QUANTUM MECHANICS

QUANTUM MECHANICS

SECOND EDITION

Amit Goswami
University of Oregon

Project Team

Editor *Lynne M. Meyers*
Developmental Editor *Brittany J. Rossman*
Marketing Manager *Keri L. Witman*
Publishing Services Coordinator *Julie Avery Kennedy*

WCB

Wm. C. Brown Publishers

President and Chief Executive Officer *Beverly Kolz*
Vice President, Director of Editorial *Kevin Kane*
Vice President, Sales and Market Expansion *Virginia S. Moffat*
Vice President, Director of Production *Colleen A. Yonda*
Director of Marketing *Craig S. Marty*
National Sales Manager *Douglas J. DiNardo*
Executive Editor *Michael D. Lange*
Advertising Manager *Janelle Keeffer*
Production Editorial Manager *Renée Menne*
Publishing Services Manager *Karen J. Slaght*
Royalty/Permissions Manager *Connie Allendorf*

A Times Mirror Company

Production Services by Lachina Publishing Services

Library of Congress Catalog Card Number: 96-83911

ISBN 0-697-15797-0

Printed in the United States of America by Times Mirror Higher Education Group, Inc., 2460 Kerper Boulevard, Dubuque, IA 52001

10 9 8 7 6 5 4 3 2 1

To Professor Manoj K. Pal

Contents

PREFACE xvii

1 AN INTRODUCTION TO THE SCHRÖDINGER EQUATION 1

1.1 Planck Takes the Quantum Leap 2

1.2 The Bohr Atom 5

1.3 From Einstein's Photons to de Broglie's Matter Waves 8

1.4 Schrödinger's Wave Equation 12

1.5 The Probability Interpretation 13

1.6 Particle in a One-Dimensional Box 17

1.7 The Phase of the Wave Function 21

1.8 The Probability Current 22

1.9 Relation Between the Phase of the Wave Function and the Probability Current 24

1.10 Outlook 25

Problems 25

Additional Problems 27

References 27

2 THE MOTION OF WAVE PACKETS 28

2.1 Gaussian Wave Packets 29
Normalization 30
Uncertainties of Gaussian Functions 30
Group and Phase Velocities 33

2.2 Some Applications of the Uncertainty Relation 34
Why We Cannot Squish Atoms 34
Time-Energy Uncertainty Relation 35
Heisenberg's Resolution of Zeno's Paradox 36

2.3 Spreading of a Gaussian Wave Packet with Time 39
Classical Limit 43
Wave Functions Corresponding to Definite Momentum and Position 44
Wave-Particle Duality 45

Problems 48

Additional Problems 49

References 50

3 SCHRÖDINGER EQUATION AS EIGENVALUE-EIGENFUNCTION EQUATION 51

3.1 The Momentum Operator 52
More on Operators 54

3.2 General Form of the Schrödinger Equation: The Hamiltonian Operator 57
Schrödinger Equation as Eigenvalue-Eigenfunction Equation 59
The Case of the Free Particle: Momentum Eigenfunctions 61
Particle in a Box Revisited 64

3.3 The Postulates of Quantum Mechanics and the Meaning of Measurement 64
Postulate 1: Description of the System 65
Postulate 2: Description of Physical Quantities 65
Postulate 3: The Expansion Postulate 66
Postulate 4: The Measurement Postulate 67
Postulate 5: The Reduction Postulate 68
Postulate 6: Time Evolution of a System 68

Problems 71

Additional Problems 73

References 74

4 THE SOLUTION OF THE SCHRÖDINGER EQUATION IN ONE DIMENSION 75

4.1 Barrier Penetration: Square Barrier 76
Application to Alpha Decay 79

4.2 The (Attractive) Square Potential Well 82
Bound States: Solutions for $E < 0$ 84

4.3 Parity 88

4.4 The Delta Function Potential 91

4.5 Solution of the Schrödinger Equation in Momentum Space 92

Problems 95

Additional Problems 98

References 100

5 LOOKING THROUGH THE HEISENBERG-BOHR MICROSCOPE 101

5.1 The Heisenberg-Bohr Microscope 101

5.2 The Einstein-Bohr Debate 103

5.3 The Double-Slit Experiment 107
A Physical Example of the Metaphorical Flashlight 111

5.4 The Delayed-Choice Experiment 113

Problems 115

References 116

6 THE DIRAC DESCRIPTION OF QUANTUM MECHANICAL STATES 117

6.1 Kets and Bras and Their Representations 118
Position and Momentum Representations 121

6.2 Operators 123
Example: The Position Operator 125

6.3 Commuting Operators and the Labeling of States 127
Commutators and Uncertainty Relations 129

6.4 The Time Evolution of a Quantum System: The Hamiltonian Operator 131

Problems 133

Additional Problems 134

References 135

7 THE ONE-DIMENSIONAL HARMONIC OSCILLATOR 136

7.1 Solution of the Schrödinger Equation for the Linear Oscillator 138
More on the Eigenfunctions 143
Comparison with the Classical Oscillator 146

7.2 The Operator Method for the Harmonic Oscillator 149
Normalized Eigenstates 152
x-Representation 153
Matrix Elements 154

Problems 157

Additional Problems 159

References 161

8 EQUATIONS OF MOTION AND CLASSICAL CORRESPONDENCE 162

8.1 The Heisenberg Equation of Motion: Schrödinger and Heisenberg Pictures 163
Ehrenfest Theorem and the Classical Limit of Quantum Mechanics 165
An Example of Classical Correspondence: The Harmonic Oscillator 169

8.2 The WKB Approximation 171
The Turning Points 173
Connection Formulas: The Case of Bound States 173
Barrier Penetration 178

Problems 179

Additional Problems 180

References 181

9 SYSTEMS OF TWO DEGREES OF FREEDOM 182

9.1 Motion of a System in Two Dimensions: Free Particles 182

9.2 Particle in a Two-Dimensional Square Box 187
The Two-Dimensional Harmonic Oscillator 189

9.3 Particle Confined to Motion on a Ring: An Introduction to the Angular Momentum Operator 190
Coordinate Transformation 191
The Uncertainty Relation Between Angular Momentum and Rotation Angle 193

9.4 Two-Particle Systems 193
Identical Particles and Exchange Degeneracy 195
The Exchange Operator and Symmetric and Antisymmetric Wave Functions 197

9.5 An Introduction to the Real-World Quantum Mechanics 198
Problems 199
Additional Problems 200
References 201

10 QUANTUM PARADOXES AND THE COPENHAGEN INTERPRETATION 202
10.1 The EPR Paradox 204
The Copenhagen Interpretation 208
Hidden Variables 208
10.2 The Paradox of Schrödinger's Cat 209
The Ensemble or Statistical Interpretation 211
The Many-Worlds Interpretation 211
10.3 When Is a Measurement? 212
Macrorealism 215
Outlook 216
Problems 216
Additional Problems 217
References 217

11 ANGULAR MOMENTUM 218
11.1 The Angular Momentum Operator and Its Eigenvalue Problem in Position Representation 220
The Eigenvalue Problem 223
Coordinate Transformation 223
The Vector Model 229
Spherical Harmonics 230
Parity 232
The Rigid Rotator 233
11.2 The Operator Method for Angular Momentum 234
Use of Raising and Lowering Operators to Generate Spherical Harmonics 238
Problems 239
Additional Problems 241
References 242

12 MOTION IN CENTRAL POTENTIAL 243

12.1 Reduction of Two-Body to One-Body Problem 244

12.2 Rotational Symmetry and Separation of Variables 245
The Hamiltonian in Spherical Coordinates 248
The "Radial Momentum" Operator 248
The Radial Equation 249

12.3 Solution of the Radial Equation for Bound States 250
Example: Particle in a Spherical Box 252

12.4 The (Spherical) Square-Well Potential 255
The Case of the S-State, $l = 0$ 255
Application to Deuteron 257
Solution for Arbitrary l 260

12.5 The Case of the Linear Potential for $l = 0$ 262

Problems 265

Additional Problems 266

References 267

13 THE HYDROGEN ATOM 268

13.1 Solution of the Radial Equation for the Coulomb Potential: The Eigenvalues for $E < 0$ 269
Numerical Estimates 271
Comments on the Eigenvalues 273

13.2 Hydrogenic Wave Functions 274
Associated Laguerre Polynomials 275
Normalization 276
Comments on the Wave Functions 277

13.3 Circulating Current and Magnetic Moment 280

Problems 282

Additional Problems 284

References 284

14 ELECTRONS IN THE ELECTROMAGNETIC FIELD 285

14.1 Motion of an Electron in a Uniform Static Magnetic Field 287
The Normal Zeeman Effect 288
The Interaction $\sim B^2$ 290

14.2 The Gauge Invariance of the Schrödinger Equation and Related Problems 291
The Aharonov-Bohm Effect 292
Flux Quantization 295
The Philosophy of Macrorealism and SQUID 296

14.3 Angular Momentum Measurement: The Stern-Gerlach Device 297

Problems 300

Additional Problems 300

References 301

15 SPIN AND MATRICES 302

15.1 Basic States, Operators, and Matrices 302
Stern-Gerlach Filters in Series with a Tilt Between Them 305
Operators and Matrices 307
Spinors 310

15.2 Analogy of Electron Spin and Polarization of Light 311

15.3 The Matrices of S_x and S_y and Solution of Eigenvalue Equations 314
The Eigenvalue Equation 315
Pauli Matrices 316

15.4 The Transformation Matrix 317

15.5 Infinite Dimensional Matrices 322

Problems 324

Additional Problems 325

References 327

16 MATRIX MECHANICS: TWO-STATE SYSTEMS 328

16.1 The Ammonia Molecule 330

16.2 The Ammonia Maser 335
The Ammonia Molecule in an External Static Electric Field 335
How Ammonia Molecules Interact with a Time-Varying Electric Field 338

16.3 Spinning Particle in a Magnetic Field: Magnetic Resonance 342

16.4 Spinning Particle in a Magnetic Field: Spin Precession 345
Spinor in an Arbitrary Direction 349

Problems 349

Additional Problems 351

References 352

17 THE ADDITION OF ANGULAR MOMENTA 353

17.1 Adding Orbital and Spin Angular Momentum 355

17.2 Adding Two Spins 357
Spin and the Pauli Principle 360

17.3 A General Method for Angular-Momentum Coupling: Clebsch-Gordan Coefficients 361
Example: Coupling of Two-Spin ½ Particles 365
Spectroscopic Notation 367

17.4 The Singlet State and the EPR Paradox Revisited 368
A Bell Inequality 370

Problems 374

Additional Problems 375

References 376

18 APPROXIMATION METHODS FOR STATIONARY STATES 377

18.1 Time-Independent Perturbation Theory for Nondegenerate States 378
Example: The Anharmonic Oscillator 382
Stark Effect in Hydrogen 383
Electric Polarizability 385

18.2 Degenerate Perturbation Theory 386
The Ammonia Molecule in an Electric Field by Perturbation Theory 387
The Weak-Field Case 388
The Strong-Field Case 389

18.3 The Variational Method 390
Example: The Anharmonic Oscillator by the Variational Method 391

Problems 393

Additional Problems 396

References 396

19 QUANTUM SYSTEMS: ATOMS WITH ONE AND TWO ELECTRONS 397

19.1 The Fine Structure of Hydrogen 398
Calculation of the Spin-Orbit Interaction 399
Calculation of Energy Shift Due to H_{kin} 400
The Fine Structure 400

19.2 **The Anomalous Zeeman Effect** 401

19.3 **The Helium Atom** 404
Screening and Effective Charge in Perturbation Theory 408
Accounting for Screening by a Variational Calculation 409
The Lowest Excited States of Helium: Exchange Energy 411

Problems 413

Additional Problems 414

References 415

20 QUANTUM SYSTEMS: ATOMS AND MOLECULES 416

20.1 **The Hartree Equation for Atoms** 417
Antisymmetrization of the Wave Function 420

20.2 **The Quantum Mechanical Explanation of the Periodic Table** 421
Spectroscopic Description or Term Value 423

20.3 **Introduction to Molecular Structure** 425
Electronic Motion, Vibration, and Rotation 429
Molecular Orbitals 430
The H_2 Molecule: An Example of the Valence-Bond Method 433

Problems 436

Additional Problems 436

References 437

21 QUANTUM SYSTEMS: FERMI AND BOSE GASES 438

21.1 **The Periodic Potential Model of the Solid** 440
Symmetry Under Finite Displacements 441
Kronig-Penny Model 442

21.2 **The Nuclear Shell Model** 446
The Three-Dimensional Harmonic Oscillator 446
The Nuclear Shell Model 451

21.3 **Boson Gas: Photon Statistics and Planck's Law** 452
Planck's Radiation Distribution Law 455

Problems 457

Additional Problems 458

References 459

22 TIME-DEPENDENT PERTURBATION THEORY AND APPLICATION TO ATOMIC RADIATION AND SCATTERING 460

22.1 Time-Dependent Perturbation Theory: General Formulation and First-Order Theory 461
First-Order Theory 462
Example: Time-Dependent Perturbation of the Harmonic Oscillator 462

22.2 Periodic Perturbation: Fermi's Golden Rule 463

22.3 Interaction of the Electromagnetic Field with Atoms 468

22.4 Calculation of Transition Rate for Spontaneous Emission 471
The Dipole Approximation 472
Summing Over Polarization 473
The Dipole Selection Rules 474
The Radial Integral: The $2P \rightarrow 1S$ Transition 476

22.5 Exponential Decay Law, Lifetime, and Line Width 477
Phase-Space Factor for Decay Rates of Unstable Particles 479

22.6 Born Approximation and Introduction to Scattering 481
The Case of the Yukawa or the Screened Coulomb Potential 483

Problems 484

Additional Problems 486

References 487

23 SCATTERING THEORY 488

23.1 Scattering Amplitude, Differential and Total Cross Sections, Center-of-Mass, and Laboratory Frames 490
Probability Currents for Elastic Scattering 491
Differential Cross Section 492
Total Cross Section 493
Transformation from the Center-of-Mass Frame to Laboratory Frame 494

23.2 Continuum Quantum Mechanics: Partial Waves 497
Free-Particle Schrödinger Equation in Spherical Coordinates 497
Expansion of a Plane Wave into Partial Waves 499
Partial Wave Expansion of the Scattering Amplitude 501
Total Cross Section and the Optical Theorem 503

23.3 Scattering by a Square-Well Potential at Low Energies 504
Breit-Wigner Formula 508
Low-Energy Neutron-Proton Scattering: Scattering Length and Effective Range 508
Scattering Lengths for Neutron-Proton Scattering 511

23.4 Inelastic Scattering 513
Example: Scattering from a Black Disc 515

23.5 Outlook 516

Problems 516

Additional Problems 518

References 519

24 THE UNFINISHED CHAPTER: THE MEANING AND INTERPRETATION OF QUANTUM MECHANICS 520

24.1 Measurement Theory 521
The Density Matrix 522
Example of the Density Matrix Formalism: A Partially Polarized Beam 524
The Interaction of a System and a Measuring Apparatus 525

24.2 The Principle of Macroscopic Distinguishability 528

24.3 Realistic Ontologies 530
Irreversibility 530
Nonlocal Hidden Variables 532
Many-Worlds Ontology 533

24.4 Does Consciousness Collapse the Wave Function?—Idealistic Extension of Heisenberg's Ontology 533
Heisenberg's Ontology and an Idealistic Extension of the Copenhagen Interpretation 534
Mind over Matter or What? 536
Wigner's Friend 536
When Is a Measurement? 537
The Watched Pot Does Boil 537
Reconciliation with the Many-Worlds Interpretation and the Cosmological Question 538
Summary of the Idealistic Interpretation of Quantum Measurement 539

24.5 A Final Outlook 540

References 540

APPENDIX: THE DELTA FUNCTION 543
Properties of the Delta Function 544
Representations of the Delta Function 545

BIBLIOGRAPHY 547

INDEX 549

Preface

The intended audience for this book is college seniors and first-year graduate students. Since quantum mechanics is the most challenging course in physics for undergraduates, a textbook at this level needs to be stimulating (a touch of Feynman and Bohm), helpful in a practical way (e.g., Liboff), modern in scope (Gasiorowicz, Das and Mellissinos), elegant (Merzbacher), exhaustive and explanatory (Chester), and, most importantly, user-friendly, which, in one way or another, none of the above-mentioned books is. I hope to combine all these qualities in a book that is also user-friendly.

What makes a book user-friendly? For quantum mechanics, it means teaching how to ask the right questions, always starting a topic with what the students know, treating the nitty-gritties carefully, satisfying the students' interest in the meaning and interpretation of quantum mechanics, treating the modern topics that the students will pursue in their careers in physics, and, last but not least, giving a lot of worked-out examples.

My goals then are (1) to write a book that is stimulating like Feynman's, which excites the students about the essence of quantum mechanics; *and* (2) to give the students a thorough grounding in the fundamental aspects of quantum mechanics to help them free themselves from their classical prejudices and learn to think and calculate quantum mechanically (I include many worked-out examples and occasional summaries as practical help for the student); *and* (3) to include a lot about quantum systems in a balanced fashion that will help the student make a bridge to whichever aspect of modern physics he or she wants to pursue; *and* (4) to go to substantial depth into meaning and interpretational questions about quantum mechanics without exceeding the mathematical level of the beginning student.

To achieve these goals, I begin with the Schrödinger wave equation and its solution for one-dimensional problems, because this is mathematically more familiar to the student. Next come the Dirac bras and kets, but treated in a gentle, hand-holding way. Midway through the book spin and matrix mechanics are introduced, following the exciting style of Feynman. The book ends with three chapters on quantum systems—one chapter each on radiation and scattering theories, and one chapter on quantum measurement theory. Consequently, the book begins simply, yet excitingly. It has a second beginning (which is great for semester systems) with matrix mechanics, which recaptures the student's interest. Finally, the book ends on a high note with both modern physics topics and interpretational questions.

The unifying approach of the book is to present quantum mechanics not only as a schema for successful calculations and predictions but also as a basis for a new and exciting world view. The most unusual aspect of the book is an ongoing presentation of the radicalness of quantum mechanics as compared to classical physics: The interference experiments, the paradoxes, nonlocality, macroscopic quantum devices, symmetry, and so forth.

I hope it is already clear that a book like this does not grow in a vacuum. I have taken help from all the books that I myself have found worthwhile (including but not limited to those already mentioned); the books especially useful in writing a particular chapter have been referenced at the end of that chapter. I am also grateful to my students Mark Mitchell and John Svitak, and to Arthur Pavlovic of West Virginia University, Philip N. Parks of Michigan Technological University, Sanford Kern of Colorado State University, and Walter Carnahan of Indiana State University, the reviewers who toiled to make the book error-free. I also thank my editor, Jeffrey L. Hahn, for his encouragement. Finally, heartfelt thanks are due my wife, Maggie, for a careful editing of the book.

PREFACE TO THE SECOND EDITION

I am grateful to the many users of this book who I credit for creating the need for a second edition. You like the user-friendliness; you like the emphasis on meaning. Thank you.

The temptation to add more material was a serious one that I actively resisted because most of you are very happy with the current size of the book. *So the only substantial change I offer is the addition of many more problems at the end of the chapters.* Also, of course, errors have been corrected (I hope, every one of them!). My thanks for the generous feedback on these errors. I have also added clarification wherever needed. I hope these changes make the book even more user-friendly.

Finally, I'd like to thank my colleague, Professor Robert Zimmerman, for a careful reading of the revised manuscript, the staff at Wm. C. Brown for their patience, and Jan Blankenship and my wife Maggie for their help.

1

An Introduction to the Schrödinger Equation

What is quantum mechanics? It's a new way of interpreting data and predicting the behavior of submicroscopic objects, perhaps of all objects, based on the idea of an essential discontinuity, the quantum, in the affairs of the world. Compared to the world view of classical physics, quantum mechanics gives us a radically new view of the world. Indeed, it will revolutionize the way you think about the world.

The discovery of quantum mechanics proceeded along two important but separate tracks. The first track was based on the realization that the allowed values of energy exchange between subatomic bodies are discrete—they involve quantum jumps. This track began with Max Planck's work on black-body radiation, got a big boost from Niels Bohr's theory of the hydrogen atom, and was carried to fruition by Werner Heisenberg's discovery of the matrix version of quantum mechanics.

The second track began with Albert Einstein's discovery of the wave-particle duality of light. Then Louis de Broglie generalized the attribute of wave-particle duality to matter, Erwin Schrödinger discovered the wave equation for matter waves, now called the Schrödinger equation, and Max Born interpreted the de Broglie-Schrödinger waves as waves of probability. This was the wave-mechanics version of quantum mechanics. Finally, Paul Dirac showed that the two versions, matrix mechanics and wave mechanics, are entirely equivalent.

However different the two tracks may look, there is one thing they have in common—they troubled many people, they continue to shock people, they should shock you! The Planck-Bohr-Heisenberg discoveries introduced quantum jumps, discontinuities, in physics. The Einstein-de Broglie-Schrödinger-

Born version introduced the paradox of wave-particle duality and more; it introduced probability into physics. Thus the bulwark of classical concepts—continuity, unambiguous language description, and causal determinism—were all squarely challenged by the new physics.

We will begin our discussion with the work of Planck and Bohr, but switch over to the de Broglie-Schrödinger track of quantum mechanics because it is based on mathematics that is more familiar to you. We will take up matrix mechanics only after you are thoroughly grounded in wave mechanics.

1.1 PLANCK TAKES THE QUANTUM LEAP

Toward the end of the nineteenth century it was becoming clear from studies, for example of black-body radiation, that classical physics does not always work. According to the classical Rayleigh-Jeans formula for the radiation energy density $u(\nu, T)$ in a cavity

$$u(\nu, T) = \frac{8\pi\nu^2}{c^3} kT \tag{1.1}$$

where ν is the frequency, c is the velocity of light, k is Boltzmann's constant, and T is the absolute temperature. The intensity of radiation goes up with the square of the frequency. Thus you could legitimately expect to get a suntan sitting in front of your fireplace. Actually, it is worse. Because, if you integrate equation (1.1) for all frequencies, you get infinity, the integral just blows up, it no longer makes sense. This is called the ultraviolet catastrophe; catastrophe it is indeed! The idea of the quantum was discovered by Max Planck in the year 1900, saving physics from the ultraviolet catastrophe.

The Rayleigh-Jeans formula is a product of two terms: The number of degrees of freedom for frequency ν given as

$$\frac{8\pi\nu^2}{c^3}$$

and the average energy, $\bar{E}$ per degree of freedom, given by kT. The derivation of the first factor need not concern us here. The second factor is derived as follows in classical physics:

$$\bar{E} = \frac{\int_0^\infty E \exp(-E/kT)\, dE}{\int_0^\infty \exp(-E/kT)\, dE} = kT \tag{1.2}$$

Why does the Rayleigh-Jeans formula lead to catastrophe? Because there are too many degrees of freedom at high frequency; the number of degrees of freedom is proportional to ν^2 and they all contribute kT to the energy density. Planck's genius lay in the observation that if we change the integrals in equation (1.2) to summations, that is, if we assume that an oscillator can take on only discrete values of energy, the average energy per degree of freedom can be made to decrease for high frequencies.

Assume then with Planck that the basic quantum of energy is given by

$$\epsilon = h\nu, \qquad h = \text{Planck's constant} \tag{1.3}$$

and that the energy of an oscillator can only be $n\epsilon$ ($n = 0,1,2,\ldots$), multiples of ϵ. Then,

$$\bar{E} = \frac{\sum_{n=0}^{\infty} n\epsilon e^{-n\epsilon/kT}}{\sum_{n=0}^{\infty} e^{-n\epsilon/kT}}$$

Substitute $x = \exp(-\epsilon/kT)$. We get

$$\bar{E} = \frac{\epsilon \sum_{n=0}^{\infty} nx^n}{\sum_{n=0}^{\infty} x^n}$$

Now use the following identities:

$$\frac{1}{1-x} = 1 + x + x^2 + \cdots$$

$$\frac{x}{(1-x)^2} = x + 2x^2 + 3x^3 + \cdots$$

You will get the desired result

$$\bar{E} = \frac{h\nu e^{-h\nu/kT}/(1-e^{-h\nu/kT})^2}{1/(1-e^{-h\nu/kT})} = \frac{h\nu}{(e^{h\nu/kT}) - 1}$$

Thus the quantum hypothesis modifies the Rayleigh-Jeans law to Planck's radiation law for the energy density of black-body radiation:

$$u(\nu, T) = \frac{8\pi h\nu^3}{c^3} \frac{1}{e^{h\nu/kT} - 1} \tag{1.4}$$

Indeed, in agreement with experimental data, it quenches the contribution to the energy density coming from high frequencies (fig. 1.1)! And more importantly, the ultraviolet catastrophe is resolved. To see this, let's integrate $u(\nu, T)$ to obtain the total energy radiated:

$$\begin{aligned} U(T) &= \frac{8\pi h}{c^3} \int_0^\infty d\nu \frac{\nu^3}{e^{h\nu/kT} - 1} \\ &= \frac{8\pi h}{c^3} \left(\frac{kT}{h}\right)^4 \int_0^\infty \frac{(h\nu/kT)^3\, d(h\nu/kT)}{e^{h\nu/kT} - 1} \\ &= \frac{8\pi k^4}{h^3 c^3} T^4 \int_0^\infty \frac{x^3}{e^x - 1} dx \end{aligned}$$

The integral is easily evaluated:

$$\begin{aligned} \int_0^\infty \frac{x^3}{e^x - 1} dx &= \int_0^\infty dx\, x^3 e^{-x} \sum_{n=0}^{\infty} e^{-nx} \\ &= \sum_{n=0}^{\infty} \frac{1}{(n+1)^4} \int_0^\infty dy\, y^3 e^{-y} = 6 \sum_{n=1}^{\infty} \frac{1}{n^4} = \frac{\pi^4}{15} \end{aligned}$$

Finally, we obtain

$$U(T) = aT^4$$

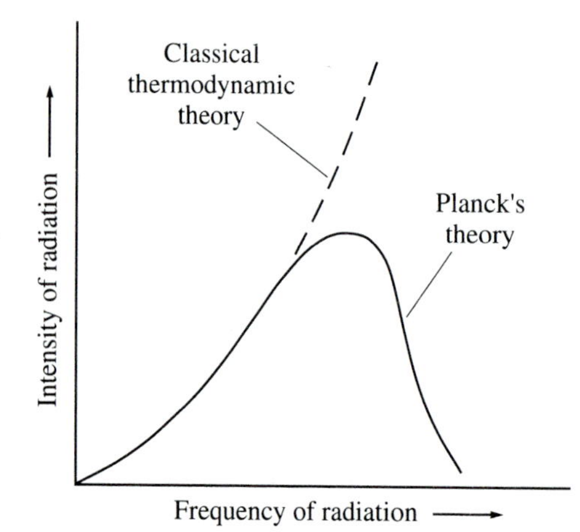

FIGURE 1.1
Black-body radiation distribution curves—comparison of Planck's and classical theories. Only Planck's theory with quantum jumps agrees with experiment.

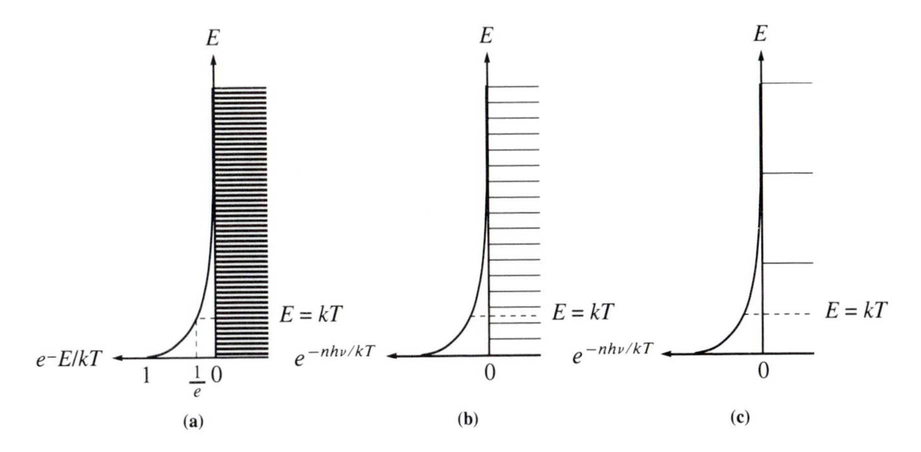

FIGURE 1.2
See text for explanation. (Adapted with permission from P. Fong, Elementary Quantum Mechanics, *©1962 by Addison-Wesley Publ. Co. Reprinted by permission of* Addison-Wesley Publishing Company, Inc., *Reading, MA.)*

with

$$a = \frac{8\pi^5 k^4}{15h^3c^3}$$

This is, of course, the Stefan-Boltzmann law for radiation from a black body.

What is going on is this. Look at figure 1.2 where we have plotted the energy of the oscillators of the radiation field along the vertical axis. The exponential curves on the left of the energy axis give the relative probability distribution among the energy levels. For the classical case of a continuous energy spectrum, figure 1.2(a), the average value of energy is roughly where the probability is reduced to $1/e$; thus it is approximately kT. Now in figure 1.2(c), the energy levels are so much separated that there are no levels at all near the classical average energy kT. Indeed, there is hardly any probability that any level other than the ground level, $E = 0$, will be occupied; thus the average energy is ~ 0 rather than kT. This is the situation at high frequency, and thus the contribution of high frequencies is quenched. The situation is quite different when the energy levels are closely spaced (at low frequencies, fig. 1.2b); the average energy in this case approaches kT, the classical case. Thus quantum theory approaches the classical result when the spacing of the energy levels approaches zero; this is an example of the *correspondence principle*—the idea that under certain situations quantum mechanics gives the same result as classical mechanics.

1.2 THE BOHR ATOM

The classical crisis with the model of the atom was not dissimilar to the case of black-body radiation. Here, Ernest Rutherford had proposed a sensible model of a nuclear atom based on his discovery of the atomic nucleus. The electrons, he assumed, must revolve around the nucleus; but then they must radiate and lose energy and eventually spiral into the nucleus. A classical nuclear atom is unstable! In 1913, Niels Bohr made the atom stable by injecting quantum ideas into Rutherford's theory.

The Bohr atom was based on two radical assumptions:

1. The electrons move around the nucleus in *quantized* circular orbits—their angular momentum is restricted to values that are an integral multiple of $\hbar$ ($= h/2\pi$),

$$mvr = n\hbar, \qquad n = 1,2,\ldots \tag{1.5}$$

   where m is the mass of the electron, v its velocity, and r is the radius of the orbit. Moreover, the orbits are assumed to be stationary; the electrons do not radiate while in these quantized orbits. Rather, the orbits form stationary, quantized states of energy for the electrons.
2. The electrons radiate only when they jump from a higher orbit to a lower one, but this jump is a quantum jump. Here Bohr used the idea of quantized light, photons, that Einstein proposed in 1905 in connection with his theory of the photoelectric effect. According to Einstein, light of frequency ν consists of quanta of energy $h\nu$. Bohr proposed that when an electron jumps from an initial stationary state of energy E_i to a final state of energy E_f, a photon is emitted:

$$E_i - E_f = h\nu \tag{1.6}$$

   This is the Bohr frequency condition. Notice that the Bohr atom is stable because once the electron is in its lowest stationary orbit (the ground state), it has no other place to jump "down."

Let's derive the Bohr formula for the quantized energy of a stationary state for the hydrogen atom based on what is called semiclassical reasoning. Use Newton's second law for the orbiting electron; the force acting on it is the Coulomb force of the nucleus, and this must equal mass times the centripetal acceleration:

$$\frac{e^2}{r^2} = \frac{mv^2}{r}$$

Combine this with the angular momentum quantization condition, equation (1.5). You now can solve for v and r:

$$v = \frac{2\pi e^2}{hn}$$

$$r = \frac{1}{4\pi^2}\frac{n^2h^2}{e^2m} \tag{1.7}$$

Hence, the quantized energy of the electron in the stationary state labeled by the quantum number n is given as

$$E_n = \frac{1}{2}mv^2 - \frac{e^2}{r}$$

$$= -\frac{2\pi^2e^4m}{h^2n^2} \tag{1.8}$$

This is the Bohr formula.

Now we can apply the Bohr frequency condition, equation (1.6), to derive the frequencies of the line spectra of hydrogen

$$\nu = \frac{2\pi^2me^4}{h^3}\left(\frac{1}{n_f^2} - \frac{1}{n_i^2}\right)$$

where n_i and n_f are the initial and final values of the quantum number n for a particular quantum jump. That this formula agrees with the observed line spectra in hydrogen, especially the Balmer series (for which n_f is 2, and n_i is 3,4, . . .), is history, and we will not discuss it here.

The energy level diagram based on the Bohr formula, equation (1.8), consisting of the ground state and some of the "excited" states, is shown in figure 1.3. The ground state energy is -13.6 electron volts (eV). Also, the radius of the Bohr orbit corresponding to the ground state, denoted as a_0, can be calculated using equation (1.7), with $n = 1$, and is given as

$$a_0 = \frac{h^2}{4\pi^2me^2} = 0.529 \times 10^{-8} \text{ cm} \tag{1.9}$$

This is called the Bohr radius. (We will use c.g.s. units throughout.)

Finally, to impress on you how radical Bohr's ideas are, let's examine one aspect of the quantum jump. The quantum jump is envisioned as a truly discontinuous affair. It's like a jump from one rung of a ladder to another without going through the intervening space.

Back to history. Following Bohr's work, Heisenberg realized that if he took all the quantum jumps in an atom and arranged them in an array, these quan-

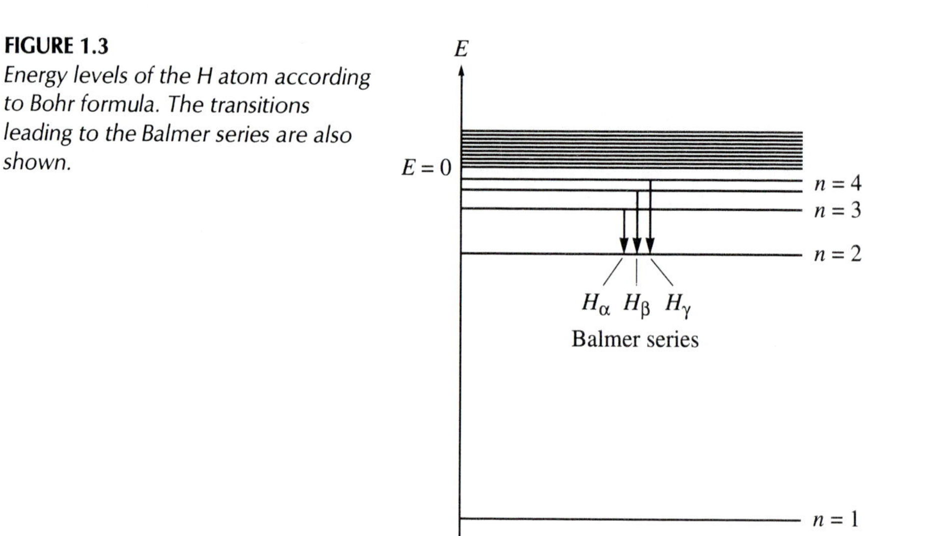

FIGURE 1.3
Energy levels of the H atom according to Bohr formula. The transitions leading to the Balmer series are also shown.

tities would satisfy an equation that no classical physicist had ever seen. It was soon realized that Heisenberg's arrays are really matrices and that Heisenberg had intuited matrix mechanics—the matrix version of quantum mechanics. For the purpose of prudent pedagogy, we will postpone the discussion of matrix mechanics until later, and instead take up the subject of how wave mechanics was conceived.

1.3 FROM EINSTEIN'S PHOTONS TO DE BROGLIE'S MATTER WAVES

De Broglie must have been a lover of music, for he realized that Bohr's stationary orbits of electrons in confinement in atoms must have something in common with stationary waves of music that we routinely create on guitar strings. The latter are well known to exist in discrete harmonics. De Broglie wondered if the discreteness of atomic orbits could be due to the discreteness of the harmonics of electron waves in captivity.

Einstein's introduction of the photon—quantized light—gave us the paradox of wave-particle duality for light. As if the wave-particle paradox of light did not provoke enough consternation, de Broglie raised the question, Could matter, electrons, be wavelike, have a duality similar to that of light? On a guitar string, the stationary waves form a discrete pattern of harmonics similar to the discrete Bohr orbits of the atom. But stationary waves are a property of waves in confinement. Asked de Broglie, Could atomic electrons be confined waves and therefore produce a discrete stationary wave pattern? For example, maybe the lowest atomic orbit is one in which one electron wavelength fits the circumference of the orbit, and the higher orbits fit two or more electron wavelengths (fig. 1.4).

Of course in science, experiments are the final arbiter. And de Broglie's idea of the electron's wave nature was demonstrated brilliantly in the Davisson-Germer experiment where the result of passing a beam of electrons through a crystal was photographed—the result was a diffraction pattern (fig. 1.5).

If electrons are waves, they should obey equation (1.3), which holds for photons, connecting one of their wave properties to a particle property. Let's write equation (1.3) in the slightly altered form:

$$E = \hbar\omega \tag{1.10}$$

where $\hbar = h/2\pi$ and $\omega = 2\pi\nu$ (ω is the angular frequency). Now a photon is a zero rest mass particle, and according to the theory of relativity, light obeys the special relationship

$$E = pc$$

between its energy E and momentum p, where c is the velocity of light. Additionally, according to the theory of waves, the angular frequency ω is related to the wave number k, via the relation

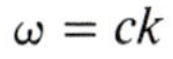

$$\omega = ck$$

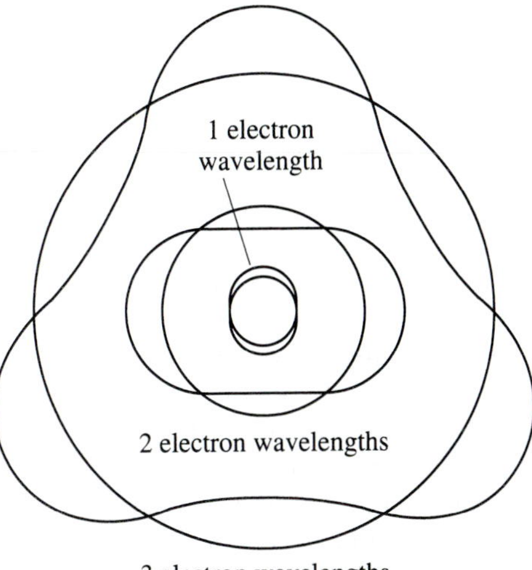

FIGURE 1.4
Stationary waves of electrons in the confinement of an atom. The electron wave fits an integral number of wavelengths in each of the successive Bohr orbits.

FIGURE 1.5
Electron diffraction pattern as revealed experimentally. (Courtesy: Stan Micklavzina.)

Combining the above equations, we get

$$pc = \hbar ck$$

or

$$p = \hbar k \tag{1.11}$$

De Broglie proposed that equation (1.11) holds even for nonzero mass matter waves. Since the wave number $k = 2\pi/\lambda$, where λ is the wavelength, equation (1.11) can also be written as

$$\lambda = h/p \tag{1.12}$$

In other words, the wavelength of the wave of a particle of matter is inversely proportional to the particle's momentum, and the constant of proportionality is Planck's constant.

With a great amount of insight, de Broglie suggested that the same relationship, equation (1.11), holds for all matter, all particles, not just electrons, with the same constant of proportionality. Thus Planck's constant h fixes the scale of things in the quantum domain of nature once and for all.

If matter is a wave, there should be a wave equation to describe a matter wave—so quipped one physicist to another at the end of a seminar in 1926 on de Broglie's waves. The physicist who said this, Peter Debye, promptly forgot about it, but the second physicist, Erwin Schrödinger, proceeded to discover the wave equation for matter, now named after him as the Schrödinger equation (see section 1.4).

The picture of the atom generated by de Broglie's idea of the matter wave, backed by the mathematical theory of the Schrödinger equation, is remarkable; it explains in simple terms the three most important properties of atoms, namely, their stability, their identity with one another, and their ability to regenerate themselves. You have already seen how stability arises—remember, that was the great contribution of Bohr. The identity of atoms of a particular species is simply a consequence of the identity of wave patterns in confinement; the structure of the stationary patterns is determined by the manner in which the electrons are confined, but not by their environment. The music of the atom is the same wherever you find it, on earth or in the Andromeda galaxy. And furthermore, the stationary pattern, depending only on the conditions of its confinement, has no recollection of past history, no memory; thus derives the property of regeneration, or repetition of the same performance over and over.

■ **PROBLEM** Derive Bohr's angular momentum quantization condition for the Bohr atom from de Broglie's relation.

SOLUTION If de Broglie waves of wavelength λ fit a Bohr orbit of radius r to satisfy the stationarity condition, we must have

$$2\pi r = n\lambda$$

where n is an integer 1,2,. . . . And since λ is $h/p = h/mv$, we get

$$\frac{2\pi r}{h/mv} = n$$

or

$$mvr = \frac{nh}{2\pi} \tag{1.5}$$

which is the Bohr angular momentum quantization relation. ■

■ **PROBLEM** Scattering of high-energy protons is like diffraction of light waves from a slit. High-energy protons of energy 200 GeV (energies in "high-energy physics" are most often given in this unit of GeV, 1 GeV = 10^9 eV) are scattered by a hydrogen target by an angle θ given by

$$p \sin\theta \cong 1.2 \text{ GeV}/c$$

Estimate the radius of the proton.

SOLUTION If R is the dimension of the "slit," then we have

$$\theta \cong \sin\theta \cong \lambda/R$$

From the experimental relationship, since $p = 200$ GeV$/c$ for relativistic protons, we have

$$\theta \cong \sin\theta \cong 1.2/200 = 6 \times 10^{-3}$$

From the de Broglie relation, $\lambda = h/p$, and thus

$$R = \lambda/\theta = h/p\theta \cong 10^{-13} \text{ cm}$$

The radius of the proton is on the order of 10^{-13} cm. ■

1.4 SCHRÖDINGER'S WAVE EQUATION

How did Schrödinger arrive at his equation? He creatively intuited it. How else could he have proposed an equation that would break all tradition of such wave equations?

Wave equations usually connect a second-order time derivative of a "wave function" $\psi(x,t)$ with its second-order spatial derivative, but Schrödinger's equation (presently, and in the first eight chapters, we will consider only the one-dimensional case) used a first-order derivative in time:

$$i\hbar \frac{\partial\psi(x,t)}{\partial t} = -\frac{\hbar^2}{2m}\frac{\partial^2\psi(x,t)}{\partial x^2} \tag{1.13}$$

This is necessary, said Schrödinger, for the solutions of the equation to display de Broglie waves in the nonrelativistic limit.

To see this, consider a solution (i.e., a wave function) in terms of a plane de Broglie wave

$$\psi = \exp(ikx - i\omega t)$$

But k is related to p, equation (1.11), and ω is related to E, equation (1.10), thus

$$\psi = \exp[i(px - Et)/\hbar]$$

Substituting in the wave equation (1.13), we get

$$E = p^2/2m$$

which is the correct nonrelativistic relationship between energy and momentum.

You may wonder, so what if we started with a wave equation with a second-order time derivative? Following the above procedure, we will then end up with the relativistic energy-momentum relationship (try it and see!). That isn't so bad! Actually, this is what Schrödinger himself attempted at first. But very soon he realized that such an equation does not work for electrons; it does not give the correct spectrum for the hydrogen atom. Incidentally, the correct relativistic equation for electrons of which the Schrödinger equation above is the nonrelativistic limit does retain the first-order time derivative; it was discovered by Dirac.

However, there is a price to pay for having a first-order time derivative in the wave equation. The solutions of the Schrödinger equation above are not real physical waves; they are complex functions with both a real and an imaginary part. This gives rise to the problem of interpretation: What does the wave function mean physically? The interpretation was given by Max Born (see section 1.5).

The appearance of $\hbar$ in the Schrödinger equation is, of course, crucial. This is how Schrödinger imposed the "quantum condition" on the wave equation of matter.

1.5 THE PROBABILITY INTERPRETATION

The Schrödinger equation gives probability waves, so said Max Born. It just tells us probabilistically where the likelihood of finding the particle will be greater; there the wave will be strong, its amplitude will be large. In other places, if the probability of finding the particle is small, the wave will be weak, having a small amplitude.

Imagine that you are looking for traffic jams from a helicopter above the streets of Los Angeles. If the cars were described by Schrödinger's waves, we would say that the wave was strong at the location of traffic jams, and in between jams the wave was weak.

Electron waves are probability waves, and here's what Born said that means mathematically: For electron waves, the modulus square of the wave function $\psi(x,t)$ at a point gives us the probability $P(x,t)$ of finding the electron at that point:

$$P(x,t) = |\psi(x,t)|^2 \tag{1.14}$$

The electron's wave function is a probability amplitude; its square gives a probability distribution. Also, since the particle must be somewhere, the probability integrated over all x (all space in one dimension) must equal 1:

$$\int_{-\infty}^{\infty} P(x,t)\, dx = 1 \tag{1.15}$$

A wave function satisfying equation (1.15) is said to be *normalized*. Equation (1.15) also makes it clear that $P(x,t)$ can be thought of as a probability density.

Now, if we use equation (1.14) to calculate the probability distribution of an electron whose wave function is given by the plane wave $\exp(ikx - i\omega t)$, we get

$$P(x,t) = |\exp(ikx - i\omega t)|^2 = 1$$

The probability is the same for the particle to be anywhere, at all times. But real particles are found to be localized; thus a plane wave is quite inadequate for describing them correctly (although, as it turns out, we do use them; see chapter 3).

The wave functions that give the proper probability distributions for a quantum object, such as shown in figure 1.8(a) below, are wave packets; literally they are superpositions of waves done in such a way that there is hardly any wave outside of a given region (fig. 1.6). The mathematics of this procedure involves the Fourier transform (see chapter 2). Since the Schrödinger equation is a linear equation, a superposition of waves forming a wave packet is also an acceptable solution for it. This is called the *superposition principle*.

This superposition principle, a consequence of the linearity of the Schrödinger equation, is behind the many spectacular wave aspects of matter (you will see!) predicted by quantum mechanics and verified by experimental data.

Probability distributions, those bell-shaped Gaussian curves, should be familiar to you. If not, here's one you can get with a little experiment of tossing a coin a few times (your arms may get tired, but it's worth it). You know that if you flip a coin, there is a fifty-fifty chance that it will land heads up. If you flip a coin 30 times, it will land heads up 15 times, on the average. Now do 100 or so of these 30-flip runs, and plot the statistical distribution of your results by counting the number of runs that show between 11 and 13 heads, between 14 and 16 heads, and so forth. The resulting graph will look like a bell curve, centered around 15, and tapering off on both sides (fig. 1.7).

So if we take the absolute square of the wave function of a quantum system and plot it, we get a bell curve. Now suppose we plot such a bell curve as a function of the electron's position. The curve should center around the ex-

FIGURE 1.6
Superposition of many sine waves produces a localized wave packet. (Reprinted with permission from P. W. Atkins, Quanta: A Handbook of Concepts. *Oxford, U.K.: Clarendon Press, 1974.)*

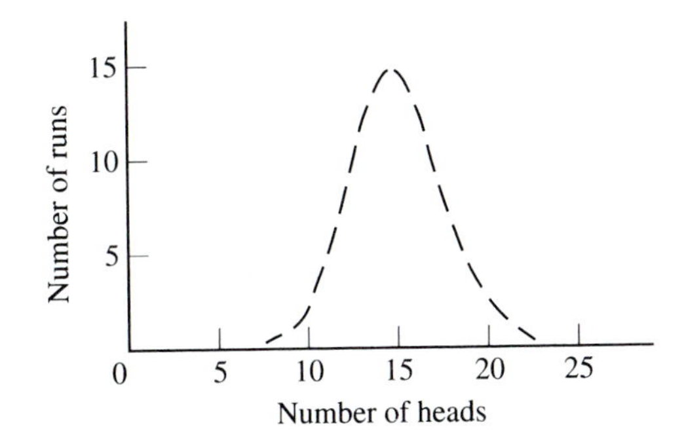

FIGURE 1.7
The result of a coin-toss experiment is a Gaussian curve.

pected position, call it x_0, of the electron, and taper off on both sides (fig. 1.8a). The width of the curve, as you can see, indicates an uncertainty in the electron's position. Call it Δx. If we do the same thing with the probability curve for the electron's momentum wave function (which is the Fourier transform of $\psi(x,t)$; see chapter 2), we will get the uncertainty of the electron's momentum; call it Δp (fig. 1.8b).

Now here is a most important thing. Heisenberg discovered that the product of the position and momentum uncertainties Δx and Δp, respectively, of a quantum object is greater than or equal to Planck's constant $\hbar$

$$\Delta x \cdot \Delta p \geq \hbar \tag{1.16}$$

This is an example of Heisenberg's uncertainty relations (there are other examples; see chapter 2). Richard Feynman once said, "Electron waves are probability waves in the ocean of uncertainty"; now you know what that means. For quantum objects described by probability waves, we can never simultaneously ascertain values of both momentum and position with arbitrary accuracy.

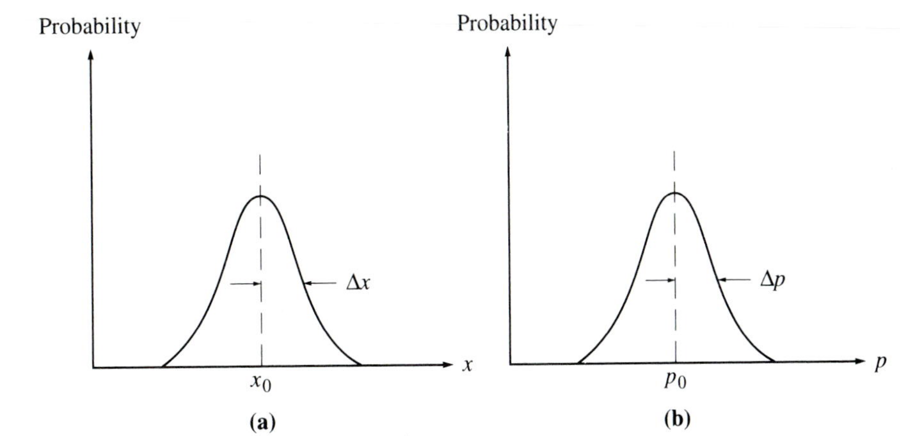

FIGURE 1.8
Probability and uncertainty. (a) The position probability distribution $|\psi(x)|^2$ is plotted as a function of x, showing the uncertainty Δx; (b) A similar plot for the momentum probability distribution.

There is an analog of the wave packets of figures 1.8(a) and 1.8(b) and the uncertainty relation, equation (1.16), in the theory of analog and digital synthesizers of music. The analog synthesizer builds a localized sound wave out of a superposition of simple sine waves of different wave number k with a range Δk (say). The sine waves have a conjugate wave form called an impulse wave; the digital synthesizer builds a wave out of different impulse waves of various x, with a range, say, Δx. Sound waves obey a relationship between Δx and Δk that is similar to the uncertainty relation, equation (1.16), namely,

$$\Delta x \cdot \Delta k \geq 1$$

For example, for percussion music, Δx is small; thus if you want to synthesize percussion sounds with an analog synthesizer, you must use a large range of Δk. The opposite is true of the sound from wind instruments. To simulate the music of a flute, for which Δk is small, with a digital synthesizer, you must use a large range of impulse waves.

As a demonstration of the uncertainty relation, suppose we could do an experiment that measures the momentum with total accuracy; the momentum uncertainty is then zero. Now what is the minimum position uncertainty? Use the uncertainty relation

$$\Delta x = \hbar/\Delta p = \hbar/0 = \infty$$

You see the uncertainty in position now approaches infinity; that is, the position is completely uncertain. There is an essential incommensurability between momentum and position. If we measure one so that the corresponding uncertainty is zero, the other quantity becomes completely uncertain, any value has equal probability. Thus it is impossible to simultaneously know both momentum and position of a quantum object.

The last statement means that in quantum mechanics, it is impossible to predict exact trajectories, because such a prediction requires the exact knowledge of both initial momentum and initial position.

Thus Bohr's semiclassical orbits are, strictly speaking, not a good description for what the electron is doing in the atom (see also chapter 2). We really cannot say that the electron is such and such distance away from the nucleus when in this or that energy state. Instead, we settle for the average value or the *expectation value*. For a function $f(x)$ connected with the system, this is defined as

$$\langle f(x) \rangle = \int \psi^*(x) f(x) \psi(x)\, dx \tag{1.17}$$

where $\psi(x)$ is the normalized wave function of the system and $\psi^*(x)$ is the complex conjugate of ψ. More on this later.

Right now, at least two other conceptual questions remain unresolved. What is the resolution of the wave-particle duality that de Broglie-Schrödinger's theory has introduced? And since waves are continuous, does the introduction of the Schrödinger wave equation get rid of the discontinuities, the quantum jumps? In chapter 2 when we deal with the mathematics of wave packets, we shall also address these questions.

1.6 PARTICLE IN A ONE-DIMENSIONAL BOX

But how does the Schrödinger formalism work? Does it work? In this section we will consider a simple example of quantization according to the Schrödinger formalism of quantum mechanics.

A "free" quantum particle satisfies the Schrödinger equation

$$i\hbar \frac{\partial \psi(x,t)}{\partial t} = -\frac{\hbar^2}{2m} \frac{\partial^2 \psi(x,t)}{\partial x^2} \tag{1.13}$$

Let's try the solution

$$\psi(x,t) = \phi(x)\exp(-iEt/\hbar)$$

Substitution in equation (1.13) gives us the *time-independent* equation

$$\frac{d^2\phi(x)}{dx^2} + k^2\phi(x) = 0$$

where we have put $E = \hbar^2k^2/2m$.

Now let's consider the solution of this equation for a particle that is inside an impenetrable box of dimension a (fig. 1.9). The box here imposes special boundary conditions

$$\phi(0) = 0 \quad \text{and} \quad \phi(a) = 0$$

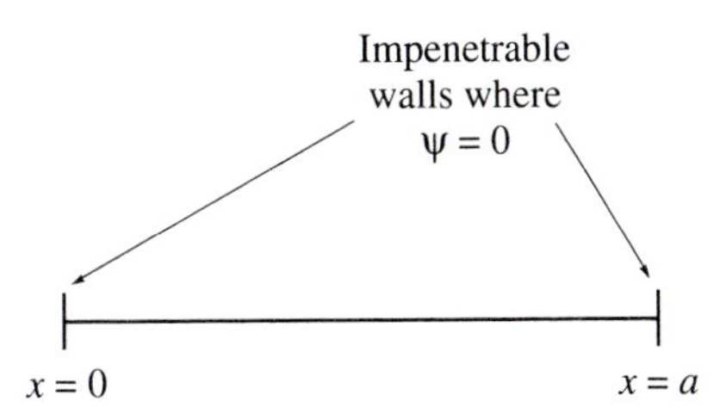

FIGURE 1.9
A one-dimensional box of confinement. The wave function must vanish at the walls.

Since the particle can be reflected at the walls, the most general wave solution representing it must contain waves going in both directions:

$$\phi(x) = A\exp(ikx) + B\exp(-ikx)$$

where $k = \sqrt{2mE/\hbar^2}$ and A and B are constants. Since $\phi(0) = 0$, we must have $B = -A$, and thus the solution becomes

$$\phi(x) = C\sin kx$$

with $C = 2iA$. Now impose the boundary condition $\phi(a) = 0$. This gives

$$\sin ka = 0$$

or $ka = n\pi$, with $n = 1,2,\ldots$. Thus k is quantized; let's call it k_n:

$$k_n = n\pi/a$$

and so energy, which is $\hbar^2k^2/2m$, is also quantized; call it E_n:

$$E_n = \frac{n^2\pi^2\hbar^2}{2ma^2} \tag{1.18}$$

with $n = 1,2,3,\ldots$. Thus quantization of energy does follow from the conditions of confinement, or boundary conditions, imposed on the Schrödinger waves.

The wave function of a state of quantized energy E_n in a box is therefore found to be

$$\phi_n(x) = C\sin\frac{n\pi x}{a}$$

Let's normalize this wave function:

$$\int_0^a |\phi_n(x)|^2\,dx = 1$$

This gives $C = \sqrt{2/a}$, and the normalized wave function for the particle within the confinement of the box is given as

$$\phi_n(x) = \sqrt{\frac{2}{a}}\sin\frac{n\pi x}{a} \tag{1.19}$$

The complete solution for a stationary state of a particle in a box is then given as

$$\begin{aligned}\psi_n(x,t) &= \phi_n(x)e^{-iE_nt/h} \\ &= \sqrt{\frac{2}{a}}\sin\frac{n\pi x}{a}\,e^{-iE_nt/h} \text{ for } 0 < x < a \\ &= 0, \text{ otherwise}\end{aligned}$$

and the probability of finding the particle at the position x, when in a stationary state $\psi_n(x,t)$, is independent of time:

$$P_n(x,t) = |\psi_n(x,t)|^2 = \frac{2}{a}\sin^2\frac{n\pi x}{a}$$

Figure 1.10 shows the wave functions for the lowest three energy states. Notice that the ground state has the least number of nodes (places where the wave function vanishes), and the excited states are characterized by a greater and greater number of nodes in the wave function.

Since the Schrödinger equation is linear, a superposition of $\psi_n(x,t)$'s is also a solution. Consider the solution

$$\psi(x,t) = \sqrt{\frac{1}{2}}\,[\psi_{n'}(x,t) + \psi_{n''}(x,t)] \qquad n' \neq n''$$

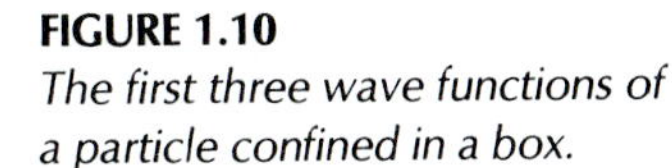
FIGURE 1.10
The first three wave functions of a particle confined in a box.

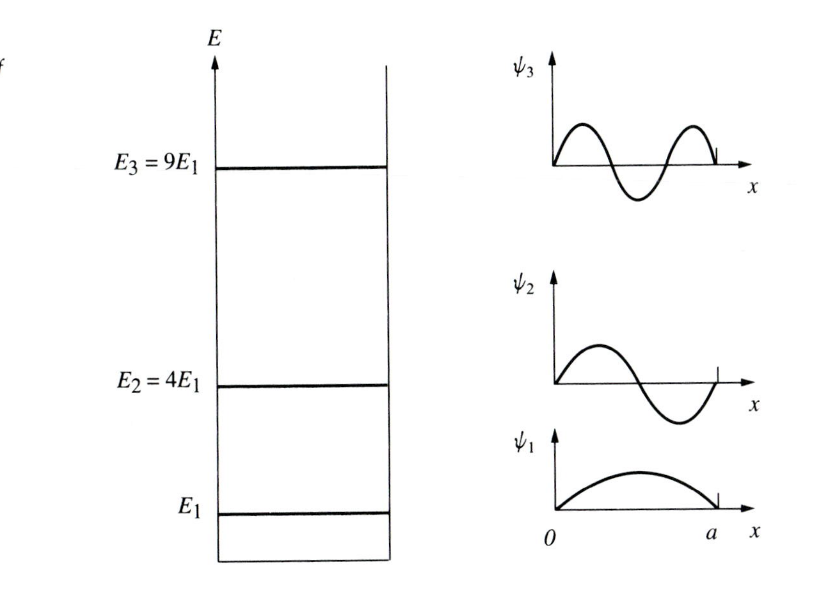

If we calculate the probability $P(x,t)$ with this wave function we get

$$\begin{aligned} P(x,t) &= |\psi(x,t)|^2 \\ &= \frac{1}{a}\left[\sin^2\frac{n'\pi x}{a} + \sin^2\frac{n''\pi x}{a}\right. \\ &\qquad \left. + 2\sin\frac{n'\pi x}{a}\sin\frac{n''\pi x}{a}\cos\{(E_{n'} - E_{n''})t/\hbar\}\right] \end{aligned} \tag{1.20}$$

There is now an oscillatory behavior, the probability changes with time, the state is not stationary. Furthermore, the probability changes with a frequency

$$\nu = \frac{E_{n'} - E_{n''}}{h}$$

which corresponds to the Bohr frequency condition for transition between the stationary states $\psi_{n'}$ and $\psi_{n''}$.

The following three additional points are useful to our understanding of a particle in a one-dimensional box.

1. According to quantum mechanics, there is a minimum energy, the ground-state energy $E_1 = \pi^2\hbar^2/2ma^2$, that the particle in a box must have, in contrast to classical physics where all energy values including $E = 0$ are permitted. You can easily see that this is a consequence of the uncertainty principle. We are confining the particle within the length a; thus there is an uncertainty in its position given by a. According to the uncertainty relation, equation (1.16), there has to be a corresponding momentum uncertainty of $\hbar/a$, and this implies that the particle has to have a minimum energy; its energy can never be zero because that would contradict the uncertainty relation. This minimum energy is called the *zero-point energy*.
2. The energy separation between successive quantized levels increases with the decrease of a, the confining length. Conversely, as a increases, the energy separations decrease; and when a is much larger than atomic dimensions, the energy separation is so small that we approach the classical correspondence limit. More explicitly, the $1/a^2$ dependence of the energy levels means that the energy level spacings are very large (~MeV) when the box is an atomic nucleus and decrease to ~eV when the box is an atom. When the box is a metallic crystal, the spacings become very small, $\sim 10^{-6}$ eV, approaching a continuum. Note that for large a, the zero-point energy also tends toward zero.
3. We have the qualitative situation in hand, but how about the quantitative situation? Let's calculate the energy levels of an electron ($m \approx 9.1 \times 10^{-28}$ g) confined to a box of atomic size ($a \approx 10^{-8}$ cm). Substituting in

the energy level formula, equation (1.18), we get $E_n \approx 40n^2$ eV. The energy difference between the ground and the first excited state is then 120 eV, and a photon emitted from a transition between them would have a frequency $\sim 3 \times 10^{16}$ Hz, which is on the same order as that in many atomic transitions.

■ **PROBLEM** A particle in a one-dimensional box (of length a) is in the ground state. Calculate the probability of finding the particle in the interval $\Delta x = 0.01a$ at the point $x = a/2$.

SOLUTION The normalized wave function for the ground state ($n = 1$) is given as

$$\psi_1 = \sqrt{\frac{2}{a}} \sin \frac{\pi x}{a} e^{-iE_1 t/\hbar}$$

Thus the required probability is given as

$$\int_{(a/2)-(\Delta x/2)}^{(a/2)+(\Delta x/2)} |\psi_1|^2 \, dx$$

However, since Δx is small, we can approximate the integral as

$$\begin{aligned} |\psi_1(x = a/2)|^2 \Delta x &= \frac{2}{a} \sin^2 \frac{\pi}{2} (0.01a) \\ &= 0.02 = 2\% \end{aligned}$$

■

1.7 THE PHASE OF THE WAVE FUNCTION

It may seem from the probability interpretation that the phase of the wave function is unimportant, since it is the modulo square $|\psi(x,t)|^2$ that we interpret as the probability; however, this is not the case because of the superposition principle.

Consider two solutions $\psi_1(x,t)$ and $\psi_2(x,t)$ of the Schrödinger equation. According to the superposition principle, the linear combination

$$\psi(x,t) = \alpha\psi_1(x,t) + \beta\psi_2(x,t) \tag{1.21}$$

where α and β are arbitrary complex numbers, is also a solution of the Schrödinger equation. Behold! The absolute square of the superposition crucially depends on the relative phase of ψ_1 and ψ_2. Such a superposition is called a *coherent superposition* and its use in the probability calculation is a guarantee that electron waves give interference similar to light waves.

You now can see a crucial difference between the use of probability in classical physics (for particles) and that in quantum physics. In classical physics, probabilities add, but in quantum physics, the probability amplitudes add (as in eq. [1.20]) giving interference.

However, an overall phase factor in the coherent superposition plays no role; we may as well write the coherent superposition as

$$\psi(x,t) = A(\psi_1(x,t) + \exp(i\theta)B\psi_2(x,t))$$

where A, B, and θ are real numbers, since this does not change the physics, namely, the probability and the interference.

In general, then, the choice of the absolute phase of a wave function is arbitrary.

1.8 THE PROBABILITY CURRENT

Coming back to the probability interpretation, we must require that

$$\int_{-\infty}^{+\infty} P(x,t=0)\,dx = 1 \tag{1.22}$$

since at some initial $t = 0$, the particle must be somewhere in space. Apart from normalization, this necessitates that we restrict the wave functions used in quantum mechanics to the class of square integrable functions for which

$$\int_{-\infty}^{\infty} dx\,|\psi(x,t)|^2 < \infty \tag{1.23}$$

that is, $\psi(x,t)$ must approach 0 as $x \to \infty$ at least as fast as $x^{-1/2-\epsilon}$, with $\epsilon > 0$, although arbitrarily small. Additionally, we will impose some continuity conditions on ψ.

To see why it is necessary to impose continuity conditions, let's try to prove that if the condition of equation (1.22) is satisfied for the initial state wave function, it is satisfied by the wave function at all times.

The Schrödinger equation for the free particle gives

$$i\hbar\frac{\partial\psi(x,t)}{\partial t} = -\frac{\hbar^2}{2m}\frac{\partial^2\psi(x,t)}{\partial x^2} \tag{1.24}$$

The complex conjugation of this equation gives the equation satisfied by ψ^*:

$$-i\hbar \frac{\partial \psi^*(x,t)}{\partial t} = -\frac{\hbar^2}{2m} \frac{\partial^2 \psi^*(x,t)}{\partial x^2} \tag{1.25}$$

Now, the time derivative of the probability $P(x,t) = \psi^*(x,t)\psi(x,t)$ is given as

$$\frac{\partial}{\partial t} P(x,t) = \frac{\partial \psi^*}{\partial t} \psi + \psi^* \frac{\partial \psi}{\partial t}$$

Substituting from equations (1.24) and (1.25), we get

$$\begin{aligned} \frac{\partial}{\partial t} P(x,t) &= \frac{1}{i\hbar}\left(\frac{\hbar^2}{2m} \frac{\partial^2 \psi^*}{\partial x^2} \psi - \frac{\hbar^2}{2m} \psi^* \frac{\partial^2 \psi}{\partial x^2}\right) \\ &= -\frac{\partial}{\partial x}\left[\frac{\hbar}{2im}\left(\psi^* \frac{\partial \psi}{\partial x} - \frac{\partial \psi^*}{\partial x} \psi\right)\right] \end{aligned}$$

We now define the *probability current* or *flux* by

$$j(x,t) = \frac{\hbar}{2im}\left[\psi^*(x,t) \frac{\partial \psi(x,t)}{\partial x} - \frac{\partial \psi^*(x,t)}{\partial x} \psi(x,t)\right] \tag{1.26}$$

Then we get

$$\frac{\partial}{\partial t} P(x,t) + \frac{\partial}{\partial x} j(x,t) = 0 \tag{1.27}$$

which is a continuity equation analogous to the continuity equation between the charge and current densities of electrodynamics. Hence $P(x,t)$ and $j(x,t)$ are sometimes referred to as the *probability density* and *probability current density*, respectively. Integrating the continuity equation, we get

$$\begin{aligned} \frac{\partial}{\partial t} \int_{-\infty}^{\infty} dx\, P(x,t) &= -\int_{-\infty}^{\infty} dx \frac{\partial}{\partial x} j(x,t) \\ &= 0 \end{aligned}$$

since for square integrable functions $j(x,t) \to 0$ at $x = \pm\infty$.

Thus, $\int_{-\infty}^{\infty} dx\, P(x,t)$ does not change in time; that is, once we have normalized the wave function at some initial time, the wave function will be normalized at later times.

Note that the continuity equation (1.27) is also a conservation law expressing the fact that a change in the particle density in a region of space is compensated for by a net change in flux *into* that region:

$$\frac{\partial}{\partial t}\int_a^b dx\, P(x,t) = -\int_a^b dx \frac{\partial}{\partial x} j(x,t)$$
$$= j(a,t) - j(b,t)$$

Now you can see why we need to impose continuity conditions: Both ψ and its derivative, $\partial\psi/\partial x$, must be continuous everywhere; otherwise, $j(x,t)$ would be singular at places, and these places would act as sources or sinks of current. In other words, matter creation or destruction would take place, which is impossible in nonrelativistic physics.

■ **PROBLEM** Find the probability current for the plane wave $\exp[i(kx - \omega t)]$.

SOLUTION For $\psi = \exp[i(kx - \omega t)]$, we have $\psi^*\partial\psi/\partial x = ik$; and $\psi\partial\psi^*/\partial x = -ik$. Substituting in the expression for the current, equation (1.26), we get

$$j(x,t) = (\hbar/2im)(ik - (-ik))$$
$$= \hbar k/m = p/m = v$$

where v is the velocity of the object; an anticipated result. ■

1.9 RELATION BETWEEN THE PHASE OF THE WAVE FUNCTION AND THE PROBABILITY CURRENT

Let's write $\psi(x,t)$ in the form

$$\psi(x,t) = \sqrt{P(x,t)}\exp\left(\frac{iS(x,t)}{\hbar}\right)$$

where P and S are real quantities. P is obviously the probability density, but what is the meaning of S, the phase of the wave function? To find out, calculate

$$\psi^* \frac{\partial\psi}{\partial x} = \sqrt{P}\frac{\partial}{\partial x}\sqrt{P} + \frac{i}{\hbar}P\frac{\partial S}{\partial x}$$

Substitute this value of $\psi^*\partial\psi/\partial x$ and its complex conjugate $\psi\partial\psi^*/\partial x$ into the expression for the probability current:

$$j(x,t) = \frac{\hbar}{2im}\left(\psi^* \frac{\partial \psi}{\partial x} - \frac{\partial \psi^*}{\partial x}\psi\right)$$
$$= \frac{1}{m} P \frac{\partial S}{\partial x} \tag{1.28}$$

Thus it is the spatial variation of the phase of the wave function that determines the strength of the current. The more the phase varies with distance, the greater the current.

Finally, we have imposed two mathematical conditions on the wave function ψ: (1) that it be square integrable and (2) that both $\psi(x)$ and $\partial\psi(x)/\partial x$ be continuous. We will impose a third condition, (3) that $\psi(x)$ be single-valued everywhere. Condition (3) is necessary so that a wave function can *uniquely* represent a given physical situation, especially so that the probability of finding a quantum object at a given place is given uniquely.

1.10 OUTLOOK

There are two aspects of learning quantum mechanics. The first and foremost is, as Richard Feynman used to say, to learn to calculate. However, the quantum mechanical way of calculating is quite different from classical ways; one finds that one has to get used to a new kind of thinking. And thus I would add a second dictum to Feynman's: Learn to think quantum mechanically. This will involve a certain amount of investment into the exploration of the meaning of quantum mechanics, but it's worth it. And if the exploration for the meaning occasionally shocks you, you can always take consolation from a comment that Niels Bohr made, "Those who are not shocked when they first come across quantum theory cannot possibly have understood it."

PROBLEMS

1. Use Planck's radiation frequency distribution law to obtain the energy density as a distribution in wavelength λ. Maximize the energy density to calculate the value of wavelength λ_{max} for which the energy density is maximum. Demonstrate that $\lambda_{max} = b/T$ and calculate the value of the constant b.
2. Classically, an electron in a circular orbit radiates at the same frequency as the frequency of its motion; that is, $\nu = v/2\pi r$, where v is the orbital velocity. Calculate the frequency associated with the electronic transition from the $(n + 1)$th Bohr orbit to the nth Bohr orbit and exhibit the correspondence principle in the limit of large n.

3. Calculate the de Broglie wavelength of (a) a 10-eV electron (b) a 10-MeV proton.
4. Schrödinger initially arrived at a wave equation for a particle of mass m that reads as follows in one dimension:

$$\frac{1}{c^2}\frac{\partial^2\psi}{\partial t^2} = \frac{\partial^2\psi}{\partial x^2} - \frac{m^2c^2}{\hbar^2}\psi$$

Show that a plane wave solution of this equation is consistent with the relativistic energy momentum relationship. (The equation above is the one-dimensional analog of what is now called the Klein-Gordon equation.)
5. Normalize the wave function

$$\psi(x) = \frac{C}{a^2 + x^2}$$

and find the expectation values of x and x^2 for this wave function.
6. Calculate the average value of the position of a particle whose wave function is given by

$$\psi(x) = \frac{1 + ix}{1 + ix^2}$$

7. A mass of 10^{-5} kg is moving with a speed of 10^{-2} cm/s in a box of length 1 cm. Treating this as the quantum mechanical particle-in-a-box problem, calculate the approximate value of the quantum number n. Discuss how the correspondence principle works in this case.
8. Consider an electron in a one-dimensional box with impenetrable walls of width 10^{-8} cm. Calculate (a) the three lowest allowed values of the energy of the electron and (b) the frequency of light that would cause the electron to jump from the ground to the third excited energy level. When the electron deexcites, what are the frequencies of the photons emitted? Plot the probability distribution for all three states and comment on where the electron is most likely to be found.
9. Consider an electron in the box of figure 1.9. What is the probability that the electron's position $x \geq a/2$?
10. Calculate the probability current for the plane wave $\exp[-i(kx + \omega t)]$. Interpret your result.
11. We have postulated that the quantum mechanical wave functions satisfy certain mathematical conditions of good behavior. Examine the following functions for this good behavior, pointing out why some of them do not qualify: (a) $\psi = x$ for $x \geq 0$, $\psi = 0$ elsewhere; (b) $\psi = x^2$; (c) $\psi = \exp(-x^2)$; and (d) $\psi = \exp(-x)$.

ADDITIONAL PROBLEMS

A1. Assume that the de Broglie formula $\lambda = h/p$ holds for photons for which $E = h\nu$. Show that photons then must have zero rest mass and that they must move with velocity c.

A2. Directly demonstrate the superposition property of the Schrödinger equation.

A3. A particle of mass m is in a one-dimensional impenetrable box extending from $x = 0$ to $x = a$. At $t = 0$, its wave function is given as:

$$\psi(x,0) = \frac{1}{\sqrt{a}}\left[\sin\frac{\pi x}{a} + \sin\frac{2\pi x}{a}\right]$$

(a) What is the probability for finding the particle in the ground state at $t = 0$? in the first excited state?

(b) What is the probability for finding the particle between $x = 0$ and $x = a/2$?

(c) Calculate (a) and (b) for time t.

A4. Consider a particle in the box of figure 1.9. Suppose the wave function of the particle is given as

$$\psi(x) = x\left(1 - \frac{x^2}{a^2}\right), \text{ for } 0 \le x < a$$

$$= 0, \text{ for } x > a$$

Find the probability that the particle will be in the ground state.

A5. Calculate the probability current density $j(x)$ for the wave function

$$\psi(x) = u(x)\exp[i\phi(x)],\ u,\ \phi \text{ real.}$$

REFERENCES

R. P. Feynman, R. B. Leighton, and M. Sands. *Feynman Lectures in Physics.*
P. Fong. *Elementary Quantum Mechanics.*
N. Herbert. *Quantum Reality.*
M. Jammer. *The Conceptual Development of Quantum Mechanics.*
R. L. Liboff. *Quantum Mechanics.*
E. H. Wichmann. *Quantum Physics.*

2

The Motion of Wave Packets

We want the solution of the Schrödinger wave equation to represent a localized particle. Localization requires us to construct a wave packet of Schrödinger waves and we do so via the mathematics of the Fourier transform: A wave packet centered at the position x_0 is given by the Fourier integral transformation

$$\psi(x, x_0) = \int_{-\infty}^{\infty} dk\, a(k, k_0) \exp[ik(x - x_0)] \tag{2.1}$$

where $a(k, k_0)$ is the Fourier transform (technically the exponential Fourier transform) of ψ. To see that ψ has the proper property of localization, notice that at $x = x_0$ the exponential goes to 1; therefore, the contributions to the integral coming from different k's add in phase (constructive interference). On the other hand, if $x - x_0$ is large, the contributions of different k oscillate and tend to cancel and we get zero for the integral (destructive interference) (fig. 2.1).

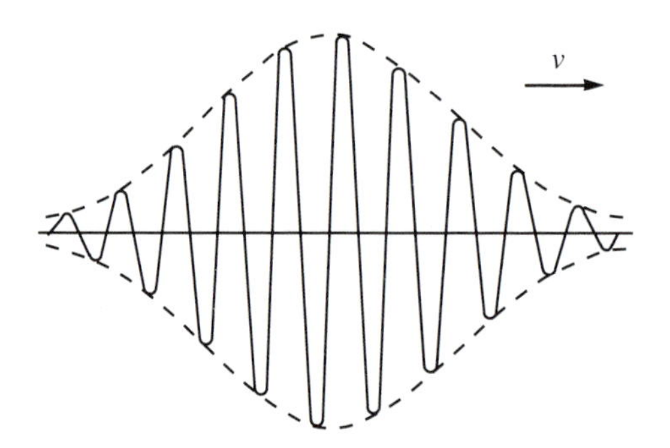

FIGURE 2.1
Schematic plot of (the real part of) a one-dimensional wave packet with a Gaussian envelope.

In this chapter, we will examine the properties of the wave packet as defined by equation (2.1), including its time evolution as determined by the Schrödinger equation. Our study will elucidate the nature of the wave-particle duality in quantum mechanics, and also the nature of the uncertainty relations.

2.1 GAUSSIAN WAVE PACKETS

Choose a Gaussian wave packet in k-space:

$$a(k) = \exp\left[-\frac{(k-k_0)^2}{2(\Delta k)^2}\right] \tag{2.2}$$

The modulo square of this wave packet is a bell curve (cf. fig. 1.8b) with an uncertainty or width given by Δk. (To see this, note that $|a(k)|^2$ starts to become small when $k - k_0 \geq \Delta k$.)

Substituting for $a(k)$ in equation (2.1), $\psi(x)$ is given by the integral

$$\psi(x) = \int_{-\infty}^{+\infty} dk \exp\left[-\frac{(k-k_0)^2}{2(\Delta k)^2} + ik(x-x_0)\right]$$

We want to convert this to the standard error integral

$$\int_{-\infty}^{+\infty} \exp(-z^2)\, dz = \sqrt{\pi} \tag{2.3}$$

To this effect we write the integral for ψ as

$$\psi(x,x_0) = \exp\left[ik_0(x-x_0) - \frac{(x-x_0)^2(\Delta k)^2}{2}\right]$$
$$\times \int_{-\infty}^{+\infty} dk \exp - \left[\frac{(k-k_0)^2}{2(\Delta k)^2} - 2\,\frac{k-k_0}{\sqrt{2}\Delta k}\,\frac{i(x-x_0)\Delta k}{\sqrt{2}} + \frac{i^2(x-x_0)^2(\Delta k)^2}{2}\right]$$

Now put

$$z = \frac{k-k_0}{\sqrt{2}\Delta k} - \frac{i(x-x_0)(\Delta k)}{\sqrt{2}}$$

The integral is then given as

$$\int_{-\infty}^{+\infty} dz \exp(-z^2)\sqrt{2}\,\Delta k = \sqrt{2\pi}\,\Delta k$$

We finally get

$$\psi(x,x_0) = \sqrt{2\pi}\,\Delta k \exp[ik_0(x - x_0) - \tfrac{1}{2}(x - x_0)^2(\Delta k)^2] \tag{2.4}$$

Thus, starting with a Gaussian function in k-space, the Fourier integral leads to a Gaussian function in x-space.

The Fourier mathematics makes it clear that the physical de Broglie-Schrödinger wave of a quantum object can be represented either by $\psi(x)$ in coordinate space or by $a(k)$ in wave-number space (which is the same as momentum space since $p = \hbar k$). Thus $a(k)$ is the wave function in wave-number space; $|a(k)|^2$ gives the probability that the momentum of the object is $\hbar k$.

Normalization

To normalize $\psi(x,x_0)$, we must require

$$\int_{-\infty}^{+\infty} dx\, \psi^*\psi = 1$$

Since

$$\int_{-\infty}^{+\infty} dx \exp[-(x - x_0)^2(\Delta k)^2] = \frac{\sqrt{\pi}}{\Delta k}$$

the normalization constant must be

$$\left[\frac{\Delta k}{\sqrt{\pi}}\right]^{1/2}$$

Thus the normalized wave packet is given as

$$\psi(x,x_0) = (\sqrt{\Delta k}/\pi^{1/4})\exp[ik_0(x - x_0) - \tfrac{1}{2}(x - x_0)^2(\Delta k)^2] \tag{2.5}$$

Uncertainties of Gaussian Functions

The probability corresponding to the wave packet defined by equation (2.5) is given as

$$P(x) = |\psi(x,x_0)|^2 \sim \exp[-(x - x_0)^2(\Delta k)^2]$$

Clearly $P(x)$ drops off to $1/e$ times its peak value when

$$(x - x_0)^2 = \frac{1}{(\Delta k)^2}$$

Consequently, the uncertainty in x is $\Delta x \sim 1/\Delta k$. Or, in general, we have

$$\Delta x \cdot \Delta k \geq 1$$

Now use the de Broglie relation $p = \hbar k$ and obtain the uncertainty relation between momentum and position:

$$\Delta x \cdot \Delta p \geq \hbar \tag{2.6}$$

There is also an alternative way to define the uncertainties that gives a slightly different result for the uncertainty relation. We can define the uncertainty Δx from

$$(\Delta x)^2 = \langle x^2 \rangle - \langle x \rangle^2 \tag{2.7}$$

where the notation $\langle f(x) \rangle$ denotes an expectation value as defined by equation (1.17). Let's simplify matters by choosing the center of the wave packet to be the origin so that $\langle x \rangle$ $(= x_0)$ is zero. Hence

$$(\Delta x)^2 = \langle x^2 \rangle = \int_{-\infty}^{\infty} dx\, x^2 |\psi(x, x_0 = 0)|^2 = \frac{\Delta k}{\sqrt{\pi}} \int_{-\infty}^{+\infty} dx\, x^2 \exp[-(\Delta k)^2 x^2]$$

Use the standard integral

$$\int_{-\infty}^{+\infty} dx\, x^{2n} e^{-x^2} = \frac{1 \cdot 3 \cdot 5 \cdots (2n-1)\sqrt{\pi}}{2^n} \qquad n = 1,2, \ldots$$

This gives

$$(\Delta x)^2 = \frac{1}{2(\Delta k)^2}$$

or

$$\Delta x = \frac{1}{\sqrt{2}\,\Delta k}$$

By the same token, we will define the expectation value of a function of k (or p) in the wave number or momentum space via the equation

$$\langle f(k) \rangle = \frac{\int_{-\infty}^{+\infty} dk\, a^*(k) f(k) a(k)}{\int_{-\infty}^{\infty} dk\, a^*(k) a(k)}$$

where we have to divide by the normalization integral of $a(k)$ since $a(k)$ in equation (2.2) is not normalized. Now we can easily calculate $\langle k^2 \rangle$ using equation (2.2) for $a(k)$ (assuming $k_0 = 0$; that is, assuming the packet to be centered at the origin in k-space) and using the standard integral above. We will get

$$\langle k^2 \rangle = \frac{(\Delta k)^2}{2} = \frac{\langle p^2 \rangle}{\hbar^2}$$

Now define the uncertainty in momentum in analogy to equation (2.7):

$$(\Delta p)^2 = \langle p^2 \rangle - \langle p \rangle^2 = \langle p^2 \rangle \tag{2.8}$$

since $\langle p \rangle = 0$, that is, the packet is centered at the origin in momentum space. This gives

$$\Delta p = \frac{\hbar \Delta k}{\sqrt{2}}$$

The product of Δx and Δp as calculated above is $\hbar/2$; this is the minimum uncertainty found for a Gaussian wave function. In general, we get

$$\Delta x \cdot \Delta p \geq \hbar/2 \tag{2.9}$$

Thus with the definitions in equations (2.7) and (2.8) for the uncertainties, the uncertainty relation is found to be numerically a little different from equation (2.6). So which is the correct one? We will show later from very general considerations (see chapter 6) that equation (2.9) is the correct expression for the uncertainty relation between x and p. So we conclude that the definitions of the uncertainties that led to equation (2.6), although qualitatively correct, must give way to the more accurate definitions based on mean-square deviations as in equations (2.7) and (2.8).

Group and Phase Velocities

Consider the position-localized wave packet centered at the origin including its time dependence

$$\psi(x,t) = \int_{-\infty}^{+\infty} a(k,k_0)\exp(ikx - i\omega t)\, dk \tag{2.10}$$

where, in general, the frequency ω is a complicated function of the wave number k, signifying that not only the position of the center of the packet but also the shape changes with time. If ω is a slowly varying function of the wave number k, we can Taylor-expand ω about $k = k_0$:

$$\omega = \omega_0 + \left(\frac{d\omega}{dk}\right)_{k=k_0}(k - k_0) + O(k - k_0)^2$$

Therefore we can write $\psi(x,t)$ as

$$\psi(x,t) \sim \exp(ik_0x - i\omega_0 t)\int_{\Delta k} a(k,k_0)\exp\left[i\left(x - \frac{d\omega}{dk}t\right)(k - k_0)\right] dk$$

The integral in the above equation is a modulating factor that modulates the wave of frequency ω_0 and wave number k_0. The phase velocity of the wave is given as

$$v_p = \frac{\omega_0}{k_0} = \frac{E}{p} = \frac{p}{2m} = \frac{v}{2}$$

It does not agree with the particle velocity. On the other hand, the modulating factor of the wave depends on x and t via the combination

$$x - \frac{d\omega}{dk}t$$

Thus the group velocity of the wave packet is given as

$$v_g = \frac{d\omega}{dk} = \frac{dE}{dp} \tag{2.11}$$

since $E = \hbar\omega$ and $p = \hbar k$.

Using $E = p^2/2m$, we get

$$\frac{dE}{dp} = \frac{p}{m} = v$$

Accordingly, the group velocity of the packet is exactly equal to the velocity of the particle it represents. It is then to the group velocity of the wave packet that we assign physical meaning.

2.2 SOME APPLICATIONS OF THE UNCERTAINTY RELATION

In this way, our attempt to simulate localization, a particle property, with waves via the construction of wave packets leads naturally to the uncertainty relation. What is the implication of the uncertainty principle for the do's and don'ts of the world—in particular, for the submicroscopic world?

Why We Cannot Squish Atoms

In the science fiction movie *The Fantastic Voyage* the main science fictiony idea was to squish people and things to miniature sizes. To the writers this might have seemed to be quite feasible; after all, everybody knows that the bulk of the atom is empty space anyway! To the uncertainty connoisseur, however, it is clear that squishing atoms is utterly impossible. Squishing means bringing the electron closer to the nucleus; this makes the uncertainty of the electron's position smaller, and correspondingly the uncertainty of the momentum must increase. So says the uncertainty relation. This would endow the electrons with runaway velocities, and the atoms would disintegrate.

We can do even better than this qualitative argument. We can show that the energy of the atom must be bounded from below due to the requirements of the uncertainty relation. Consider the hydrogen atom. What is the uncertainty of the electron's position? It is the radial distance from the nucleus, call it r. Call the momentum uncertainty p, then

$$pr \sim \hbar$$

We can use this relation in the expression for the energy to eliminate p. We get

$$E = \frac{p^2}{2m} - \frac{e^2}{r} = \frac{\hbar^2}{2mr^2} - \frac{e^2}{r}$$

The minimum of energy is given when

$$\frac{\partial E}{\partial r} = -\frac{\hbar^2}{mr^3} + \frac{e^2}{r^2} = 0$$

Solving for r, we get

$$r = \frac{\hbar^2}{me^2}$$

The corresponding value of E is found to be

$$E = -\frac{me^4}{2\hbar^2}$$

This agrees with the Bohr formula, equation (1.7), for $n = 1$. This is, of course, partly fortuitous, since we could easily have put in an extra factor such as $1/2$ (cf. eq. [2.9]) in the uncertainty relation we began with.

Time-Energy Uncertainty Relation

We will now manipulate the position-momentum uncertainty relation to derive an uncertainty relation between energy and time. We have

$$\Delta x \cdot \Delta p \geq \hbar$$

Write this as

$$\frac{m\Delta x}{p} \frac{p\Delta p}{m} \geq \hbar$$

The first factor is equal to $\Delta x/v$ or Δt, the uncertainty in the time-localizability of the wave packet, and it appears in conjunction with ΔE, the uncertainty in energy (since $E = p^2/2m$, $\Delta E = p\Delta p/m$). Thus we get

$$\Delta t \cdot \Delta E \geq \hbar \tag{2.12}$$

This is the desired time-energy uncertainty relation.

We can derive the time-energy uncertainty relation from the wave packet formalism, too. Represent $\psi(t)$ by the Fourier integral

$$\psi(t) = \frac{1}{\sqrt{2\pi}} \int_{-\infty}^{\infty} d\omega \, a(\omega) e^{-i\omega t}$$

$a(\omega)$ is the Fourier transform of ψ in frequency space and it can be obtained from $\psi(t)$ via the inverse Fourier integral

$$a(\omega) = \frac{1}{\sqrt{2\pi}} \int_{-\infty}^{\infty} dt \, \psi(t) e^{i\omega t}$$

As before, the Fourier transform of a Gaussian distribution function is a Gaussian distribution function. Thus, if we take

$$\psi(t) \sim \exp\left[-\frac{t^2}{2(\Delta t)^2}\right]$$

then

$$a(\omega) \sim \int_{-\infty}^{\infty} dt\, e^{-[t^2/2(\Delta t)^2]+i\omega t}$$
$$\sim e^{-\omega^2(\Delta t)^2/2}$$

The width of $|a(\omega)|^2$ is $\Delta\omega \approx 1/\Delta t$. A wave train whose duration is short is necessarily made up of a larger range of frequencies; this is why high fidelity in AM radio broadcasting requires a broad frequency bandwidth. Finally, since $E = \hbar\omega$, equation (2.12) follows.

However, the alternative derivation that we gave for the position-momentum uncertainty relation falls through in the case of time energy—we cannot define the expectation value of a function of time as we can for a function of position. There seems to be a fundamental difference between position and time in quantum mechanics, relativity notwithstanding. Because of this, many authors still debate the validity of the time-energy uncertainty relation.

But physically, the time-energy uncertainty relation is significant. It tells us that in order for a quantum state to have a well-defined value of energy, the state must be really stationary—it must last a long, long time. Indeed, when a quantum state is short-lived, such as an excited state of an atom, its energy is smeared just a little, as revealed by the *natural* width that all spectral lines exhibit.

Heisenberg's Resolution of Zeno's Paradox

One question that might have led Heisenberg to the uncertainty principle is, Can an object occupy a given place and be moving at the same time? This is actually an ancient question first raised by an ancient Greek philosopher, Zeno, in his paradox, "The Arrow." Here's a recap.

> The arrow cannot fly. It cannot fly because an object that is behaving itself in a uniform manner is either continually moving or continually at rest. The arrow is certainly behaving itself in such a uniform manner. Now watch the arrow as it travels along its flight path. Clearly, at any instant, the arrow is occupying a given place. Therefore, if it is occupying a given place, it must be at rest there. The arrow must be at rest at the instant we picture, and since the instant we have chosen is any instant, the arrow cannot be moving at any instant. Thus the arrow is always at rest and cannot fly.†

Aristotle saw through part of the problem that this paradox exposed about movement and our perception of it. You can understand Aristotle's resolution best by thinking through a modern example of how a filmmaker creates motion out of still frames; if he or she is careful to remove nonuniformities, the still pic-

†Quoted in F. A. Wolf, *Taking the Quantum Leap*, p. 17.

tures, when run at 24 frames per second, certainly generate the perception of continuous motion to our eyes. But that's just an illusion. Now imagine that we run the frames at such a rate that the time interval between two consecutive frames is infinitesimal, practically frozen "still points" of time dissolving into one another. If we can imagine that we are taking snapshots of the flying arrow for each of these still points of time, one frame will dissolve to the next to show the arrow moving continuously.

Unfortunately, this involves an infinite number of frames passing before our eyes each second—a real problem of infinities!

Heisenberg more clearly saw Zeno's real concern, that trying to localize the position of a moving arrow does violence to the fact of its movement. Heisenberg, via his uncertainty principle, recognized that an object cannot be said to occupy a given place and be moving at the same time in a predictable fashion. Any attempt to take a snapshot of an object results only in giving us its position, but we lose information about its state of motion.

In general, then, the uncertainty relation makes it impossible to talk about sharply defined trajectories of objects, because it is impossible to avoid fundamental errors (due to the uncertainty principle) in the simultaneous ascertaining of initial conditions—initial values of both position and velocity.

By the same token, you should give up thinking of the atom in terms of Bohr orbits. If you are reluctant, expecting to map out the orbit of the electron by measurement, think again. Suppose you want to do an experiment that ascertains the position of an atomic electron right on the nth orbit. Then you must do your position measurement so that any error is much less than the radial separation between the nth and $(n-1)$th orbit (using eq. [1.7])

$$\Delta x \ll r_n - r_{n-1} \cong 2\hbar^2 n/me^2$$

But this will cost you an uncertainty in p given as

$$\Delta p \gg me^2/2n\hbar$$

Translate this into an uncertainty in energy:

$$\Delta E = p\Delta p/m \gg (e^2/\hbar n)\cdot(me^2/2\hbar n) = me^4/2\hbar^2 n^2$$

where we have used equation (1.7). In other words, the uncertainty in energy is much greater than the binding energy of the electron for the nth orbit. Thus any attempt at trying to map out the electron's orbit by measurement will knock the electron right out of its orbit.

Hence the wise doctrine of Heisenberg: "What does not exist in the theory cannot be observed in an experiment describable by the same theory." See chapter 5 for further discussion on this point.

■ **PROBLEM** Billiard balls of mass 100 g and diameter 10 cm travel with velocity $v \cong 1$ m/s over a distance $l \cong 1$ m. (a) Assume that the relative uncertainty of position $\Delta x/l$ is equal to the relative uncertainty of momentum $\Delta p/mv$ and that the uncertainty product $\Delta x \cdot \Delta p$ has the smallest value permitted. Estimate Δx. (b) If one ball is kept fixed and the other balls are scattered off of it, show that the uncertainty relation gives the well-known formula for diffraction by a circular screen.

SOLUTION (a) We have from the uncertainty principle, $\Delta p \cdot \Delta x = \hbar/2$, and from the conditions of the problem, $\Delta p = \Delta x(p/L)$. Combining, we get $(\Delta x)^2 = \hbar l/2p$. Thus

$$\Delta x = (\hbar l/2p)^{1/2} = (\hbar l/2mv)^{1/2} \approx 2 \times 10^{-15} \text{ cm}$$

which is smaller than the diameter of the proton. Clearly, you can't blame Heisenberg for your difficulties at the pool table.

(b) Suppose for the scattered ball, $\Delta x = d$. Since $\Delta x \cdot \Delta p \approx \hbar$, we have

$$\Delta p/p = \hbar/(p\,\Delta x) = \hbar/pd = \Lambda/d,$$

where Λ is the de Broglie wavelength of the ball. But $\Delta p/p = \sin\theta$, where θ is the angle of scattering of the ball. It follows that

$$d \sin\theta = \Lambda$$

the desired diffraction formula. How far from the "diffraction screen" do you have to be to see the intensity maximum at the center of the geometrical shadow (called the Poisson spot)? Consult figure 2.2. Clearly,

$$R = (d/2)/\sin\theta \approx d^2/\Lambda$$

Since $\Lambda = 2\pi\hbar/mv \approx 6 \times 10^{-21}$ cm, this gives

$$R \approx 10^{32} \text{ cm},$$

which is greater than the current size of the universe! ■

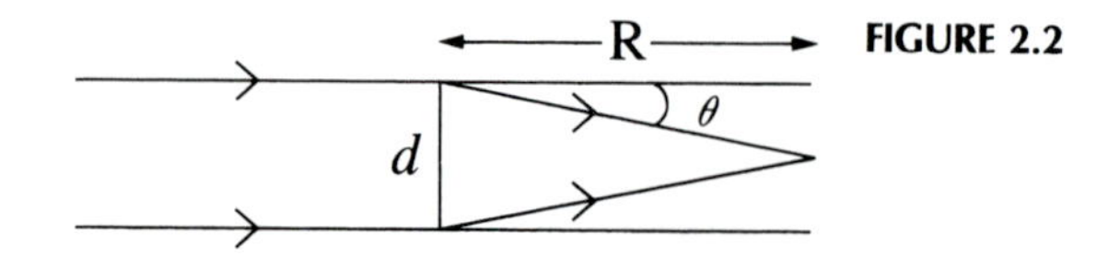

FIGURE 2.2

2.3 SPREADING OF A GAUSSIAN WAVE PACKET WITH TIME

It may now seem that the wave packet way of making a localized particle out of a dispersed wave is the solution to the problem of wave-particle duality—the question as to how quantum objects can be both wave and particle. Is the charge of the electron all spread out because it disperses as a wave? No, we are tempted to say; our trouble arises from thinking of waves as plane waves. As soon as we think of a wave packet, we can see that the charge is localized inside the packet.

But does it stay localized? Since the wave packet must satisfy the Schrödinger equation at all times, what does this fact say about the spreading of the wave packet? If the packet remains localized, we have solved the wave-particle puzzle. But what if the packet spreads?

Let's calculate. The calculation takes two steps.

First Step Given any wave function $\psi(x, x_0, t = 0)$, we can always calculate its Fourier transform $a(k, k_0)$ by inverting the Fourier integral. If

$$\psi(x, x_0, 0) = \int_{-\infty}^{+\infty} dk\, a(k, k_0) e^{ik(x-x_0)}$$

then

$$a(k, k_0) = \frac{1}{2\pi} \int_{-\infty}^{+\infty} dx\, \psi(x, x_0, 0) e^{-ik(x-x_0)} \tag{2.13}$$

which is the wave function in wave-number space.

The proof of this inversion is instructive because it will enable us to introduce a very useful function. Substitute for ψ in equation (2.13):

$$a(k, k_0) = \frac{1}{2\pi} \int_{-\infty}^{+\infty} dx \int_{-\infty}^{+\infty} dk'\, a(k', k_0) e^{ik'(x-x_0)} e^{-ik(x-x_0)}$$

where we have changed the dummy integration variable in ψ from k to k' to avoid conflict of notation. Now assume that we can interchange the order of the two integrations. This gives us

$$a(k, k_0) = \int_{-\infty}^{+\infty} dk'\, a(k', k_0) \left[\frac{1}{2\pi} \int_{-\infty}^{+\infty} dx\, e^{i(k'-k)(x-x_0)} \right]$$

The integral within the square bracket is called the Dirac delta function:

$$\delta(k' - k) = \frac{1}{2\pi} \int_{-\infty}^{+\infty} dx\, e^{i(k'-k)x} \tag{2.14}$$

For the equation for $a(k,k_0)$ above to hold, the delta function must have the unusual property that it vanishes when $k' \neq k$, and tends to infinity when $k' = k$ (fig. 2.3), in such a way that

$$\int_{-\infty}^{\infty} dk'\, a(k')\delta(k'-k) = a(k) \tag{2.15}$$

It is in the form of this kind of integral that the delta function can be shown to have any meaning, although in quantum mechanics we will often manipulate the delta function by itself, taking it for granted that in any meaningful equation the function will always occur in an integral such as the above (see the appendix for further exposition on the δ function).

So given a $\psi(x,x_0,0)$ at some initial time, we can always calculate its Fourier transform $a(k,k_0)$. The crucial point is to recognize that $a(k,k_0)$ does not depend on time; thus the ψ at a later time must have the same $a(k,k_0)$ for its Fourier transform. That is,

$$\psi(x,x_0,t) = \int_{-\infty}^{\infty} dk\, a(k,k_0) e^{i\{k(x-x_0)-\omega t\}} \tag{2.16}$$

Second Step The second step consists of carrying out the above procedure explicitly. Our initial normalized wave packet is

$$\psi(x,x_0) = (\sqrt{\Delta k}/\pi^{1/4})\exp[ik_0(x-x_0) - \tfrac{1}{2}(x-x_0)^2(\Delta k)^2]$$

Therefore, using equation (2.13), $a(k,k_0)$ is given as

$$a(k,k_0) = \frac{\sqrt{\Delta k}}{2\pi\pi^{1/4}} \int_{-\infty}^{\infty} dx\, e^{-(x-x_0)^2(\Delta k)^2/2} e^{-i(k-k_0)(x-x_0)}$$

This integral is similar to one we did before; the trick is to complete the square in x and use the error integral, equation (2.3). The final result is

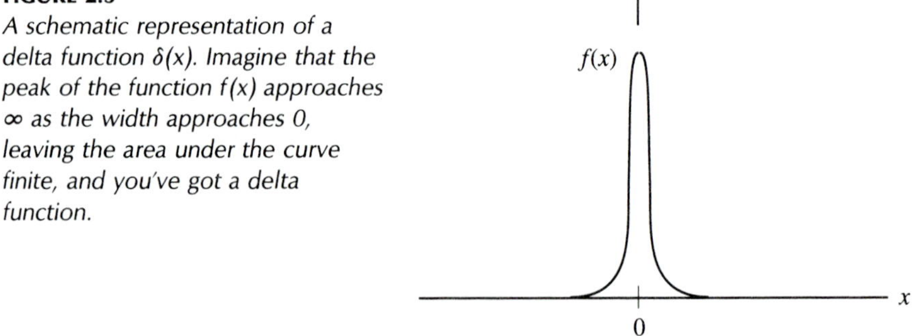

FIGURE 2.3
A schematic representation of a delta function $\delta(x)$. Imagine that the peak of the function $f(x)$ approaches ∞ as the width approaches 0, leaving the area under the curve finite, and you've got a delta function.

$$a(k,k_0) = \frac{1}{\sqrt{2\pi\Delta k}\,\pi^{1/4}} \exp\left[-\frac{(k-k_0)^2}{2(\Delta k)^2}\right]$$

We are back where we began, since this is the same function we began with, equation (2.2), except for the constant. This is, of course, to be expected. We now substitute this value of $a(k,k_0)$ into $\psi(x,x_0,t)$, equation (2.16), to obtain the latter. For simplicity, we will put $x_0 = 0$ (that is, center the packet at the origin).

$$\psi(x,t) = \frac{1}{\sqrt{2\pi\Delta k}\,\pi^{1/4}} \int_{-\infty}^{\infty} dk \exp\left[-\frac{(k-k_0)^2}{2(\Delta k)^2} + ik\left(x - \frac{\hbar k t}{2m}\right)\right]$$

where we have substituted $\omega = \hbar k^2/2m$. Once more, completing the square and using the error integral leads us to the desired result (the steps are messy, however, and will be left to the reader as an exercise):

$$\psi(x,t) = \frac{1}{\gamma\sqrt{\Delta k}\,\pi^{1/4}} \exp\left[-\frac{\left(x - \frac{\hbar k_0 t}{m}\right)^2}{2\gamma^2}\right] \exp\left[ik_0\left(x - \frac{\hbar k_0 t}{2m}\right)\right] \tag{2.17}$$

where

$$\gamma^2 = \frac{1}{(\Delta k)^2} + \frac{i\hbar t}{m}$$

Now let's calculate the width of the wave packet. The probability is

$$\begin{aligned} P(x,t) &= |\psi(x,t)|^2 \\ &= \frac{1}{|\gamma|^2 \Delta k\, \pi^{1/2}} \exp\left[-\frac{1}{(\Delta k)^2} \frac{\left(x - \frac{\hbar k_0 t}{m}\right)^2}{|\gamma|^4}\right] \end{aligned} \tag{2.18}$$

We come to the conclusion that the width of the packet changes with time. At $t = 0$, the width was $1/\Delta k$, but at time t the width is, as seen from equation (2.18),

$$\begin{aligned} \Delta x(t) &= \Delta k|\gamma|^2 \\ &= \Delta k\left[\frac{1}{(\Delta k)^4} + \frac{\hbar^2 t^2}{m^2}\right]^{1/2} \\ &= \frac{1}{\Delta k}\left[1 + \frac{\hbar^2 t^2 (\Delta k)^4}{m^2}\right]^{1/2} \end{aligned}$$

Introducing the notation

$$T = \frac{m}{\hbar(\Delta k)^2} \tag{2.19}$$

for the time constant of the spreading of the packet, we get for the width at time t

$$\Delta x(t) = \Delta x(t=0)\left[1 + \left(\frac{t}{T}\right)^2\right]^{1/2} \tag{2.20}$$

where $\Delta x(t=0)$ is, of course, $1/\sqrt{2}\Delta k$.

Behold! You cannot keep the wave-particle duality of quantum objects under wraps just by introducing wave packets. The packets spread with time. For $t \ll T$, the packet does not grow noticeably, but for $t \gg T$, the original form is completely bloated. For an electron, we can estimate the time constant T for the spreading of the packet by putting $1/\Delta k$ $(= \Delta x(t=0))$ equal to the electron's Compton wavelength $\hbar/mc$:

$$\begin{aligned} T &= \frac{m}{\hbar}\frac{1}{(\Delta k)^2} = \frac{m}{\hbar}\left(\frac{\hbar}{mc}\right)^2 \\ &= \frac{\hbar}{mc^2} \sim 10^{-21}\ \text{s} \end{aligned}$$

Obviously, even though we may localize an electron within its Compton wavelength at some initial time, its wave packet will spread all over town in a matter of seconds.

It is this spreading of the wave packet that promotes jokes like the following. Perhaps you will like the quantum mechanical way of materializing a Thanksgiving turkey: Prepare your oven and wait; there is a nonzero probability that a turkey from a nearby grocery store will materialize in your oven. For best results, start sufficiently ahead of time. And remind yourself if you get impatient that this is how poets wait for the idea of a poem to materialize in their heads.

Unfortunately for the turkey lover, for macroscopic masses, the spreading is small (it better be! remember the correspondence principle?). For $m = 10^{-6}$ g, and initial width $1/\Delta k = 10^{-2}$ cm, $T \sim 10^{17}$ s, the spreading is macroscopically appreciable only in a time comparable to the lifetime of the universe.

Notice that the maximum of the Gaussian that $P(x,t)$ (eq. [2.18]) represents has shifted from the initial center $x = 0$ to

$$x = \frac{\hbar k_0}{m} t$$

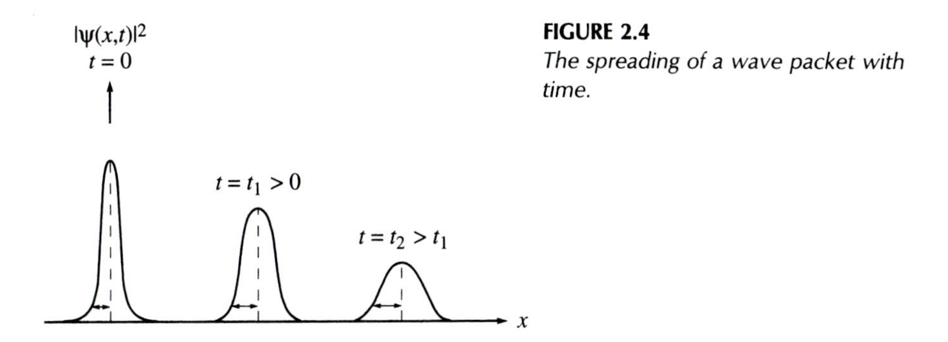

FIGURE 2.4
The spreading of a wave packet with time.

You can identify $\hbar k_0/m$ as the group velocity v_g of the wave group that $\psi(x,t)$ represents.

Figure 2.4 sketches the behavior of a Gaussian wave packet with time including the change in the width and the shift of the center.

Notice also that since Δx increases with time and Δp remains constant, the uncertainty product $\Delta x \cdot \Delta p$ must also grow with time.

For $t \gg T$, we have

$$\Delta x(t) \cong \frac{\hbar}{m\Delta x(0)}\, t$$

Thus, the narrower the packet is initially, the quicker it spreads. You can see in this the hidden influence of the uncertainty principle. If the confinement length is small, the uncertainty in velocity $\Delta v \approx \hbar/(m\Delta x(0))$ is large, meaning that the packet will contain many waves of high velocity, much greater than the average group velocity p_0/m. Because of the velocity fluctuation, the distance covered by the particle must also be uncertain by a proportionate amount of

$$\Delta v t = \frac{\hbar}{m\Delta x(0)}\, t$$

Classical Limit

Let's look at the classical limit, $\hbar \to 0$, to see if all quantum effects disappear.

If we let $\hbar \to 0$ in equation (2.18) for the probability $P(x,t)$, since $|\gamma|^2 \to 1/(\Delta k)^2 = (\Delta x(t=0))^2$, we get

$$P(x,t) = \frac{1}{\sqrt{\pi}\,\Delta x(t=0)} \exp\left[-\left(x - \frac{p_0 t}{m}\right)^2 \Big/ (\Delta x(t=0))^2\right]$$

This describes a classical free particle whose momentum is precisely p_0, but whose position is distributed about the origin according to a Gaussian of width $\Delta x(t=0)$. Now take the limit

$$\Delta x(t=0) \to 0$$

This is facilitated by one of the representations of the δ function given in the appendix

$$\delta(x) = \lim_{\alpha \to 0} \frac{1}{\alpha\sqrt{\pi}} e^{-x^2/\alpha^2}$$

We get

$$P(x,t) \to \delta(x - p_0 t/m)$$

This means that the probability is zero for finding the particle anywhere except on the classical trajectory of the particle

$$x = p_0 t/m$$

In other words, we are back to classical determinism, as expected.

Note that the order in which the limits are taken is important. Check it out.

Another important thing to note is that although $P(x,t)$ has a well-defined classical limit, the wave function $\psi(x,t)$ does not (i.e., its phase does not).

Wave Functions Corresponding to Definite Momentum and Position

In equation (2.1), suppose that the wave function in momentum space $a(k, k_0)$ is a δ function:

$$a(k, k_0) = \delta(k - k_0)$$

Then

$$\psi(x, x_0) = \int_{-\infty}^{\infty} dk\, \delta(k - k_0) e^{ik(x-x_0)}$$
$$= e^{ik_0(x-x_0)}$$

Now ψ in position space is a plane wave. Since the modulo square of a plane wave is 1, it follows that the probability of finding the particle in any position is the same. In other words, the position uncertainty is infinite. It follows from the uncertainty relation that the momentum uncertainty must be zero. That is, when the momentum space wave function is a δ function, $\delta(k - k_0)$, the momentum has a definite value $\hbar k_0$. And in position space, the particle is now represented by a plane wave.

Likewise, now we can see that the position space wave function corresponding to a definite position is to be represented by a δ function, $\delta(x - x_0)$.

To summarize, in position space, the wave function that corresponds to a definite value of momentum is the plane wave, and the wave function that corresponds to a definite value of position is a delta function.

Wave-Particle Duality

Here are a couple of puzzling questions that can finally be discussed in full. Does the quantum picture of the electron waving around the atomic nucleus imply that the electron's charge and mass are smeared all over the atom? Or does the fact that a free electron spreads out, as a wave must according to the theory of Schrödinger (the wave of a free particle is a plane wave, $\exp(ikx)$, which has the probability of 1 to appear at any position x), mean that the electron is everywhere with its charge now smeared all over space? In other words, how do we reconcile the wave picture of the electron with the fact that it has particle-like localized properties? The answers are subtler than we thought when we began this chapter.

You may have thought that with wave packets, at least, we should be able to confine the electron in a small place. Alas, things don't stay that simple! A wave packet that satisfies the Schrödinger equation at a given moment of time has to spread with the passage of time as shown above. How do we reconcile *that* with the picture of a localized particle? The answer is that we must include the act of observation in our reckoning.

If we want to measure the electron's charge, we must intercept it with something, like a cloud of vapor in a cloud chamber. As a result of this measurement, we must assume that the electron's wave collapses, so now we are able to see the electron's track through the cloud of vapor. Thus another wise saying of Heisenberg: "The path of the electron comes into existence only when we observe it." So you see, when we measure it, we always find the electron localized as a particle. The electron does spread out according to Schrödinger's theory, but probabilistically, in *potentia*, to use Heisenberg's word (who took it from Aristotle). The electron cloud defines the volume within which the electron may be located with nonzero probability—within the cloud the electron exists in formless potentia. Only when we measure it is the electron manifested and the wave is "reduced" to a particle.

When Schrödinger introduced his wave equation, he and others thought that maybe they had gotten rid of quantum jumps—discontinuity—from physics, since wave motion is continuous. But the particle nature of quantum objects has to be reconciled with the wave nature, and thus wave packets have to be introduced. Finally, with the recognition of the spreading of the wave in potentia, and with the realization that it is our observation that must *instantly* collapse the size of the packet, we also see that the collapse has to be discontinuous. Quantum jumps are back once again.

It sure seems that we cannot have quantum mechanics without quantum jumps. Schrödinger made a famous visit to Bohr in Copenhagen where he made his protests against quantum jumps for days, ultimately conceding the point with this emotional outburst, "If I knew that one has to accept this damned quantum jump, I'd never have gotten involved with quantum mechanics."

Coming back to the atom, if we measure the position of the electron while in an atomic stationary state, we will again collapse its probability cloud to find it in a particular position, not smeared everywhere. If we make a large number of measurements to look for the electron, we will find it more often at those places where the probability of finding it is high (as predicted by the Schrödinger equation). Indeed, after a large number of measurements, if we plot the distribution of the measured positions, it will look quite like the distribution given by the solution of the Schrödinger equation (fig. 2.5). The details of such solutions will be given in chapter 13.

And how about Zeno's arrow from this perspective? If the arrow is a submicroscopic atomic projectile, when we make an initial observation of it, we find it localized in a wave packet. But after the observation, the packet spreads; the spread of the packet is the cloud of our uncertainty about the packet. If we observe again, the packet localizes once more, but always spreading between our observations (fig. 2.6). Even for the macroscopic arrow, quantum mechanics predicts essentially the same picture, the only difference being that the spreading of the wave packet is small between observations.

Now we are really getting to the crux of the matter. Whenever we measure it, the electron always appears at some one place, as a particle. The probability distribution simply identifies the place or places where it is likely to be found when we do measure—no more than that. On the other hand, when we are not measuring it, the electron does spread, in potentia, and it does exist in more than one place at the same time, as a wave, as a cloud—no less than that.

The wave and particle natures of the electron are not dualistic, simply opposite polarities, said Niels Bohr, but they are complementary properties that are revealed to us in complementary experiments. In truth, electrons are neither wave nor particle; their true nature transcends both descriptions, said Bohr. Since this *complementarity principle* is a subtle, even puzzling, concept, we will postpone a more complete treatment until chapter 5.

To summarize, according to the *Copenhagen interpretation* of quantum mechanics developed through the ideas of Born, Heisenberg, and Bohr, we cal-

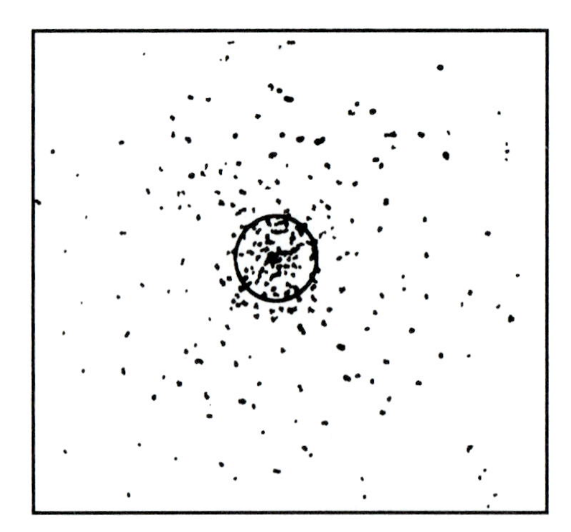

FIGURE 2.5
The results of repeated measurements of the position of a hydrogenic electron in the lowest energy state. The solid circle indicates where the predicted radial probability distribution reaches its peak. Note that the electron's wave function usually collapses, upon measurement, where the probability for finding it is high.

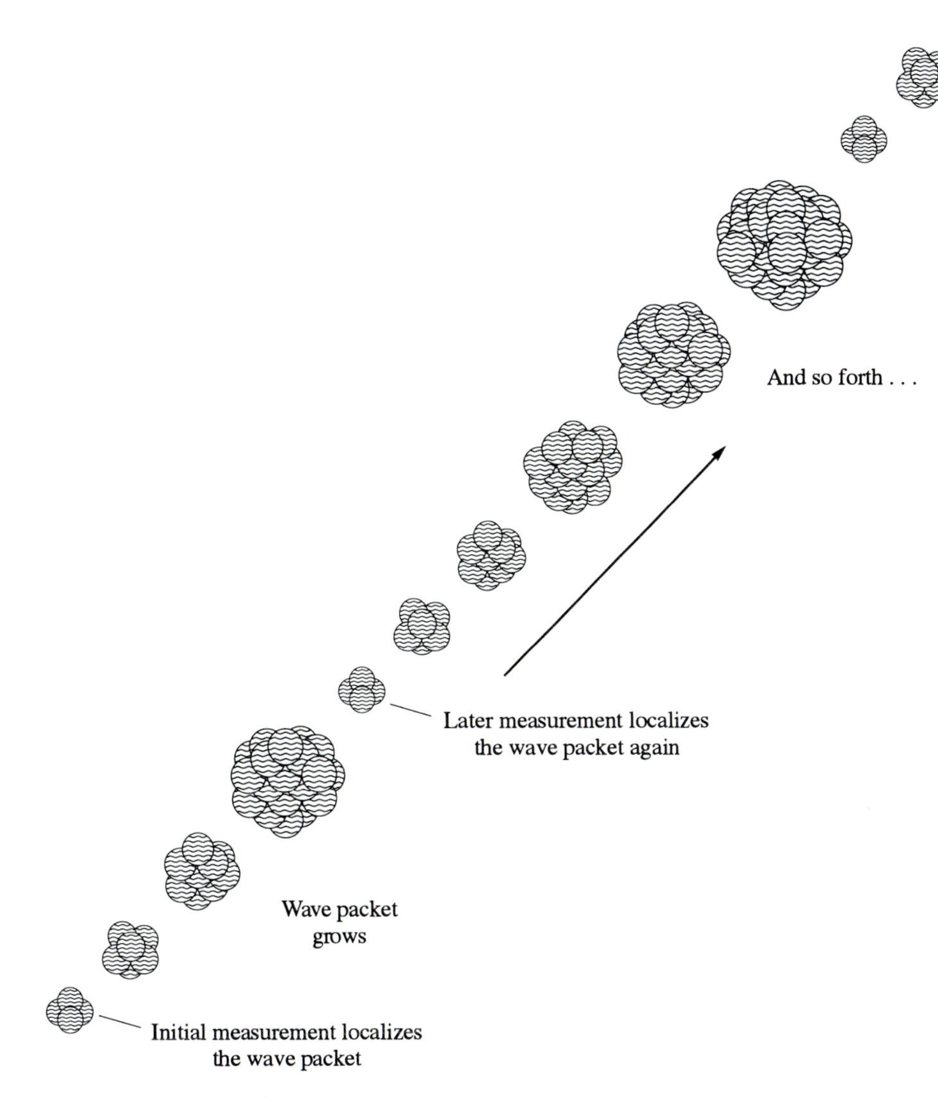

FIGURE 2.6
An artist's rendition of the time evolution of a quantum object. The wave packet of the object grows in size in between measurements. Measurement collapses the packet. (Adapted with permission from R. H. March, Physics for Poets, *2d ed. New York: McGraw-Hill, 1978.)*

culate quantum objects probabilistically, we determine their attributes somewhat uncertainly, and we understand them complementarily. Quantum mechanics with the Copenhagen interpretation presents a new and exciting world view that challenges old concepts such as deterministic trajectories of motion and causal continuity. If initial conditions do not forever determine the motion, if instead, every time we observe, there is a new beginning, the world is creative at the base level.

PROBLEMS

1. Find the Fourier transform of the following function:

$$\psi(x) = 1 - |x| \qquad |x| < 1$$
$$= 0 \qquad |x| \geq 1.$$

2. Calculate the wave function $\psi(x)$ using the Fourier integral, equation (2.1), when the Fourier transform $a(k)$ is given as

$$a(k) = 0 \qquad k < -k_0$$
$$= N \qquad -k_0 < k < k_0$$
$$= 0 \qquad k_0 < k$$

Normalize $\psi(x)$ to find the value of N. Show by defining Δx in a suitable (but reasonable) manner that $\Delta k \cdot \Delta x > 1$ independent of the value of k_0.

3. Suppose at $t = 0$, a system is in the state given by the wave function

$$\psi(x,0) = \frac{1}{\sqrt{L}} \qquad |x| < L/2$$
$$= 0 \qquad \text{otherwise}$$

If, at the same instant, the momentum of the particle is measured, what are the possible values that will be found, and with what probability?

4. The relation between wavelength and frequency for a wave guide is given as

$$\lambda = \frac{c}{\sqrt{\nu^2 - \nu_0^2}}$$

Calculate the group velocity of the wave.

5. Show that the phase velocity and group velocity for the plane wave of a relativistic electron are given as $v_p = c^2/v$ and $v_g = v$, respectively. How would you interpret these results?

6. Use the uncertainty relation to find an estimate of the ground-state energy of a harmonic oscillator. The energy of the oscillator is given as

$$E = \frac{p^2}{2m} + \frac{1}{2} m\omega^2 x^2$$

where ω is the oscillator frequency.

7. Use the uncertainty relation to estimate the ground-state energy of a particle sitting on a table of height H under the influence of gravity.
8. You are on top of a ladder of height L and are throwing marbles of mass m to the floor, aiming at a crack using equipment of the highest precision, of course. Even so, quantum mechanics says, you will miss the crack. Why? Estimate the typical distance by which you will miss the crack. (*Hint*: Take the typical transverse distance to be $\Delta x + (\Delta p/m)$[time of fall].)
9. Perform the integral that led to equation (2.17).
10. An excited state of an atom has a lifetime of 10^{-9} s. What is the natural line width of the energy level?
11. Electrons of kinetic energy 100 eV travel a distance of 1 km. If the size of the initial wave packet is given as 1 micron, what is the size at the end of their travel?
12. An arrow of mass 200 g travels for 10 s with a velocity of 10 m/s. If the initial packet has a size of 1 cm, what is the size of the packet at its destination?

ADDITIONAL PROBLEMS

A1. One way to measure the distance of an object is to send a pulse of radar microwaves to the object and measure the time between radiation of the original and reception of the reflected pulses. Suppose you want to measure the range R with a precision of 1000 m. What width Δt of the pulse should you use? How wide would be the bandwidth of frequencies passed by the amplifier into the receiver?

A2. Electrons in beta decay are emitted from the nuclei with energies as high as a few MeV. Assuming the electron energy to be 1 MeV and the nuclear size to be 10^{-13} cm, show that electrons cannot be confined in the nucleus before the decay. (*Hint*: Use the uncertainty principle.)

A3. Suppose the position uncertainty of an object is a. What is its momentum uncertainty? Hence estimate the energy of the object.

A4. Suppose you measure the position of an electron, initially at rest, within an accuracy of 10^{-11} m. Estimate the momentum imparted to the electron by the measurement process.

A5. Gravitational waves disperse in water according to the law

$$v_p = \left(\frac{\lambda g}{2\pi}\right)^{1/2}$$

Calculate the group velocity.

A6. Consider an object with the wave function $\psi(x) = c(a^2 + x^2)^{-1}$. Normalize the wave function. (a) Calculate the expectation value $\langle x^n \rangle$, $n = 1,2,3,\ldots$ where x is the position of the object. Calculate the root mean square deviation Δx. (b) Calculate the expectation value $\langle p^n \rangle$, $n = 1,2,3,\ldots$ where p is the momentum of the object. Calculate Δp. Now calculate $\Delta x \cdot \Delta p$. (*Hint*: Use the Fourier transform of $\psi(x)$.)

A7. Suppose instead of a Gaussian, we use,

$$a(k) = (\hbar/\delta)^{1/2} \exp[-\hbar(k - k_0)/\delta]$$

which is properly normalized and peaks about $k = k_0$. Show that the Fourier transform of this function, which is the position wave function at time $t = 0$, $\psi(x,0)$, is given as

$$\psi(x,0) = (2\hbar^3/\pi\delta^3)^{1/2} \exp[ik_0 x][x^2 + \hbar^2/\delta^2]^{-1}$$

Show that $\psi(x,0)$ is normalized. Now calculate the uncertainty product based on mean-square deviations and compare with the uncertainty product based on the spreads of the two wave functions. (You may have to consult a table of integrals for this problem.)

REFERENCES

S. Gasiorowicz. *Quantum Physics.*
W. Heisenberg. *The Physical Principles of Quantum Theory.*
W. Heisenberg. *Physics and Philosophy.*
J. L. Powell and B. Crasemann. *Quantum Mechanics.*

3

Schrödinger Equation as Eigenvalue-Eigenfunction Equation

So far we have considered the Schrödinger equation for free particles. This is nice for getting acquainted with quantum mechanics, but by far the more important problems of physical nature arise from the action of force fields on objects. How do we generalize the Schrödinger equation?

The generalization is facilitated when we interpret the free-particle Schrödinger equation using the mathematical concept of an "operator." What is an operator? If you exercise, your muscles build up by the action of the exercise; exercise changes the muscles. The action of operators on functions is a bit like that of exercise on muscles: They change the functions. In quantum mechanics, physical quantities such as x and p are represented by operators because they bring about changes in the wave functions upon which they act. The difference between the exercise operator and our quantum mechanical operators is that the latter are described in precise mathematical terms. In this chapter, we will study some of the most important operators in quantum mechanics, and show how the Schrödinger formalism is a method for calculating wave functions for which one or more of these operators have measurable characteristic values or "eigenvalues."

In chapter 2, we calculated expectation values $\langle x^2 \rangle$ with the position wave function $\psi(x)$ and $\langle p^2 \rangle$ with the momentum wave function $a(p)$. Tacitly, we assumed that we know how x "operates" on $\psi(x)$ and how p operates on $a(p)$, namely, by multiplication. Somehow the operator nature of x and p is trivial in these examples. Now let's ask, "How does p operate on $\psi(x)$?" as an introduction to a nontrivial quantum mechanical operator.

3.1 THE MOMENTUM OPERATOR

We will begin with the classical equations

$$p = mv = m\frac{dx}{dt}$$

Now consider the expectation value of p in position space via this equation:

$$\begin{aligned}\langle p\rangle &= m\frac{d}{dt}\langle x\rangle \\ &= m\frac{d}{dt}\int_{-\infty}^{\infty} dx\,\psi^*(x,t)x\psi(x,t) \\ &= m\int_{-\infty}^{\infty} dx\left(\frac{\partial\psi^*(x,t)}{\partial t}x\psi(x,t) + \psi^*(x,t)x\frac{\partial\psi(x,t)}{\partial t}\right)\end{aligned}$$

Note that all change in x with time is being determined by the change in ψ; there is no $\partial x/\partial t$ term above. This is how Schrödinger's mechanics works. Now use the Schrödinger equation

$$i\hbar\frac{\partial\psi}{\partial t} = -\frac{\hbar^2}{2m}\frac{\partial^2\psi}{\partial x^2}$$

and its complex conjugate equation to substitute for $\partial\psi/\partial t$ and $\partial\psi^*/\partial t$. This gives

$$\langle p\rangle = -\frac{i\hbar}{2}\int_{-\infty}^{\infty} dx\left[\frac{\partial^2\psi^*}{\partial x^2}x\psi - \psi^* x\frac{\partial^2\psi}{\partial x^2}\right]$$

Since we are dealing with square integrable functions, the integral of all derivatives vanishes. Thus the trick for evaluating the above integral is to express it, as far as we can, as a derivative. To this end, we write

$$\begin{aligned}\frac{\partial^2\psi^*}{\partial x^2}x\psi &= \frac{\partial}{\partial x}\left[\frac{\partial\psi^*}{\partial x}x\psi\right] - \frac{\partial\psi^*}{\partial x}\psi - \frac{\partial\psi^*}{\partial x}x\frac{\partial\psi}{\partial x} \\ &= \frac{\partial}{\partial x}\left[\frac{\partial\psi^*}{\partial x}x\psi\right] - \frac{\partial}{\partial x}(\psi^*\psi) + \psi^*\frac{\partial\psi}{\partial x} \\ &\quad - \frac{\partial}{\partial x}\left[\psi^* x\frac{\partial\psi}{\partial x}\right] + \psi^*\frac{\partial\psi}{\partial x} + \psi^* x\frac{\partial^2\psi}{\partial x^2}\end{aligned}$$

Accordingly, the integrand above is given as

$$\frac{\partial}{\partial x}\left[\frac{\partial \psi^*}{\partial x} x\psi - \psi^* x \frac{\partial \psi}{\partial x} - \psi^*\psi\right] + 2\psi^* \frac{\partial \psi}{\partial x}$$

As stated above, the integral of the derivative vanishes, and we are left with

$$\langle p \rangle = \int_{-\infty}^{\infty} dx\, \psi^*(x,t)\left(-i\hbar \frac{\partial}{\partial x}\right)\psi(x,t)$$

This suggests that in position space, momentum is represented by the differential operator

$$\hat{p} = -i\hbar \frac{\partial}{\partial x} \tag{3.1}$$

To emphasize its difference from a multiplicative function, whenever there may be an ambiguity we will denote an operator by a hat on its head, as on p in equation (3.1).

We will now verify that in momentum space, all this still leads to our old assumed result—that $\hat{p}$ is simply multiplicative. To this end, we rewrite the Fourier integral for $\psi(x)$ and its inverse $a(p = \hbar k)$ as follows:

$$\psi(x) = \frac{1}{\sqrt{2\pi\hbar}} \int_{-\infty}^{\infty} dp\, a(p) e^{ipx/\hbar} \tag{3.2}$$

and

$$a(p) = \frac{1}{\sqrt{2\pi\hbar}} \int_{-\infty}^{\infty} dx\, \psi(x) e^{-ipx/\hbar} \tag{3.3}$$

The advantage of this "renormalization" can be seen as follows. Suppose ψ is normalized:

$$\int_{-\infty}^{\infty} dx\, \psi^*(x)\psi(x) = 1$$

Then, according to equations (3.2) and (3.3), we have

$$\int_{-\infty}^{\infty} dp\, a^*(p)a(p) = \int_{-\infty}^{\infty} dp\, a^*(p) \frac{1}{\sqrt{2\pi\hbar}} \int_{-\infty}^{\infty} dx\, \psi(x)e^{-ipx/\hbar}$$
$$= \int_{-\infty}^{\infty} dx\, \psi(x) \frac{1}{\sqrt{2\pi\hbar}} \int_{-\infty}^{\infty} dp\, a^*(p)e^{-ipx/\hbar}$$
$$= \int_{-\infty}^{\infty} dx\, \psi(x)\psi^*(x) = 1$$

Thus with this renormalized definition, if we start with a normalized function, we end up with a normalized Fourier transform. This is known as Parseval's theorem.

Back to the consideration of the form of $\hat{p}$ in momentum space. We have

$$\langle \hat{p} \rangle = \int_{-\infty}^{\infty} dx\, \psi^*(x) \frac{\hbar}{i} \frac{d\psi(x)}{dx}$$
$$= \int_{-\infty}^{\infty} dx\, \psi^*(x) \frac{\hbar}{i} \frac{d}{dx} \frac{1}{\sqrt{2\pi\hbar}} \int_{-\infty}^{\infty} dp\, a(p)e^{ipx/\hbar}$$
$$= \int_{-\infty}^{\infty} dp\, a(p)p \frac{1}{\sqrt{2\pi\hbar}} \int_{-\infty}^{\infty} dx\, \psi^*(x)e^{ipx/\hbar}$$
$$= \int_{-\infty}^{\infty} dp\, a(p)pa^*(p)$$

as anticipated. When $\hat{p}$ operates on $a(p)$, it is simply multiplicative, it doesn't have a hat. (This result also justifies our calling $a(p)$ the wave function in momentum space.) Take notice because this is a general rule: An operator is multiplicative in its own "home" space where it has no hat.

More on Operators

Generally speaking, an operator acts on a function and changes it into another unless it is operating on functions in its home space. In quantum mechanics, measurable attributes associated with objects, called *observables*, are represented by operators. Let's now discuss four important aspects of quantum mechanical operators.

First, operators in quantum mechanics are generally *linear* operators (though not without exception). What is a linear operator? A linear operator is defined by the following two properties:

$$\hat{L}(\psi_1(x) + \psi_2(x)) = \hat{L}\psi_1(x) + \hat{L}\psi_2(x) \tag{3.4a}$$

and

$$\hat{L}c\psi(x) = c\hat{L}\psi(x) \tag{3.4b}$$

where c is an arbitrary complex number. It is easy to see that both x and p satisfy the criteria of linearity. Check it out!

■ **PROBLEM** Consider two operators $\hat{O}_1$ and $\hat{O}_2$ defined by the following operations:

$$\hat{O}_1 \psi(x) = \psi(x) + x$$

$$\hat{O}_2 \psi(x) = \frac{d\psi(x)}{dx} + 2\psi(x)$$

Check for the linearity of $\hat{O}_1$ and $\hat{O}_2$.

SOLUTION Let's check for the linearity of $\hat{O}_1$ first.

$$\begin{aligned} \hat{O}_1 [\psi_1(x) + \psi_2(x)] &= \psi_1(x) + \psi_2(x) + x \\ &\neq \hat{O}_1 \psi_1(x) + \hat{O}_1 \psi_2(x) \end{aligned}$$

Thus $\hat{O}_1$ is not a linear operator. On the other hand,

$$\begin{aligned} \hat{O}_2 [\psi_1(x) + \psi_2(x)] &= \frac{d[\psi_1(x) + \psi_2(x)]}{dx} + 2[\psi_1(x) + \psi_2(x)] \\ &= \hat{O}_2 \psi_1(x) + \hat{O}_2 \psi_2(x) \end{aligned}$$

It is also easy to see that

$$\hat{O}_2 c\psi(x) = c\hat{O}_2 \psi(x)$$

Thus $\hat{O}_2$ is linear. ■

Second, operators do not necessarily commute, in contrast to ordinary complex numbers. For this reason they are sometimes referred to as q-numbers as opposed to ordinary "c-numbers" that obey a commutative algebra.

In particular, the operators $\hat{x}$ and $\hat{p}$ do not commute. Let's denote the commutator $\hat{x}\hat{p} - \hat{p}\hat{x}$ by the *commutator bracket* $[\hat{x}, \hat{p}]$.

$$\begin{aligned} [\hat{x}, \hat{p}] \psi(x,t) &= x \frac{\hbar}{i} \frac{\partial \psi(x,t)}{\partial x} - \frac{\hbar}{i} \frac{\partial}{\partial x} x\psi(x) \\ &= i\hbar \psi(x,t) \end{aligned}$$

Thus $\hat{x}$ and $\hat{p}$ do not commute; instead they satisfy the *commutation relation*

$$[\hat{x}, \hat{p}] = i\hbar \tag{3.5}$$

This means that we have to be careful about the order of operators in considering operator products in quantum mechanics. It's a little bit like learning to discriminate the order in which you ask, "Do you like me?" and "Do you love me?" In the latter case the romantic learns very quickly that the order matters. So it is, even if you are not a quantum romantic, the order of operators matters.

An important thing about the commutation relation, equation (3.5), is that it holds true not only in position space but also in momentum space. From what has been said above it must already be clear to you that the position operator *cannot* be a multiplicative operator in momentum space. An easy way to see that its form in momentum space must be

$$\hat{x} = i\hbar \frac{\partial}{\partial p}$$

is to check out that with this form, the same commutation relation, equation (3.5), obtains in momentum space as well.

Third, the momentum operator $-i\hbar\partial/\partial x$ has an "i" in it. How do we know that its expectation value is real?

By definition

$$\langle \hat{p} \rangle = \int_{-\infty}^{\infty} dx\, \psi^*(x) \frac{\hbar}{i} \frac{d\psi}{dx}$$

The complex conjugate of this equation is given as

$$\langle \hat{p} \rangle^* = \int_{-\infty}^{\infty} dx \psi(x)(-) \frac{\hbar}{i} \frac{d\psi^*}{dx}$$

The difference is

$$\langle \hat{p} \rangle - \langle \hat{p} \rangle^* = \frac{\hbar}{i} \int_{-\infty}^{\infty} dx \frac{d}{dx} (\psi^* \psi)$$

The integral is zero for square integrable functions, and thus $\langle \hat{p} \rangle$ is real.

Occasionally, we use wave functions in quantum mechanics that are not square integrable (such as $\exp(ikx)$), but in such a case we restrict ourselves to a bounded region $0 \leq x \leq L$, and impose *periodic boundary conditions* on ψ:

$$\psi(x) = \psi(x + L)$$

With this manipulation, we can ensure that $\langle p \rangle$ is real:

$$\langle p \rangle - \langle p \rangle^* = \frac{\hbar}{i} \int_0^L dx \frac{d}{dx} (\psi^* \psi) = \frac{\hbar}{i} (|\psi(L)|^2 - |\psi(0)|^2)$$
$$= 0$$

Operators with real expectation values are called *hermitian operators*. That is, for any hermitian operator $\hat{O}$,

$$\langle \hat{O} \rangle_\psi = \langle \hat{O} \rangle_\psi^* \tag{3.6}$$

where the notation indicates the wave function in which the expectation value will be calculated. This gives an alternative definition of a hermitian operator:

$$\int dx\, \psi^* \hat{O} \psi = \int dx\, (\hat{O}\psi)^* \psi \tag{3.7}$$

In quantum mechanics, all *observables*, measurable attributes of systems, are represented by hermitian operators.

What is the meaning of the expectation value of an observable $\hat{O}$? In terms of measurement, conceptualize that you have measured the quantity for a large number of identical systems described by the wave function ψ, performing identical experiments. Suppose all these equivalent experiments give you the values $o_1, o_2, o_3, \ldots o_n$. Then the expectation value of the observable $\hat{O}$ is given as

$$\langle \hat{O} \rangle = (1/n) \Sigma_n o_n$$

Fourth, time is not an operator in quantum mechanics (here is the difference between x and t in quantum mechanics that we hinted at in chapter 2, only more technically put) but energy, the conjugate variable of time via the uncertainty relation, is. Now we turn to this energy operator.

3.2 GENERAL FORM OF THE SCHRÖDINGER EQUATION: THE HAMILTONIAN OPERATOR

Let's go back to the free-particle Schrödinger equation

$$i\hbar \frac{\partial \psi}{\partial t} = -\frac{\hbar^2}{2m} \frac{\partial^2 \psi}{\partial x^2}$$

We now note that since

$$\hat{p} = \frac{\hbar}{i}\frac{\partial}{\partial x}$$

we can write the Schrödinger equation above in the form

$$i\hbar \frac{\partial \psi(x,t)}{\partial t} = \frac{\hat{p}^2}{2m}\psi(x,t)$$

But the operator acting on ψ on the right-hand side can be recognized as the operator for the kinetic energy of a free particle, which is all the energy that a free particle has. It is customary to call the operator representing the total energy of a system the *Hamiltonian*, denoted by $\hat{H}$. Thus the Schrödinger equation for the free particle can also be written as

$$i\hbar \frac{\partial \psi(x,t)}{\partial t} = \hat{H}\psi \tag{3.8}$$

This form of writing the Schrödinger equation opens us to the general form of the equation even when the particle is not free, when it moves in a potential $V(x)$. The Hamiltonian, or the total energy operator, in this case, can be anticipated from classical mechanics to be

$$\hat{H} = \frac{\hat{p}^2}{2m} + V(x) \tag{3.9}$$

Thus the Schrödinger equation for a particle moving in a potential $V(x)$ is given as

$$\begin{aligned} i\hbar \frac{\partial \psi(x,t)}{\partial t} &= \left[\frac{\hat{p}^2}{2m} + V(x)\right]\psi(x,t) \\ &= -\frac{\hbar^2}{2m}\frac{\partial^2 \psi(x,t)}{\partial x^2} + V(x)\psi(x,t) \end{aligned} \tag{3.10}$$

Henceforth, equation (3.8) (in conjunction with eq. [3.9]) or equation (3.10) will be referred to as the Schrödinger equation (in one dimension).

Notice that $\hat{H}$ is a linear, hermitian operator. It is the most important operator in quantum mechanics. In order to apply quantum mechanics to a system, in order to write down the system's Schrödinger equation, we need to know the Hamiltonian of the system. To do quantum mechanics is to know your Hamiltonian.

So how do we know our Hamiltonian? For most cases we just take it from classical mechanics, relying on the correspondence principle, and saving our ingenuity only for those cases where there is no classical analog (e.g., spin).

One other thing. In converting a classical Hamiltonian to a quantum mechanical one, we must replace all observables by their operator counterparts in quantum mechanics. Occasionally, we may encounter a case where the product of x and p occurs in the classical Hamiltonian. Since in quantum mechanics the order of the operators matters, in which order do we put x and p? In such a case, we simply symmetrize the product:

$$xp \rightarrow \frac{1}{2}(xp + px)$$

One more thing in passing. If it bothers you that $V(x)$ was introduced in the Schrödinger equation (eq. [3.10]) without the quantum scale factor $\hbar$, the reason is that it is the total energy that is quantized, and in position space the $\hbar$ in the momentum accomplishes the task. For momentum space, indeed, $V(x)$ would contain $\hbar$, and not the kinetic energy term.

Schrödinger Equation as Eigenvalue-Eigenfunction Equation

The equation

$$i\hbar \frac{\partial\psi(x,t)}{\partial t} = \left[-\frac{\hbar^2}{2m}\frac{\partial^2}{\partial x^2} + V(x)\right]\psi(x,t) \tag{3.10}$$

is called the *time-dependent* Schrödinger equation. Let's separate the time and space variables in this equation by a standard mathematical procedure. Substitute

$$\psi(x,t) = u(x)T(t)$$

This gives

$$i\hbar u(x)\frac{dT(t)}{dt} = \left[-\frac{\hbar^2}{2m}\frac{d^2u(x)}{dx^2} + V(x)u(x)\right]T(t)$$

Divide by $u(x)T(t)$. We get

$$\frac{1}{T(t)}\, i\hbar \frac{dT(t)}{dt} = \frac{1}{u(x)}\left[-\frac{\hbar^2}{2m}\frac{d^2}{dx^2} + V(x)\right]u(x)$$

Because the variable dependence has been separated, the equation above can be satisfied only if both sides are equal to a constant. Call this constant E. We get

$$i\hbar \frac{dT(t)}{dt} = ET(t)$$

which is easily integrated to give

$$T(t) = Ce^{-iEt/\hbar}$$

where C is an integration constant. You can see that E is none other than the total energy of the system. The equation for $u(x)$

$$\left[-\frac{\hbar^2}{2m}\frac{d^2}{dx^2} + V(x)\right]u(x) = Eu(x) \tag{3.11}$$

is called the *time-independent* Schrödinger equation. We note that the quantity within the square brackets is the Hamiltonian operator $\hat{H}$. Consequently, equation (3.11) can also be written as

$$\hat{H}u(x) = Eu(x) \tag{3.12}$$

If an operator acting on a function gives the function back multiplied by a constant, the function is called the *eigenfunction* of the operator, and the constant its *eigenvalue*. Accordingly, equation (3.12) is an eigenvalue-eigenfunction equation for the operator H, since $u(x)$ defines a special class of functions, energy eigenfunctions, that are such that H operating on them gives back the same function multiplied by the eigenvalue E of H. Thus whereas the time-dependent Schrödinger equation describes the development of the system in time, the time-independent Schrödinger equation is an eigenvalue equation for the Hamiltonian of the system, the eigenvalues being the allowed values for the total energy of the system.

Of particular importance is that the eigenvalue E of H as determined from the solution of equation (3.11) is real. Presently, we will prove this using the continuity relation, equation (1.27). The probability associated with the wave function

$$\psi(x,t) = u(x)e^{-iEt/\hbar} \tag{3.13}$$

is given as

$$P(x,t) = \psi^*(x,t)\psi(x,t) = u^*(x)u(x)e^{-i(E-E^*)t/\hbar}$$

Substituting in the continuity equation

$$\frac{\partial P(x,t)}{\partial t} + \frac{\partial j(x,t)}{\partial x} = 0 \tag{1.27}$$

and integrating over x, we get

$$-\frac{i}{\hbar}(E - E^*)\int_{-\infty}^{\infty} dx\, u^*(x)u(x) = -\int_{-\infty}^{\infty} dx\, \frac{\partial j(x,t)}{\partial x}$$

For square integrable functions, the right-hand side above vanishes, and it follows that $E = E^*$, and E is real.

If E is real, the time-dependent phase of $\psi(x,t)$ in equation (3.13) is purely oscillatory. Not only that, but the probability density $P(x,t)$ is now time independent, so that ψ is stationary. It should also be noted that the expectation value of any time-independent operator O in a state described by such a stationary wave function is independent of time and can be calculated from the time-independent wave function alone (this is another reason we call the states represented by these wave functions stationary states).

$$\langle \hat{O} \rangle = \int dx\, \psi^*(x,t)\hat{O}\psi(x,t) = \int dx\, u^*(x)\hat{O}u(x) \tag{3.14}$$

If $\hat{O}$ is $\hat{H}$, the Hamiltonian itself, and u is normalized, then

$$\langle \hat{H} \rangle = \int dx\, u^*(x)\hat{H}u(x) = E\int dx\, u^*(x)u(x) = E$$

The eigenvalue E is thus also the expectation value of the total energy of the system.

We will see below that the great importance of the time-independent Schrödinger equation is that its solution not only gives us a particular solution of the full Schrödinger equation (3.10) but promises to give us all solutions of physical interest.

Note that the different eigenvalues E of a system will each belong to its own eigenfunction. If the eigenvalue is discrete with quantum number n, we denote the corresponding eigenfunction as $u_n(x)$; on the other hand, if the eigenvalue E is continuous (that is, there is no restriction on the values it can take), we label the eigenfunction by $u_E(x)$.

We now can write the most general solution of the Schrödinger equation (3.10) as a coherent superposition of $u(x)T(t)$:

$$\psi(x,t) = \sum_n A_n u_n(x)e^{-iE_nt/\hbar} + \int dE\, A_E u_E(x)e^{-iEt/\hbar} \tag{3.15}$$

The Case of the Free Particle: Momentum Eigenfunctions

We started quantum mechanics with the free-particle Schrödinger equation. Let's look at it again from the vantage point of an eigenvalue-eigenfunction equation. The Hamiltonian is simply

$$H = \frac{p^2}{2m} = -\frac{\hbar^2}{2m}\frac{\partial^2}{\partial x^2}$$

The time-independent Schrödinger equation gives

$$\frac{d^2u(x)}{dx^2} = -\frac{2mE}{\hbar^2}u(x) = -k^2u(x)$$

where $k^2 = 2mE/\hbar^2$.

Notice that for each eigenvalue E, there are two solutions, two wave functions, $\exp(ikx)$ and $\exp(-ikx)$. This situation where there is more than one eigenfunction for a given eigenvalue is called *degeneracy*. The free-particle solutions are *degenerate*. This degeneracy has to do with their being simultaneous eigenfunctions of another operator, the momentum operator.

The eigenvalue-eigenfunction equation for the momentum operator is easily constructed:

$$\hat{p}u_p(x) = -i\hbar\frac{du_p(x)}{dx} = pu_p(x)$$

where p denotes the eigenvalue. The solution is

$$u_p(x) = Ce^{ipx/\hbar} = Ce^{ikx} \tag{3.16}$$

where, as usual, C is a constant of integration, which is to be evaluated from normalization. The eigenvalues p must be real so that the wave function remains finite at both $x = +\infty$ and $x = -\infty$. There is no other constraint on p, so that p defines a continuous spectrum.

Now you can see the explanation of the degeneracy of the energy eigenfunctions of the free-particle problem; they are also simultaneously eigenfunctions of momentum. The solution $\exp(ikx)$ belongs to the eigenvalue $+p$ $(+\hbar k)$ of momentum (the particle moving in the positive x-direction), and $\exp(-ikx)$ belongs to the eigenvalue $-p$ $(= -\hbar k)$, but the energy is the same in both cases.

Finally, how do we normalize the free particle wave function when clearly the normalization integral considered in the usual fashion diverges? One way to deal with this problem is to impose periodic boundary conditions. We demand that the wave function take on the same value on any two points separated by the basic length L. The wave function, equation (3.16), is now constrained; this loss of freedom is manifest in that k is allowed to take on only a sequence of discrete values

$$k = \frac{2\pi n}{L} \qquad (n = 0, \pm 1, \pm 2, \ldots)$$

The normalization of the eigenfunctions is now achieved by requiring

$$|C|^2 \int_{-(1/2)L}^{(1/2)L} dx\, |e^{ikx}|^2 = 1$$

This gives $C = 1/\sqrt{L}$ and the free-particle wave function "box normalized" by means of the periodic boundary condition is given as

$$u(x) = \frac{1}{\sqrt{L}} e^{ikx} = \frac{1}{\sqrt{L}} e^{ipx/\hbar} \tag{3.17}$$

Consider now two such eigenfunctions belonging to two different values of the wave number k and k'. The integral

$$\int_{-L/2}^{L/2} dx\, u_{k'}^*(x) u_k(x) = \frac{1}{L} \int_{-L/2}^{L/2} dx\, e^{i(k-k')x}$$

is called the orthogonality integral. When this integral vanishes, as is the case above, the two functions u_k and $u_{k'}$ are said to be orthogonal. We can combine the orthogonality and the normalization conditions of these wave functions into a single *orthonormality relation*:

$$\int_{-L/2}^{L/2} dx\, u_{k'}^*(x) u_k(x) = \delta_{kk'} \tag{3.18}$$

where $\delta_{kk'}$ is the Kronecker delta defined as

$$\begin{aligned} \delta_{kk'} &= 1, \quad \text{for } k = k' \\ &= 0, \quad \text{for } k \neq k' \end{aligned}$$

The orthonormality integral for the eigenfunctions defined by equation (3.16) can be formally carried out using the delta function. We note

$$\begin{aligned} \int_{-\infty}^{\infty} dx\, u_{p'}^*(x) u_p(x) &= |C|^2 \int_{-\infty}^{\infty} dx\, e^{i(p-p')x/\hbar} \\ &= 2\pi\hbar |C|^2 \delta(p - p') \end{aligned}$$

Clearly, with the choice

$$C = 1/\sqrt{2\pi\hbar}$$

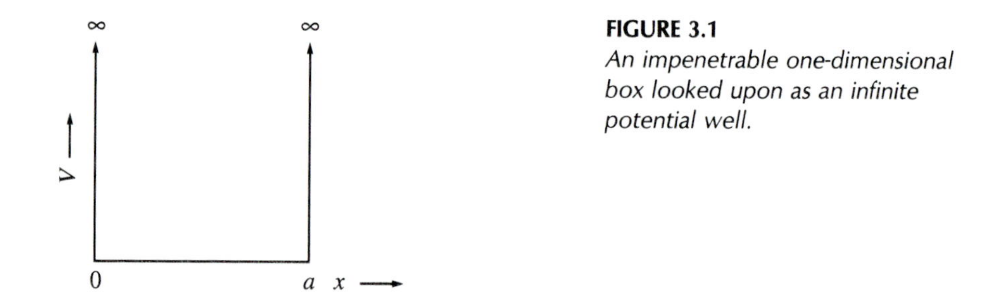

FIGURE 3.1
An impenetrable one-dimensional box looked upon as an infinite potential well.

the orthonormality integral is similar to equation (3.18) except that the Kronecker δ appropriate for discrete indices is replaced by the Dirac delta function appropriate for continuous indices. Finally, the delta function normalized free-particle wave function is given as

$$u_p(x) = \frac{1}{\sqrt{2\pi\hbar}} e^{ipx/\hbar} \tag{3.19}$$

Particle in a Box Revisited

Let's also take a second look at the particle-in-a-box problem of chapter 1. The Hamiltonian is the free-particle Hamiltonian inside the box, $0 \le x \le a$, and at the walls it is the Hamiltonian of a particle in an infinite potential, $V(0) = V(a) = \infty$ (fig. 3.1). Thus, at the walls the wave function must vanish. The eigenfunction of H belonging to the eigenvalue

$$E_n = \frac{n^2\pi^2\hbar^2}{2ma^2} \quad \text{with } n = 1,2,3,\ldots$$

was then shown in chapter 1 to be

$$u_n(x) = \left(\frac{2}{a}\right)^{1/2} \sin\frac{n\pi x}{a}$$

Also note that these eigenfunctions satisfy the orthonormality relation:

$$\int_0^a dx\, u_m(x)u_n(x) = \delta_{mn}$$

3.3 THE POSTULATES OF QUANTUM MECHANICS AND THE MEANING OF MEASUREMENT

We now have the essence of the mathematical formalism that is the basis of quantum mechanics. The next question is, How does the theory relate to results of experiments?

Before we consider the answer to this question, it is useful to review all the postulates of quantum mechanics posited so far.

Postulate 1: Description of the System

For every object moving in an external potential, there exists a wave function ψ that determines all that can be known about the object. Additionally, we assume that the wave function is single valued and square integrable and that both ψ and $\partial\psi/\partial x$ are continuous. (Sometimes, we have to relax the continuity of $\partial\psi/\partial x$. For example, for the infinite potential problem, $\partial\psi/\partial x$ is not continuous at the boundary.)

Postulate 2: Description of Physical Quantities

Physical observables connected with quantum objects are represented by hermitian operators that operate on the wave functions. An operator $\hat{O}$ is hermitian if its expectation value is real. Additionally, the eigenvalues of hermitian operators are real and thus physically realizable. We have already proven the reality of eigenvalues for the Hamiltonian and momentum operators; a more general proof can be given as follows. Write the eigenvalue equation for a hermitian operator $\hat{O}$:

$$\hat{O}\psi_n = o_n\psi_n$$

where ψ_n is its eigenfunction belonging to the eigenvalue o_n. Now multiply by ψ_n^* from the left and integrate:

$$\int dx\,\psi_n^*\hat{O}\psi_n = o_n\int dx\,\psi_n^*\psi_n$$

Taking the complex conjugate we get

$$\int dx\,(\hat{O}\psi_n)^*\psi_n = o_n^*\int dx\,\psi_n^*\psi_n$$

In view of the definition of a hermitian operator, equation (3.7), it follows that

$$o_n = o_n^* \tag{3.20}$$

The eigenvalues of hermitian operators are real.

Before stating the next postulate, which is new, a general mathematical proof of the orthogonality of the eigenfunctions of hermitian operators is instructive. Consider two eigenfunctions of $\hat{O}$, ψ_1 and ψ_2, belonging to the eigenvalues o_1 and o_2, respectively:

$$\hat{O}\psi_1 = o_1\psi_1$$

$$\hat{O}\psi_2 = o_2\psi_2$$

Multiply the first equation by ψ_2^* and integrate:

$$\int dx\, \psi_2^* \hat{O} \psi_1 = o_1 \int \psi_2^* \psi_1\, dx$$

Now take the complex conjugate of the second equation above, multiply it on the right by ψ_1, integrate, and use the hermiticity of $\hat{O}$, equation (3.7) in a more general form; i.e., $\int dx\, \psi_1^* \hat{O} \psi_2 = \int dx\, (\hat{O}\psi_1)^* \psi_2$:

$$\int dx\, \psi_2^* \hat{O} \psi_1 = o_2^* \int dx\, \psi_2^* \psi_1$$

Now subtract one from the other of the last two equations. Since the eigenvalues are real, we get

$$(o_1 - o_2) \int dx\, \psi_2^* \psi_1 = 0 \tag{3.21}$$

This proves the orthogonality of eigenfunctions that belong to two different eigenvalues. However, equation (3.21) can be trivially satisfied if the two eigenvalues are the same, namely, if two different eigenfunctions, such as ψ_2 and ψ_1 above, have the same eigenvalue. Here a reminder about linear independence is necessary. A set of functions is linearly independent if the linear equation

$$\sum_i c_i \psi_i = 0$$

implies that all the c_i are zero. Thus two or more eigenfunctions belonging to the same eigenvalue are not considered to be different unless they are linearly independent, not expressible as a linear combination of the others. In the case of a genuine degeneracy, it is always possible to find a set of linear combinations of the degenerate eigenfunctions that are orthogonal. The procedure is called *Schmidt orthogonalization* and will be left as an exercise. Now let's introduce postulate 3.

Postulate 3: The Expansion Postulate

The set of functions ψ_i that are eigenfunctions of the eigenvalue equation of the operator $\hat{O}$ belonging to the eigenvalues o_i

$$\hat{O}\psi_i = o_i \psi_i$$

constitute, quite generally, an infinite set of linearly independent, orthogonal functions. In addition, if the functions are normalized, then they are said to form a complete, orthonormal set:

$$\int dx\, \psi_i^* \psi_j = \delta_{ij} \tag{3.22}$$

We will assume that *any wave function Ψ of quantum mechanics can be expressed as a coherent superposition of a complete set*:

$$\Psi = \sum_j c_j \psi_j \tag{3.23}$$

This is the expansion postulate.

The advantage of such a general postulate is enormous. Make an analogy with a music synthesizer. Its versatility lies in the ability to synthesize a given piece of music starting from piano waveforms or guitar waveforms or even from percussion waves if that's what the music calls for. The idea of the complete set expansion has the same versatility. You can use it to expand a given wave function using the eigenfunctions of whichever attribute you want. And this versatility is especially useful when you want to measure a particular attribute.

Since the set is orthonormal, the expansion coefficients c_j are easily evaluated. Multiply the left-hand side of equation (3.23) by ψ_k^* and integrate over x:

$$\begin{aligned}\int dx\, \psi_k^* \psi &= \sum_j c_j \int dx\, \psi_k^* \psi_j \\ &= \sum c_j \delta_{jk} = c_k\end{aligned}$$

where we have used the orthogonality, equation (3.22), of the set ψ_j.

Using the complete set expansion, equation (3.23), now we can express the expectation value of the operator $\hat{O}$ in the state represented by the wave function ψ above as follows:

$$\begin{aligned}\langle \hat{O} \rangle_\psi &= \sum_j \sum_k c_k^* c_j \int dx\, \psi_k^* \hat{O} \psi_j \\ &= \sum_j \sum_k c_k^* c_j o_j \int \psi_k^* \psi_j \, dx = \sum_j |c_j|^2 o_j\end{aligned} \tag{3.24}$$

Now we are ready to make the promised connection of the quantum formalism with the results of measurements.

Postulate 4: The Measurement Postulate

If a measurement of a physical observable represented by the operator O is carried out on the state of a system represented by the wave function ψ above, then $|c_j|^2$ in equation (3.24) gives the probability that the result will be the eigenvalue o_j. This is the all-important measurement postulate.

Question: In what situation will a measurement of the observable represented by $\hat{O}$ give a specific eigenvalue o_k with a 100% probability? Only if

$$|c_j|^2 = \delta_{jk}$$

that is, only if the system is in the state represented by the eigenfunction ψ_k belonging to the eigenvalue o_k.

Postulate 5: The Reduction Postulate

Furthermore, after a particular measurement that has given the eigenvalue o_k of $\hat{O}$, the system must be left in a state described by the eigenfunction ψ_k, otherwise an immediate repetition of the measurement would not necessarily yield the same result. Thus to guarantee the reproducibility of measurements, we must posit that *a coherent superposition reduces to an eigenfunction upon measurement.* This is called the *reduction postulate.*

Before measurement of a particular observable attribute of a quantum object in an arbitrary quantum state represented by a wave function ψ, a multitude of possibilities exists for the outcome of the measurement, each with a certain probability. By virtue of the expansion postulate, we look upon ψ as a coherent superposition of the eigenfunctions of the attribute; ψ is now a probability-weighted possibility structure in potentia. The act of measurement actualizes one of these possibilities by collapsing the coherent superposition into a particular eigenfunction of the observable corresponding to a particular eigenvalue.

We will also note in passing that although we have used discrete eigenvalues for defining our complete set above, all the mathematics is easily generalized when the eigenvalues span a continuous spectrum (by replacing sums by integrals and Kronecker deltas by delta functions) or both a discrete and a continuous spectrum.

The Fourier integral, equation (3.2), can be looked upon as an example of a complete set expansion of a wave function $\psi(x)$ in terms of the continuous momentum eigenfunctions $\exp(ipx/\hbar)$. Again $|a(p)|^2$ then represents the probability that a measurement of momentum in the state represented by ψ will give the value p.

Postulate 6: Time Evolution of a System

Between measurements, the time evolution of the system is given by the time-dependent Schrödinger equation, equation (3.8). The Hamiltonian operator, representing the total energy of the system, is crucial here. More on this in chapter 6.

The important point to note is that the time evolution via the Schrödinger equation is a continuous and deterministic affair. It is measurement, observation, that introduces discontinuity and probability into quantum physics. This play of continuous time development and discontinuous change brought about by measurement is schematically shown in figure 3.2.

FIGURE 3.2
Time evolution of the wave function and the effect of measurement. (Adapted with permission from P. Fong, Elementary Quantum Mechanics, *©1962 by Addison-Wesley Publ. Co. Reprinted by permission of Addison-Wesley Publishing Company, Inc., Reading, MA.)*

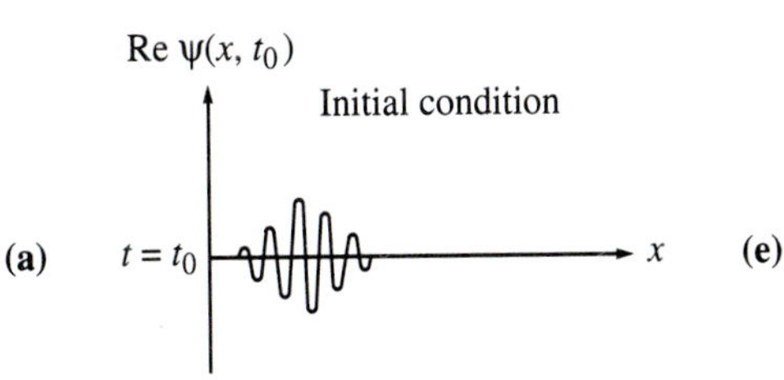

From t_0 to t_{1-}, ψ changes according to the Schrödinger equation.

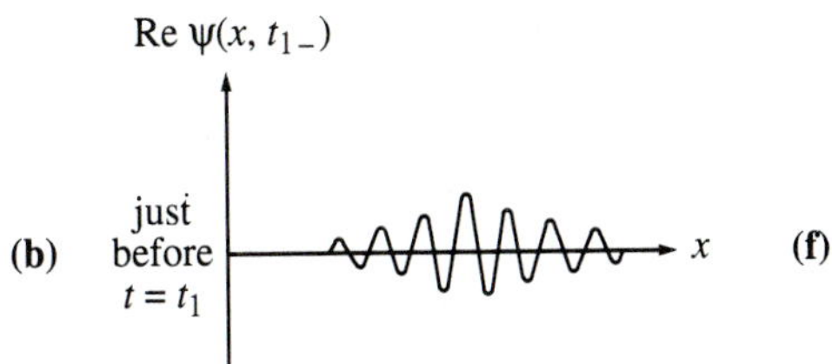

At $t = t_1$, a measurement of position is carried out.

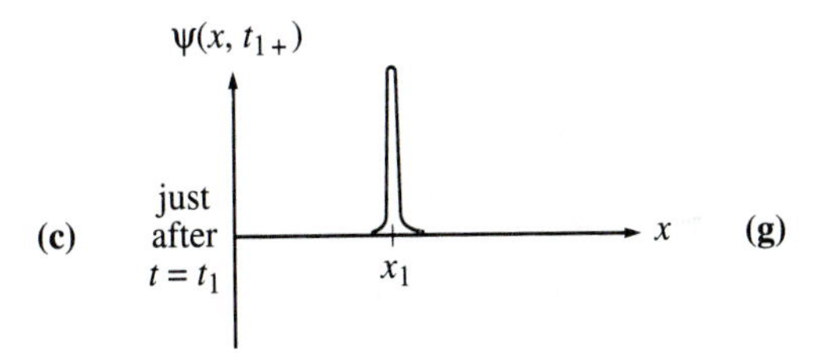

From t_{1+} to t_{2-}, ψ changes according to the Schrödinger equation.

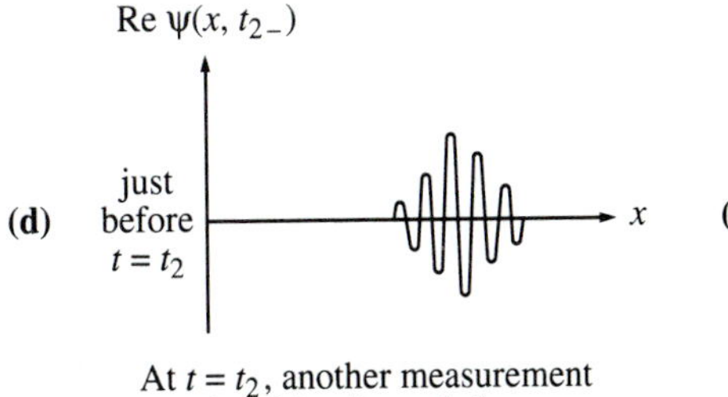

At $t = t_2$, another measurement of position is carried out.

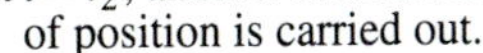

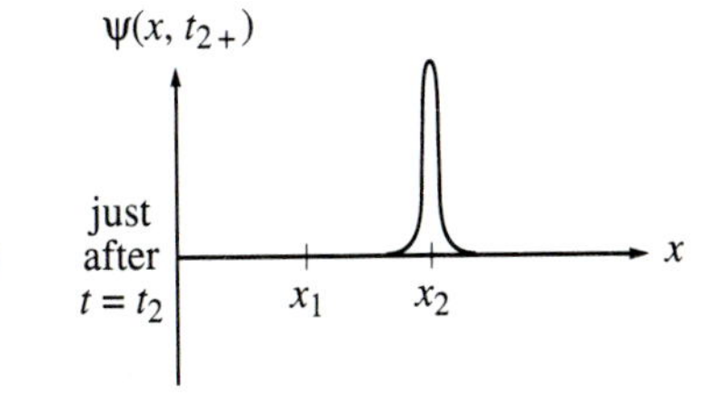

From t_{2+} to t_{3-}, ψ changes according to the Schrödinger equation.

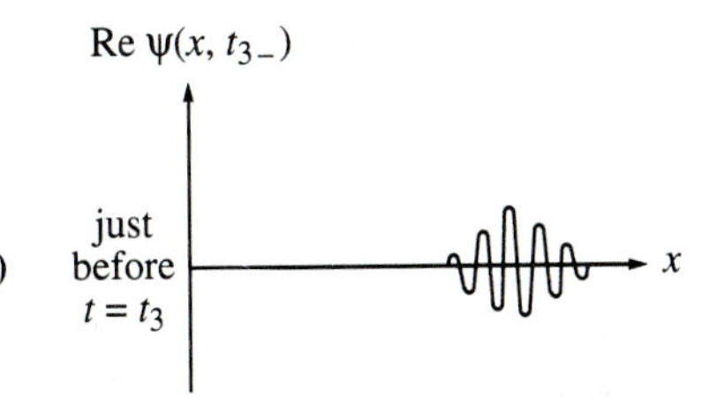

At $t = t_3$, a measurement of momentum is carried out.

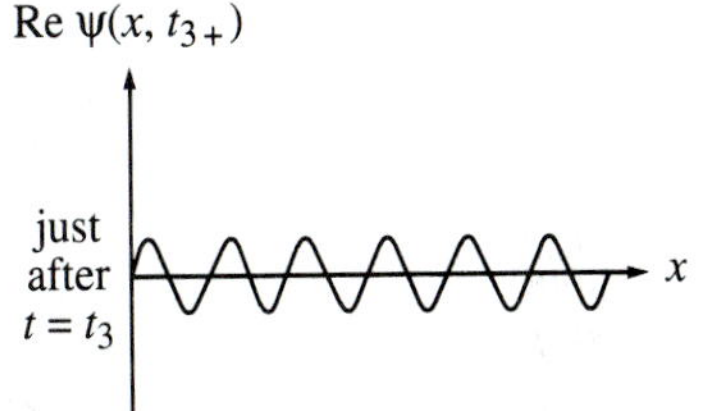

From t_{3+} to t_4, ψ changes according to the Schrödinger equation.

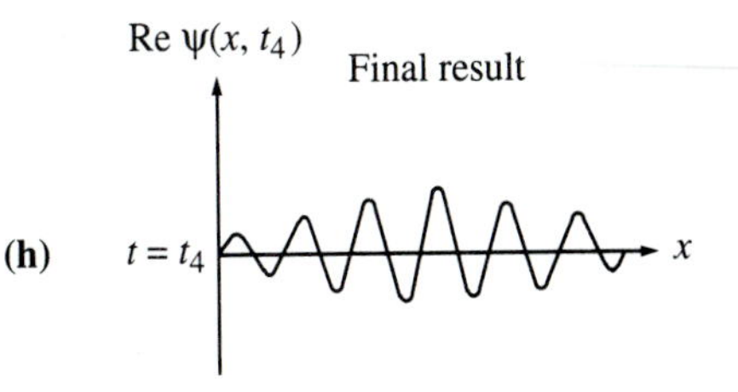

■ **PROBLEM** The complete set expansion of an initial wave function $\psi(x,0)$ of a system in terms of energy eigenfunctions ψ_n of the system has three terms; that is, $n = 1$, 2, and 3. The measurement of energy on the system represented by $\psi(x,0)$ gives the values E_1 and E_2 with probability 1/4, and E_3 with probability of 1/2. Write down the most general expansion of $\psi(x,0)$ and $\psi(x,t)$ consistent with the data.

SOLUTION The initial wave function $\psi(x,0)$ is given as

$$\psi(x,0) = \frac{1}{2}\psi_1 + \frac{1}{2}e^{i\gamma}\psi_2 + \frac{1}{\sqrt{2}}e^{i\delta}\psi_3$$

since the phases are not determined by the probability data. As to $\psi(x,t)$, here is the advantage of expanding in terms of energy eigenstates. These states are stationary with time dependence $\exp(-iE_n t/\hbar)$. Once you determine the expansion coefficients at some initial time, the wave functions at all subsequent times are given by the same expansion coefficients. Thus $\psi(x,t)$ is given as

$$\psi(x,t) = \frac{1}{2}e^{-iE_1 t/\hbar}\psi_1 + \frac{1}{2}e^{i\gamma}e^{-iE_2 t/\hbar}\psi_2 + \frac{1}{\sqrt{2}}e^{i\delta}e^{-iE_3 t/\hbar}\psi_3$$ ■

■ **PROBLEM** Consider a particle in a one-dimensional box (with walls at $x = 0$ and $x = a$) in its ground state. Suddenly, the walls of the box are removed. What is the probability that the particle will be found to have its momentum between p and $p + dp$? Is the conservation of energy violated?

SOLUTION Since the walls are removed quickly, the wave function $\psi(x)$ can be taken to be the same after removal as immediately before removal (this is sometimes called the sudden approximation). Thus,

$$\psi(x) = u_0(x) = \left(\frac{2}{a}\right)^{1/2} \sin\frac{\pi x}{a}$$

which vanishes at $x = 0$ and $x = a$. After the removal of the walls, the relevant eigenfunctions are free-particle eigenfunctions:

$$\psi_p(x) = \frac{1}{\sqrt{2\pi\hbar}}e^{ipx/\hbar}$$

We expand $\psi(x)$ above in terms of the complete set $\psi_p(x)$:

$$\psi(x) = \int_{-\infty}^{\infty} dp\, a(p)\frac{1}{\sqrt{2\pi\hbar}}e^{ipx/\hbar}$$

where $a(p)$ is the probability amplitude that the momentum of the particle is p. Now $a(p)$ is given as

$$a(p) = \frac{1}{\sqrt{2\pi\hbar}} \int_{-\infty}^{\infty} dx\, \psi(x) e^{-ipx/\hbar}$$

Substituting for $\psi(x)$, we get

$$a(p) = \frac{1}{\sqrt{2\pi\hbar}} \sqrt{2/a} \int_0^a dx \sin\frac{\pi x}{a}\, e^{-ipx/\hbar}$$

The integral is easily evaluated giving

$$a(p) = \frac{\pi}{a\sqrt{\pi\hbar a}} \frac{1}{(\pi/a)^2 - (p/\hbar)^2} (1 + e^{-ipa/\hbar})$$

Consequently, the probability that the particle has momentum between p and $p + dp$ after the removal of the walls is given as

$$|a(p)|^2 dp = \frac{2\pi}{\hbar a^3} \frac{1 + \cos(pa/\hbar)}{[(p/\hbar)^2 - (\pi/a)^2]^2}\, dp$$

This is the answer to the first part of the problem.

As to the second part of the problem, the whole question of conservation of energy in quantum mechanics is interesting. First of all, and this is related to the uncertainty principle (see chapter 5), the energy of a system, after a measurement is made, changes uncontrollably due to the interaction with the measuring apparatus. Between measurements, since no transfer of energy is taking place, the information about energy is carried by the probabilities $|c_j|^2$ of finding the energy eigenvalue E_j, and these probabilities are independent of time. Thus the fact that the energy is distributed in a time-independent fashion over the different eigenstates is the quantum mechanical equivalent of the law of energy conservation. ■

PROBLEMS

1. Test the following operators for linearity:

(a) $\hat{O}_1\psi(x) = x^2\psi(x)$ (b) $\hat{O}_2\psi(x) = \exp(\psi(x))$

(c) $\hat{O}_3\psi(x) = \psi^*(x)$ (d) $\hat{O}_4\psi(x) = x^2\, d\psi(x)/dx$

2. A one-dimensional impenetrable box of length a contains an electron that suffers a small perturbation and emits a photon of frequency

$$\nu = 3E_1/h$$

where E_1 is the energy of the ground state. From this would it be correct to conclude that the initial state of the electron is the $n = 2$ box state? Why or why not?

3. Evaluate the following commutators by operating on a wave function: (a) $[x, d/dx]$ (b) $[x^2, d/dx]$.

4. Show that in the momentum representation, if x is represented by the operator $i\hbar\partial/\partial p$, then the commutation relation $[x, p] = i\hbar$ holds.

5. Verify the following properties of the commutator of operators: (a) $[\hat{A}, \hat{B}] = -[\hat{B}, \hat{A}]$ (b) $[\hat{A}, \hat{B}_1 + \hat{B}_2] = [\hat{A}, \hat{B}_1] + [\hat{A}, \hat{B}_2]$ (c) $[\hat{A}\hat{B}, \hat{C}] = [\hat{A}, \hat{C}]\hat{B} + \hat{A}[\hat{B}, \hat{C}]$ (d) $[\hat{A}, \hat{B}\hat{C}] = [\hat{A}, \hat{B}]\hat{C} + \hat{B}[\hat{A}, \hat{C}]$. (*Note*: These commutators are useful. You should make them part of your quantum paraphernalia.)

6. Calculate the commutator $[xp^2, px^2]$ using the fundamental commutator $[x, p] = i\hbar$.

7. A certain system is described by the Hamiltonian operator

$$H = -\frac{d^2}{dx^2} + x^2$$

Show that $Ax\exp(-x^2/2)$ is an eigenfunction of H and determine the eigenvalue; also find A by normalization.

8. The unnormalized ground-state wave function of a particle is given as

$$\psi_0(x) = \exp(-\alpha^4 x^4/4)$$

with eigenvalue $E_0 = \hbar^2\alpha^2/m$. What is the potential in which the particle moves?

9. Show that if the potential energy $V(x)$ is changed by an additive constant amount everywhere, the stationary-state wave functions remain unchanged. What happens to the energy eigenvalues?

10. Calculate the uncertainty product $\Delta p \cdot \Delta x$ using the box wave functions, equation (1.19).

11. A particle in an impenetrable potential box with walls at $x = 0$ and $x = a$ has the following wave function at some initial time:

$$\psi(x) = \frac{1}{\sqrt{5a}}\sin\frac{\pi x}{a} + \frac{3}{\sqrt{5a}}\sin\frac{3\pi x}{a}$$

What are the possible results of the measurement of energy on the system and with what probability would they occur? What is the form of the wave function after such a measurement? Suppose immediately after a measurement, energy is remeasured. What are now the relative probabilities of the possible outcomes?

12. A measurement of energy in problem 11 gives E_1, the ground state energy. What is the probability of finding the electron's momentum between $\hbar k$ and $\hbar(k + dk)$ in a subsequent measurement?

13. An electron is in the ground state of a box with sides at $x = 0$ and $x = a$. Suddenly a wall is moved from $x = a$ to $x = 2a$. What is the probability that the electron will be found (a) in the ground state and (b) in the first excited state of the new box?

14. Let $\psi_n(x)$ denote the orthonormal stationary states of a system corresponding to energy E_n. Suppose the normalized wave function of the system at time $t = 0$ is $\psi(x,0)$, and a measurement of energy yields the value E_1 with probability 1/2, E_2 with probability 3/8, and E_3 with probability 1/8. (a) Write down the most general expansion for $\psi(x,0)$ consistent with the data. (b) What is the corresponding expansion for $\psi(x,t)$, the wave function of the system at time t? (c) Show explicitly that the expectation value of the Hamiltonian H does not change with time.

15. If C and D are two arbitrary hermitian operators, work out which, if any, of the following combinations (i) CD, (ii) D^2, (iii) $CD - DC$, (iv) CDC, and (v) $CD + DC$

(a) are hermitian.

(b) have real nonnegative expectation value.

(c) have pure imaginary expectation value.

16. ψ_1 and ψ_2 are two linearly independent and normalized (but not orthogonal) eigenfunctions of H belonging to the same eigenvalue (i.e., they are degenerate). (a) Show that $c_1\psi_1 + c_2\psi_2$ is also an eigenfunction of H belonging to the same eigenvalue as ψ_1 or ψ_2. (b) Construct two linear combinations of ψ_1 and ψ_2 that are orthogonal to each other.

ADDITIONAL PROBLEMS

A1. If $\hat{A}$ and $\hat{B}$ are linear operators, is the product $\hat{A}\hat{B}$ linear?

A2. Show that the commutator $[x^2, p] = 2i\hbar x$ and the commutator $[x^n, p] = ni\hbar x^{n-1}$.

A3. A particle moves in one dimension in a potential $V(x)$. Show that the expectation value of the momentum in a stationary state with a discrete eigenvalue is zero. (*Hint*: First evaluate $[H, x]$.)

A4. Show from the Schrödinger equation by comparing the signs of $d^2\psi/dx^2$ and ψ that the wave function is concave toward the x-axis in regions whenever $|V(x) - E| > 0$ and concave away from the x-axis whenever $|V(x) - E| < 0$.

A5. Show that there is no degeneracy in the bound states for one-dimensional motion.

A6. Consider the Schrödinger equation for a complex potential

$$V_1(x) + iV_2(x)$$

in which both V_1 and V_2 are real. Is the Hamiltonian hermitian? Follow through the derivation of the continuity equation (see chapter 1) with this potential and find the additional terms that appear when $V_2 \neq 0$. Provide a physical interpretation of the result. What possible use can you think of for such complex potentials?

A7. In the infinite well potential, we set $V \to \infty$ at the walls and $V = 0$ inside. Suppose, instead, that the potential inside $V = C$, where C is a constant. Calculate the energy levels.

REFERENCES

M. Chester. *Primer of Quantum Mechanics.*
Fayyazuddin and Riazuddin. *Quantum Mechanics.*
P. Fong. *Elementary Quantum Mechanics.*
S. Gasiorowicz. *Quantum Physics.*
A. Sudbery. *Quantum Mechanics and the Particles of Nature.*

4

The Solution of the Schrödinger Equation in One Dimension

In this chapter we will consider some simple one-dimensional potential models for you to gain experience with the solution of the time-independent Schrödinger equation. Although the real world is three dimensional, it turns out that our models will often simulate the real world well enough to make their study worthwhile.

The simplest potential models we choose also emphasize the difference between classical and quantum descriptions of the motion of objects. One potential model we will discuss is the so-called square potential: The potential is constant everywhere except at the two points where it changes quite discontinuously from zero to a finite constant value. If we think in terms of forces, the force goes to infinity at the points of discontinuity, and is zero elsewhere. This is not how forces in classical systems behave. But in quantum mechanics, the square potentials give a pretty good description of some systems such as atomic nuclei.

Recall that the time-independent Schrödinger equation for a system is an eigenvalue equation for its Hamiltonian. As such, for attractive potentials we can have both positive and negative energy solutions.

If the total energy E is less than zero, it means that the system is bound to the attractive potential. In quantum mechanics, we are then looking at a localized eigenfunction; the wave function must vanish as the distance from the origin tends toward infinity. As in the case of the box, an infinite potential well, such a restriction imposes discreteness on the energy spectrum. Such quantized energy levels are found in atoms, molecules, nuclei, and so forth, and thus our models are expected to lead to some insight into the physics of these systems. The $E > 0$ solutions corresponding to an attractive potential are important for the study of scattering problems—scattering of a projectile from a target that can be represented by a potential.

When the potential is repulsive, obviously only solutions with $E > 0$ are possible. But what if E is greater than zero, yet less than the value of the potential? Classically the potential then is an impassable "barrier," such as is presented to billiard balls on a pool table by the walls of the table. But this is the strange thing about quantum mechanics: There is a finite probability that quantum billiard balls can penetrate those walls and fall off the table. It is not easy to play quantum pool! Since this is an exciting and new phenomenon, this is where we will begin our discussion.

4.1 BARRIER PENETRATION: SQUARE BARRIER

A square (also called rectangular) barrier (fig. 4.1) is defined as

$$\begin{aligned} V(x) &= 0, && \text{for } x < -a \\ &= V_0, && \text{for } -a < x < a \\ &= 0, && \text{for } x > a \end{aligned} \tag{4.1}$$

Inside the barrier the Schrödinger equation is given as

$$-\frac{\hbar^2}{2m}\frac{d^2u}{dx^2} + V_0 u(x) = Eu(x)$$

Let's write the equation in the form

$$\frac{d^2u(x)}{dx^2} + \frac{2m}{\hbar^2}(E - V_0)u(x) = 0 \tag{4.2}$$

We will consider only the interesting situation of $0 < E < V_0$. Define the real quantity α

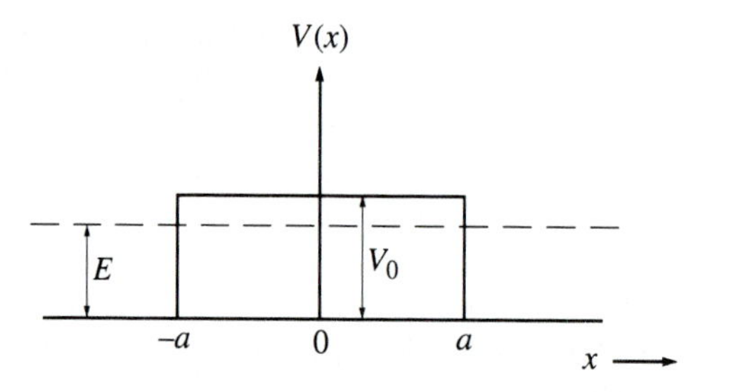

FIGURE 4.1
A square or rectangular barrier.

$$\alpha^2 = \frac{2m(V_0 - E)}{\hbar^2} \tag{4.3}$$

Introducing α^2 in equation (4.2) we get

$$\frac{d^2u(x)}{dx^2} - \alpha^2 u(x) = 0 \tag{4.4}$$

The inside solution of the Schrödinger equation is thus given as

$$u(x) = Ae^{\alpha x} + Be^{-\alpha x}, \qquad |x| < a \tag{4.5}$$

Outside the barrier, the Schrödinger equation is the free-particle equation. If we assume that the particle is incident from the left, then the appropriate solution in the outside region $x < -a$ is given as

$$u(x) = e^{ikx} + Re^{-ikx}, \qquad x < -a \tag{4.6}$$

where k is the wave number of a free particle

$$k^2 = \frac{2mE}{\hbar^2} \tag{4.7}$$

The flux from the left incident on the barrier is then given as

$$\begin{aligned} j &= \frac{\hbar}{2im} [(e^{-ikx} + R^* e^{ikx})(ike^{ikx} - ikRe^{-ikx}) - \text{complex conjugate}] \\ &= \frac{\hbar k}{m} (1 - |R|^2) \end{aligned} \tag{4.8}$$

Here the first term is the flux of $\exp(ikx)$, which we identify as the incident wave. The second term of the flux $\hbar k|R|^2/m$ must be interpreted as the flux due to the reflected wave, $R\exp(-ikx)$, reflected off the barrier and traveling to the left. Note the negative sign for the reflected current; it is the current of e^{-ikx}—that's why.

For $x > a$, we write the solution in the form

$$u(x) = Te^{ikx}, \qquad x > a \tag{4.9}$$

where we omit the $\exp(-ikx)$ solution in this region since it would describe a wave coming from $+\infty$, whereas the way we have set up the experiment, we can only have a wave going toward $+\infty$ for $x > a$; this is the transmitted wave. The flux due to the transmitted wave is given by

$$j = \frac{\hbar k}{m} |T|^2 \tag{4.10}$$

The next important step is to impose the boundary conditions obtained by our requirements of continuity of u and du/dx at $x = -a$ and $x = a$, respectively. This gives us the needed four equations for our four unknowns:

$$\begin{aligned} e^{-ika} + Re^{ika} &= Ae^{-\alpha a} + Be^{\alpha a} \\ ik(e^{-ika} - Re^{ika}) &= \alpha(Ae^{-\alpha a} - Be^{\alpha a}) \\ Ae^{\alpha a} + Be^{-\alpha a} &= Te^{ika} \\ \alpha(Ae^{\alpha a} - Be^{-\alpha a}) &= ikTe^{ika} \end{aligned} \tag{4.11}$$

The solution is tedious but straightforward. We are interested only in the expression for T for which we get

$$T = e^{-2ika} \frac{2k\alpha}{2k\alpha \cosh 2\alpha a - i(k^2 - \alpha^2)\sinh 2\alpha a} \tag{4.12}$$

Consequently, $|T|^2$, the ratio of transmitted current to incident current, is given by

$$|T|^2 = \frac{(2k\alpha)^2}{(k^2 + \alpha^2)^2 \sinh^2 2\alpha a + (2k\alpha)^2} \tag{4.13}$$

There! We get a finite probability for the transmission of the particle even though its energy is less than the potential barrier. This is a purely quantum effect signifying the wave property of quantum objects.

Hence quantum objects are said to be able to *tunnel* through a classically impassable energy barrier, although you should not take this language literally. To be sure, there is a finite probability for the particle to be inside the classically forbidden barrier region where its kinetic energy is negative, but the point is that nobody can "see" a particle *actually* go through a classically forbidden region. Particle detectors can detect only objects of kinetic energy greater than zero; if you insert a detector inside the barrier to see the particle, you are not only making a hole in the potential but also in your objective, because the object will no longer belong to a classically forbidden region where you wanted to find it! Another way to say this is that our effort to observe the object with any measuring instrument will impart to it an uncontrollable amount of energy. This is how the uncertainty principle works in such measurement situations!

Similarly, we cannot say what kind of time the particle takes to tunnel through a barrier. It may do it now or later; quantum mechanics gives us only average behavior—the average time it takes to travel to the other side. It seems best to think of quantum tunneling as "as if" tunneling—tunneling in potentia. Alternatively, we can say that quantum objects go across a barrier for which they do not have enough energy by doing a quantum "jump"—Bohr style; they never go through the intervening space, and they do not do it causally taking a finite amount of time. Now they are here, on this side of the barrier; and then they are there, on the other side.

When $\alpha a \gg 0$, the *transmission coefficient* $|T|^2$ becomes

$$|T|^2 = \left(\frac{4k\alpha}{k^2 + \alpha^2}\right)^2 e^{-4\alpha a} \approx e^{-4\alpha a} \tag{4.14}$$

As a result, the probability of barrier transmission is extremely sensitive to both a, the range of the barrier, and to α, or $V_0 - E$, the barrier height that the object has to jump. It has to overcome the hurdles of both a long jump and a high jump!

Application to Alpha Decay

The barrier transmission phenomenon is observed in many situations involving quantum systems. Here we will consider the case of alpha radioactivity of nuclei, the original application worked out by Gamow, Gurney, and Condon.

First we need to figure out how to generalize the above calculation to the case of a more physical barrier where the shape is not so square (fig. 4.2). Notice that $|T|^2$, equation (4.14), can be written as

$$|T|^2 = e^{-2\sqrt{2m/\hbar^2(V_0-E)}\cdot 2a}$$

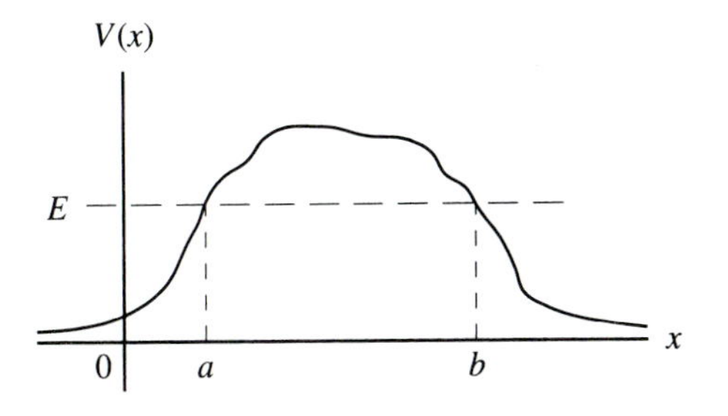

FIGURE 4.2
A realistic potential barrier has a more jagged look than a square barrier.

For a broad range of barriers, the appropriate generalization for an order of magnitude estimate can now be guesstimated:

$$|T|^2 \approx \exp\left[-2\int_a^b dx \sqrt{\frac{2m}{\hbar^2}\,(V(x)-E)}\right] \tag{4.15}$$

We can approximate the situation for alpha emission from a nucleus by the potential barrier shown in figure 4.3. Inside the nucleus, the alpha particle is a free particle, $E > 0$ (if the alpha were bound, how would the nucleus decay?), and has to tunnel through the Coulomb barrier

$$V(r) = \frac{ZZ'e^2}{r}$$

where Z and Z' are the atomic numbers of the daughter nucleus and the alpha particle, respectively, into which the parent nucleus is splitting. Here r is the radial distance, but we can use equation (4.15) for $|T|^2$, treating x as r, since it's just a dummy integration variable. Write $|T|^2$ as $\exp(-G)$. This defines G as

$$G = 2\sqrt{2m/\hbar^2}\int_R^b dr\left(\frac{ZZ'e^2}{r} - E\right)^{1/2}$$

where R is the radius of the daughter nucleus (fig. 4.3) and the upper limit of the integral, b, is taken to be the classical "turning point" where the integrand vanishes, since

$$E = \frac{ZZ'e^2}{b}$$

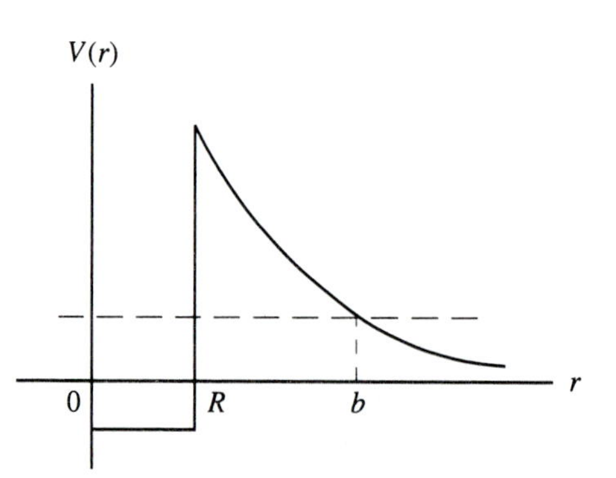

FIGURE 4.3
Model potential barrier for alpha decay of nuclei.

The integral above is given as

$$\int_R^b dr\left(\frac{1}{r} - \frac{1}{b}\right)^{1/2} = \sqrt{b}\left[\cos^{-1}\left(\frac{R}{b}\right)^{1/2} - \left(\frac{R}{b} - \frac{R^2}{b^2}\right)^{1/2}\right]$$

For low energies and high barriers, $b \gg R$, we get

$$G \approx 2\left(\frac{2mZZ'e^2b}{\hbar^2}\right)^{1/2}\frac{\pi}{2}$$

But $b = ZZ'e^2/E = 2ZZ'e^2/mv^2$, where v is the velocity of the alpha particle inside the nucleus. Hence

$$G = \frac{2\pi ZZ'e^2}{\hbar v}$$

To calculate the alpha particle *escape probability* per second, we have to multiply the transmission coefficient $\exp(-G)$ by the rate of the alpha hitting the barrier, which is $\sim v/R$. For 1 MeV alpha, and using $R = 1.2 \times 10^{-13}A^{1/3}$ cm with $A = 216$, we estimate $v/R \approx 10^{21}\ \text{s}^{-1}$. Consequently,

$$\text{escape probability per second} = \tau^{-1} = 10^{21}e^{-G}$$

where τ denotes the decay time. Noting that $Z' = 2$ and that the mass of alpha $m \approx 4 \times 10^3\ \text{MeV}/c^2$, we get

$$G = \frac{2\pi ZZ'e^2}{\hbar\sqrt{2E/m}} \approx \frac{4Z}{\sqrt{E}(\text{in MeV})}$$

Therefore, we obtain

$$\log_{10}\frac{1}{\tau} = C_1 - C_2\frac{Z}{\sqrt{E}(\text{MeV})} \tag{4.16}$$

where C_1 and C_2 are two constants, never mind our estimates for them. This formula, first derived by Gamow, Gurney, and Condon, fits the data on alpha decay quite remarkably. It is also remarkable that we can derive the formula from a one-dimensional calculation.

We will not discuss the case of $E > V_0$, leaving it to you as a problem. The solution is very similar to the case of $E > 0$ solution for the attractive square-well potential considered below. Note, however, that the solutions are oscillatory in all three regions, and that the boundary conditions can be satisfied only if R, the amplitude of the reflected wave, is nonzero. This is again contrary to the classical expectation where the whole beam is transmitted.

4.2 THE (ATTRACTIVE) SQUARE POTENTIAL WELL

Consider now the square potential well (fig. 4.4) defined as follows:

$$\begin{aligned} V(x) &= 0, && \text{for } x < -a \\ &= -V_0, && \text{for } -a \leq x \leq a, \text{ where } V_0 \text{ is a constant } >0 \\ &= 0, && \text{for } x > a \end{aligned} \tag{4.17}$$

Let's first consider $E > 0$. The Schrödinger equation is now

$$\frac{d^2u(x)}{dx^2} + \frac{2m}{\hbar^2}(E + V_0)u(x) = 0 \tag{4.18}$$

inside the well; outside the well, we have the free-particle wave equation. Thus we can immediately write down the solutions:

$$\begin{aligned} u(x) &= e^{ikx} + Re^{-ikx} && x < -a \\ &= Ae^{ik'x} + Be^{-ik'x} && -a < x < a \\ &= Te^{ikx} && x > a \end{aligned} \tag{4.19}$$

where the wave number k' inside the well is defined by

$$k'^2 = \frac{2m(E + V_0)}{\hbar^2} \tag{4.20}$$

and the free-particle wave number k is given by equation (4.7). Notice how our solutions for the different regions are again dictated by the experimental situation, which we assume is the same as for $V > 0$ described in section 4.1: The

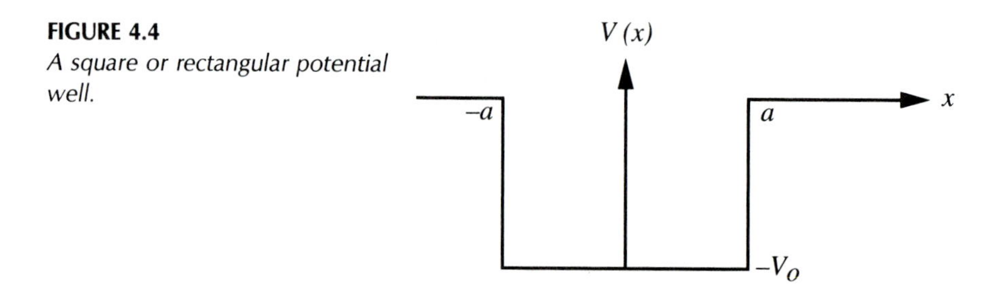

FIGURE 4.4
A square or rectangular potential well.

wave is coming in from the left with an incoming current $\hbar k/m$; it is partly reflected with current $\hbar k|R|^2/m$ and partly transmitted with current $\hbar k|T|^2/m$.

Again matching u and du/dx at $x = -a$ and at $x = a$ gives us the four equations we need for solving for the four unknowns we have:

$$
\begin{aligned}
e^{-ika} + Re^{ika} &= Ae^{-ik'a} + Be^{ik'a} \\
ik(e^{-ika} - Re^{ika}) &= ik'(Ae^{-ik'a} - Be^{ik'a}) \\
Ae^{ik'a} + Be^{-ik'a} &= Te^{ika} \\
ik'(Ae^{ik'a} - Be^{-ik'a}) &= ikTe^{ika}
\end{aligned}
\tag{4.21}
$$

We now are interested in both R and T. We get

$$
\begin{aligned}
R &= ie^{-2ika}\,\frac{(k'^2 - k^2)\sin 2k'a}{2kk'\cos 2k'a - i(k'^2 + k^2)\sin 2k'a} \\
T &= e^{-2ika}\,\frac{2kk'}{2kk'\cos 2k'a - i(k'^2 + k^2)\sin 2k'a}
\end{aligned}
\tag{4.22}
$$

Several things can be noticed about this solution.

1. There is usually some reflection, a quantum wave effect as noted before. In fact, reflection dominates at low energy. Only when $E \gg V_0$ does reflection tend to zero, and it's all transmission.
2. Since there is no time dependence, the continuity equation (1.27) requires that the current is the same everywhere. This means that the current to the left of the well is the same as the transmitted current to the right:

$$\frac{\hbar k}{m}(1 - |R|^2) = \frac{\hbar k}{m}|T|^2 \tag{4.23}$$

   With the solutions given by equation (4.22) for R and T, you can easily verify that equation (4.23) holds. Of course, the current is also conserved inside the well. Check it out!
3. There are, however, special conditions under which the reflected current goes to zero exactly, namely, when

$$\sin 2k'a = 0$$

   In other words, $2k'a = n\pi$, with $n = 1, 2, 3, \ldots$. This translates into

$$\frac{2m(V_0 + E)a^2}{\hbar^2} = \frac{n^2\pi^2}{4}$$

Thus whenever the incident energy is given as

$$E = -V_0 + \frac{n^2\pi^2\hbar^2}{8ma^2}$$

there is only transmission. This is called transmission *resonance*, and it was experimentally observed by Ramsauer and Townsend when they scattered low-energy electrons ($E \approx 0.1$ eV) off noble atoms.

Physically, what causes the resonance is this. The maxima in transmission occur when $2k'a = n\pi$. Note that the de Broglie wavelength of the wave inside is given as $\lambda = 2\pi/k'$. This means that the transmission is boosted whenever the distance $4a$ back and forth that the wave traverses inside the well due to reflection is equal to an integer number of its de Broglie wavelengths. Then the (multiply) reflected waves and the incident wave are in phase and constructively reinforce one another.

Bound States: Solutions for $E < 0$

Since $E < 0$, we have $E = -|E|$, and the equation inside the well becomes

$$\frac{d^2u(x)}{dx^2} + k'^2u(x) = 0 \tag{4.24}$$

where we have defined the inside wave number k' now via

$$k'^2 = \frac{2m(V_0 - |E|)}{\hbar^2} \tag{4.25}$$

On the other hand, the Schrödinger equation outside the well is given as

$$\frac{d^2u(x)}{dx^2} - \beta^2u(x) = 0 \tag{4.26}$$

where $\beta^2 = 2m|E|/\hbar^2$. The general solution of equation (4.26) is given as

$$u(x) = Ce^{\beta x} + De^{-\beta x} \tag{4.27}$$

But here's the catch. The condition of good behavior at infinity allows only one-half of the solution for both $x < -a$ and $x > a$, and not the same half, of course. Thus the permitted bounded solutions in these two regions are given as

$$u(x) = Ce^{\beta x} \qquad x < -a$$
$$= De^{-\beta x} \qquad x > a \tag{4.28}$$

Since we have to match these real functions at the boundaries of the potential, it is convenient to choose the solution inside as a coherent superposition of real functions as well.

$$u(x) = A\cos k'x + B\sin k'x \qquad -a < x < a \tag{4.29}$$

We now do the matching at $x = \pm a$. This gives

$$\begin{aligned}
Ce^{-\beta a} &= A\cos k'a - B\sin k'a \\
\beta Ce^{-\beta a} &= k'(A\sin k'a + B\cos k'a) \\
De^{-\beta a} &= A\cos k'a + B\sin k'a \\
-\beta De^{-\beta a} &= -k'(A\sin k'a - B\cos k'a)
\end{aligned} \tag{4.30}$$

What we have is a set of four simultaneous linear homogeneous equations for four unknowns and the condition for a solution to exist is that the determinant of the equations vanish:

$$\begin{vmatrix}
\cos k'a & -\sin k'a & -e^{-\beta a} & 0 \\
k'\sin k'a & k'\cos k'a & -\beta e^{-\beta a} & 0 \\
\cos k'a & \sin k'a & 0 & -e^{-\beta a} \\
-k'\sin k'a & k'\cos k'a & 0 & \beta e^{-\beta a}
\end{vmatrix} = 0$$

This leads to the equation

$$\left(\tan k'a - \frac{\beta}{k'}\right)\left(\tan k'a + \frac{k'}{\beta}\right) = 0 \tag{4.31}$$

Since k' and β are functions of the energy E, equation (4.31) imposes restrictions on the values of energy E that permit a solution for A, B, C, and D; in other words, energy quantization. Moreover, there are two types of solution, one obtained when $\tan k'a = \beta/k'$, the other when $\cot k'a = -\beta/k'$. Let's study them sequentially.

First, if

$$\tan k'a = \beta/k' \tag{4.32}$$

going back to equation (4.30), we get from the first two equations

$$\beta C e^{-\beta a} = \beta \cos k'a(A - B \tan k'a)$$
$$= \beta \cos k'a(A - B\beta/k')$$

and

$$\beta C e^{-\beta a} = k' \cos k'a(A \tan k'a + B)$$
$$= k' \cos k'a(\beta A/k' + B)$$

It follows that B must be equal to zero. Thus in this case, the interior solution is

$$u(x) = A \cos k'x \tag{4.33}$$

where A has to be determined from normalization.

The solution (4.33) is an even function of x; that is, if we change the sign of x, $u(x)$ remains unchanged.

The eigenvalue condition, equation (4.32), can be solved graphically as follows. Define

$$\lambda = \frac{2mV_0a^2}{\hbar^2}$$

and $\xi = k'a$. Then the eigenvalue condition, equation (4.32), can be written as

$$\frac{\sqrt{\lambda - \xi^2}}{\xi} = \tan \xi$$

Then the eigenvalues can be obtained as the points of intersection of the two functions $\tan \xi$ and $(\lambda - \xi^2)^{1/2}/\xi$ plotted as a function of ξ, as shown in figure 4.5. The following comments are now appropriate:

1. As expected, the eigenvalues are discrete.
2. There are a greater number of eigenvalues (that is, more bound states) the larger λ is (that is, the larger V_0a^2 is). There are more eigenvalues if the well is deep or broad or both deep and broad.
3. However, there is always at least one bound state, no matter how shallow or how narrow the well is. This is a special characteristic of the even solution of one-dimensional attractive potentials and is not borne out for realistic three-dimensional potentials.

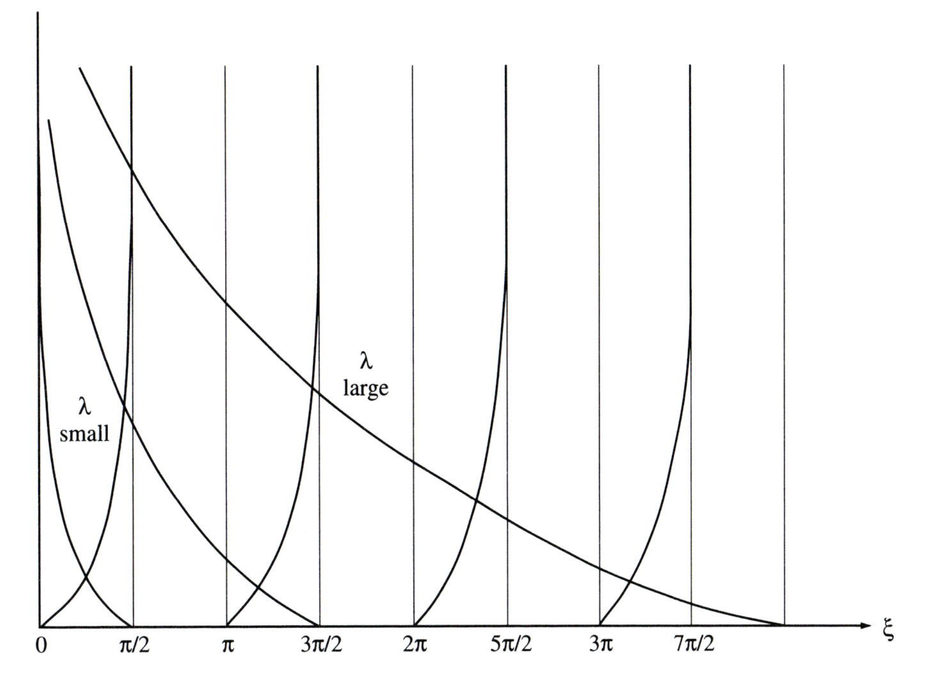

FIGURE 4.5
Graphical solution of the eigenvalue equation for the one-dimensional square well (even case). The intersections of the curves for tanξ (the rising curves) and $(\lambda - \xi^2)^{1/2}/\xi$ (the falling curves) give the locations of the discrete eigenvalues for different values of λ.

Second, when

$$\cot k'a = -\beta/k' \tag{4.34}$$

we find that $A = 0$ and thus the eigenfunctions are

$$u(x) = B \sin k'x \tag{4.35}$$

where B again must be determined from normalization. Note that these eigenfunctions are odd functions of x; they change sign when x changes sign.

The eigenvalue condition, equation (4.34), for this solution can be written as

$$\frac{\sqrt{\lambda - \xi^2}}{\xi} = -\cot \xi$$

The graphs of the functions on the two sides of the above equation are depicted in figure 4.6. Interestingly, the one unwanted feature (item 3 above) of the even solutions is now gone. We find that there will be no intersection of the two curves unless $\lambda \geq \pi^2/4$; that is,

$$\frac{2mV_0a^2}{\hbar^2} \geq \frac{\pi^2}{4} \tag{4.36}$$

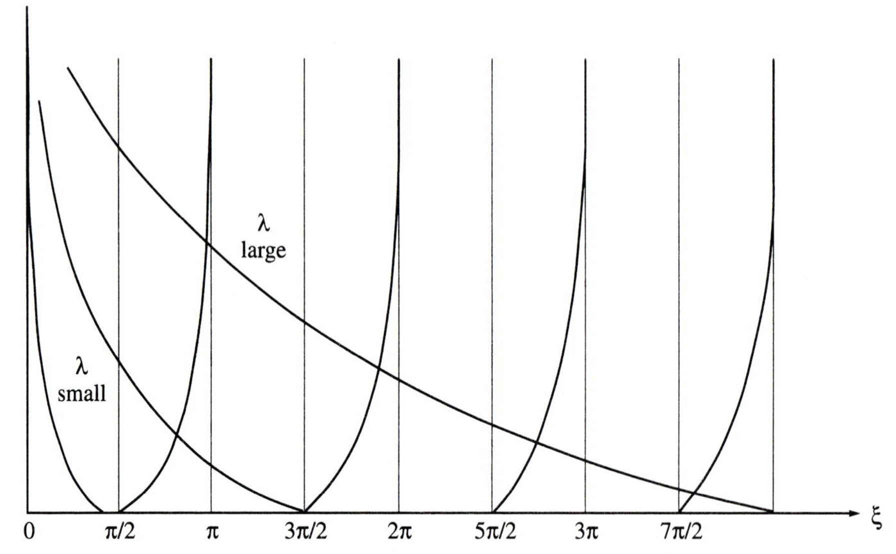

FIGURE 4.6
Graphical solution of the eigenvalue equation for the one-dimensional square well (odd case). The intersections of the rising curves for $-\cot\xi$ *and the falling curves for* $(\lambda - \xi^2)^{1/2}/\xi$ *give the locations of the eigenvalues for various* λ.

This is a range-depth condition for the existence of a bound state, and this is what we obtain for the realistic three-dimensional square-well case (see chapter 12). Why? In the three-dimensional case, the radial coordinate r cannot be less than zero, but otherwise, with a redefinition of the wave function so that it vanishes at $r = 0$, the Schrödinger equation is the same as the one-dimensional one when we consider the ground state. Thus the radial problem can be simulated as a one-dimensional motion in a square well with an infinite potential wall at $x = 0$. Now the wave function has to vanish at $x = 0$, the solution has to be odd, and the range-depth relation will prevail (see problem 14).

4.3 PARITY

For the square-well potential above, we found two types of solution: one class of eigenfunctions even in x, the other class of eigenfunctions odd in x. This, we will now show, is due to an interesting property of the Hamiltonian for the problem.

Consider the general Hamiltonian

$$H = \frac{p^2}{2m} + V(x) \tag{4.37}$$

but now suppose that $V(x)$ has the special property

$$V(x) = V(-x)$$

Now consider the Schrödinger equation

$$H\psi(x) = E\psi(x)$$

Because of the above special *symmetry* property of $V(x)$, it is clear that whenever $\psi(x)$ is a solution, so is $\psi(-x)$, and so are the linear combinations of the two:

$$\psi_e(x) = \frac{1}{\sqrt{2}}[\psi(x) + \psi(-x)]$$

$$\psi_o(x) = \frac{1}{\sqrt{2}}[\psi(x) - \psi(-x)] \tag{4.38}$$

where the subscripts e and o denote even and odd, respectively. The factor $1/\sqrt{2}$ has been added for normalization. Note that if $\psi(x) = \psi_e(x)$ then $\psi_o(x)$ is zero, and vice versa.

We will define a special symmetry operator called the *parity* operator P as follows:

$$P\psi(x) = \psi(-x) \tag{4.39}$$

that is, when operating on a wave function $\psi(x)$, P changes the sign of x in the argument of ψ. If $\psi(x)$ describes the state of a system, $P\psi(x)$ describes its mirror image. It also follows from this definition that

$$P^2\psi(x) = \psi(x)$$

Now consider the eigenvalue equation for P:

$$P\psi(x) = \lambda\psi(x)$$

λ being the eigenvalue. It follows that

$$P^2\psi(x) = \lambda^2\psi(x) = \psi(x)$$

In view of this, λ—the eigenvalues of P—can only be ± 1. Note that the even combination, ψ_e above, is the eigenfunction of parity belonging to the eigenvalue $+1$, and the odd combination ψ_o is the eigenfunction of parity belonging to the eigenvalue -1. Henceforth they will be referred to as the even and odd parity eigenfunctions, respectively.

Notice that the Hamiltonian H, equation (4.37), itself is reflection symmetric; it remains unchanged if we change $x \to -x$, if $V(x) = V(-x)$. Because of this, we have $PH(x) = H(-x)P = H(x)P$. H commutes with P. It follows that

$$PH\psi(x) = HP\psi(x) = EP\psi(x)$$

If $\psi(x)$ is an eigenfunction of H, so is $P\psi(x)$. $\psi(x)$ is an eigenfunction of both H and P.

It should now be clear that the even and odd bound state solutions of the attractive square-well potential obtained in section 4.2 are nothing but the even and odd parity solutions, respectively. For the eigenfunction, equation (4.33), we have

$$P \cos k'x = \cos(-k'x) = \cos k'x$$

Thus we have a solution of even parity. For the odd solutions, equation (4.35),

$$P \sin k'x = \sin(-k'x) = -\sin k'x$$

Therefore, these are eigenfunctions of odd parity.

It is often possible to simplify the solution of the Schrödinger equation considerably by taking the symmetries of its Hamiltonian into account. Let's demonstrate this for the bound-state solutions of the attractive square well. Since the eigenfunctions are simultaneous eigenfunctions of H and P, we can look for a specific parity solution to begin with. Then, inside the well, we would have

$$u(x) = A \cos kx \quad \text{for even parity}$$

$$u(x) = B \sin kx \quad \text{for odd parity}$$

The outside solutions are given as before, in equation (4.28). It is now convenient to join the two boundary conditions at each of $x = \pm a$ by the one condition that the logarithmic derivative of the wave function

$$\frac{d}{dx}(\ln \psi) = \frac{1}{\psi}\frac{d\psi}{dx}$$

be continuous at these points (because then the constant coefficients cancel out). This immediately gives the two eigenvalue conditions obtained in section 4.2. Check it out!

4.4 THE DELTA FUNCTION POTENTIAL

Let's now consider another useful and simple potential model, the delta function potential, $V(x) = -\Lambda\delta(x)(\Lambda > 0)$, which is interesting also because the mathematics has one little subtlety in this case.

For concreteness, let's consider an attractive potential (as above) and the bound state solution, $E < 0$, in such a potential. Let's write the Schrödinger equation in the form

$$\frac{d^2u(x)}{dx^2} - \kappa^2 u(x) = -\frac{2m\Lambda}{\hbar^2}\,\delta(x)u(x) \tag{4.40}$$

where

$$\kappa^2 = 2m|E|/\hbar^2 \tag{4.41}$$

For everywhere other than $x = 0$, the Schrödinger equation reads

$$\frac{d^2u}{dx^2} - \kappa^2 u = 0$$

Since the solutions have to go to zero at $x = \pm\infty$, we have

$$\begin{aligned} u(x) &= Ae^{\kappa x} \qquad x < 0 \\ &= Be^{-\kappa x} \qquad x > 0 \end{aligned} \tag{4.42}$$

The continuity of the wave function at $x = 0$ gives $A = B$. Now here is the special catch for the delta function potential—the derivative of the wave function is not continuous (here we have to relax one of our mathematical expectations just as we have relaxed the square integrability condition for the plane wave). This can be seen by integrating the Schrödinger equation (4.40) from $-\epsilon$ to ϵ (where ϵ is positive but arbitrarily small), which also will enable us to evaluate the discontinuity:

$$\begin{aligned} \lim_{\epsilon\to 0}\int_{-\epsilon}^{\epsilon} dx\,\frac{d^2u(x)}{dx^2} &= \left(\frac{du}{dx}\right)_{0+} - \left(\frac{du}{dx}\right)_{0-} \\ &= -\lim_{\epsilon\to 0}\frac{2m\Lambda}{\hbar^2}\int_{-\epsilon}^{\epsilon} dx\,\delta(x)u(x) \\ &= -2m\Lambda u(0)/\hbar^2 \end{aligned}$$

Clearly, the delta function potential itself acts as a boundary condition.

Now if we substitute the solutions, equation (4.42), in the equation above, we get an eigenvalue condition

$$-\kappa A - \kappa A = -\frac{2m\Lambda A}{\hbar^2}$$

Consequently, we get $\kappa = m\Lambda/\hbar^2$. Using equation (4.41), we get the energy eigenvalues

$$E = -\frac{m\Lambda^2}{2\hbar^2} \tag{4.43}$$

There is only one bound state. The corresponding normalized eigenfunction is

$$u(x) = \sqrt{\kappa}\, e^{-\kappa|x|} \tag{4.44}$$

This gives you the idea how to deal with delta function potentials. The scattering problem $E > 0$ will be left as an exercise.

4.5 SOLUTION OF THE SCHRÖDINGER EQUATION IN MOMENTUM SPACE

We noted in chapter 3 that the roles of momentum and position are reversed in momentum space where p becomes a multiplicative operator and x becomes the differential operator $i\hbar\partial/\partial p$. Thus to work with the time-independent Schrödinger equation

$$H\psi = (p^2/2m + V(x))\psi = E\psi$$

in momentum space, we must replace all x by the operator $i\hbar\partial/\partial p$.

As a specific example, take the case of the linear potential:

$$\begin{aligned} V(x) &= Cx \quad \text{with } C > 0, \text{ for } x > 0 \\ &= \infty \quad \text{for } x < 0 \end{aligned} \tag{4.45}$$

The potential is plotted in figure 4.7. According to the recipe in the beginning of this section, the Schrödinger equation in momentum space is now given as

$$\left[\frac{p^2}{2m} + Ci\hbar\frac{d}{dp} - E\right]a(p) = 0, \qquad E > 0 \tag{4.46}$$

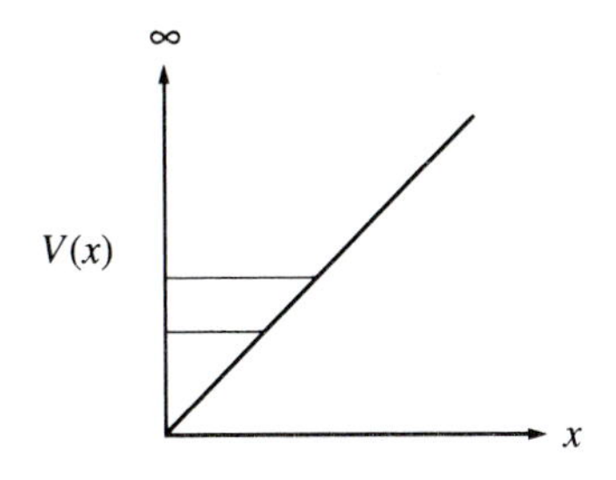

FIGURE 4.7
The one-dimensional linear potential with an infinite barrier at x = 0.

This can be written as

$$\frac{i}{C\hbar}(p^2/2m - E) = \frac{1}{a(p)}\frac{da(p)}{dp}$$

Integrating, we get

$$\ln(a(p)/A) = \frac{i}{C\hbar}\left[\frac{p^3}{6m} - Ep\right]$$

where A is an integration constant. Exponentiating, we get

$$a_E(p) = A\exp\left[\frac{i}{C\hbar}\left(\frac{p^3}{6m} - Ep\right)\right] \tag{4.47}$$

where we now label the eigenfunction with the eigenvalue E. We next evaluate the integration constant A by normalization:

$$\int_{-\infty}^{\infty} dp\, a_E^*(p)a_{E'}(p) = \delta(E - E')$$

which is a generalization of equation (3.22). Substituting for $a_E(p)$ from equation (4.47), we get

$$\begin{aligned}\delta(E - E') &= |A|^2\int_{-\infty}^{\infty} dp\exp\left[\frac{i}{C\hbar}(E - E')p\right]\\ &= |A|^2 2\pi\delta\left[\frac{E - E'}{C\hbar}\right] = 2\pi|A|^2 C\hbar\delta(E - E')\end{aligned}$$

using equation (A.7) in the appendix. Thus $2\pi|A|^2C\hbar = 1$. Thus the normalized momentum-space wave function for this problem is given as

$$a(p) = \frac{1}{\sqrt{2\pi C\hbar}} \exp\left[\frac{i}{C\hbar}\left(\frac{p^3}{6m} - Ep\right)\right] \tag{4.48}$$

So you see, the solution of the problem in momentum space is quite simple. However, the boundary condition for the problem is given for position space, namely, the wave function in position space must vanish at $x = 0$. Accordingly, we do need to transform back to position space via the Fourier integral, equation (3.2):

$$\psi(x) = \frac{1}{\sqrt{2\pi\hbar}} \int_{-\infty}^{\infty} dp\, e^{ipx/\hbar} a(p)$$

Substituting for $a(p)$ from equation (4.48), we obtain

$$\psi(x) = \frac{1}{2\pi\hbar C^{1/2}} \int_{-\infty}^{\infty} dp \exp\left[\frac{i}{C\hbar}\left(\frac{p^3}{6m} - Ep\right) + \frac{ipx}{\hbar}\right]$$

Applying the boundary condition, $\psi(0) = 0$, we get

$$\int_{-\infty}^{\infty} dp \exp\left[\frac{i}{C\hbar}\left(\frac{p^3}{6m} - Ep\right)\right] = 0$$

The exponential function can be split into a sine and a cosine function; the integral of the sine function, which is odd in p, is identically 0. Thus for the integral of the even cosine function, we get

$$\int_0^{\infty} dp \cos[(p^3/6m - Ep)/C\hbar] = 0$$

Substituting $q = p/(2mC\hbar)^{1/3}$, the integral becomes a standard one whose value is expressed in terms of a standard mathematical function:

$$\int_0^{\infty} dq \cos[q^3/3 - E(2m/C^2\hbar^2)^{1/3} q] = \pi Ai[-E(2m/C^2\hbar^2)^{1/3}]$$

where $Ai(x)$ denotes the Airy function, a standard special function. Thus the boundary condition that $\psi(x)$ goes to 0 at $x = 0$ gives us the discrete energy eigenvalues E_n as the zeroes of an Airy function

$$Ai[-E_n(2m/C^2\hbar^2)^{1/3}] = 0$$

Calling these zeroes a_n, we get

$$E_n = -(C^2\hbar^2/2m)^{1/3}a_n \tag{4.49}$$

The first few zeroes of the Airy function a_n are

$$a_1 = -2.338, \qquad a_2 = -4.088, \qquad a_3 = -5.521, \qquad a_4 = -6.787$$

In this way, we see that the position-space wave function for the present case is an Airy function. We will have more to say about the Airy functions later (see chapters 8 and 12); the three-dimensional analog of the linear potential is used in models of quark confinement in elementary particle physics.

Finally, make special note of the fact that the Schrödinger eigenvalue problem can be set up and solved either in position space or in momentum space. This must suggest to you that behind these diverse "representations" (position or momentum) lurks an abstract unified description of the states of a quantum system that is independent of the particular representation used. We will pick up this subject again in chapter 6. Meanwhile, in chapter 5, we will tackle the question of the meaning of quantum mechanics.

PROBLEMS

1. Calculate the transmission coefficient for barrier penetration for electrons of energy 5 eV, assuming that the barrier is one dimensional and rectangular with width 10^{-7} cm and height 6 eV.
2. Imagine a billiard table with sinusoidal walls (just to make the potential barrier interesting) (fig. 4.8). Estimate the probability of a billiard ball going across the wall of the table by quantum tunneling.
3. Consider the step potential (fig. 4.9) defined by $V(x) = 0$ for $x < 0$ and $V(x) = V_0$ for $x > 0$. Let a particle of mass m and energy E be incident on it. (a) For $E > V_0$, calculate the probability of reflection and transmission. (b) For $E < V_0$, calculate the probability of reflection; also show that there is finite probability that the particle penetrates into the region $x > 0$.

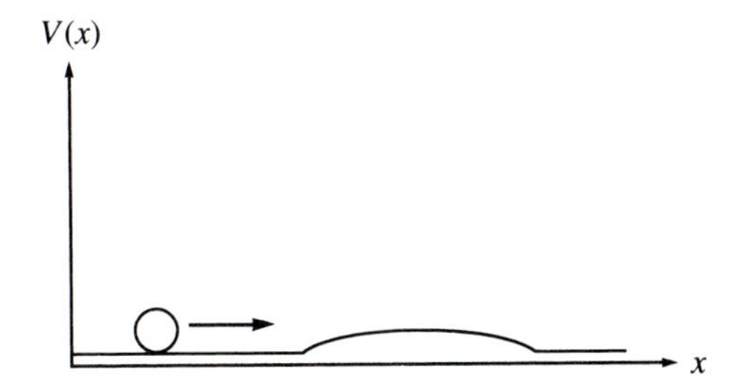

FIGURE 4.8
A sinusoidal potential barrier.

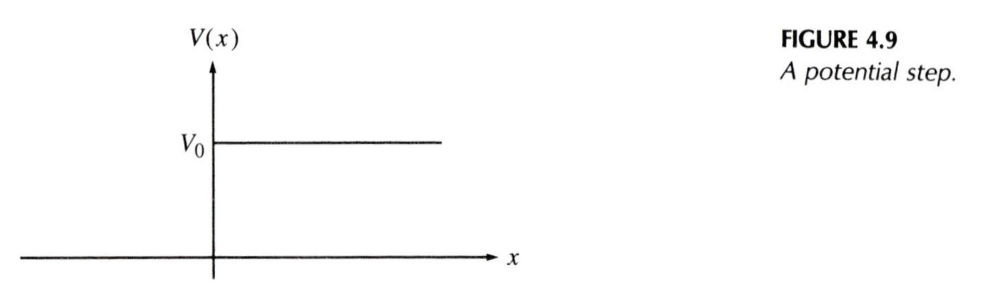

FIGURE 4.9
A potential step.

4. Consider the rectangular barrier (fig. 4.1). Consider the $E > V_0$ case and calculate the reflection and transmission coefficients.
5. A one-dimensional potential barrier is shown in figure 4.10. Calculate the transmission coefficient for particles of mass m and energy E ($V_1 < E < V_0$) incident on the barrier from the left.
6. A beam of 10 electron-volt electrons is incident on a potential barrier of height 30 eV and width 0.5 Å. Calculate the transmission coefficient.
7. Consider a one-dimensional box centered at $x = 0$. At the walls, $x = \pm a$, the potential $V_0 \to \infty$ (fig. 4.11). Find the energy eigenvalues and eigenfunctions for the problem. Why is the solution you obtain different from the box problem worked out in chapter 1?
8. Consider the Schrödinger equation for the potential shown in figure 4.12 and set up the boundary conditions matching equations for the case when particles of mass m and energy $E < V_0$ are incident from the right. Do not attempt to solve the equations.
9. A particle of mass m is confined in the double potential well shown in figure 4.13. The potential is zero for $-b - a/2 < x < -a/2$, and for $a/2 < x < b + a/2$. The potential is V_0 for $-a/2 \leq x \leq a/2$. Elsewhere the potential is ∞. Consider the bound-state problem. (a) Write down the Schrödinger equation and its solution for each of the regions appropriate for the problem. (b) Set up the boundary condition matching equations. (c) Is there a symmetry in the problem? Simplify the matching equations of part (b) using the symmetry. (d) Solve the matching equations and obtain transcendental equations for the eigenvalues.
10. Consider a repulsive δ function potential for the $E > 0$ solution: $V(x) = \Lambda\delta(x)$, $\Lambda > 0$. Find the reflection and transmission coefficients.

FIGURE 4.10

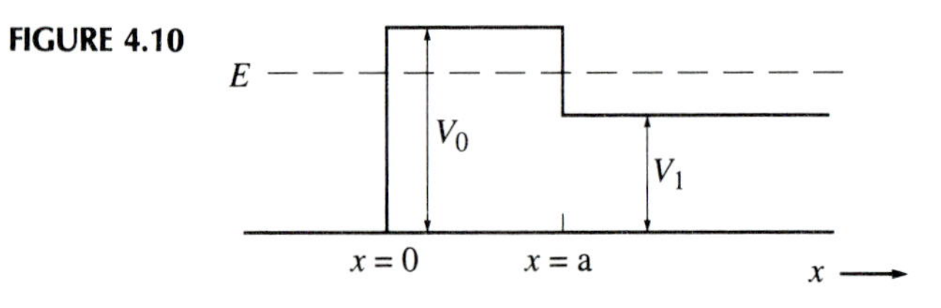

FIGURE 4.11

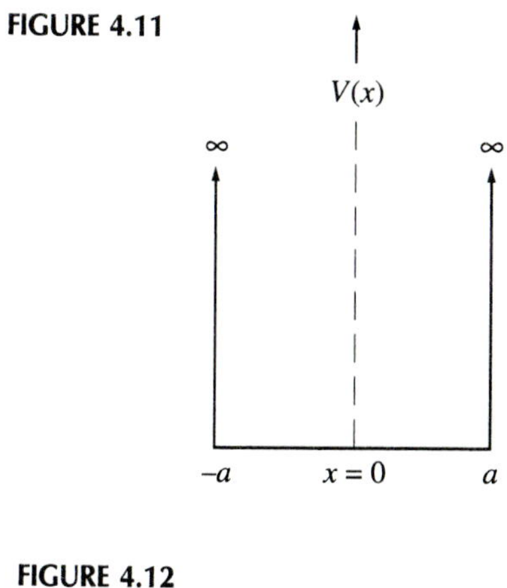

FIGURE 4.12

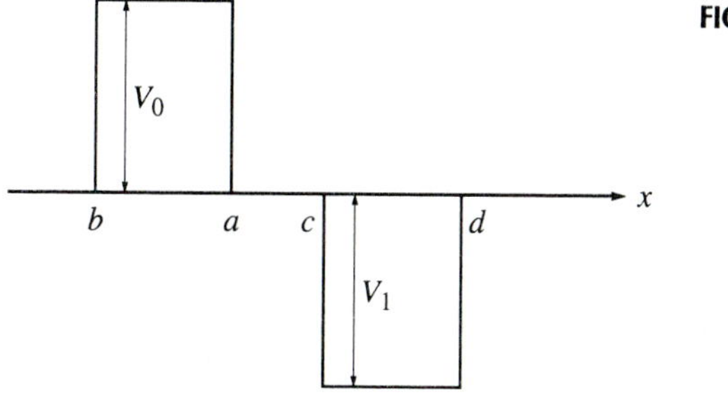

11. Calculate the transmission coefficient for a particle of mass m incident on the potential

$$V(x) = \Lambda[\delta(x+a) + \delta(x-a)]$$

12. Consider a particle of mass m bound by the attractive δ function potential worked out in the text. Now imagine that the potential suddenly changes to $V=0$ for $|x|<a$, $V\to\infty$ for $|x|>a$. Find the probability that the particle will be in the lowest even parity state of this new potential.

FIGURE 4.13

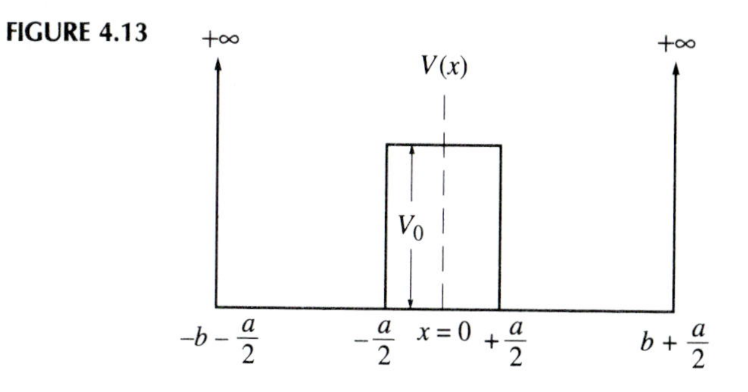

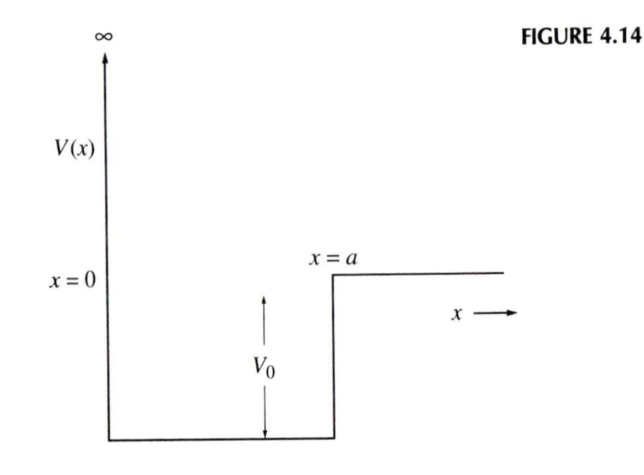

FIGURE 4.14

13. Consider a particle of mass m in the momentum dependent potential $V = Cp$, where C is a constant. Solve the Schrödinger equation in the momentum representation for $E > 0$. Then find the coordinate space-wave function.
14. Prove that the parity operator defined by $P\psi(x) = \psi(-x)$ is hermitian, and show that its eigenfunctions are orthogonal.
15. Consider the motion of a particle of mass 0.8×10^{-24} gm in a well, shown in figure 4.14, with range 1.4×10^{-13} cm. If the binding energy of the system is 2.2 MeV, find the depth of the potential V_0 in MeV. This is a simulation of the deuteron nucleus in one dimension.

ADDITIONAL PROBLEMS

A1. Consider the problem of possible cold fusion. A 1 cm^3 of palladium contains roughly 10^{20} hydrogen atoms. If each hydrogen atom loses its electron to the conduction band of Pd, we have 10^{20} protons. Consider the central collision of two protons as a one-dimensional e^2/x Coulomb barrier penetration problem (fig. 4.15). In the figure, E_b is the binding energy of the nucleons and $a \approx 10^{-13}$ cm is the proton radius. Assume that the palladium is at room temperature, $T_R \approx 300$ K, so that the energy of the protons is $E \approx kT_R$, where k is the Boltzmann constant. Calculate the transmission coefficient $|T|^2$ for a proton tunneling through the Coulomb barrier of another proton using the same procedure as employed for alpha decay in the text. Estimate the mean period between proton collisions in terms of the proton density n in palladium and the velocity v of the protons. How many such cold fusion processes do you expect in a year?

A2. Show that the bound-state wave functions can always be chosen as real. Hence show that the probability current density $j(x, t)$ is zero for any bound state. Is there any physical interpretation of this result?

FIGURE 4.15

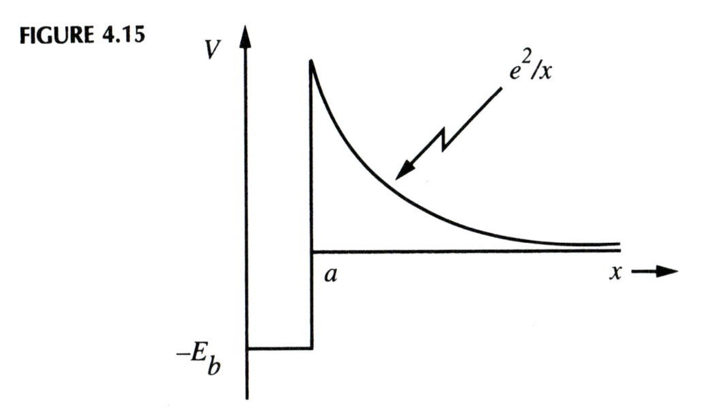

A3. Consider an electron in the ground state of an infinite one-dimensional square well potential of width a. Suddenly, the infinite square well is changed into a finite-walled one-dimensional well of width $15a$ and depth V_0. Assume the infinite well to have been positioned at the center of the wider finite well. Estimate the probability that the electron ends up in the ground state of the finite potential.

A4. Consider the bound-state motion of a particle in the potential

$$V(x) = V_1 \text{ for } x < 0,$$

$$V(x) = 0, \text{ for } 0 < x < a,$$

$$V(x) = V_2 \text{ for } x > a$$

where V_1 and V_2 are both constants >0 and $V_1 \geq V_2$. Write down the Schrödinger equations in the three regions and derive a transcendental equation for determining the bound-state energy eigenvalues. Discuss the special cases (a) $V_1 = V_2 \to \infty$ and (b) $V_1 \to \infty$.

A5. Two identical potential wells of width a and depth V_0 are separated by a distance d comparable to a (fig. 4.16). Discuss the motion of a particle in this potential. Pay special attention to the limits (a) $d \to 0$ and (b) $a \to \infty$.

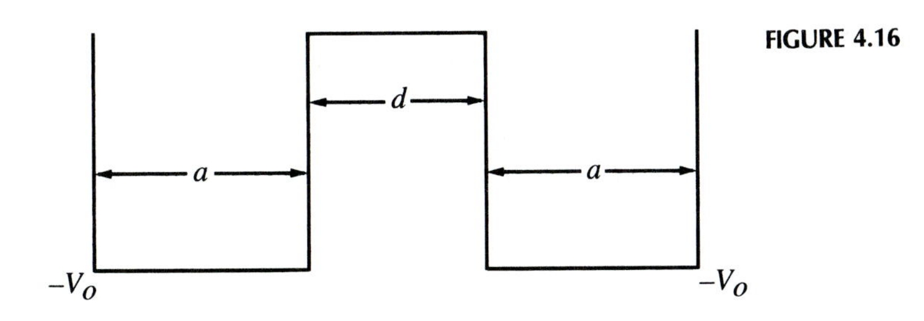

FIGURE 4.16

A6. Consider the motion of a particle in the exponential potential

$$V = V_0 \exp(-|x|/a)$$

(a) Find a substitution of variable that will reduce the Schrödinger equation to the Bessel equation

$$\frac{d^2u}{dz^2} + \frac{1}{z}\frac{du}{dz} + \left(1 + \frac{n^2}{z^2}\right)u = 0$$

whose solutions are the Bessel functions. Look in any book on mathematical special functions.

(b) By classifying the stationary states according to their parity, deduce the equation that determines the bound-state eigenvalues of the system.

A7. Prove that the wave function of a definite parity in position space has identical parity in momentum space.

A8. Suppose we put an electron in the potential well you calculated for problem 15 above. Is there a bound state? What conclusion can you draw from this about the possibility of an electron existing inside a nucleus?

REFERENCES

A. Z. Capri. *Nonrelativistic Quantum Mechanics.*
E. Fermi. *Nuclear Physics.*
P. Fong. *Elementary Quantum Mechanics.*
S. Gasiorowicz. *Quantum Physics.*
E. Merzbacher. *Quantum Mechanics.*
H. Smith. *Introduction to Quantum Mechanics.*

5

Looking Through the Heisenberg-Bohr Microscope

Gradually, I hope, you are beginning to see how quantum mechanics works, how to use the Schrödinger equation to calculate and predict the often strange behavior of quantum objects. But what does all this mean?

Heisenberg and Bohr, the two pioneers in the exploration of the meaning of quantum mechanics, gave us two principles, uncertainty and complementarity. More than anything else, these principles, especially when we examine how they operate in a given experimental situation, help us understand quantum mechanics. And understanding is what we will attempt in this chapter.

We will begin with the Heisenberg-Bohr microscope, the analysis that Heisenberg did of the attempt to see an electron through a microscope (Bohr's contribution was to further elucidate Heisenberg's analysis). But looking through the Heisenberg-Bohr microscope is more than just examining this particular analysis. It's learning to look through the maze of paradoxes that quantum measurements present to us. This is what the chapter is about.

5.1 THE HEISENBERG-BOHR MICROSCOPE

If you are a believer in classical determinism (this tendency is not uncommon), when you first encounter the uncertainty principle, you will attempt to find ways to violate this principle by finding situations and ways to determine both the position and the momentum of a quantum object with arbitrary accuracy. Thus, for example, you may think of using a "gamma-ray microscope" for measuring the position of your object. Position determination is limited only by diffraction effects of the light you use; diffraction effects depend on the wavelength of the light: The shorter the wavelength, the less likely the diffraction effect is to spoil your measurement. But to measure the momentum accurately, you have to measure the position again in a short while, and thereby compute the velocity of the object. And here arises the problem.

In classical physics, people thought that the act of measurement did not affect the object they measured. This is because, at least in principle, one can make the energy and momentum of one's probe arbitrarily small. But in quantum theory, you cannot make your probe any smaller than a photon. When you scatter a gamma photon off your object, say an electron, in order to see it, the scattering affects the electron's momentum and spoils the momentum measurement. And behold: The momentum of the photon is inversely proportional to its wavelength. Thus the shorter the wavelength of the photon, the larger its momentum, and the greater is the scattering uncertainty introduced into the electron's momentum. Clearly then, the harder you try to measure the position accurately, the worse is your chance to capture the momentum correctly. Isn't this what the uncertainty principle is saying? And if you try to do better on the momentum measurement and use long wavelength light, diffraction effects will get worse and ruin your position measurement. There is no way to win, is there?

To see how the uncertainty principle is quantitatively vindicated on such occasions of measurement as above, consider, following Heisenberg and Bohr, the Heisenberg-Bohr microscope (fig. 5.1). If θ is the half-angle subtended by the aperture of the microscope, as shown, then the uncertainty of position measurement is given by the minimum distance by which the microscope can resolve two objects. This minimum distance is calculated in any standard optics book, and is given as

$$\Delta x = \frac{\lambda}{2 \sin \theta}$$

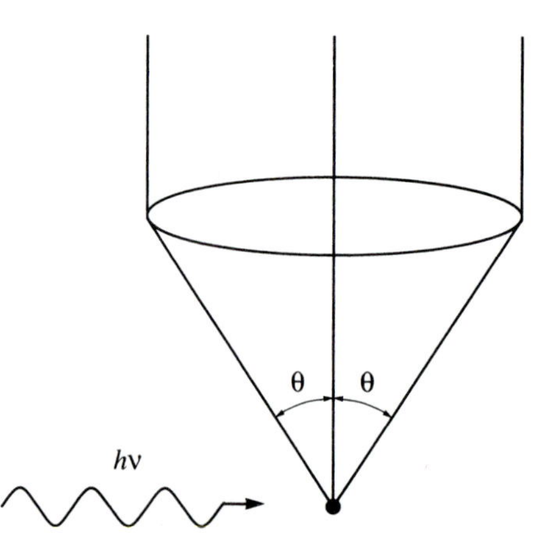

FIGURE 5.1
The Heisenberg-Bohr microscope.

(Don't be embarrassed if you are not familiar with this formula; the story is that young Heisenberg himself was not aware of it, and Bohr had to point it out to him.) Assume that the x-component of the momentum of the incoming photon is known precisely. Since the photon can be scattered at any angle between 0 and θ, the x-component of its momentum after scattering will be anywhere between 0 and $p \sin \theta$, where p is its total momentum. Since momentum is conserved, the uncertainty in the electron's recoil momentum must be greater than or equal to the momentum uncertainty of the scattered photon (the equality being obtained for the case when the initial momentum of the electron is known precisely):

$$\Delta p_x \geq p \sin \theta$$

Thus, the product of the two uncertainties is given as

$$\Delta x \Delta p_x \geq (\lambda/2 \sin \theta) p \sin \theta = h/2$$

where we have used $\lambda = h/p$ in the last step.

This idea that the process of measuring affects that which is measured in the submicroscopic domain of nature, and hence the uncertainty principle (imposed as a limit to the attainable accuracy of our measurements), is correct, but it does not do full justice to the profundity of the principle. To gain further insight, we'll look at some more gedanken experiments on the subject considered by Einstein and Bohr in their famous debate.

5.2 THE EINSTEIN-BOHR DEBATE

It is well known that Einstein never accepted the indeterministic aspect of quantum mechanics. And yet, although Einstein was most likely wrong in his opposition, posterity can be grateful for his position. Einstein's debate with Bohr on this subject, which began at the Solvay Congress in 1929, remains one of the most elucidating introductions to the ubiquity of the uncertainty principle.

Consider the single-slit diffraction experiment of light passing through a slit of finite width making a diffraction pattern on a screen (fig. 5.2). According to the wave theory of light, the angular spread θ of the pattern is inversely proportional to the width a of the slit:

$$\sin \theta = \frac{\lambda}{a}$$

where λ is the wavelength of the incident light. Now if we look at this experiment in terms of position and momentum measurement of photons, the position uncertainty along the x-direction (perpendicular to the incident beam direction) is given by the width a of the slit. What can we say about the com-

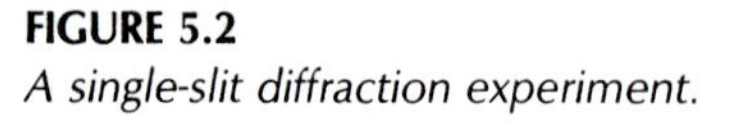

FIGURE 5.2
A single-slit diffraction experiment.

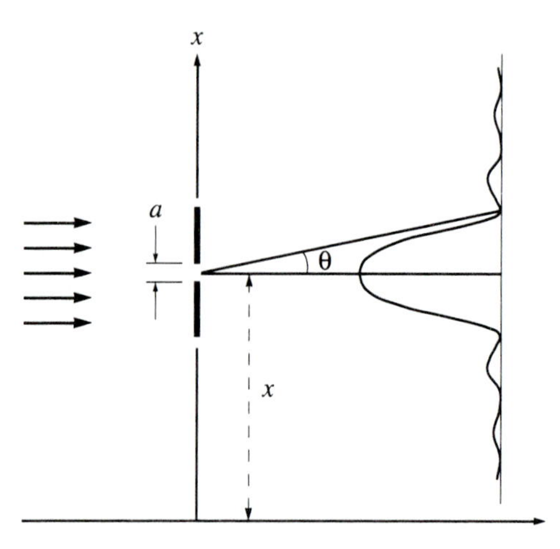

ponent of momentum in the x-direction? All we know is that the photon will arrive at the screen somewhere within the diffraction pattern, but we don't know where; thus the uncertainty in momentum is given by the angular spread of the pattern. But the angular spread is inversely proportional to the width of the slit. In this way we see that

$$\Delta p_x \approx p \sin \theta = \frac{h}{\lambda}\frac{\lambda}{a} = \frac{h}{a} = \frac{h}{\Delta x}$$

the momentum uncertainty is inversely proportional to the position uncertainty, in agreement with the uncertainty principle. Trying to reduce the position uncertainty by reducing the width of the slit leads to a greater spread of the diffraction pattern, and this corresponds to an increase of the momentum uncertainty in exact agreement with the uncertainty relation.

But Einstein argued as follows. Suppose we can measure the momentum of the photon as it leaves the slit. By making the slit as narrow as we like, we can reduce the position uncertainty as much as we wish; we are now measuring the momentum as accurately as we can, thus we should be able to make the product of the two uncertainties arbitrarily small, it would seem.

However, the screen with the slit in this single-slit diffraction experiment is rigidly connected to the earth, so how can we measure its recoil, which will give us the momentum of the photon as it leaves? Easy, said Einstein. Suppose the screen carries a shutter that opens the slit for a short interval Δt under the pressure of the incident radiation. Then the momentum transferred by the photon to the edges of the shutter can be calculated by applying the laws of energy and momentum conservation, giving in its turn the accurate value of the x-component of the momentum of the photon as it leaves the slit.

Not so fast, said Bohr. The energy and momentum transfer between the shutter and the photon are themselves subject to the time-energy uncertainty relation, and this makes them unanalyzable.

Behold: The shutter, since it exposes the slit of width, let's say, Δx for a time Δt, must move with the velocity $v = \Delta x/\Delta t$. A momentum transfer Δp to the shutter therefore involves an energy exchange with the photon

$$\Delta E = v\Delta p \sim (1/\Delta t)\Delta x\Delta p = \hbar/\Delta t$$

Thus $\Delta E\Delta t \sim \hbar$, the time-energy uncertainty relation holds, and indeed the momentum transfer cannot be calculated.

I have another one for you, said Einstein to Bohr, not being an easy person to shut up. He then posed a gedanken experiment involving the double-slit interference experiment. But Bohr rose to the occasion and used the double-slit experiment to elucidate the quantum uncertainty and complementarity principles with such clarity that Bohr's analysis has become a classic of quantum measurement theory, and we will discuss it in great detail in section 5.3.

Einstein came back to Bohr the following year with yet another intriguing example of what he thought was a violation of the uncertainty principle. This is the famous "Einstein box" experiment (fig. 5.3) involving a box full of radiation that contains a shutter controlled by a clock. Suppose we arrange the shut-

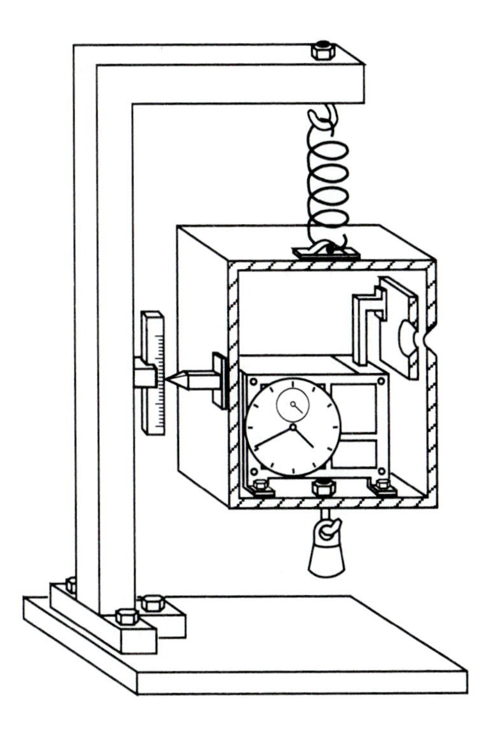

FIGURE 5.3
Einstein's box gedanken experiment schematically depicted. (Adapted from Niels Bohr, Atomic Physics and Human Knowledge. *New York: Wiley, 1958. Reprinted with permission from North Holland Publishing Company, Amsterdam.)*

ter mechanism to open a hole in the box for an arbitrarily small time interval. We can measure the energy of the radiation escaping by accurately weighing the box both before and after the escape of the energy pulse. Thus both the energy of the photon and its time of arrival at a distant detector could be predicted with arbitrary accuracy, violating the time-energy uncertainty relation.

Bohr's solution to the puzzle is classic. Examine the weighing process in great detail, said he. Any weighing must involve reading a pointer for which the accuracy is limited to Δx. This implies a momentum uncertainty for the box given by $\Delta p = \hbar/\Delta x$. The momentum uncertainty must be smaller than the impulse due to the change in mass, $g\Delta mt$, where t is the time needed to weigh and g is the acceleration of gravity:

$$\frac{\hbar}{\Delta x} < g\Delta mt \tag{5.1}$$

Bohr then invoked the equivalence principle (gravitational mass = inertial mass), Einstein's own brainchild, according to which a change in the vertical position Δx in a gravity field implies a change in the clock rate. Assume an atomic clock. Then we can calculate the change in the clock rate by equating the change in the potential energy $mg\Delta x$ (where m is the mass of the photon) with the change in the energy of the wave motion of light $h\Delta\nu$, where ν is the light frequency. But $\Delta\nu = \Delta t/t^2$ and $m = h\nu/c^2$. Thus we get

$$\frac{\Delta t}{t} = \frac{g\Delta x}{c^2} \tag{5.2}$$

Combining equations (5.1) and (5.2), we get

$$\Delta t > \frac{\hbar}{c^2\Delta m}$$

Or since $c^2\Delta m$ is equal to ΔE, we have

$$\Delta E\Delta t > \hbar$$

So Bohr showed, using Einstein's own theory against him, that the time-energy uncertainty relation is preserved after all in the Einstein box experiment.

Bohr won this debate, no doubt. Yet quantum philosophy, like all philosophy, contains two great facets: *epistemology*, how we know things, and *ontology*, the nature of being of things. The above demonstrations show that Bohr certainly won the epistemological debate, but Einstein also had an ontological concern.

As a wave, an electron spreads out, even as it passes through a single hole of a diffraction experiment. If there is a screen a distance away, its wave packet may cover the entire screen. However, as the electron makes a flash on the fluorescent screen, its wave must instantaneously contract down to a point. Einstein was worried that such quantum measurements violate the locality principle of relativity—the speed of light limitation of the travel of all signals in space-time—and smack of action-at-a-distance.

The answer to this ontological concern has already been touched upon in chapter 2. Quantum waves are not ordinary waves traveling in space, they are waves in potentia, as Heisenberg pointed out. But Einstein felt that the way to avoid action-at-a-distance is to adopt an ensemble interpretation of probability—that Schrödinger waves, instead of representing a single quantum object, represent a whole ensemble of them distributed in space (i.e., $|\psi(x)|^2$ is the probability that some particle of the ensemble is found at x). This is a very different ontology! To examine the ontological question of quantum objects, let's turn to the double-slit experiment.

5.3 THE DOUBLE-SLIT EXPERIMENT

The double-slit experiment, as you know, is a setup where a well-defined beam of objects, say electrons, will pass through a screen with two narrow slits in it (fig. 5.4a) before falling on a fluorescent screen. Since electrons are waves, the beam will be split into two sets of waves by the two-slitted screen. These waves will then interfere with one another, and the result of the interference will show on the fluorescent screen as a pattern of alternate bright and dark fringes (fig. 5.4b). Importantly, the spacing of the fringe pattern is a measure of the wavelength of the waves.

Remember though that the electron waves are probability waves. Thus we must say that it is the probability of an electron arriving at the light areas that is high, and that the probability of an electron arriving at the dark areas is low, hence the pattern. You see, we must not get carried away and think from the interference pattern that the electron waves are classical waves, because the electrons do arrive at the fluorescent screen in a particlelike manner, one localized flash per electron. It is the totality of spots made by a large number of electrons that looks like the wave interference pattern. However, doesn't this sound more like the ensemble interpretation?

To settle this question, suppose we make the electron beam very weak, so weak that at any one moment, only one electron arrives at the slits on the average. Do we still get an interference pattern? Quantum mechanics, interpreted according to the Copenhagen interpretation, unambiguously says yes. And experimental data seem to agree with this viewpoint. Einstein's ensemble interpretation has no easy way of interpreting this data.

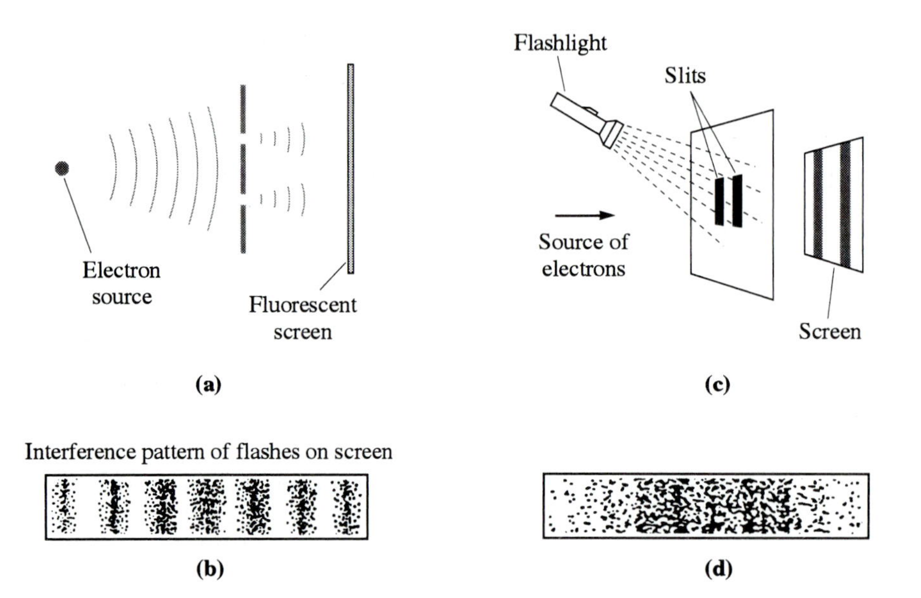

FIGURE 5.4
(a) The double-slit experiment (schematic) for electrons. It is a big clue for the nature of being (ontology) of the electron. (b) The electron double-slit interference pattern showing the wave nature of electrons. (c) When we try to find out which slit the electron is passing through by focusing flashlights on the slits, the electrons show their particle nature—there are only two interference fringes exactly where we would expect them if the electrons were miniature baseballs. (d) With a dimmer flashlight, some of the interference returns.

But it is not easy to accept the Copenhagen view. We cannot get interference without a split beam! You may yourself cry in the agony of intellectual confusion, Doesn't it take two waves to interfere? Can a single electron split, pass through both slits, and interfere with its own better half? Yes, yes. Quantum mechanics says yes to all these questions. As Paul Dirac, one of the pioneers of quantum mechanics, put it, "Each photon [or electron in the present case] interferes only with itself."

If all of this sounds a bit strange and confusing, maybe now is the time to remember the words of Niels Bohr: "Those who are not shocked when they first come across quantum theory cannot possibly have understood it."

Let's be patient and see how the mathematics of interference works. Let's consider a point x on the screen and denote the probability wave amplitudes for the electron to reach x from slit 1 and 2, respectively, by $\psi_1(x)$ and $\psi_2(x)$. If only slit 1 is open, the probability of the electron arriving at x is simply $|\psi_1(x)|^2$. If slit 2 is open instead, the same probability is $|\psi_2(x)|^2$. But when both slits are open, the total wave function contributing to x is the coherent superposition

$$\psi(x) = \psi_1(x) + \psi_2(x) \tag{5.3}$$

and we must calculate the probability by squaring this total ψ. For the probability $P(x)$ this gives

$$P(x) = |\psi_1(x) + \psi_2(x)|^2 = |\psi_1(x)|^2 + |\psi_2(x)|^2 + 2\,\mathrm{Re}[\psi_1^*(x)\psi_2(x)] \tag{5.4}$$

There is now an interference term in addition to the single-slit terms $|\psi_1(x)|^2$ and $|\psi_2(x)|^2$. It is because in quantum mechanics the probability rule is that the amplitudes add before squaring (not the already-squared amplitudes, as in classical physics) that we get the interference term in equation (5.4).

But try to imagine the coherent superposition—that an electron is passing 50% through one slit and 50% through the other slit. You will probably get exasperated and begin to disbelieve this strange consequence of quantum mechanics. Does the electron really pass through both slits at the same time? Why should you take that for granted? Why, you may ask, don't we find out by making a measurement (Einstein had suggested something similar to Bohr)? The simplest "gedanken" way of making such a measurement is to look with a flashlight (following Richard Feynman). We can focus a flashlight on the slits to see which hole the electron is really passing through.

So we turn the light on, and as we see an electron passing through a particular slit, we look to see where the flash appears on the fluorescent screen. But what we find is that every time an electron goes through a slit, its flash appears just behind the slit it passed through. So after a time the combined effect on the screen looks like figure 5.4(c); the interference pattern has disappeared.

What's happening now is the play of the uncertainty principle, as Bohr explained. As soon as we locate the electron and determine which slit it passed through, we lose the information about the electron's momentum. An electron, as we have already seen in the analysis of the single-slit diffraction experiment, is very delicate; the collision with the photon that we are using to observe it affects it so its momentum changes by an unpredictable amount. As soon as we lose the information about the electron's momentum, we also must lose information about its wavelength, according to de Broglie's relation. But if there were interference fringes, from their spacing we would be able to measure the wavelength. Thus the fringe structure cannot exist anymore—the interference pattern is destroyed.

The point is that the position and momentum measurements on the electron are really complementary, as Bohr first pointed out forcefully; they are mutually exclusive processes. We can concentrate on the momentum and measure the wavelength and thence the momentum of the electron from the interference pattern, but then we cannot tell which slit the electron went through. Or we can concentrate on the position and lose the interference pattern, the information about the wavelength and momentum. When we localize (find out which slit the electron goes through), we reveal the particle aspect of the electron; and when we don't localize (don't worry about which slit the electron is passing through), the electron shows its wave aspect. This is what Bohr is saying with his complementarity principle.

So are quantum objects both wave and particle, but we can see only one attribute with a particular experimental arrangement? This "both/and" thinking is certainly correct, but there is a further subtlety. We must also say, the electron is neither a wave (because we never *see* the electron spread out like a wave, a single electron is always found at one flash on the screen) nor a particle (how else would it appear on the screen at places forbidden for classical particles?). What can be "both/and" and "neither/nor"? So the complementarity principle is saying that the electron's true nature transcends both wave and particle descriptions.

To see more clearly how complementarity operates, suppose in the experiment above that we use weak batteries to make the flashlight we shine on the electrons somewhat dimmer. When we repeat the experiment that led to figure 5.4(c) with dimmer and dimmer flashlights, we find that some of the interference pattern begins to appear again, becoming more and more prominent as we make the flashlight dimmer and dimmer (fig. 5.4d). When the flashlight is turned off completely, the full interference pattern comes back.

What is happening now is this. As the flashlight goes dim, the number of photons scattering off the electrons decreases, so some of the electrons entirely escape being "seen" by the light. Those electrons that are still seen appear behind slit 1 or slit 2, just where we would expect them. But the unseen electrons split and interfere with themselves to make the wave-interference pattern on the screen when enough of them have arrived there. In the limit of strong light, only the particle nature of the electrons is seen; in the limit of no light, only the electrons' wave nature is seen. In the case of various intermediate situations of dim light, both aspects show up to a similarly varied degree. Thus complementarity—we can see either the particle nature or the wave nature of a quantum object at any given time—is not a property of the whole ensemble, but must hold for each individual object!

Finally, what about Einstein's concern about action-at-a-distance? Heisenberg, while discussing a situation (similar to the double slit; see fig. 5.6) where a light wave packet strikes a half-silvered mirror and splits into two packets that go their separate ways, had this to say:

> If now an experiment yields the result that the photon is, say, in the reflected part of the packet, then the probability of finding the photon in the other part of the packet immediately becomes zero. The experiment in the position of the reflected packet then exerts a kind of action (reduction of the wave packet) at

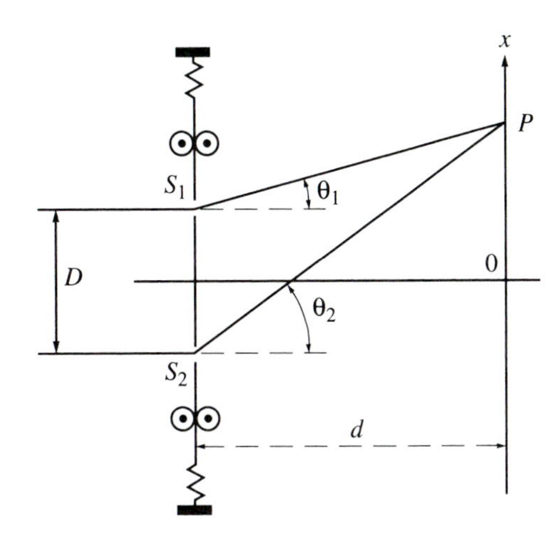

FIGURE 5.5
The double-slit experiment with a movable screen (schematic). The momentum of the double-slitted screen is measured before and after the electron passes through, to determine which slit the electron went through in order to reach the point P on the fluorescent screen.

the distant point occupied by the transmitted packet, and one sees this action is propagated with a velocity greater than light. However, it is also obvious that this kind of action can never be utilized to transmit a signal so that it is not to conflict with the postulates of the theory of relativity.†

A Physical Example of the Metaphorical Flashlight

A real-life example of the metaphorical flashlight with which we can "see" which slit an electron is passing through in the double-slit experiment can be constructed, following Einstein, if the double-slitted screen is mounted so that it can move vertically along the same plane. Then it's possible to measure its vertical (x-) momentum. Let the electron gun be at ∞. Consider an electron that hits the fluorescent screen at P (fig. 5.5). Some of the momentum of the electron is transferred to the double-slitted screen, but the amount of momentum transferred depends on which slit the electron passes through. If the electron passes through slit S_1 (see fig. 5.5), then the momentum transferred to the screen is

$$p_1 = -p \sin \theta_1$$

where p is the electron's momentum. If, however, the electron passes through slit S_2, the momentum transferred is given as

$$p_2 = -p \sin \theta_2$$

Thus by measuring the momentum transferred to the double-slitted screen, we can tell which slit the electron passed through. But how does it ruin the interference pattern as the uncertainty (and complementarity) principle demands?

†W. Heisenberg, *The Physical Principles of the Quantum Theory*, p. 39.

The point is that we must apply the uncertainty principle not only to the motion of submicroscopic objects such as electrons but also to macro measurement apparatuses such as the two-slitted screen (as Bohr pointed out to Einstein in their original Solvay debate). Thus the uncertainty Δp of the screen's momentum must be small enough for us to tell the difference between p_1 and p_2 above:

$$\Delta p \ll |p_1 - p_2|$$

But we can know the position of the screen only within an accuracy of Δx given as

$$\Delta x \geq \frac{h}{|p_1 - p_2|} \tag{5.5}$$

Now if D is the separation between slits and d is the distance between the slit and the fluorescent screen, and further if we assume that the angles θ_1 and θ_2 are small (i.e., $d/D \gg 1$), then

$$\sin\theta_1 \approx \theta_1 = \frac{x - D/2}{d}$$

$$\sin\theta_2 \approx \theta_2 = \frac{x + D/2}{d}$$

where x denotes the position of the electron impact point P on the fluorescent screen (fig. 5.5). This gives

$$|p_1 - p_2| = p\ |\sin\theta_1 - \sin\theta_2| \approx p\ |\theta_1 - \theta_2|$$
$$\approx \frac{h}{\lambda}\frac{D}{d}$$

where λ is the de Broglie wavelength of the electron. Substituting in equation (5.5), we get

$$\Delta x \geq \frac{\lambda d}{D}$$

But $\lambda d/D$ is the fringe separation! Clearly, if the vertical position of the slits can be known only within an accuracy that is greater than or at best equal to the separation between the fringes, the fringes will be impossible to observe. This is a direct consequence of the uncertainty principle applied to the motion of the double-slitted screen.

A very important point then is this. We must apply quantum mechanics not only to submicroscopic systems, but also to macrosystems of measurement.

5.4 THE DELAYED-CHOICE EXPERIMENT

One unique characteristic of the complementarity principle is this: What attribute the quantum object reveals depends on how *we* choose to observe it. Nowhere is the importance of conscious choice in the shaping of quantum reality better demonstrated than in the delayed-choice experiment with a beam splitter suggested by John Wheeler (fig. 5.6). A light beam is split into two beams of equal intensity by using a half-silvered mirror M_1; these two beams are then reflected by two regular mirrors A and B to a crossing point P on the right. If we choose to detect the particle mode of the photons, we put detectors or coun-

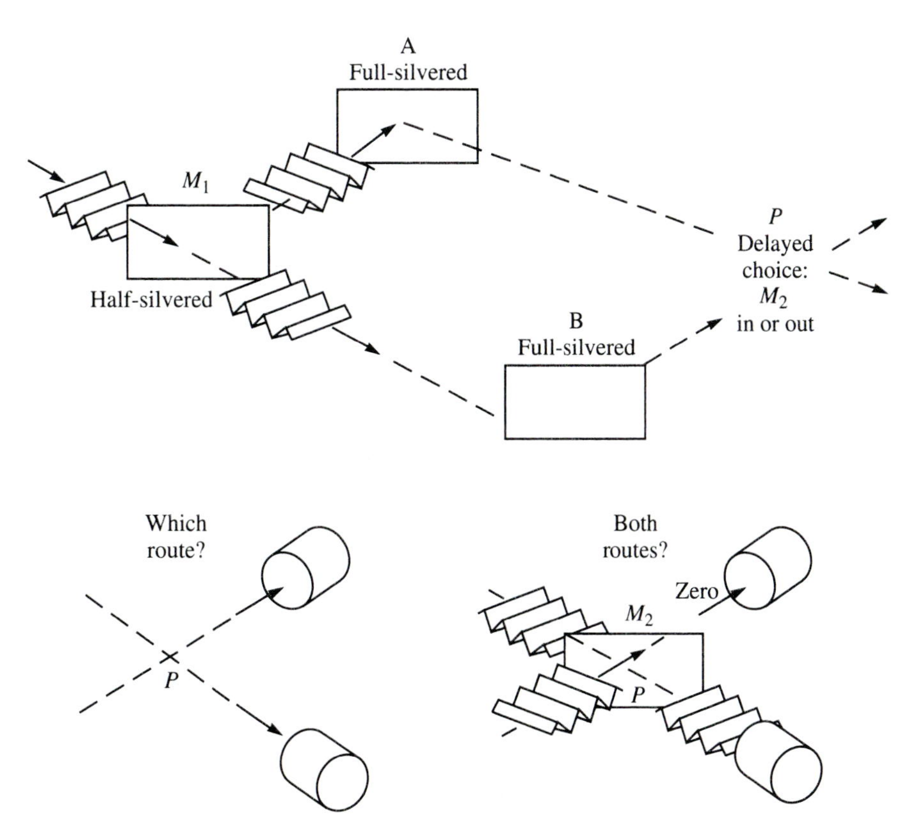

FIGURE 5.6
The delayed choice experiment is shown schematically in the upper picture. Lower left: the arrangement for seeing the particle nature of photons. Both detectors click, but one at a time, signifying which route the photon takes. Lower right: the arrangement for seeing the wave nature of photons. One of the detectors never clicks, signifying wave-interference cancellation. The photon must travel both routes at the same time in this case. Delayed choice means to put the mirror at P (or take the mirror out if it is already there) after the photon has left the mirror M_1.

ters past the point of crossing P, as shown at the lower left in figure 5.6. One or the other counter will tick, defining the localized path of a photon, to show its particle aspect.

To detect the wave aspect of the photon we take advantage of the phenomenon of wave interference. If we put a second half-silvered mirror M_2 at P (fig. 5.6, bottom right), the two waves created by beam splitting at M_1 will be forced by M_2 to interfere constructively on one side of P (where the counter ticks) and destructively on the other side (where now the counter never ticks). But notice that when we are detecting the wave mode of the photons, we must agree that each photon is traveling by both routes A and B; otherwise how can there be interference?

But the subtlest aspect of the experiment is yet to come! In the delayed-choice experiment, the experimenter decides at the very last moment, in the very last pico (10^{-12}) second, whether or not to insert the half-silvered mirror at P, whether or not to measure the wave aspect. In effect, this means that the photons have already done their travel past the point of splitting if you think of them as classical objects! Even so, inserting the mirror at P always shows the wave aspect, and not inserting the mirror, the particle aspect. Was each photon moving in one path or two? The photons seem to respond to even our delayed choice instantly and retroactively(!). A photon travels one path or both paths, exactly in harmony with our choice in actual laboratory experiments. How does the photon know? Or is it action-at-a-distance or retroactive action? or both? The answer is simple: There is no manifest photon until we see it, and thus how we see it determines its attributes. There is no paradox in the delayed-choice experiment, if you give up the idea that there is a material world laid in concrete even when we are not observing it. Says Wheeler,

> nature at the quantum level is not a machine that goes its inexorable way. Instead what answer we get depends on the question we put, the experiment we arrange, the registering device we choose. We are inescapably involved in bringing about that which appears to be happening.†

The story is that in Copenhagen, at the Bohr Institute, they used to make all students of quantum mechanics read a story by the Danish writer Poul Moller. In the story, the hero points out the paradox of thinking—we watch our thoughts as they parade by, but whose thoughts are they? Ours! The observer and the observed get mixed up. The same thing happens in quantum physics. There is no sharp line where the observed ends and the observer begins.

Finally, the delayed choice experiment has been carried out in the laboratory in the 1980s by several experimental groups. See, for example, Hellmuth, et al., "Realization of the delayed choice experiment," in D. M. Greenberger

---

†J. A. Wheeler, "Law Without Law," in J. A. Wheeler and W. H. Zurek (eds.), *Quantum Theory and Measurement*, p. 185.

FIGURE 5.7

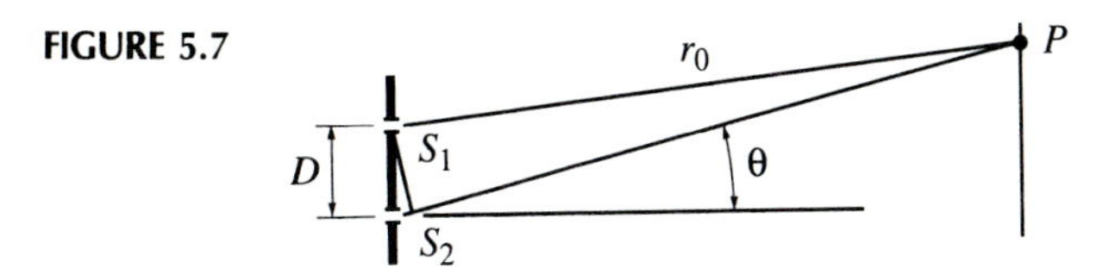

(ed.), *New Techniques and Ideas in Quantum Measurement Theory*, pp. 108–114 (New York: New York Academy of Sciences). In June 1995, the experiment even made the pages of *Newsweek* magazine as an example of "quantum weirdness."

PROBLEMS

1. Richard Feynman said, "[the double-slit experiment] has in it the heart of quantum mechanics; in reality it contains the *only* mystery." Can you tell what mystery Feynman is talking about?
2. While performing the double-slit experiment, if we close one of the slits, the interference pattern disappears at once. Is this subject to the Einstein criticism of action-at-a-distance? Why or why not?
3. In the double-slit arrangement shown in figure 5.7, suppose the amplitudes at slits 1 and 2 are given as

$$\psi_1 = |\psi_1| e^{i(\phi - \omega t)}, \qquad \psi_2 = |\psi_2| e^{i(\phi - \omega t)}$$

   What are the amplitudes at the point P assuming that the screen is far enough away that the same angle θ can be used to describe both rays? Calculate the intensity distribution on the screen as a function of θ.
4. Suppose we do an experiment in which electrons pass through two consecutive slits (fig. 5.8). If we consider an electron that has passed through the second slit, its position uncertainty Δx is $\approx a$, the width of the slit. Since we know that the electron has passed through both slits, the uncertainty in the direction of the momentum is $\Delta\theta \leq a/d$. Then the uncertainty in the x-momentum must be

$$\Delta p \approx \frac{ap}{d}$$

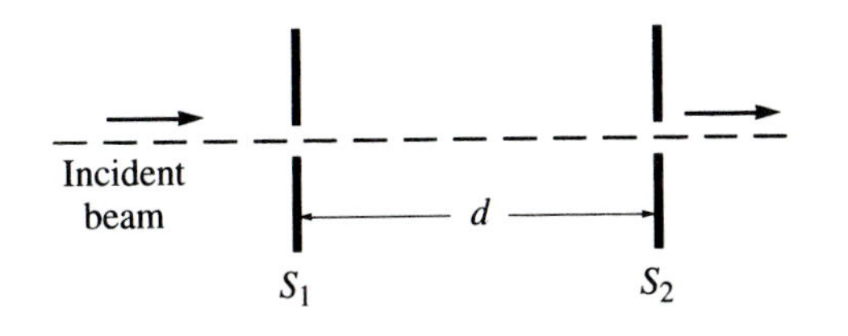

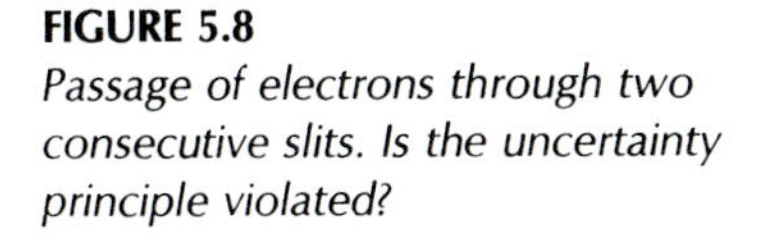
FIGURE 5.8
Passage of electrons through two consecutive slits. Is the uncertainty principle violated?

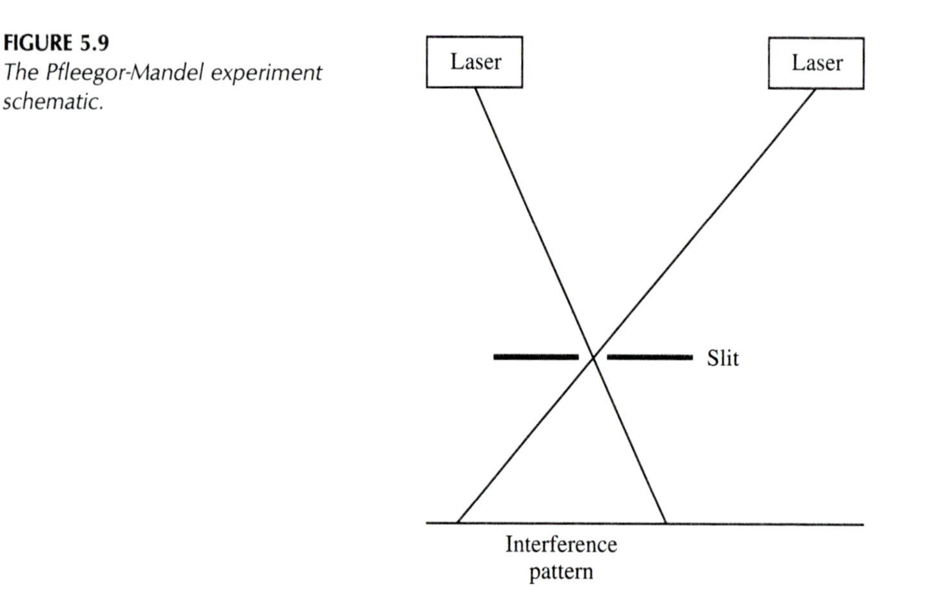

FIGURE 5.9
The Pfleegor-Mandel experiment schematic.

Arguing in this way, we find

$$\Delta x \cdot \Delta p \approx \frac{a}{d}\, ap$$

Thus it appears that by making a arbitrarily small and d arbitrarily large, we can make the uncertainty product as small as we wish. Defend the uncertainty principle.

5. In the Pfleegor-Mandel experiment, photons from two different lasers of identical frequency pass through a slit and combine on a screen to make an interference pattern (fig. 5.9). The interference pattern persists even when the laser beams are so weak that only one photon from either source passes through the slit at a time. Show by using the uncertainty principle at the detection site that the interference pattern in this case is due to the fact that we cannot know which laser a photon is coming from, thus giving us a coherent superposition.

REFERENCES

C. Cohen-Tannoudji, B. Diu, and F. Laloe. *Quantum Mechanics.*
R. P. Feynman, R. B. Leighton, and M. Sands. *The Feynman Lectures in Physics.*
W. Heisenberg. *The Physical Principles of the Quantum Theory.*
M. Jammer. *The Philosophy of Quantum Mechanics.*
R. L. Pfleegor and L. Mandel. "Interference of Independent Photon Beams." *Physical Review, 159,* 1084, 1967.
J. A. Wheeler and W. Zurek (eds.). *Quantum Theory and Measurement.*

6

The Dirac Description of Quantum Mechanical States

In this chapter we will develop a powerful, abstract way of representing and working with the notion of a *state* of a quantum mechanical system. This abstract description was invented by Paul Dirac. Dirac's motivation was to find a formulation of quantum mechanics that integrated Schrödinger's wave mechanics and Heisenberg's matrix mechanics.

First, what is a state? The idea is that if we know the state of a quantum mechanical system, we should know all about the system. We can be a little more specific. A quantum mechanical system is in a definite state when everything you want to know about it that can be known without violating the laws of quantum mechanics is known. The state has to be described in such a way that the availability of this knowledge is conveyed.

In the preceding chapters, we have been describing the states of quantum systems by wave functions. We have found that the description of quantum mechanical states is radically different from that of states in classical physics. For one thing, for quantum states we can specify less than we can for classical states. For classical states, for example, we can specify both position and momentum, and hence our prediction of future events is deterministic. For quantum states, we cannot simultaneously specify both position and momentum of the system, by the dictate of the uncertainty principle; because of this paucity of specification we can only predict probabilities, not certainties, of future events.

A second striking difference in the description of quantum states is that the wave functions describing them obey the superposition principle. This often leads to a situation where a quantum state is described by a coherent superposition of quite contradictory facets—for example, the electron is at slit 1 as well as slit 2 in the double-slit experiment. Such a description of a state would be an anachronism in classical physics.

Any abstract formulation of quantum mechanical states must incorporate the above aspects we already know. Hence a good place to start is the expansion postulate, which incorporates the idea of coherent superposition in a most succinct manner. Remember the objective is to develop an abstract description of the state that is independent of the specific representation that a wave function uses.

6.1 KETS AND BRAS AND THEIR REPRESENTATIONS

Any ordinary vector $\mathbf{A}$ can be expanded in terms of basis vectors $\mathbf{e}_i$:

$$\mathbf{A} = \sum_i A_i \mathbf{e}_i$$

Since the unit vectors $\mathbf{e}_i$ are orthonormal,

$$\mathbf{e}_i \cdot \mathbf{e}_j = \delta_{ij}$$

we have

$$A_i = \mathbf{e}_i \cdot \mathbf{A}$$

In this way, A_i, the scalar product of the vector $\mathbf{A}$ and basis vector $\mathbf{e}_i$, is the projection of $\mathbf{A}$ along $\mathbf{e}_i$.

Compare the above equations with those of the expansion of a wave function $\psi(x)$ in terms of a complete orthonormal set of eigenfunctions $\phi_i(x)$ of a hermitian operator O:

$$\psi(x) = \sum_i c_i \phi_i(x) \tag{6.1}$$

We can see the following analogy:

$$\text{eigenfunctions } \phi_i \rightarrow \text{basis vectors } \mathbf{e}_i$$

$$\text{wave function } \psi \rightarrow \text{vector } \mathbf{A}$$

$$\text{expansion coefficient } c_i \rightarrow \text{scalar product } \mathbf{e}_i \cdot \mathbf{A}$$

Nevertheless there are differences; in quantum mechanics we deal with complex continuous functions, and the vector space they correspond to is complex and infinite dimensional. Thus the same notation as that of linear vectors will not do.

Following Dirac, let's denote the basis vectors of our quantum mechanical vector space, called the *Hilbert space*, by the notation $|\phi_i\rangle$. They will be called the *base kets* or basic states, or simply basis. The state corresponding to the wave function ψ will now be represented by a *state vector* $|\psi\rangle$ in this vector space. (Note that we are labeling the state vectors in a way that anticipates the wave functions they correspond to. This is not the only way of labeling state vectors, as we shall see.)

Fortunately, the similarities between the linear vector space and the Hilbert space go quite deep, so that we can anticipate operations in the Hilbert space from the corresponding operation in the linear vector space. Thus the sum of two state vectors is again a state vector and so is the product of a state vector and an arbitrary complex number.

The most important quantity is the scalar product (sometimes also called inner product). In the above analogy, it is given by the complex numbers c_i:

$$c_i = \int dx\, \phi_i^* \psi$$

We introduce the notation $\langle\phi_i|\psi\rangle$, called the *bracket*, to denote the scalar product in our vector space:

$$c_i = \langle\phi_i|\psi\rangle = \int dx\, \phi_i^*(x)\psi(x) \tag{6.2}$$

To complete the description of the notation, the notation $\langle\phi|$ is called a *bra*. It's the *dual* partner to a ket, the other half of the bra-ket.

In this new notation, the complete set expansion, equation (6.1), can be written as

$$|\psi\rangle = \sum_i \langle\phi_i|\psi\rangle|\phi_i\rangle \tag{6.3}$$

which is the same as

$$|\psi\rangle = \sum_i |\phi_i\rangle\langle\phi_i|\psi\rangle$$

In other words, the key to the complete set expansion is the *completeness* rule that the base kets must satisfy:

$$\sum_i |\phi_i\rangle\langle\phi_i| = 1 \tag{6.4}$$

In other words, *the ket-bra sum must be equal to 1* for base kets. From the orthonormality relation of the eigenfunctions $\phi_i(x)$

$$\int dx\, \phi_i^*(x)\phi_j(x) = \delta_{ij}$$

and the definition, equation (6.2), of the bracket, it also follows that

$$\langle \phi_i | \phi_j \rangle = \delta_{ij} \tag{6.5}$$

Now recall that physically, the c_i in equation (6.2) is a probability amplitude. Since $c_i = \langle \phi_i | \psi \rangle$, we can now physically interpret the scalar product as the probability amplitude that we start with a ket $|\psi\rangle$ and end up with the bra $\langle \phi_i |$. Now here is a difference between the ket and the bra: A ket is the initial state, a possible cause, a starting position; on the other hand, a bra is the final condition, that's where the system ends up, it is the possible effect. And $|c_i|^2$ is the probability of finding the system in the eigenstate $|\phi_i\rangle$ starting from the state $|\psi\rangle$. If ψ is normalized, the sum of all the probabilities must be 1, and so we have

$$\sum_i |c_i|^2 = 1 = \sum_i c_i^* c_i = \sum_i \langle \phi_i | \psi \rangle^* \langle \phi_i | \psi \rangle$$

But from the normalization of ψ, we can write, using equation (6.3),

$$\langle \psi | \psi \rangle = 1 = \sum_i \langle \psi | \phi_i \rangle \langle \phi_i | \psi \rangle$$

Comparing the above two equations, we can see that in order for both equations to hold, we must have the relationship

$$\langle \psi | \phi_i \rangle = \langle \phi_i | \psi \rangle^* \tag{6.6}$$

Reversing the bra and ket is the same thing as complex conjugation. Physically, for the probability amplitudes of going from one state to another, we must have (forward amplitude)* = reverse amplitude. If this were not true, probability would not be conserved.

The eigenfunctions $\phi_i(x)$ in the complete set expansion, equation (6.1), are the eigenfunctions of some hermitian operator $\hat{O}$ belonging to the eigenvalue o_i:

$$\hat{O}\phi_i(x) = o_i \phi_i(x) \tag{6.7}$$

We will now generalize this equation for our abstract state vectors. Corresponding to every operator in position space, there is, in the Hilbert space, an abstract operator that transforms one state vector into another. And if such an operator operating on a state vector gives the same vector back multiplied by a number, the state vector is called the eigenstate of the operator, and the number the eigenvalue. Thus the vector equation that corresponds to the eigenvalue equation (6.7) is written as

$$\hat{O}|\phi_i\rangle = o_i|\phi_i\rangle \tag{6.8}$$

Of course, only the eigenvectors of a hermitian operator will define a complete set of base kets. Now let's work with two such sets of base kets, the eigenkets of the position and momentum operators, respectively.

Position and Momentum Representations

The base kets of position space are written as $|x\rangle$. They are the eigenstates of the position operator, and we label them by the eigenvalue x of this operator. This is another useful way of labeling. Since x is a continuous variable, the two properties of base kets noted in equations (6.4) and (6.5) have to be rewritten, replacing the summation by an integral and the Kronecker delta by the Dirac delta function:

$$\int dx\, |x\rangle\langle x| = 1 \tag{6.9}$$

$$\langle x'|x\rangle = \delta(x' - x) \tag{6.10}$$

Now consider our initial definition of the scalar product, equation (6.2), and apply it to the scalar product $\langle\psi|\psi\rangle$. We have

$$\langle\psi|\psi\rangle = \int dx\, \psi^*(x)\psi(x) \tag{6.11}$$

But using equation (6.9), we can also write

$$\begin{aligned}\langle\psi|\psi\rangle &= \int dx\, \langle\psi|x\rangle\langle x|\psi\rangle \\ &= \int dx\, \langle x|\psi\rangle^*\langle x|\psi\rangle \end{aligned} \tag{6.12}$$

where we have used equation (6.6). Comparing equations (6.11) and (6.12), it is clear that

$$\langle x|\psi\rangle = \psi(x)$$

It fits! When the system is the state $|\psi\rangle$, the probability amplitude for finding it at the position x is the wave function $\psi(x)$. We are back to where we began. Also, since we can write

$$|\psi\rangle = \int dx\, |x\rangle\langle x|\psi\rangle = \int dx\, |x\rangle\psi(x) \tag{6.13}$$

it is clear that $\langle x|\psi\rangle$ (or $\psi(x)$) is also the "component" or the projection of the state vector $|\psi\rangle$ onto the base kets of position space. In other words, $\psi(x)$ is the *position representation* of the state vector $|\psi\rangle$. It is the expression of the abstract entity $|\psi\rangle$ using the language of position space. You can think of each of the $\psi(x)$ as a word-expression of $|\psi\rangle$ in this language.

Similarly, we denote the base kets in the momentum representation as $|p\rangle$—the eigenstates of the momentum operator. Again we label the states by the corresponding eigenvalue. Expanding the state vector $|\psi\rangle$ in these momentum space base kets, we get

$$|\psi\rangle = \int dp\, |p\rangle\langle p|\psi\rangle \tag{6.14}$$

where $\langle p|\psi\rangle$ = the amplitude for ending with momentum p, starting with state $|\psi\rangle = a(p)$, the momentum wave function. In this way $a(p)$ can be seen as the momentum representation of the state vector $|\psi\rangle$.

Let's multiply equation (6.14) by the bra $\langle x|$ from the left. This gives

$$\langle x|\psi\rangle = \int dp\, \langle x|p\rangle\langle p|\psi\rangle \tag{6.15}$$

Now we know what $\langle x|\psi\rangle$ and $\langle p|\psi\rangle$ are: $\psi(x)$ and $a(p)$, respectively; they are two representations of the state vector $|\psi\rangle$, two different language expressions. What is $\langle x|p\rangle$? It is the *transformation coefficient* from one basis to another. Metaphorically, the collection of $\langle x|p\rangle$ is the language dictionary that enables us to translate one language expression of a concept into another. By definition, $\langle x|p\rangle$ is also the amplitude "starting with momentum eigenstate $|p\rangle$ and ending with position x"; in other words, it is the momentum wave function in the position representation. We already know what that is!

$$\langle x|p\rangle = (2\pi\hbar)^{-1/2} e^{ipx/\hbar}$$

Substituting for $\langle x|\psi\rangle$, $\langle p|\psi\rangle$, and $\langle x|p\rangle$ in equation (6.15), we get

$$\psi(x) = \frac{1}{\sqrt{2\pi\hbar}} \int dp\, a(p) e^{ipx/\hbar}$$

You recognize the Fourier expansion of $\psi(x)$! Thus the Fourier expansion is nothing but an application of the ket-bra sum rule.

The derivation of the inverse Fourier expansion will be left as an exercise for you, but the advantage of the Dirac notation is obvious. In the wave function notation, or Fourier notation, we denote the position representation of a state as $\psi(x)$ and the momentum representation as $a(p)$; but this notation obscures the fact that they are in fact representing the same thing. So the symbological coup of the Dirac notation is that it enables us to see the universality that lies behind the different representations, the state vector itself. The state vector is the unity behind the diversity of all the different representations.

To summarize, we have uncovered three different meanings or uses attributed to the scalar product of two state vectors $\langle\phi|\psi\rangle$:

1. It is the amplitude for the event that starts with $|\psi\rangle$ and ends with $\langle\phi|$. The absolute square of this amplitude $|\langle\phi|\psi\rangle|^2$ is the probability density for the event.
2. If $|\phi\rangle$ is a member of a basic set, then the scalar product $\langle\phi|\psi\rangle$ is also the representation of the state vector $|\psi\rangle$ in the basis $|\phi\rangle$.
3. If $|\psi\rangle$ and $|\phi\rangle$ are members of two different bases, then their scalar product $\langle\phi|\psi\rangle$ is the coefficient of transformation between the two bases.

6.2 OPERATORS

It is important initially for you to remember that the operators we are talking about now work on the abstract state vectors and are quite different from the algebraic operators we previously introduced, those that operate on wave functions. The latter are representations of the abstract operators in a particular space. For example, the operator $-i\hbar\partial/\partial x$ is the form in position space of the quantum mechanical operator that corresponds to the observable that we call momentum.

However, since both kinds of operators will occur frequently, and often in the same situation, and because it is not particularly important to introduce a sharp distinction between them, we are not going to introduce any special notation to distinguish them.

By definition, we can write the result of an operation denoted by the operator $\hat{A}$ on a state vector $|\psi\rangle$ by the equation

$$|\phi\rangle = \hat{A}|\psi\rangle \tag{6.16}$$

The results of the operation $\hat{A}$ on $|\psi\rangle$ lead to the new state represented by the state vector $|\phi\rangle$. Now suppose we expand $|\psi\rangle$ in a certain basis $|i\rangle$ using the ket-bra sum-rule:

$$|\psi\rangle = \sum_i |i\rangle\langle i|\psi\rangle \tag{6.17}$$

Now let's do two things to equation (6.16). First, multiply it from the left by $\langle j|$, where the state $|j\rangle$ belongs to the same set as $|i\rangle$; and second, insert the expansion, equation (6.17), for $|\psi\rangle$. This gives

$$\langle j|\phi\rangle = \sum_i \langle j|\hat{A}|i\rangle\langle i|\psi\rangle$$

This is an algebraic equation: All the quantities in it are numbers including $\langle j|\hat{A}|i\rangle$. Thus $\langle j|\hat{A}|i\rangle$ is the algebraic representation of the abstract operator $\hat{A}$ in the basis $|i\rangle$; we will refer to it as the *matrix element* of $\hat{A}$ between the states $|i\rangle$ and $|j\rangle$ in anticipation of matrix mechanics, which we will introduce later. It is also understood that in a matrix element $\langle j|\hat{A}|i\rangle$, $\hat{A}$ operates to the right giving a new ket, and we take the scalar product of this new ket by multiplying the bra from the left.

Let's make some general comments about quantum mechanical operators. The most important operators in quantum mechanics are those that represent observables and they are linear and hermitian. Let the set $|\alpha_i\rangle$ denote a complete set of eigenstates of such an operator $\hat{A}$, where we have used the real eigenvalues α_i of $\hat{A}$ to label its eigenstates. How do we express the linearity and hermiticity of $\hat{A}$?

Expand an arbitrary state vector $|\psi\rangle$ in terms of the set $|\alpha_i\rangle$:

$$|\psi\rangle = \sum_i c_i|\alpha_i\rangle$$

Then by virtue of linearity, we get

$$\hat{A}|\psi\rangle = \sum_i c_i\hat{A}|\alpha_i\rangle = \sum_i c_i\alpha_i|\alpha_i\rangle$$

The hermitian conjugate of the operator $\hat{A}$ is denoted by $\hat{A}^\dagger$ and can be defined by the relation

$$\langle\phi|\hat{A}^\dagger|\psi\rangle = \langle\psi|\hat{A}|\phi\rangle^* \tag{6.18}$$

Now you can see why we tacitly assume that an operator sandwiched between a bra and ket is operating on the ket on its right; because on the left, on the bra, it's really its hermitian conjugate $\hat{A}^\dagger$ that operates. In the above definition, you can see that $[(\hat{A})^\dagger]^\dagger = \hat{A}$ operates on the bra to the left, and then the right-hand side follows as a result of the (forward)* = backward rule, equation (6.6). In Dirac's clever and compact notation, you just have to remember this subtlety.

An operator $\hat{A}$ is said to be hermitian if $\hat{A}^\dagger = \hat{A}$. Thus for a hermitian operator, we have

$$\langle\phi|\hat{A}|\psi\rangle = \langle\psi|\hat{A}|\phi\rangle^* \tag{6.19}$$

Let's prove once again the most important aspect of hermitian operators, namely, that their eigenvalues are real. Expand the states $|\phi\rangle$ and $|\psi\rangle$ in terms of the basis $|\alpha_i\rangle$:

$$|\phi\rangle = \sum_i b_i|\alpha_i\rangle \qquad b_i = \langle\alpha_i|\phi\rangle$$

$$|\psi\rangle = \sum_i c_i|\alpha_i\rangle \qquad c_i = \langle\alpha_i|\psi\rangle$$

Therefore,

$$\begin{aligned}\langle\phi|\hat{A}|\psi\rangle &= \sum_i\sum_j \langle\phi|\alpha_i\rangle\langle\alpha_i|\hat{A}|\alpha_j\rangle\langle\alpha_j|\psi\rangle \\ &= \sum_i b_i^*\alpha_i c_i\end{aligned}$$

where we have used the facts that $\langle\phi|\alpha_i\rangle = \langle\alpha_i|\phi\rangle^*$ and $\langle\alpha_i|\hat{A}|\alpha_j\rangle = \alpha_j\langle\alpha_i|\alpha_j\rangle = \alpha_j\delta_{ij} = \alpha_i$. Similarly,

$$\langle\psi|\hat{A}|\phi\rangle^* = \left(\sum_i c_i^*\alpha_i b_i\right)^* = \sum_i c_i\alpha_i^* b_i^*$$

It follows that $\alpha_i^* = \alpha_i$. The eigenvalues of hermitian operators are real.

Another class of important operators in quantum mechanics is called *unitary operators*, which are defined as

$$UU^\dagger = U^\dagger U = 1$$

Note that unitary operators conserve the scalar product of state vectors.

Example: The Position Operator

The position representation $|x\rangle$ is defined by the eigenstates of the position operator $\hat{x}$

$$\hat{x}|x\rangle = x|x\rangle$$

In its own space, operating by x is the same thing as multiplying by x; at home x doesn't have a hat.

The expectation value of x in the state $|\psi\rangle$ is given by

$$\begin{aligned}\langle\hat{x}\rangle = \langle\psi|\hat{x}|\psi\rangle &= \int dx\int dx'\,\langle\psi|x\rangle\langle x|\hat{x}|x'\rangle\langle x'|\psi\rangle \\ &= \int dx\,\psi^*(x)x\psi(x)\end{aligned}$$

where we have used $\langle x|\hat{x}|x'\rangle = x'\delta(x-x')$. Another expected result!

What is the form of $\hat{x}$ in the momentum representation? Actually, we already know it, but let's do a formal proof. Consider the expectation of $\hat{x}$ in the state $|\psi\rangle$ again, but this time expand $|\psi\rangle$ in the momentum representation. This gives

$$\langle \hat{x} \rangle = \langle \psi | \hat{x} | \psi \rangle = \int dp \, \langle \psi | p \rangle \langle p | \beta \rangle \tag{6.20}$$

where $|\beta\rangle = \hat{x}|\psi\rangle$, and our job is to evaluate it in the p-representation.

Consider $\langle x|\beta\rangle$. Expand it using the ket-bra sum rule for the p-basis:

$$\langle x|\beta\rangle = \int dp \, \langle x|p\rangle\langle p|\beta\rangle = (2\pi\hbar)^{-1/2} \int dp \, e^{ipx/h} \langle p|\beta\rangle \tag{6.21}$$

But we can also expand $\langle x|\beta\rangle$ as

$$\begin{aligned}
\langle x|\beta\rangle &= \langle x|x|\psi\rangle = x\langle x|\psi\rangle \\
&= x \int dp \, \langle x|p\rangle\langle p|\psi\rangle \\
&= (2\pi\hbar)^{-1/2} x \int dp \, e^{ipx/h} a(p) \\
&= (2\pi\hbar)^{-1/2} \int dp \, (-i\hbar d/dp \, e^{ipx/h}) a(p)
\end{aligned}$$

Integrating by parts, we get

$$\langle x|\beta\rangle = (2\pi\hbar)^{-1/2} \left[\frac{\hbar}{i} \, [e^{ipx/h} a(p)]_{-\infty}^{\infty} - \frac{\hbar}{i} \int dp \, e^{ipx/h} \frac{da(p)}{dp} \right]$$

So long as we consider bound states, $a(p) \to 0$ as $p \to \pm\infty$, and the first term above goes to zero. We are left with

$$\langle x|\beta\rangle = (2\pi\hbar)^{-1/2} i\hbar \int dp \, e^{ipx/h} \frac{da(p)}{dp}$$

Comparing with equation (6.21), we find

$$\langle p|\beta\rangle = i\hbar \frac{d}{dp} a(p)$$

Substituting in equation (6.20), we get

$$\langle x \rangle = \int dp\, a^*(p)(i\hbar d/dp)a(p)$$

Thus $\hat{x} = i\hbar d/dp$ in the momentum representation.

Now we will show that this operator $i\hbar d/dp$ is hermitian using the definition of the hermitian conjugate, equation (6.18), and using the momentum representation:

$$\langle A^\dagger \rangle = \int dp\, a^*(p) A^\dagger a(p) = \langle A \rangle^* = \left[\int dp\, a^*(p) A a(p) \right]^*$$
$$= \int dp\, [Aa(p)]^* a(p)$$

Putting $A = i\hbar d/dp$ in the above equation, we get

$$\int dp \left(i\hbar \frac{da(p)}{dp} \right)^* a(p) = \int dp\, (-i\hbar) \frac{d}{dp} (a^* a) - \int dp\, (-i\hbar) a^* \frac{da(p)}{dp}$$
$$= \int dp\, a^* (i\hbar d/dp) a(p)$$

This shows that the hermitian conjugate of $i\hbar d/dp$ is $i\hbar d/dp$ because $[d/dp]^\dagger = -d/dp$ and $[i\hbar]^\dagger = -i\hbar$.

We can use similar arguments to show that the momentum operator is likewise simply multiplicative in the momentum representation, but is represented by the hermitian operator $-i\hbar d/dx$ in the coordinate representation.

Generally speaking, for the matrix element $\langle i|A|j\rangle$ of an operator A in a basis $|i\rangle$, the operator is the meat of the sandwich, no doubt, but the bread, the basis, does change the flavor of the sandwich.

6.3 COMMUTING OPERATORS AND THE LABELING OF STATES

In chapter 3, we found that the free particle energy eigenfunctions have a twofold degeneracy; any eigenfunction of energy is also a simultaneous eigenfunction of momentum belonging to eigenvalues $\pm p$.

In such a case, we have a problem in labeling the eigenstates. The energy eigenvalue E alone is not enough; we have to specify if the eigenvalue of momentum is $+p$ or $-p$ in order to completely specify the state's labels. (For a real-world example, think of the hydrogen atom. The spectra of the hydrogen atom exhibit degeneracy. We need more than the quantum number for energy (the principal quantum number) to label the eigenstates of the hydrogen atom.)

In chapter 4, we found a similar situation with the solution of the square well centered at $x = 0$. In this case, although there is no degeneracy, the eigenfunctions are nevertheless simultaneous eigenfunctions of the Hamiltonian and the parity operator, and it is convenient to have that information for labeling purposes.

Thus it is important to inquire into the general condition under which simultaneous eigenstates of two operators occur. Suppose $|\psi_\alpha\rangle$ is an eigenstate of the operator A belonging to the eigenvalue α:

$$A|\psi_\alpha\rangle = \alpha|\psi_\alpha\rangle$$

The state $|\psi_\alpha\rangle$ will be a simultaneous eigenstate of a second operator B if

$$B|\psi_\alpha\rangle = \beta|\psi_\alpha\rangle$$

where β is the eigenvalue of B. But if both equations hold, then the operation of AB and BA on the state $|\psi_\alpha\rangle$ produces the same result:

$$\begin{aligned} AB|\psi_\alpha\rangle &= A\beta|\psi_\alpha\rangle = \beta A|\psi_\alpha\rangle = \beta\alpha|\psi_\alpha\rangle \\ &= \alpha\beta|\psi_\alpha\rangle = \alpha B|\psi_\alpha\rangle = B\alpha|\psi_\alpha\rangle = BA|\psi_\alpha\rangle \end{aligned}$$

It follows that $AB = BA$, as far as the whole complete set of these eigenstates $|\psi_\alpha\rangle$ is concerned; and since any state can be expanded in terms of the complete set $|\psi_\alpha\rangle$, the equality $AB = BA$ holds for any state. Hence the order in which A and B operate on a state does not matter. In other words, A and B commute:

$$[A, B] = 0$$

Thus, in the case of the square well, the Hamiltonian and the parity operators commute, and we get simultaneous eigenfunctions in the problem. In the case of the free particle, likewise, momentum and H commute.

Now we can see how the labeling problem in general can be solved in the presence of degeneracy. If a certain observable operator has degenerate eigenstates with identical eigenvalues, the states can be distinguished by finding a second operator that commutes with the first. The eigenvalue of the second operator will provide the distinction. If it happens that two commuting operators A and B have two simultaneous eigenstates with the same two eigenvalues, then there must be a third commuting operator C, which has a different eigenvalue for the two states (or some linear combinations of them), that can distinguish between them; after all, if no such observable exists, then the two states are not physically distinguishable, in which case they must be identical. The same could be said for two simultaneous eigenstates of three commuting operators; if there is still a degeneracy, we find a fourth observable that has a different eigenvalue for the degenerate eigenstates; and so forth.

When we have found a *complete* commuting set of operators, then no two of their simultaneous eigenstates can have the same eigenvalues for all of them, and now a quantum mechanical state has been specified uniquely, and its labeling is complete. We have gotten the largest amount of information about a state that quantum mechanics allows. Thus a quantum mechanical state gives a set of eigenvalues (measurable attributes) for a complete set of commuting observables.

And when we have found such a state defined by the eigenvalues of a complete commuting set of observables, what can we say about the result of the measurement of an observable outside of this set (that is, an observable that does not commute with any one of the members of the complete set)? Nothing as far as a particular measurement on one particular object is involved! We can, of course, calculate expectation values or the average result of many measurements of such an observable.

Thus for an atomic state that is an eigenstate of H, and for which H does not commute with either the momentum or the position operator, what can we say about the result of a particular measurement of an electron's position and momentum? Comprehend what Robert Oppenheimer said:

> If we ask, for instance, whether the position of the electron remains the same, we must say "no"; if we ask whether the position of the electron changes with time, we must say "no"; if we ask whether the electron is at rest, we must say "no"; if we ask whether it is in motion, we must say "no."†

On the other hand, if two observables commute, since they have simultaneous eigenstates and eigenvalues, they can be simultaneously measured. They are compatible. Operationally, this means that if two observables A and B commute and are compatible, then if A is measured, and then B, and then A again, the second measurement of A will give the same value as the first. The measurement of B does not mess up the measurement of A. This is the key thing: To be compatible, the measurement of one observable must survive the measurement of the other.

Commutators and Uncertainty Relations

Conversely, it is also clear that if two observables do not commute, they do not have simultaneous eigenfunctions or eigenvalues. This means we cannot measure them simultaneously, because that would imply their having simultaneous eigenvalues. For two such operators A and B, the measurement of B will mess

†J. R. Oppenheimer, *Science and Common Understanding*, p. 69.

up any previous measurement of A. In fact, the measurement of B will destroy the eigenstate of A that the measurement of A created and replace it with an eigenstate of B. We have already demonstrated the veracity of such statements for the position and momentum operators and attributed that to the existence of the uncertainty relation between them. Thus the question, Does the existence of a nonvanishing commutator imply an uncertainty relation? We will now derive such a general uncertainty relation.

Consider two observables A and B that do not commute. A measure ΔA of the uncertainty of A for a state $|\psi\rangle$ of the system is given by

$$(\Delta A)^2 = \langle A^2 \rangle - \langle A \rangle^2 = \langle \psi | A^2 | \psi \rangle - \langle \psi | A | \psi \rangle^2$$
$$= \langle \psi | (A - \langle \psi | A | \psi \rangle)^2 | \psi \rangle$$

since the second term is simply multiplicative. Now define

$$A' = A - \langle \psi | A | \psi \rangle$$

Then $(\Delta A)^2 = \langle \psi | A'^2 | \psi \rangle$. In the same way, if we define $B' = B - \langle \psi | B | \psi \rangle$, then $(\Delta B)^2 = \langle \psi | B'^2 | \psi \rangle$, where ΔB is the uncertainty of B. Then the product

$$(\Delta A)^2 (\Delta B)^2 = \langle \psi | A'^2 | \psi \rangle \langle \psi | B'^2 | \psi \rangle$$
$$= \langle A'\psi | A'\psi \rangle \langle B'\psi | B'\psi \rangle$$

since A' and B' are hermitian. We will now invoke the Schwartz inequality, which says that for any two complex functions f and g the following inequality holds:

$$\int f^* f \, d\tau \int g^* g \, d\tau \geq \left| \int f^* g \, d\tau \right|^2 \tag{6.22}$$

Identify f with $A'\psi$ and g with $B'\psi$. We get

$$(\Delta A)^2 (\Delta B)^2 \geq |\langle A'\psi | B'\psi \rangle|^2 \geq [\mathrm{Im}(\langle A'\psi | B'\psi \rangle)]^2$$
$$= [(\langle A'\psi | B'\psi \rangle - \langle B'\psi | A'\psi \rangle)/2i]^2$$
$$= \langle [A', B']/2i \rangle^2 = \langle [A, B]/2i \rangle^2 \tag{6.23a}$$

This gives

$$\Delta A \cdot \Delta B \geq \tfrac{1}{2} |[A, B]| \tag{6.23b}$$

Equations (6.23a) and (6.23b) are called the generalized uncertainty relations. If $A = x$ and $B = p$, $[A, B] = i\hbar$, and we have

$$(\Delta x)^2 \cdot (\Delta p)^2 \geq (\hbar/2)^2$$

In other words, $\Delta x \cdot \Delta p \geq \hbar/2$, and we have yet another derivation of the momentum-position uncertainty relation that shows that the uncertainty relation is a direct consequence of the commutation relation between x and p.

6.4 THE TIME EVOLUTION OF A QUANTUM SYSTEM: THE HAMILTONIAN OPERATOR

Recall the postulates of quantum mechanics from chapter 3. It is quite easy to translate them into the Dirac language. Thus we say, for example, that a system is represented by a state vector of unit magnitude in the Hilbert space, and that observables are represented by operators that operate on these state vectors. The expansion postulate is now given by equation (6.3):

$$|\psi\rangle = \sum_i \langle \phi_i | \psi \rangle | \phi_i \rangle \tag{6.3}$$

and the measurement postulate—that the measurement of an observable gives an eigenstate $|\phi_i\rangle$ of the observable with probability $|\langle \phi_i | \psi \rangle|^2$—and the reduction postulate—that measurement reduces the state vector $|\psi\rangle$ from the coherent superposition, equation (6.3), to an eigenstate $|\phi_i\rangle$ of the observable we are measuring—easily follows. And this brings us to the sixth postulate—How does the system evolve between measurements?

Time now enters our description of the state vectors. Suppose initially, we have a state ket $|\alpha, t_0\rangle$, where we have introduced the time label to denote the time dependence of the ket. At a later time, the ket will evolve to become $|\alpha, t_0; t\rangle$, say; our inquiry is about the relationship of $|\alpha, t_0; t\rangle$ with $|\alpha, t_0\rangle$—How does the state ket change with a displacement of time from t_0 to t? This is, of course, a most important equation of quantum mechanics. (And we already know the answer, but we will pretend we don't. Thus look at the following discussion as a formal derivation of sorts of the Schrödinger equation.)

Define a time-evolution operator denoted by $U(t, t_0)$ via the equation

$$|\alpha, t_0; t\rangle = U(t, t_0) |\alpha, t_0\rangle \tag{6.24}$$

What are some of the properties that U has to have? First of all, at time t_0,

$$|\alpha, t_0; t_0\rangle \equiv |\alpha, t_0\rangle$$

Thus $U(t_0,t_0) = 1$. Second, if we consider a sequence of times $t_0 \to t_1 \to t_2$, we must insist that the U that takes us from t_0 to t_2 must be the product of the two U's that are needed to go from t_0 to t_1, and then from t_1 to t_2:

$$U(t_2,t_0) = U(t_2,t_1)U(t_1,t_0) \tag{6.25}$$

And third, for an infinitesimal increment of time Δt, U must deviate from 1 by an increment proportional to Δt; that is,

$$U(t + \Delta t, t) = 1 - \frac{i}{\hbar} H\Delta t \tag{6.26}$$

This defines the operator H (in anticipation) and we will have to find out what it means as we go along.

Substituting equation (6.26) into equation (6.25) we get

$$\begin{aligned} U(t + \Delta t, t_0) &= U(t + \Delta t, t)U(t, t_0) \\ &= \left(1 - \frac{i}{\hbar} H\Delta t\right) U(t, t_0) \end{aligned}$$

In this way, we see that

$$U(t + \Delta t, t_0) - U(t, t_0) = -\frac{i}{\hbar} H\Delta t U(t, t_0)$$

We can write this last equation in the differential form

$$i\hbar \frac{\partial}{\partial t} U(t, t_0) = HU(t, t_0) \tag{6.27}$$

We have found the law of time evolution of U. The rest is easy. Multiply equation (6.27) on the right by $|\alpha, t_0\rangle$:

$$i\hbar \frac{\partial}{\partial t} U(t, t_0)|\alpha, t_0\rangle = HU(t, t_0)|\alpha, t_0\rangle$$

Since $|\alpha, t_0\rangle$ does not depend on time, this gives

$$i\hbar \frac{\partial}{\partial t} |\alpha, t_0; t\rangle = H|\alpha, t_0; t\rangle \tag{6.28}$$

This is the formal statement of the Schrödinger equation in the state vector language. For quantum mechanical state vectors, operation by H produces the same result as operation by $i\hbar\partial/\partial t$. If we know H, we know how the system time-evolves.

To connect up to familiar equations, let's consider the position representation. In equation (6.28), put $|\alpha, t_0; t\rangle \equiv |\psi\rangle$, multiply from the left by $\langle x|$, and use the ket-bra sum rule for the basis $|x\rangle$:

$$i\hbar \frac{\partial}{\partial t} \langle x|\psi\rangle = \int dx' \langle x|H|x'\rangle\langle x'|\psi\rangle$$

Since H, which we can identify with the Hamiltonian of the system, is a local operator, its matrix element at x can never depend on another point of space x'. Following Schrödinger we assert H to be the local operator representing the total energy of the system

$$\langle x|H|x'\rangle = H(x)\delta(x - x') \tag{6.29}$$

with

$$H(x) = (-\hbar^2/2m)\partial^2/\partial x^2 + V(x)$$

Substituting, we recover the position representation of the Schrödinger equation:

$$i\hbar \frac{\partial\psi(x,t)}{\partial t} = H\psi = -\frac{\hbar^2}{2m}\frac{\partial^2\psi(x,t)}{\partial x^2} + V(x)\psi(x,t)$$

PROBLEMS

1. Consider a particle in a one-dimensional infinite potential box of length a. Its wave function has the most general form

$$\psi(x) = \sum_{n=1}^{\infty} c_n\sqrt{2/a}\sin\frac{n\pi x}{a}$$

   Write this in the Dirac notation using the ket-bra sum for the box base kets denoted by $|n\rangle$. Show the correspondence between the two notations. Interpret the $\langle x|\psi\rangle$, $\langle x|n\rangle$, and $\langle n|\psi\rangle$ that will arise as you do the problem.
2. Using Dirac notation, prove that the eigenstates of a hermitian operator belonging to two different eigenvalues are orthogonal.
3. If A and B are two hermitian operators, prove (a) that the product AB is hermitian only if A and B commute and (b) that $(A + B)^n$ is hermitian.
4. Prove that for any operator A, $A + A^\dagger$ and $i(A - A^\dagger)$ are hermitian.

5. Prove that if A is a hermitian operator, $\langle A^2 \rangle \geq 0$.
6. The initial state $|\psi_i\rangle$ of a quantum system is given in an orthonormal basis of three states $|\alpha\rangle$, $|\beta\rangle$, and $|\gamma\rangle$ that form a complete set:

$$\langle \alpha | \psi_i \rangle = i/\sqrt{3}, \qquad \langle \beta | \psi_i \rangle = (2/3)^{1/2}, \qquad \langle \gamma | \psi_i \rangle = 0$$

Calculate the probability of finding the system in a state $|\psi_f\rangle$ given in the same basis as

$$\langle \alpha | \psi_f \rangle = (1+i)/\sqrt{3}, \qquad \langle \beta | \psi_f \rangle = 1/\sqrt{6}, \qquad \langle \gamma | \psi_f \rangle = 1/\sqrt{6}$$

7. Suppose A and B are two commuting operators and $|\psi_{\alpha i}\rangle$ $(i = 1, n)$ are n (linearly independent) eigenstates of A all belonging to the same eigenvalue α. Show that linear combinations of these eigenstates can be chosen that are also simultaneously the eigenstates of B. (*Hint*: Expand $B|\psi_{\alpha i}\rangle$ in terms of the eigenstates of A.)
8. Prove that if H is hermitian, then the hermitian conjugate of $\exp(iH)$ defined as

$$\sum_{n=0}^{\infty} i^n H^n / n!$$

is the operator $\exp(-iH)$.
9. A unitary operator U is defined by

$$UU^\dagger = U^\dagger U = 1$$

Show that if the state $|\psi\rangle$ is normalized, so is $U|\psi\rangle$.
10. Show that if the operator H is hermitian, then $\exp(-iH)$ is unitary.
11. From the infinitesimal operator for time development, equation (6.26), show that the time-development operator corresponding to a finite time t is given by $\exp(-iHt/\hbar)$.

ADDITIONAL PROBLEMS

A1. An antihermitian operator is defined by $\hat{O}^\dagger = -\hat{O}$. Show that any operator can be expressed as a linear combination of a hermitian and an antihermitian operator.

A2. Write in Dirac notation:
(a) $\int dx\, \phi^*(x)\phi(x) = N$.
(b) $\hat{O} = \psi(x) \int dx'\, \phi^*(x')$
(c) $\partial\psi(x)/\partial x = \phi^*(x) \int dx'\, a^*(x') b(x')$

A3. Show that a two-level Hamiltonian can be written as:

$$H = E_1 |1\rangle\langle 1| + E_2 |2\rangle\langle 2| + V[|1\rangle\langle 2| + |2\rangle\langle 1|]$$

where the states $|1\rangle$ and $|2\rangle$ are orthonormal with energy eigenvalues E_1 and E_2, respectively, and V is the interaction.

A4. Let A and B be two anticommuting operators, that is

$$AB + BA = 0$$

If $B^2 = 1$, what does it mean for the eigenvalues of A?

A5. Show that for any two operators A and B,

$$e^A B e^{-A} = B + [A, B] + (1/2!)[A, [A, B]] + \cdots$$

What is the nth term of the expansion on the right-hand side? (*Hint*: Consider the Taylor expansion of the operator function $F(\lambda) = e^{\lambda A} B e^{-\lambda A}$.)

A6. Show explicitly that unitary transformations conserve scalar products of state vectors.

A7. An arbitrary unitary operator U can be decomposed as

$$U = \frac{U + U^\dagger}{2} + i\,\frac{U - U^\dagger}{2i} \equiv H_1 + iH_2$$

(a) Show that H_1 and H_2 are hermitian operators.
(b) Show $[H_1, H_2] = [U, H_1] = [U, H_2] = 0$; thus U, H_1, and H_2 have simultaneous eigenstates. Call these $|h_1 h_2\rangle$. Using these eigenstates, show that
(c) the eigenvalues of U are of unit magnitude.

REFERENCES

M. Chester. *Primer of Quantum Mechanics.*
P. A. M. Dirac. *The Principles of Quantum Mechanics.*
R. P. Feynman, R. B. Leighton, and M. Sands. *The Feynman Lectures in Physics.*
E. Merzbacher. *Quantum Mechanics.*
A. Sudbery. *Quantum Mechanics and the Particles of Nature.*

7

The One-Dimensional Harmonic Oscillator

We will now begin a love affair with the simple harmonic oscillator in quantum mechanics. The harmonic oscillator is a pretty problem; its Schrödinger equation is exactly solvable, and it is easily treated by the purely formal operator method. Additionally, even in its one-dimensional form, the one we will treat in this chapter, it simulates certain excitations of some physical systems—among them, molecules, nuclei, and solids—pretty well. And finally, this is one problem to which we will keep returning.

The simplest of simple harmonic oscillators, as you know from classical physics, is a mass on a spring (fig. 7.1). The force of the spring on the mass is a Hooke's law force $-kx$, k being the spring constant and x being the displacement of the mass from its equilibrium position. Equivalently, the mass has a potential energy of $\frac{1}{2}kx^2$. If the momentum of the mass m is p, we can write the oscillator Hamiltonian as

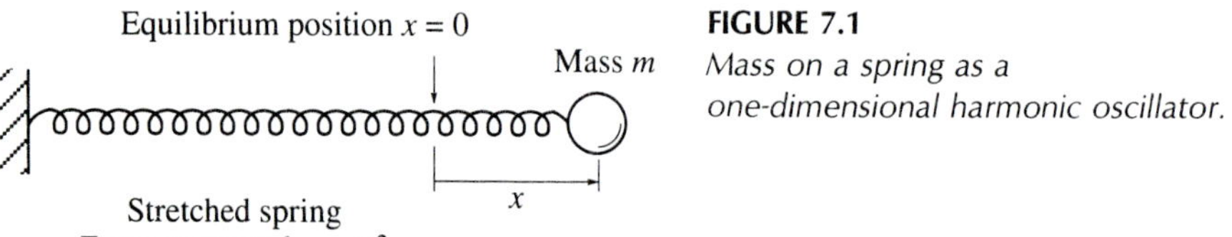

FIGURE 7.1
Mass on a spring as a one-dimensional harmonic oscillator.

$$H = \frac{p^2}{2m} + \frac{1}{2} kx^2 \tag{7.1}$$

From the outset, however, we will slightly idealize the picture shown in figure 7.1. We will assume that x spans a spectrum from $-\infty$ all the way to ∞, all permitted values that position can take. Thus if you think of the harmonic oscillator as a mass on a spring, you have to imagine that the stationary wall from which the spring is suspended is infinitely removed from the equilibrium position of the mass.

The oscillator Hamiltonian, equation (7.1), also gives us an approximate way to look at any system near equilibrium when the system is displaced by a small amount from the equilibrium position. For example, consider a system moving in the potential $V(x)$ shown in figure 7.2. Near the bottom, the potential is rather closely represented by a parabolic dependence on x, which is the oscillator potential. How do we find the oscillator constant? It is equal to

$$k = \left.\frac{d^2 V}{dx^2}\right|_{x = x_0} \tag{7.2}$$

where x_0 is the point where the potential is minimum. Can you see why? Taylor expand $V(x)$ about x_0:

$$V(x) = V(x_0) + \left(\frac{dV}{dx}\right)_{x_0} (x - x_0) + \frac{1}{2}\left(\frac{d^2 V}{dx^2}\right)_{x_0} (x - x_0)^2$$

The first term, being a constant, represents just a shift of the energy scale. The second term is zero, x_0 being the equilibrium point. Clearly, the leading term of the expansion is the oscillator potential with the oscillator constant given by equation (7.2).

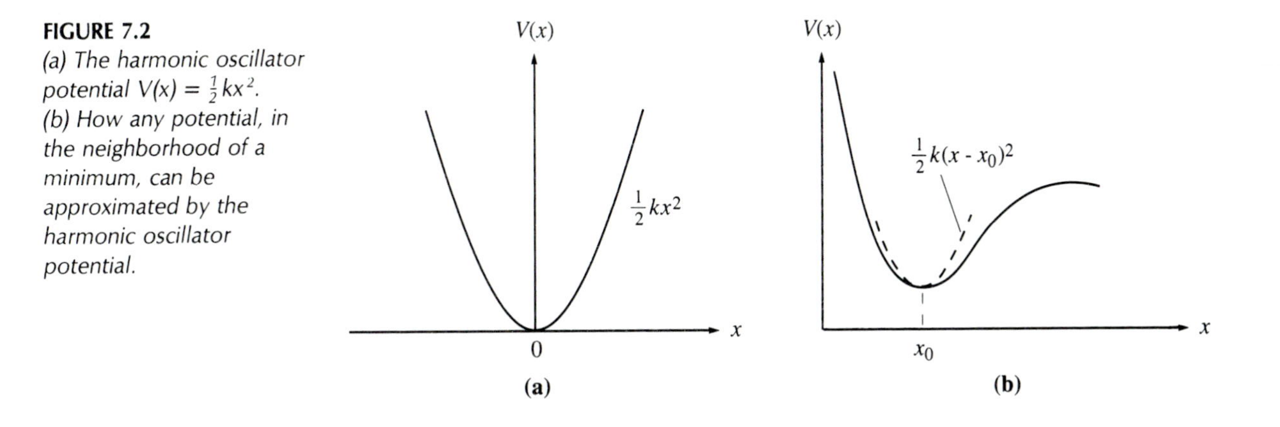

FIGURE 7.2
(a) The harmonic oscillator potential $V(x) = \frac{1}{2}kx^2$.
(b) How any potential, in the neighborhood of a minimum, can be approximated by the harmonic oscillator potential.

7.1 SOLUTION OF THE SCHRÖDINGER EQUATION FOR THE LINEAR OSCILLATOR

Since the oscillator potential is infinitely high as $x \to \infty$, the particle is always in some sort of bound state. Let's represent the eigenstates as $|E\rangle$, and write the time-independent Schrödinger equation as

$$H|E\rangle = E|E\rangle$$

We will first solve this equation in the position representation. To this end, we multiply the equation above from the left by $\langle x|$, introduce the notation $\langle x|E\rangle = u(x)$, and use equation (7.1) for H with p replaced by $-i\hbar d/dx$. In this way, we find the Schrödinger equation for the oscillator Hamiltonian to be given as

$$\frac{d^2u(x)}{dx^2} + \frac{2m}{\hbar^2}\left(E - \frac{1}{2}kx^2\right)u(x) = 0 \tag{7.3}$$

As you know by now, the easy way to solve differential equations like this is to look into the mathematical literature where somebody else has already done the work. The solutions of equation (7.3) are special functions called *Hermite functions*.

It is, however, instructive to go through the details of the steps to the solution. To this purpose, we introduce the oscillator frequency $\omega = \sqrt{k/m}$ and make a change of variables from x to a dimensionless coordinate ξ defined as

$$\xi = \sqrt{m\omega/\hbar}\, x$$

We also introduce a new dimensionless quantity ϵ defined by

$$\epsilon = 2E/\hbar\omega \tag{7.4}$$

Substituting all this into equation (7.3), we get

$$\frac{d^2u}{d\xi^2} + (\epsilon - \xi^2)u = 0 \tag{7.5}$$

First we will obtain the asymptotic solution of the equation above, call it u_∞. In this limit, $\xi \gg 0$, $\epsilon \ll \xi^2$, and the equation satisfied by u_∞ is given as

$$\frac{d^2u_\infty}{d\xi^2} - \xi^2 u_\infty = 0$$

Multiply the equation above by 2 $du_\infty/d\xi$. We get

$$\frac{d}{d\xi}(du_\infty/d\xi)^2 - \xi^2\frac{d}{d\xi}(u_\infty^2) = 0$$

This can be further simplified if we make the approximation that ξu_∞^2 is negligibly small. We have

$$\frac{d}{d\xi}\left[\left(\frac{du_\infty}{d\xi}\right)^2 - \xi^2 u_\infty^2\right] = -2\xi u_\infty^2 \cong 0 \tag{7.6}$$

The last step will have to be justified, but the advantage of the step should be obvious: We can integrate the equation! Integration gives

$$\frac{du_\infty}{d\xi} = (C + \xi^2 u_\infty^2)^{1/2}$$

where C is a constant of integration. But both u_∞ and $du_\infty/d\xi$ go to 0 as $\xi \to \infty$, so $C = 0$. We get

$$\frac{du_\infty}{d\xi} = \pm\xi u_\infty$$

or

$$\frac{1}{u_\infty}\frac{du_\infty}{d\xi} = \pm\xi$$

Integrating again, we get

$$u_\infty = \exp(-\xi^2/2) \tag{7.7}$$

where we have ignored, as unacceptable, the positive exponential solution $\exp(\xi^2/2)$ that diverges as $\xi \to \infty$. You now can satisfy yourself that our approximation in equation (7.6) is justified; the term $2\xi u_\infty^2$ is indeed small compared to the term we saved.

Having obtained the asymptotic solution, we will now seek the solution of equation (7.5) for all ξ in the form

$$u(\xi) = h(\xi)\exp(-\xi^2/2) \tag{7.8}$$

Substitution into equation (7.5) gives the equation that must be satisfied by the function $h(\xi)$:

$$\frac{d^2h}{d\xi^2} - 2\xi \frac{dh}{d\xi} + (\epsilon - 1)h = 0 \tag{7.9}$$

This last equation can be solved by the method of power series.

Before we apply the power series method, it is advantageous to use the parity symmetry of the harmonic oscillator Hamiltonian: $H(x) = H(-x)$. The eigenfunctions must have "good" parity; that is, parity is an additional label for them. In constructing the solution of the Schrödinger equation for the oscillator we may keep the solutions confined within all even or within all odd parity.

Thus we are led to expand the even parity and odd parity solutions of equation (7.9) as two different power series; calling the even solution h_e and the odd h_o, we write

$$h_e = a_0 + a_2\xi^2 + a_4\xi^4 + \cdots$$

$$h_o = a_1\xi + a_3\xi^3 + a_5\xi^5 + \cdots \tag{7.10}$$

Then

$$\frac{d^2h_e}{d\xi^2} = 2\cdot 1a_2 + 4\cdot 3a_4\xi^2 + \cdots + (n+2)(n+1)a_{n+2}\xi^n$$

$$-2\xi\frac{dh_e}{d\xi} = -2\cdot 2a_2\xi^2 + \cdots - 2\cdot na_n\xi^n$$

$$(\epsilon - 1)h_e = (\epsilon - 1)a_0 + \cdots + (\epsilon - 1)a_n\xi^n$$

The differential equation (7.9) is solved if the three series above add up to zero. This means that the coefficient of all powers of ξ must vanish:

$$2a_2 + (\epsilon - 1)a_0 = 0$$

$$(n+2)(n+1)a_{n+2} + (\epsilon - 2n - 1)a_n = 0$$

But these are recursion formulas—if you know a_{n-2}, you can calculate a_n, if you know a_{n-4}, you can calculate a_{n-2}, and eventually, if you know a_0, you can calculate a_2. Carrying this process out, we get

$$a_n = \frac{(1-\epsilon)(5-\epsilon)\cdots(2n-3-\epsilon)}{n!}\,a_0 \tag{7.11}$$

where n is even, 2, 4.... The constant a_0 is undetermined at this point.

A similar manipulation with the equation for $h_o(\xi)$ will give us the recursion relations for the odd parity solutions, $n =$ odd. We get

$$a_n = \frac{(3-\epsilon)(7-\epsilon)\cdots(2n-3-\epsilon)}{n!}\,a_1 \tag{7.12}$$

where n is odd, 3, 5 . . . . Here a_1 is also, as yet, undetermined.

For large n, both recursion relations (7.11) and (7.12) give

$$a_{n+2} = \frac{2}{n+2}\,a_n$$

Now notice that if we expand $\exp(\xi^2) = \sum \xi^{2n}/n!$, the coefficient b_n of ξ^n is

$$b_n = \frac{1}{(n/2)!}$$

And likewise,

$$b_{n+2} = \frac{1}{\left(\frac{n}{2}+1\right)!} = \frac{1}{\frac{n}{2}+1}\,b_n = \frac{2}{n+2}\,b_n$$

Evidently, both $h_e(\xi)$ and $h_o(\xi)$ behave as $\exp(\xi^2)$ for large ξ. This means that the wave function u behaves asymptotically as

$$u \sim \exp(-\xi^2/2)\exp(\xi^2) \sim \exp(\xi^2/2)$$

We find that $u \to \infty$ as x (or ξ) $\to \infty$. But this is unacceptable; what good is a wave function for which there is an overwhelming probability of finding the particle not here, but elsewhere, at ∞?

So we cannot find a general solution for u, even or odd parity, for any arbitrary value of the energy E. The question is, Are there specific values of E for which we can find well-behaved solutions that quantum mechanics demands? The answer is yes, and it leads to the quantization of energy that we anticipate.

Look back at the recursion formulas (7.11) and (7.12). If

$$\epsilon = 2n + 1, \text{ for } n = 0,1,2,\ldots$$

then a_{n+2} vanishes and the series terminates for a_n irrespective of whether n is even or odd. If the series terminates, then the solution is a polynomial in ξ times $\exp(-\xi^2/2)$, perfectly satisfactory as a quantum mechanical wave function since the exponential always guarantees convergence for $x \to \infty$.

And so we find that the energy values that a quantum mechanical oscillator is allowed are quantized. From equation (7.4), we have

$$E = \frac{1}{2}\hbar\omega\epsilon = \hbar\omega\left(n + \frac{1}{2}\right), \qquad n = 0,1,2,\ldots \tag{7.13}$$

Whenever E is equal to one of the quantized values E_n above, depending on the value of n (even or odd), one of the two series terminates (even or odd). The other can be set at zero by putting a_1 or a_0 at zero, as the case may be, and we are then left with the particular polynomial that goes with the particular eigenvalue. These polynomials are called Hermite polynomials, denoted as $H_n(\xi)$. Explicitly, corresponding to the energy eigenvalue E_n, the (unnormalized) eigenfunction is given as

$$u_n(\xi) = H_n(\xi)e^{-\xi^2/2} \tag{7.14}$$

We will have more to say about the Hermite polynomials (see below), but first, let's consider the following comments on the eigenstates.

1. The energy levels (fig. 7.3) are equally spaced. This is a characteristic of some parts of the molecular and nuclear experimental spectra, and the harmonic oscillator provides a good model description of these spectra,

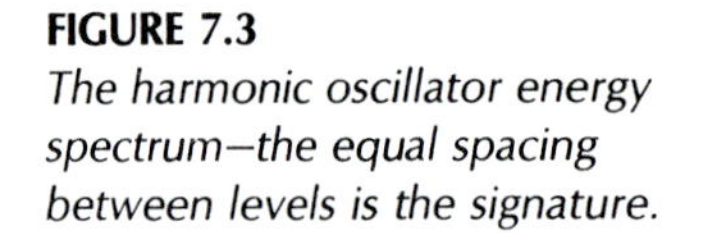
FIGURE 7.3
The harmonic oscillator energy spectrum—the equal spacing between levels is the signature.

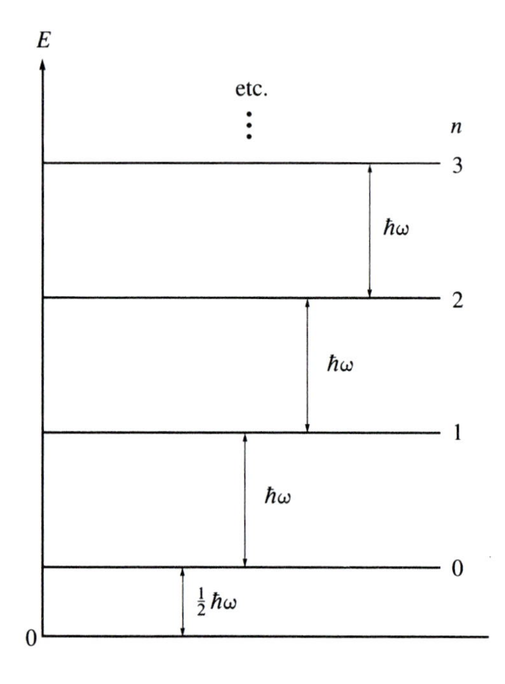

so much so that the spectra are referred to as *vibrational spectra*. There are also excitations in solids called *phonons* that fall in the same category.

2. For each eigenvalue, there is only one eigenfunction, there is no degeneracy. This property seems to be a common characteristic of bound states for one-dimensional potentials that remain finite for all finite values of x.
3. The eigenstates of the Hamiltonian are also eigenstates of the parity operator. States that correspond to zero or an even value of the quantum number n have positive parity; states labeled by odd integer n are of negative parity.
4. Even for $n = 0$, the ground state, the energy of the oscillator is $\frac{1}{2}\hbar\omega$, which is the zero-point energy. As in the case of the particle-in-a-box problem, the zero-point energy can be understood as a consequence of the uncertainty principle.
5. Since the Hermite polynomials corresponding to an eigenstate labeled by the quantum number n are polynomials of degree n, the corresponding oscillator eigenfunction has n nodes. Notice how the number of nodes goes up as the energy increases. This is understandable, since the kinetic energy is proportional to the curvature of the wave function (d^2u/dx^2). The larger the curvature (i.e., the more the wave function buckles, or bends backward and forward, in order to go through a zero) the greater is the kinetic energy.

More on the Eigenfunctions

Let's now examine the eigenfunctions a little more closely:

$$u_n(x) = C_n H_n\left(\sqrt{\frac{m\omega}{\hbar}}\, x\right) \exp\left(-\frac{m\omega}{2\hbar}\, x^2\right) \tag{7.15}$$

where we now have added a normalization constant C_n. It is instructive to determine the first few Hermite polynomials directly using the recursion relations (7.11) and (7.12). For example, for $n = 2$, we get $a_2 = -2a_0 = -2$ (since we can choose $a_0 = 1$). Thus the polynomial corresponding to $n = 2$ can be written as

$$H_2(\xi) = 1 - 2\xi^2$$

However, it is customary to define the Hermite polynomials so that the coefficient of the highest power of ξ is 2^n. Thus, for the case of $n = 2$, we choose a_0 to be -2 instead of 1, and obtain

$$H_2(\xi) = 4\xi^2 - 2$$

In this way, you will find that the first few Hermite polynomials are given as

$$H_0(\xi) = 1 \qquad H_3(\xi) = -12\xi + 8\xi^3$$

$$H_1(\xi) = 2\xi \qquad H_4(\xi) = 12 - 48\xi^2 + 16\xi^4$$

$$H_2(\xi) = -2 + 4\xi^2 \qquad H_5(\xi) = 120\xi - 160\xi^3 + 32\xi^5 \tag{7.16}$$

Verify them.

The Hermite polynomials satisfy the following integral formula:

$$\int_{-\infty}^{\infty} d\xi\, H_n(\xi)H_m(\xi)e^{-\xi^2} = \sqrt{\pi}2^n n!\delta_{mn} \tag{7.17}$$

With the help of this formula, we can easily evaluate the normalization integral of the eigenfunctions given by equation (7.15):

$$\int_{-\infty}^{\infty} dx\, u_n^2(x) = C_n^2\sqrt{\hbar/m\omega}\sqrt{\pi}2^n n!$$

This fixes C_n, which we have chosen to be real. The normalized oscillator eigenfunctions are thereby given as

$$u_n(x) = 2^{-n/2}(n!)^{-1/2}\left(\frac{m\omega}{\hbar\pi}\right)^{1/4} H_n(\sqrt{m\omega/\hbar}x)\exp\left(-\frac{m\omega}{2\hbar}x^2\right) \tag{7.18}$$

The orthogonality of the Hermite functions, equation (7.18), of different n follows from the orthogonality of the Hermite polynomials, equation (7.17). Hence these functions form a complete orthonormal set. The first five eigenfunctions of the linear harmonic oscillator are shown in figure 7.4.

There are a couple of recursion relations among the Hermite polynomials that are useful for the calculation of expectation values over harmonic oscillator wave functions. These are:

$$\frac{dH_n(\xi)}{d\xi} = 2nH_{n-1}(\xi) \tag{7.19}$$

$$2\xi H_n(\xi) = H_{n+1}(\xi) + 2nH_{n-1}(\xi) \tag{7.20}$$

■ **PROBLEM** Calculate $\langle x\rangle$ for the eigenfunction $u_n(x)$.

SOLUTION By definition $\langle x\rangle$ is given as

$$\langle x\rangle = \int_{-\infty}^{\infty} dx\, u_n(x)xu_n(x)$$

Changing variables to ξ, we get

$$\langle x \rangle = |C_n|^2(\hbar/m\omega) \int_{-\infty}^{\infty} d\xi\, e^{-\xi^2} H_n(\xi)\xi H_n(\xi)$$

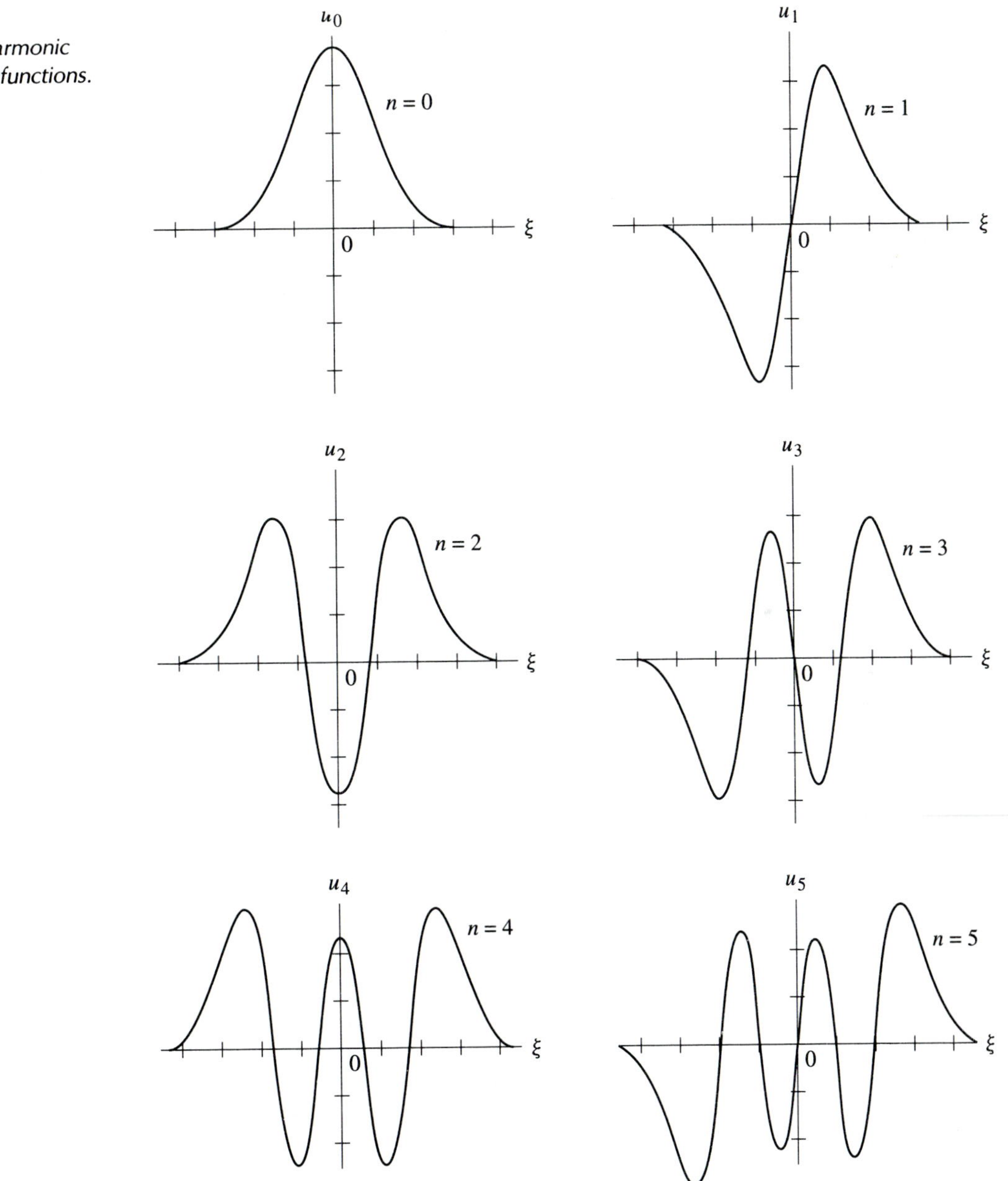

FIGURE 7.4
The first five harmonic oscillator eigenfunctions.

Now invoke the recursion relation, equation (7.20), which expresses $\xi H_n(\xi)$ in terms of H_{n+1} and H_{n-1}:

$$\langle x \rangle = \frac{1}{2}\,|C_n|^2\,\frac{\hbar}{m\omega}\left[\int_{-\infty}^{\infty} d\xi\, H_n(\xi)H_{n+1}(\xi)e^{-\xi^2} + 2n\int_{-\infty}^{\infty} d\xi\, H_n(\xi)H_{n-1}(\xi)e^{-\xi^2}\right]$$

But both integrals are zero by virtue of the orthogonality of the Hermite polynomials, equation (7.17). Consequently,

$$\langle x \rangle = 0$$ ■

Comparison with the Classical Oscillator

First, there is the question of time dependence. Classically, the simple harmonic oscillator oscillates in such a manner that the position of the particle represented by the oscillator changes from one moment to another. Quantum mechanics, on the other hand, says quite unequivocally that for any eigenstate of energy, although there is a distribution of probabilities for various positions, this distribution is frozen as far as time is concerned; this is the usual meaning of energy eigenstates being stationary. Is it possible to reconcile these two very different pictures?

The answer lies in the consideration of not one single eigenstate but a superposition of eigenstates as in a wave packet. Consider, for example, the superposition $\psi(x,t)$ of the first two oscillator eigenstates:

$$\psi(x,t) = \frac{1}{\sqrt{2}}\,[\exp(-iE_0 t/\hbar)u_0(x) + \exp(-iE_1 t/\hbar)u_1(x)]$$

If we plot $|\psi(x,t)|^2$, we get figure 7.5, where the plot is made for four different values of time. Clearly, the probability oscillates with time with just the frequency of the harmonic oscillator as expected classically. It is therefore reasonable to expect that when we take a superposition of a large number of oscillator eigenstates we will get classical behavior. This classical correspondence will be demonstrated in chapter 8.

It should be clear, however, from the outset, from one look at the eigenfunctions, equation (7.18), that the quantum solution of the harmonic oscillator is radically different from that for the classical oscillator. In classical mechanics, the oscillator is forbidden to go beyond the potential, beyond the turning points where its kinetic energy turns negative. But clearly, the quantum wave functions extend beyond the potential, and thus there is a finite probability for the oscillator to be found in a classically forbidden region.

Let's be quantitative. To be specific, let's compare the quantum and classical probabilities for the states corresponding to $n = 0$ and $n = 1$. The quan-

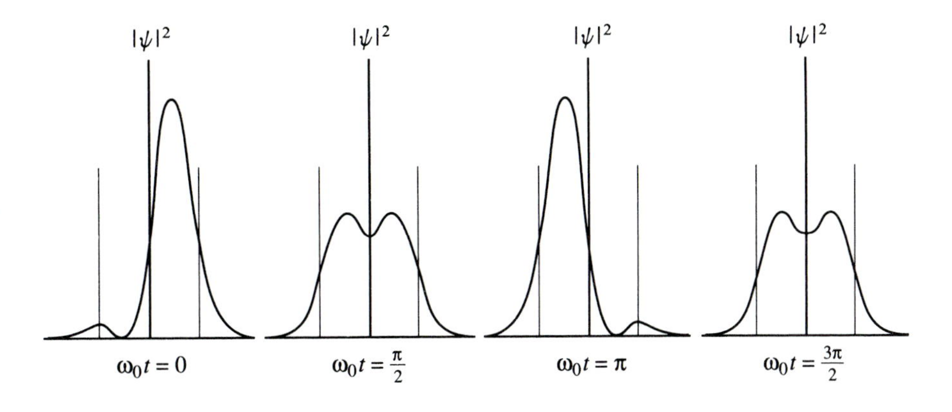

FIGURE 7.5
The probability corresponding to the superposition of the first two oscillator eigenfunctions of equal amplitude (with their time dependence included) plotted at four different times. Classical oscillatory behavior is clearly seen. The vertical lines indicate the classical limits of motion, assuming an energy $E = \langle H \rangle = \hbar\omega$.

tum probabilities are easily calculated by taking the square of the appropriate wave functions, u_0 and u_1.

How do we calculate the classical probabilities? Let the period of oscillation of the oscillator be denoted by $\tau = 2\pi/\omega$. The classical probability P_{cl} of finding the particle inside a region Δx can be defined as the fraction of time $\Delta t/\tau$ that the oscillator spends inside the spatial interval Δx:

$$P_{cl}(x)\Delta x = \frac{\Delta t}{\tau} = \frac{\omega}{2\pi}\frac{2\Delta x}{v(x)} \tag{7.21}$$

where v is the velocity. Now the classical solution of the harmonic oscillator can be written as

$$x = A \sin \omega t$$

where the amplitude A is related to the energy E,

$$A = \sqrt{2E/m\omega^2} \tag{7.22}$$

The velocity $v = A\omega \cos \omega t$, but we can easily express it as a function of x:

$$v(x) = A\omega(1 - x^2/A^2)^{1/2} \tag{7.23}$$

Inserting this into equation (7.21), we get the desired expression for the classical probability:

$$P_{cl}(x)\Delta x = \frac{1}{\pi A}\frac{1}{(1 - x^2/A^2)^{1/2}}\Delta x \tag{7.24}$$

where A is given by equation (7.22). As expected, the classical probability is nonzero only for $-A \leq x \leq A$; the oscillator is confined within the turning points. For $x > A$, from equation (7.22), it is clear that the potential energy $\frac{1}{2}m\omega^2x^2 > E$, and classically this is impossible.

The quantum and classical probabilities are compared in figures 7.6(a) and 7.6(b) for $n = 0$ and $n = 1$, respectively.

For $n = 0$, strictly speaking, the difference between the two probabilities is striking; and this is because, classically, there is no zero-point energy, $E = 0$, which implies $A = 0$, and the probability is nonzero only for $x = 0$. In contrast, the quantum mechanical probability has its maximum at $x = 0$, but does not do a vanishing act everywhere else. Surely you can see the play of the uncertainty principle here. (Note, however, that in figure 7.6 we compare classical and quantum probability densities of each state for the same total energy.)

For $n = 1$, the classical probability is maximum at the turning points, whereas the quantum probability reaches maximum much closer to the point of equilibrium.

In both cases, the quantum probability does not vanish in the classically forbidden region. But there is no contradiction here because, as stated before, any attempt to measure the oscillator in a state of negative kinetic energy entails an interaction of the measuring apparatus with the particle that changes the particle's energy because of the uncertainty principle involved in the measurement. The measurement does localize the particle, but after the measurement we can no longer say that the region of localization is classically forbidden.

For large n, the average of the quantum mechanical probability distribution is found to be given by the classical probability curve (see problem 6 at the

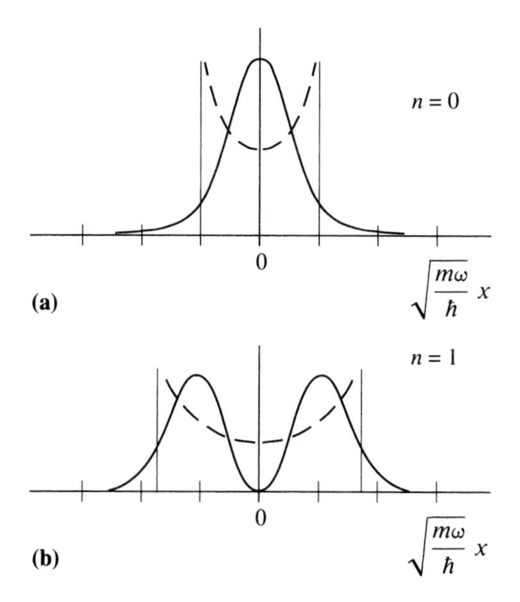

FIGURE 7.6
Comparison of quantum (solid curve) and classical probability distributions (dashed curve) for the harmonic oscillator for the first two oscillator states (for the same total energy).

end of the chapter). This is an example of Bohr's correspondence principle and will be further discussed in chapter 8.

7.2 THE OPERATOR METHOD FOR THE HARMONIC OSCILLATOR

In the operator method of solving for the eigenstates of the oscillator Hamiltonian, x and p in equation (7.1) are quantum mechanical operators in Hilbert space satisfying the commutation relation

$$[x,p] = i\hbar$$

and we do not use any particular representation space. It is important to introduce two additional operators:

$$a = \sqrt{m\omega/2\hbar}\,x + ip/\sqrt{2m\hbar\omega}$$

$$a^\dagger = \sqrt{m\omega/2\hbar}\,x - ip/\sqrt{2m\hbar\omega} \tag{7.25}$$

Note that a and $a^\dagger$ are not hermitian, but $(a^\dagger)^\dagger = a$. Furthermore, the operators a and $a^\dagger$ obey the commutation relation

$$[a, a^\dagger] = 1 \tag{7.26}$$

The commutator of a and $a^\dagger$ is unity. This commutation relation defines the algebra of these operators. These two operators will act as our springboard to move away from using the position and momentum representations and focus on the Hamiltonian in its own space.

So our next task is to write the oscillator Hamiltonian, equation (7.1), in terms of the a and $a^\dagger$ operators. To this end, note that

$$\begin{aligned}\hbar\omega a^\dagger a &= \hbar\omega\left(\sqrt{\frac{m\omega}{2\hbar}}\,x - \frac{ip}{\sqrt{2m\hbar\omega}}\right)\left(\sqrt{\frac{m\omega}{2\hbar}}\,x + \frac{ip}{\sqrt{2m\hbar\omega}}\right) \\ &= \frac{p^2}{2m} + \frac{1}{2}\,m\omega^2x^2 + \frac{i\omega}{2}\,[xp - px] \\ &= H - \frac{1}{2}\,\hbar\omega\end{aligned}$$

In this way we see that the oscillator Hamiltonian H can be written as

$$H = \hbar\omega(a^\dagger a + \tfrac{1}{2}) \tag{7.27}$$

The commutators of the operators a and $a^\dagger$ with H are now easily derived:

$$\begin{aligned}[H,a] &= [\hbar\omega a^\dagger a, a] = \hbar\omega[a^\dagger, a]a \\ &= -\hbar\omega a \end{aligned} \tag{7.28}$$

Noting that for any two operators $[A,B]^\dagger = [B^\dagger, A^\dagger]$, we get, taking the hermitian conjugate of equation (7.28),

$$[H,a^\dagger] = \hbar\omega a^\dagger \tag{7.29}$$

The meaning of the commutation relations above will be clear a little later on, but first let's obtain the energy eigenstates by solving the eigenvalue equation for the H operator by purely algebraic method using these commutation relations. We have

$$H|E\rangle = E|E\rangle \tag{7.30}$$

We will now show that if $|E\rangle$ is an eigenstate of H, so are $a|E\rangle$ and $a^\dagger|E\rangle$, but belonging to different eigenvalues. We will first demonstrate this for $a|E\rangle$. From $[H,a] = -\hbar\omega a$, we have

$$Ha|E\rangle - aH|E\rangle = -\hbar\omega a|E\rangle$$

Now use the eigenvalue equation (7.30). This gives

$$Ha|E\rangle = (E - \hbar\omega)a|E\rangle \tag{7.31}$$

Therefore, if $|E\rangle$ is an eigenstate of H with eigenvalue E, then $a|E\rangle$ is an eigenstate of H corresponding to the eigenvalue $E - \hbar\omega$. The operator a is thus a *lowering* operator (also called a *step-down operator*) that lowers the energy eigenvalue by the amount of $\hbar\omega$: $a|E\rangle \sim |E - \hbar\omega\rangle$.

What happens then if we keep operating upon an eigenstate $|E\rangle$ with the lowering operator a? The energy eigenvalue will decrease and decrease. Can it decrease indefinitely?

The answer is a resounding no! The oscillator Hamiltonian H consists of the sum of the squares of two hermitian operators. For any hermitian operator Q,

$$\langle\psi|Q^2|\psi\rangle = \langle Q^\dagger\psi|Q\psi\rangle = \langle Q\psi|Q\psi\rangle \geq 0$$

Thus $\langle\psi|H|\psi\rangle \geq 0$; the eigenvalues of H must be positive *semidefinite* (which is a fancy way of saying that eigenvalues are greater than or equal to zero). Thus H must have a state of nonnegative lowest energy, a ground state, call it $|u_0\rangle$;

and the buck stops here. Even the operation by a cannot lower the energy further. This must mean that

$$a|u_0\rangle = 0 \tag{7.32}$$

This equation then defines the ground state $|u_0\rangle$. To determine the ground state energy, we operate on $|u_0\rangle$ by H as given by equation (7.27). This gives

$$\begin{aligned} H|u_0\rangle &= (\hbar\omega a^\dagger a + \tfrac{1}{2}\hbar\omega)|u_0\rangle \\ &= \tfrac{1}{2}\hbar\omega|u_0\rangle \end{aligned} \tag{7.33}$$

where we have used equation (7.32). Clearly then $\frac{1}{2}\hbar\omega$ is the energy of the ground state—the zero-point energy.

Now we want to find out what happens if $a^\dagger$ is applied to the ground state. To this end, we multiply equation (7.29) from the right by $|u_0\rangle$. We get

$$Ha^\dagger|u_0\rangle - a^\dagger H|u_0\rangle = \hbar\omega a^\dagger|u_0\rangle$$

This last equation can be written as

$$Ha^\dagger|u_0\rangle = (\hbar\omega + \tfrac{1}{2}\hbar\omega)a^\dagger|u_0\rangle \tag{7.34}$$

Thus acting on an eigenstate of energy $\frac{1}{2}\hbar\omega$, $a^\dagger$ produces an eigenstate with a higher energy eigenvalue $\frac{1}{2}\hbar\omega + \hbar\omega$. In other words, $a^\dagger$ is a *raising operator* or *step-up operator*.

The state $a^\dagger|u_0\rangle$ must be the first excited state of the system, which we will denote by $|u_1\rangle$. Can there not be a state $|E_{\text{in}}\rangle$ intermediate in energy between $|u_0\rangle$ and $|u_1\rangle$? No, because the lowering operator a acting upon $|E_{\text{in}}\rangle$ would produce a state lower in energy than the ground state, which is impossible. Therefore in $|u_0\rangle$ and $|u_1\rangle$ we have the first two lowermost states of the oscillator Hamiltonian.

To generate states of higher excitation, just extend this reasoning and realize that we can go up the energy ladder by simply the repeated application of $a^\dagger$. Clearly, if we operate on $|u_0\rangle$ with $a^\dagger$ n times, we get

$$H(a^\dagger)^n|u_0\rangle = (n + \tfrac{1}{2})\hbar\omega(a^\dagger)^n|u_0\rangle \tag{7.35}$$

This has the familiar form of an eigenvalue equation from which we can easily pick out the energy eigenvalue

$$E_n = (n + \tfrac{1}{2})\hbar\omega, \qquad n = 0,1,2,\ldots$$

It's a little like magic! Previously, we found energy quantization by solving a differential equation and applying an explicit boundary condition. Now we see that energy quantization can be obtained purely from the operator properties of H, a, and $a^\dagger$.

It also follows from equation (7.35) that the nth energy eigenstate is given by $(a^\dagger)^n|u_0\rangle$ except for normalization. Additionally, from the form of H, equation (7.27), it is clear that the operator defined by

$$N = a^\dagger a$$

has the eigenvalue n for the nth excited energy eigenstate:

$$N(a^\dagger)^n|u_0\rangle = n(a^\dagger)^n|u_0\rangle \tag{7.36}$$

The operator N is called the *number operator*; it keeps track of the number of energy quanta in a given eigenstate. Thus our labeling the eigenstates by n, as in $|u_n\rangle$, is appropriate.

Normalized Eigenstates

If $|u_n\rangle$ denotes the normalized eigenstate corresponding to the eigenvalue n of the number operator (or corresponding to the energy E_n), we can write

$$(a^\dagger)^n|u_0\rangle = c_n|u_n\rangle$$

We will now determine the constants c_n. Note that the way we have written the equation above implies that $c_0 = 1$.

Clearly, we have

$$\begin{aligned} |c_n|^2\langle u_n|u_n\rangle = |c_n|^2 &= \langle (a^\dagger)^n u_0|(a^\dagger)^n u_0\rangle \\ &= \langle u_0|a^n(a^\dagger)^n|u_0\rangle \end{aligned} \tag{7.37}$$

We can use the commutation relation, equation (7.26), of a and $a^\dagger$ to carry a through $(a^\dagger)^n$. For a starter, note that

$$a(a^\dagger)^n = n(a^\dagger)^{n-1} + (a^\dagger)^n a$$

Consequently, we have

$$a^n(a^\dagger)^n = a^{n-1}[n(a^\dagger)^{n-1} + (a^\dagger)^n a]$$

The second term gives zero for the expectation value in equation (7.37). The first term gives $n|c_{n-1}|^2$. So we end up with a recursion relation

$$|c_n|^2 = n|c_{n-1}|^2$$

from which it follows that

$$|c_n|^2 = n!\,|c_0|^2 = n!$$

We will use the arbitrariness of the total phase to choose c_n to be real. (Actually, this choice is dictated by our desire that the x-representation of $|u_n\rangle$ will have the previously obtained normalized form, eq. [7.18].) In this way we get the normalized eigenstate of H as

$$|u_n\rangle = \frac{1}{\sqrt{n!}}\,(a^\dagger)^n|u_0\rangle \tag{7.38}$$

The orthogonality relation of the eigenstates

$$\langle u_m|u_n\rangle = 0 \qquad m \neq n$$

can be proven the same way, but this will be left as an exercise. The states $|u_n\rangle$ then make up a complete orthonormal basis, which we will refer to as the *energy or number* (n-) *representation.*

x-Representation

To find the eigenfunctions, the coordinate representation of the eigenstates, which are given as $u_n(x) = \langle x|u_n\rangle$, we start with equation (7.32) and multiply it with $\langle x|$ from the left to obtain

$$\langle x|a|u_0\rangle = 0$$

We now substitute in the defining equation for a, equation (7.25), the x-representation of the quantum mechanical operators x and p, which are x and $-i\hbar d/dx$, respectively:

$$\left(\sqrt{\frac{m\omega}{2\hbar}}\;x + \hbar\,\frac{1}{\sqrt{2m\hbar\omega}}\,\frac{d}{dx}\right)u_0(x) = 0$$

We can rewrite this in terms of the dimensionless variable ξ given by

$$\xi = \sqrt{m\omega/\hbar}\,x$$

We obtain (upon dividing out a factor of $1/\sqrt{2}$)

$$\left(\xi + \frac{d}{d\xi}\right)u_0 = 0$$

which is easily integrated to give

$$u_0 = N_0 \exp(-\xi^2/2)$$

The constant of integration N_0 can be evaluated from the normalization condition on u_0:

$$1 = \langle u_0 | u_0 \rangle = \int_{-\infty}^{\infty} dx \, |u_0(x)|^2 = |N_0|^2 \int_{-\infty}^{\infty} dx \exp(-m\omega x^2/\hbar)$$
$$= |N_0|^2 \sqrt{\frac{\hbar\pi}{m\omega}}$$

Accordingly,

$$N_0 = (m\omega/\hbar\pi)^{1/4}$$

The eigenfunctions $u_n(x)$ are now easily determined:

$$u_n(x) = \langle x | u_n \rangle = \langle x | \frac{(a^\dagger)^n}{\sqrt{n!}} | u_0 \rangle$$
$$= \frac{(m\omega/\hbar\pi)^{1/4}}{\sqrt{2^n n!}} \left(\xi - \frac{d}{d\xi} \right)^n e^{-\xi^2/2} \tag{7.39}$$

These are the previously obtained Hermite functions, equation (7.18). Check it out for the first few n's.

Matrix Elements

The lowering operator a and the raising operator $a^\dagger$ introduced above do not correspond to observables. Are they then just computational aids, mere tools for carrying out the algebra? No, as their names suggest, they indicate something quite akin to physical effects. We will now show that they have ubiquitous use in working with the energy or n-representation.

First, let's derive the matrix elements of $a^\dagger$ and a in the n-representation. (For the definition of matrix elements, see p. 124; for further elaborate exposition of matrices in quantum mechanics, see chapter 15.) From equation (7.38), we have

$$|u_{n+1}\rangle = \frac{1}{\sqrt{(n+1)!}} (a^\dagger)^{n+1} |u_0\rangle$$

Alternatively, we can write

$$\sqrt{n+1} \, |u_{n+1}\rangle = \frac{1}{\sqrt{n!}} (a^\dagger)^{n+1} |u_0\rangle = a^\dagger |u_n\rangle$$

Multiply the equation by the bra $\langle u_{n+1}|$ from the left, and you have the desired matrix element of $a^\dagger$:

$$\langle u_{n+1}|a^\dagger|u_n\rangle = \sqrt{n+1}\,\langle u_{n+1}|u_{n+1}\rangle = \sqrt{n+1} \tag{7.40}$$

since $|u_{n+1}\rangle$ is normalized. It is also clear from the orthogonality of the set $|u_n\rangle$ that in general, $\langle u_m|a^\dagger|u_n\rangle = (n+1)^{1/2}\,\delta_{m,n+1}$. Clearly, we can think of $a^\dagger$ also as a *creation operator*; it creates a state with one additional energy quantum.

For finding the matrix element of a in the n-representation, we use the fact that

$$a|u_n\rangle = \frac{1}{\sqrt{n!}}\,a(a^\dagger)^n|u_0\rangle \tag{7.41}$$

But

$$a(a^\dagger)^n = (a^\dagger)^n a + n(a^\dagger)^{n-1}$$

The first term gives zero when substituted into equation (7.41). In this way we get, upon multiplying equation (7.41) by the bra $\langle u_{n-1}|$ from the left,

$$\begin{aligned}\langle u_{n-1}|a|u_n\rangle &= \langle u_{n-1}|\,\frac{1}{\sqrt{n!}}\,n(a^\dagger)^{n-1}|u_0\rangle = \sqrt{n}\,\langle u_{n-1}|u_{n-1}\rangle \\ &= \sqrt{n}\end{aligned} \tag{7.42}$$

And, in general, $\langle u_m|a|u_n\rangle = n^{1/2}\delta_{m,n-1}$. Hence a is sometimes called a *destruction operator*; it destroys one energy quantum from the state on which it acts.

To summarize, we have three important operators for the harmonic oscillator: the creation operator $a^\dagger$ that creates an energy quantum, thus taking us up the energy ladder, the destruction operator a that destroys an energy quantum and takes us to a state where the number of quanta is one less, and finally the number operator N that maintains the status quo, always telling us how many quanta are present. For this reason, the late J. J. Sakurai used to joke, using his knowledge of Hindu mythology, that $a^\dagger$ is like *Brahma* the creator, a is like *Shiva* the destroyer, and N is the benign *Vishnu*, the preserver.

As an example of the application of these matrix elements we will now work out the matrix elements of $\hat{x}$ and $\hat{p}$ in the energy representation, in the basis defined by the complete set of eigenstates $|u_n\rangle$. You already know one way of calculating such matrix elements denoted by, let's say $\langle u_m|x|u_n\rangle$, using the x-representation of the basis $|u_n\rangle$. For example, we can write

$$\langle u_m|x|u_n\rangle = \int_{-\infty}^{\infty} dx' \int_{-\infty}^{\infty} dx'' \langle u_m|x'\rangle\langle x'|x|x''\rangle\langle x''|u_n\rangle$$
$$= \int_{-\infty}^{\infty} dx'\, u_m^*(x')x'u_n(x')$$

since $\langle x'|\hat{x}|x''\rangle = x'\delta(x'-x'')$. In this way of calculating, you have to perform a complicated integral involving the Hermite functions $u_n(x)$.

The algebraic method using the matrix elements of a and $a^\dagger$ is simpler. To this end, first, we note that

$$x = (\hbar/2m\omega)^{1/2}(a^\dagger + a)$$
$$p = i(m\hbar\omega/2)^{1/2}(a^\dagger - a) \tag{7.43}$$

Now use equations (7.40) and (7.42). We get

$$\langle u_{n+1}|x|u_n\rangle = (\hbar/2m\omega)^{1/2}\langle u_{n+1}|a^\dagger|u_n\rangle = (n+1)^{1/2}(\hbar/2m\omega)^{1/2}$$
$$\langle u_{n-1}|x|u_n\rangle = (\hbar/2m\omega)^{1/2}\langle u_{n-1}|a|u_n\rangle = n^{1/2}(\hbar/2m\omega)^{1/2} \tag{7.44}$$

All other matrix elements are zero. And for the momentum operator p, we get

$$\langle u_{n+1}|p|u_n\rangle = i(\hbar m\omega/2)^{1/2}\langle u_{n+1}|a^\dagger|u_n\rangle = i(n+1)^{1/2}(\hbar m\omega/2)^{1/2}$$
$$\langle u_{n-1}|p|u_n\rangle = -i(\hbar m\omega/2)^{1/2}\langle u_{n-1}|a|u_n\rangle = -in^{1/2}(\hbar m\omega/2)^{1/2} \tag{7.45}$$

All other matrix elements vanish. The operators x and p connect only states for which the index n differs by $\Delta n = \pm 1$. These matrix elements will be important when we consider transitions between harmonic oscillator states (see chapters 18 and 22).

Now a sneak preview. Everything there is to know about the operators x and p in the basis $|u_n\rangle$ is represented by the above matrix elements. It is customary to write in full these matrix elements as square arrays of numbers, where the number index on the left is taken as the row index and the number index on the right of the matrix element signifies the column of the array. For example, the array for $(2m\omega/\hbar)^{1/2}x$ is

$$\begin{pmatrix} 0 & \sqrt{1} & 0 & 0 & \cdot \\ \sqrt{1} & 0 & \sqrt{2} & 0 & \cdot \\ 0 & \sqrt{2} & 0 & \sqrt{3} & \cdot \\ \cdot & \cdot & \cdot & \cdot & \cdot \end{pmatrix} \tag{7.46}$$

Such arrays are called matrices and such a representation of an operator is called a matrix representation. Clearly the matrix representation of $\hat{x}$ is infinite dimensional, but there are other physical quantities for which the matrix representations are finite (see chapter 15).

What interpretation can we give of a matrix representation of an operator? The important thing to notice here is that x does not commute with H and hence x and H cannot have simultaneous eigenvalues. Instead, each value of x is in general associated with several values of H. This property of x is reflected when we describe x as a matrix; the matrix element x_{mn} belongs symmetrically to two values $(m + \frac{1}{2})\hbar\omega$ and $(n + \frac{1}{2})\hbar\omega$ of H. We will come back to matrix representations of operators in more detail in chapter 15.

PROBLEMS

1. Solve the Schrödinger equation for the linear harmonic oscillator Hamiltonian in the momentum representation.
2. Consider a coherent superposition of the $n = 1$ and $n = 2$ oscillator eigenfunctions:

$$\psi = \frac{1}{\sqrt{2}} [\exp(-iE_1 t/\hbar)u_1 + \exp(-iE_2 t/\hbar)u_2]$$

   where u_1 and u_2 are Hermite functions. Plot $|\psi|^2$ as a function of x for four different times

$$\omega_0 t = 0, \pi/2, \pi, 3\pi/2$$

   where $\omega_0 = E_0/\hbar$. Compare with the classical limits of motion.
3. Use the ground-state harmonic oscillator eigenfunction to show by direct integration that:

$$\langle p \rangle = 0$$

$$\langle p^2 \rangle = \hbar m\omega - m^2\omega^2\langle x^2 \rangle$$

4. Use the nth energy state wave function of the one-dimensional oscillator to calculate the uncertainty product $\Delta x \cdot \Delta p$. (*Hint*: Use the recursion relations, eqs. [7.19] and [7.20].)
5. What are the energy levels of a particle of mass m moving in the one-dimensional potential well defined by

$$V(x) = \infty \qquad \text{for } x < 0$$
$$= \tfrac{1}{2} m\omega^2 x^2 \qquad \text{for } x > 0$$

(*Hint*: You don't need a lengthy calculation.)

6. Expand the Dirac delta function as a discrete sum using the complete set of linear harmonic oscillator eigenfunctions.

7. Plot the probability density of the linear harmonic oscillator as a function of x for $n = 12$. Compare the result with the classical probability density (for the same total energy) and discuss the limit $n \to \infty$.

8. Consider a proton as a bound oscillator with a natural frequency of 3×10^{21} Hz. What is the energy of its ground and first excited states? What is its classical oscillation amplitude?

9. A macroscopic oscillator consists of a mass of 1 gram attached to a spring. The mass is displaced from its equilibrium position and released to exhibit simple harmonic motion of period 1 s. The mass is also found to pass through the zero displacement position with a velocity of 8 cm/s. (a) Is the oscillator in an eigenstate of the Hamiltonian? (b) Estimate the order of magnitude of the quantum number n associated with the energy $\langle E \rangle$ of the system.

10. At $t = 0$, an ensemble of particles is in the state

$$|\psi\rangle = \frac{1}{4}\,|u_0\rangle + \frac{i}{2}\,|u_1\rangle + \frac{i\sqrt{11}}{4}\,|u_2\rangle$$

where $|u_0\rangle$, $|u_1\rangle$, and $|u_2\rangle$ are the first three energy eigenstates of the linear harmonic oscillator. (a) If a single observation is made on one of the particles, what is the likely value of energy that will turn up and with what probability? (b) If the experiment is repeated many times, each time on a different particle of the ensemble, what is the likely average value of energy to turn up as a result of measurement? (c) If a particle is left undisturbed, what is its state at time t? (d) When all the particles are measured, how would you describe the state of the ensemble?

11. Prove the orthogonality of the eigenstates $|u_n\rangle$ using the operator method.

12. Use the operator method for calculating the expectation values $\langle p^2 \rangle$ and $\langle x^2 \rangle$ in the energy eigenstate $|u_n\rangle$. (*Hint*: Use the completeness relation for the oscillator eigenstate, $\Sigma\,|u_n\rangle\langle u_n| = 1$.) Hence prove the quantum analog of the virial theorem:

$$\langle V \rangle = \langle T \rangle = \langle H \rangle / 2$$

ADDITIONAL PROBLEMS

A1. A harmonic oscillator of mass m is located within a distance of ± 0.001 $(\hbar/m\omega)$ of the origin. If you measure the energy of the oscillator, what is the probability of your finding the value $\hbar\omega/2$? the value $3\hbar\omega/2$? the value $2\hbar\omega$?

A2. If $\phi(x)$ is an even function and $\psi(x)$ is odd under $x \to -x$, show that

$$\int_{-\infty}^{\infty} dx\, \phi(x)\psi(x) = 0$$

Hence show directly that the oscillator eigenfunctions u_0 and u_1 are orthogonal.

A3. Suppose that the Hamiltonian

$$H = \frac{p^2}{2m} + V(x)$$

has a set of eigenstates $|n\rangle$ with eigenvalues E_n, the lowest being the state $|0\rangle$ with eigenvalue E_0. Prove the sum rule:

$$\Sigma_n (E_n - E_0)|\langle n|x|0\rangle|^2 = \hbar^2/2m$$

(*Hint*: Consider $[x, [H, x]]$.) Verify the sum rule above for a harmonic oscillator.

A4. Suppose a particle is in the ground state of the oscillator potential

$$V_1(x) = \tfrac{1}{2} m\omega_1^2 x^2$$

Suddenly the potential changes to

$$V_2(x) = \tfrac{1}{2} m\omega_2^2 x^2$$

What is the probability that the particle will be in the ground state of the new potential?

A5. A particle of mass m is in the ground state of the modified oscillator potential

$$V(x) = \tfrac{1}{2} m\omega^2 x^2 \text{ for } x > 0,\ V(x) = \infty \text{ for } x < 0.$$

At time $t = 0$, the potential is suddenly changed to

$$V(x) = \tfrac{1}{2} m\omega^2 x^2$$

What is the probability that a subsequent measurement of energy will give the value $\hbar\omega/2$? What is the probability for measuring the energy value $3\hbar\omega/2$?

A6. Solve the bound-state Schrödinger equation for the potential

$$V(x) = -3\hbar^2/(ma^2 \cosh^2(x/a))$$

(*Hint*: Substitute $y = x/a$, use the ansatz $\psi(y) = \exp(iky)\phi(y)$, and then substitute $z = \tanh y$. Find a solution of the resulting equation for ϕ by the power series method using the appropriate boundary conditions.)

A7. The Hamiltonian of a particle is given as

$$H = E_0 a^\dagger a + E_1(a^\dagger + a)$$

$$[a, a^\dagger] = 1$$

Find the eigenvalues of H.

A8. The operator c is defined by the following relations:

$$c^2 = 0$$

$$cc^\dagger + c^\dagger c = 1$$

where $c^\dagger$ is the hermitian conjugate of c. Show that
(a) $N = c^\dagger c$ is hermitian.
(b) $N^2 = 1$.
(c) The eigenvalues of N are 0 and 1.
(d) If $|0\rangle$ and $|1\rangle$ denote the two eigenstates of N corresponding to the eigenvalues 0 and 1, respectively, show that

$$c^\dagger|0\rangle = |1\rangle$$

$$c|0\rangle = 0.$$

A9. Consider the Hamiltonian

$$H = \hbar\omega_0(c^\dagger c + \tfrac{1}{2})$$

where c, $c^\dagger$ are anticommuting as in problem A8. Denoting the eigenstates of H as $|n\rangle$, show that the only nonvanishing states are the states $|0\rangle$ and $|1\rangle$ above.

A10. Display the matrix of the operators $a^\dagger$ and a (that is, in the form of equation (7.46)) in the energy representation.

REFERENCES

A. Z. Capri. *Nonrelativistic Quantum Mechanics.*
A. Das and A. C. Melissinos. *Quantum Mechanics: A Modern Introduction.*
R. H. Dicke and J. P. Wittke. *Introduction to Quantum Mechanics.*
Fayyazuddin and Riazuddin. *Quantum Mechanics.*
E. Merzbacher. *Quantum Mechanics.*
J. L. Powell and B. Crasemann. *Quantum Mechanics.*

8

Equations of Motion and Classical Correspondence

The early development of quantum theory largely used Bohr's idea of classical correspondence—that there should be certain limits under which classical motion prevails. By looking at those limits, one then could check the validity of the ad hoc rules by which Bohr, Sommerfeld, and others carried out the calculations of quantum theory before the development of quantum mechanics.

What is our motivation for studying the classical limit of quantum mechanics now that we have a fundamental theory, not a patchwork? Well, first of all, we know classical theory does work for the motion of most macroscopic bodies. It is necessary to find out if classical mechanics can really be derived from quantum mechanics and under what limits. There is also a second important motivation. Few quantum mechanical potential problems are exactly solvable. Finding the classical limit of quantum mechanics should help us develop approximation methods for solving those problems for which the motion is approximately classical.

We will find that the classical equation of motion holds better and better whenever the potential varies slowly with distance so that the wave properties of matter do not find much manifest expression. We will develop an approximate method, called the Wentzel-Kramers-Brillouin or WKB method for short, for solving the Schrödinger equation precisely under these circumstances.

There is no analog of state vectors in classical mechanics; on the other hand, most quantum mechanical observables have direct classical analogs. Accordingly, a simple way to begin to see the relation of quantum and classical mechanics is to write the quantum equation as an equation of motion for observables for which classical analogs exist.

8.1 THE HEISENBERG EQUATION OF MOTION: SCHRÖDINGER AND HEISENBERG PICTURES

In chapter 6, we derived a differential equation for the time development operator $U(t, t_0)$, equation (6.27). If the Hamiltonian of the system does not depend on time, we can easily integrate that equation to obtain

$$U(t, t_0) = e^{-iH(t-t_0)/\hbar} \tag{8.1}$$

Since $U(t_0, t_0) = 1$, the constant of integration must be equal to 1. So the state vector at a finite time t is related to the state vector at time $t = 0$ by the equation

$$|\psi(t)\rangle = e^{-iHt/\hbar}|\psi(0)\rangle \tag{8.2}$$

Now consider the time dependence of the expectation value of an operator A in the state $|\psi(t)\rangle$ (we assume that A does not have any explicit time dependence),

$$\begin{aligned}\langle A\rangle_t &= \langle\psi(t)|A|\psi(t)\rangle \\ &= \langle\psi(0)|e^{iHt/\hbar}Ae^{-iHt/\hbar}|\psi(0)\rangle \\ &= \langle\psi(0)|A(t)|\psi(0)\rangle = \langle A(t)\rangle_0\end{aligned} \tag{8.3}$$

where we have used the hermiticity of H and have defined the operator $A(t)$ as

$$A(t) = e^{iHt/\hbar}Ae^{-iHt/\hbar} \tag{8.4}$$

The result, equation (8.4), is very interesting since it shows that we have a choice. In our development of quantum mechanics in this book, we have ascribed all the time dependence to the wave function (or state vectors), leaving the operators time independent. This is called the *Schrödinger picture*. But there is another alternative—we can use $A(t)$ as our operator in which all the time dependence is taken up in the operators, and this leaves the state vectors fixed once and for all in time. This is called the *Heisenberg picture*. And equation (8.3) assures us that physical results, such as expectation values, remain unchanged irrespective of what picture we use.

In the Heisenberg picture, the state vectors are constant fixtures, and there is no need to invoke them in our calculations. Instead, we concentrate on the equation of motion obeyed by the observable $A(t)$. From equation (8.4), differentiating with respect to time, we get

$$\frac{d}{dt}A(t) = \frac{iH}{\hbar}e^{iHt/\hbar}Ae^{-iHt/\hbar} + e^{iHt/\hbar}A(-iH/\hbar)e^{-iHt/\hbar}$$

$$= \frac{i}{\hbar}[H,A(t)] \tag{8.5}$$

This is the *Heisenberg equation of motion* that replaces the Schrödinger equation in the Heisenberg picture.

Let's summarize with the help of a metaphor. The Schrödinger picture rotates the dance floor with the dancers remaining still, while the Heisenberg picture leaves the dance floor alone and lets the dancers rotate.

Since the two pictures are completely equivalent, in principle, we can do our calculations using either picture. However, for a general operator, the operator equations of motion in the Heisenberg picture are often difficult to solve for most systems (there are exceptions), and thus, in practice, we use the Schrödinger picture in which we deal with the more conventional linear differential equation.

One interesting aspect of the Heisenberg equation is that it leads to equations that are formally quite similar to the corresponding classical equations (see below). Except, of course, Heisenberg's equations are for operators with nonzero commutators between them. What Heisenberg discovered is that it is possible to describe nature with classical equations, but we must realize that $[x,p]$ is $i\hbar$, and it is this little correction to the classical way of doing things that leads to all the wonderful phenomena of quantum mechanics.

■ **PROBLEM** Consider the harmonic oscillator Hamiltonian in the Heisenberg picture. Derive the Heisenberg equations of motion for $a(t)$ and $a^\dagger(t)$ and solve them for their time dependence. Hence determine $\hat{x}(t)$ and $\hat{p}(t)$ for the oscillator.

SOLUTION From equation (7.27), it is easy to see, using the definition of equation (8.4), that the oscillator Hamiltonian in the Heisenberg picture can be written as

$$H = \hbar\omega(a^\dagger(t)a(t) + \tfrac{1}{2})$$

Furthermore, the commutation relations of a and $a^\dagger$ with one another and with H retain the same form:

$$[a(t),a^\dagger(t)] = 1$$

$$[H,a(t)] = -\hbar\omega a(t)$$

$$[H,a^\dagger(t)] = \hbar\omega a^\dagger(t)$$

Note that since these commutation relations remain unchanged, the arguments leading to the eigenvalue spectrum of H also remain valid; thus the operator method outlined in chapter 7 works equally well irrespective of whether we use the Schrödinger or the Heisenberg picture. The Heisenberg equation of motion, equation (8.5), now gives

$$\frac{d}{dt}a(t) = -i\omega a(t)$$

$$\frac{d}{dt}a^{\dagger}(t) = i\omega a^{\dagger}(t)$$

Fortunately in this case we have only first-order equations to solve for obtaining the time dependence. We get

$$a(t) = e^{-i\omega t}a(0)$$

$$a^{\dagger}(t) = e^{i\omega t}a^{\dagger}(0)$$

Now using the defining equations of a and $a^{\dagger}$, equation (7.25), which retains the same form also for the Heisenberg operators, we easily find the time dependence of $\hat{x}$ and $\hat{p}$:

$$\hat{x}(t) = \hat{x}(0)\cos\omega t + \frac{\hat{p}(0)}{m\omega}\sin\omega t$$

$$\hat{p}(t) = \hat{p}(0)\cos\omega t - m\omega\hat{x}(0)\sin\omega t$$

This is indeed quite similar to the classical solution, except, of course, that $\hat{x}(0)$ and $\hat{p}(0)$ are operators as indicated explicitly by their hats; in fact, they are the corresponding Schrödinger picture operators. ■

Ehrenfest Theorem and the Classical Limit of Quantum Mechanics

Consider now a physical quantity whose associated operator is A, which is hermitian. To establish the connection with classical mechanics where there are no operators, but only c-numbers, we have to consider expectation values of operators. Thus the classical analog of quantum mechanics is to be sought from equation (8.5) by taking the expectation value of this equation in the state vector $|\psi(0)\rangle$. We have

$$\frac{d}{dt}\langle\psi(0)|A(t)|\psi(0)\rangle = \frac{i}{\hbar}\langle\psi(0)|[H,A(t)]|\psi(0)\rangle \tag{8.6}$$

This is the quantum equation of motion for the time development of the expectation value of an operator in the Heisenberg picture. However, by virtue of equations (8.3) and (8.4), it is easy to see that a similar equation holds for the Schrödinger picture:

$$\frac{d}{dt}\langle\psi(t)|A|\psi(t)\rangle = \frac{i}{\hbar}\langle\psi(t)|[H,A]|\psi(t)\rangle \tag{8.7}$$

It is instructive to derive equation (8.7) using the Schrödinger equation—using the Schrödinger picture from the beginning. In the Schrödinger picture, the time dependence of $\langle A\rangle$ comes entirely from the time dependence of the state vector $|\psi(t)\rangle$. Since $\langle A\rangle$ is given as

$$\langle A\rangle = \langle\psi(t)|A|\psi(t)\rangle$$

the time rate at which this average value changes is given by

$$\frac{d\langle A\rangle}{dt} = \left[\frac{\partial}{\partial t}\langle\psi|\right]A|\psi\rangle + \langle\psi|A\frac{\partial}{\partial t}|\psi\rangle$$

But from the Schrödinger equation, we have

$$\frac{\partial}{\partial t}|\psi\rangle = \frac{1}{i\hbar}H|\psi\rangle$$

$$\frac{\partial}{\partial t}\langle\psi| = -\frac{1}{i\hbar}\langle\psi|H$$

In this way, we get

$$\begin{aligned}\frac{d\langle A\rangle}{dt} &= -\frac{1}{i\hbar}\langle\psi|HA - AH|\psi\rangle \\ &= \frac{i}{\hbar}\langle[H,A]\rangle \end{aligned} \tag{8.7}$$

One of the consequences of the quantum equation of motion, equation (8.7), is that when a physical observable A commutes with the Hamiltonian H, its expectation value $\langle A\rangle$ is constant in time for any state. Then A is said to be conserved and is called a constant of the motion. Thus the conservation law in quantum mechanics is statistical; it holds for the average value of the observable.

We can apply the conclusion above to the momentum p and see that momentum is conserved only if p commutes with H, that is, if p commutes with $V(x)$. But p commutes with $V(x)$ when V is a constant independent of x; that is, when there is no force field acting on the system. It is the same way in classical mechanics.

Now we will apply our equation of motion, equation (8.7), to the cases of x and p, respectively. For x, we get

$$\begin{aligned}\frac{d\langle x\rangle}{dt} &= \frac{i}{\hbar}\langle[H,x]\rangle \\ &= \frac{i}{\hbar}\left\langle\left[\frac{p^2}{2m} + V(x),x\right]\right\rangle\end{aligned}$$

Of course, x commutes with any function of x; that is, $[V(x),x] = 0$. The other commutator is easily calculated:

$$\begin{aligned}[p^2,x] &= p[p,x] + [p,x]p \\ &= -2i\hbar p\end{aligned}$$

Consequently, we get

$$\frac{d\langle x\rangle}{dt} = \left\langle\frac{p}{m}\right\rangle \tag{8.8}$$

For the time rate of change of $\langle p\rangle$, we have

$$\begin{aligned}\frac{d\langle p\rangle}{dt} &= \frac{i}{\hbar}\left\langle\left[\frac{p^2}{2m} + V(x),p\right]\right\rangle \\ &= -\frac{i}{\hbar}\langle[p, V(x)]\rangle\end{aligned}$$

since p^2 commutes with p. To find the commutator of p and $V(x)$, let's operate on a wave function $\psi(x)$ with the commutator $[p, V(x)]$. We get

$$\begin{aligned}pV(x)\psi(x) - V(x)p\psi(x) &= -i\hbar\frac{d}{dx}[V(x)\psi(x)] - V(x)(-i\hbar)\frac{d}{dx}\psi(x) \\ &= -i\hbar\frac{dV(x)}{dx}\psi(x)\end{aligned}$$

Accordingly,

$$[p, V(x)] = -i\hbar\frac{dV(x)}{dx} \tag{8.9}$$

And consequently, we obtain

$$\frac{d\langle p\rangle}{dt} = -\left\langle \frac{dV(x)}{dx} \right\rangle \tag{8.10}$$

We see that the quantum equation of motion for $\langle x\rangle$ gives us equation (8.8), which is nothing but the quantum counterpart of the classical relationship between velocity and momentum. And the equation of motion of $\langle p\rangle$ leads us to equation (8.10), which can be easily seen as the quantum analog of Newton's second law of motion ($-dV(x)/dx$ is the x-component of a Newtonian force). This is the *Ehrenfest theorem*: Newtonian mechanics holds for the expectation values of the corresponding quantum operators.

Equations (8.8) and (8.10)—the Ehrenfest theorem—do not themselves constitute the classical limit of quantum mechanics; however, they can show us under what circumstances classical mechanics, and classical motion, are obtained as a special case of quantum mechanics, which all objects, microscopic or macroscopic, ultimately follow.

The big hurdle blocking the realization of classical motion, of course, is that in equation (8.10), we cannot in general replace

$$\left\langle \frac{dV(x)}{dx} \right\rangle$$

by

$$\frac{dV(\langle x\rangle)}{d\langle x\rangle}$$

and thus recover a classical equation for an analogous classical variable $x_{\text{cl}} = \langle x\rangle$. This replacement can be made whenever $V(x)$ is a slowly varying function of x. In this way, the Ehrenfest theorem is useful, because it tells us exactly when we can expect classical equations to be a valid limiting case of quantum equations. The classical equations pertain whenever the uncertainty in the variables (i.e., their statistical spread around the average, typical of quantum mechanics) is small and ignorable. Even so, we cannot, in general, expect the conditions for the Ehrenfest theorem to remain valid for all times since we know that wave packets spread, and uncertainties of the variables increase (remember chapter 2?), with time.

For macroscopic bodies, Ehrenfest's theorem, and a classical correspondence in the sense above, usually holds. Even for the description of the motion of microbodies in a macroscopic field (such as in a particle accelerator), we can invoke classical correspondence and treat the motion as essentially classical.

An Example of Classical Correspondence: The Harmonic Oscillator

For the harmonic oscillator Hamiltonian,

$$H = \frac{p^2}{2m} + \frac{1}{2} m\omega^2 x^2 \tag{7.1}$$

the Ehrenfest theorem, equation (8.10), gives

$$\frac{d}{dt} \langle p \rangle = -\left\langle \frac{dV}{dx} \right\rangle = -m\omega^2 \langle x \rangle$$

Additionally, equation (8.8) gives

$$\frac{d}{dt} \langle x \rangle = \frac{\langle p \rangle}{m}$$

Combining the two equations, we obtain

$$\frac{d^2 \langle x \rangle}{dt^2} + \omega^2 \langle x \rangle = 0 \tag{8.11}$$

Thus $\langle x \rangle$ satisfies the classical equation for the harmonic oscillator. The solution can be written as

$$\langle x \rangle_t = x_{\text{cl}}(t) = \sqrt{\frac{2E}{m\omega^2}} \cos(\omega t + \phi) \tag{8.12}$$

Under what condition does the quantum solution for the harmonic oscillator given in chapter 7 give the classical solution, equation (8.12), for the expectation value? Let's find out.

A general solution of the Schrödinger equation for the harmonic oscillator can be written as a complete set expansion using the oscillator eigenstates

$$|\psi(t)\rangle = \sum_{n=0}^{\infty} c_n e^{-iE_n t/\hbar} |u_n\rangle$$

$$= e^{-i\omega t/2} \sum c_n e^{-in\omega t} |u_n\rangle$$

where c_n are the coefficients of the expansion and where we have substituted $E_n = (n + \frac{1}{2})\hbar\omega$. The expectation value of $\hat{x}$ in the state $|\psi(t)\rangle$ is then given as

$$\langle x \rangle = \sum_n \sum_{n'} c_n^* c_{n'} e^{i(n-n')\omega t} \langle u_n | x | u_{n'} \rangle$$

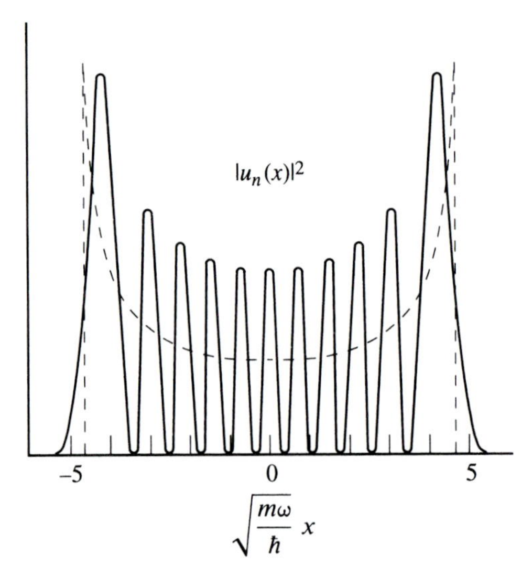

FIGURE 8.1
Comparison of classical and quantum harmonic oscillator (of the same total energy) probability distribution for large n (n = 10) as an example of how the correspondence principle works.

Substituting for the matrix element of x from equation (7.44), we get

$$\langle x \rangle = \sqrt{\frac{\hbar}{2m\omega}} \sum_{n=1}^{\infty} \sqrt{n}\,(c_n^* c_{n-1} e^{i\omega t} + c_{n-1}^* c_n e^{-i\omega t})$$

Now put $c_n = |c_n| \exp(i\phi_n)$. This gives

$$\langle x \rangle = \sqrt{\frac{2}{m\omega^2}} \sum_{n=0}^{\infty} \sqrt{n\hbar\omega}\, |c_n|\, |c_{n-1}| \cos(\omega t + \phi_{n-1} - \phi_n)$$

This does not look like the classical solution yet, but if the phase $\phi_{n-1} - \phi_n$ is ϕ independent of n, and $|c_n| \approx |c_{n-1}|$, and moreover c_n is appreciably different from zero only for large n so that $E_n \approx n\hbar\omega$, then we get

$$\langle x \rangle = \sqrt{\frac{2}{m\omega^2}} \left(\sum_{n=0}^{\infty} \sqrt{E_n}\, |c_n|^2 \right) \cos(\omega t + \phi)$$

The quantity within parentheses can be recognized as $\langle \sqrt{E} \rangle$, and thus we have recovered the classical solution, equation (8.12), for large n.

As additional confirmation of this classical correspondence, let's plot, following the procedure of chapter 7, the quantum and classical probability distributions when the oscillator quantum number n is large, say $n = 10$ (fig. 8.1). Notice that indeed for large n, the classical distribution curve describes the average of the quantum mechanical probability distribution, indicating the validity of the correspondence principle.

8.2 THE WKB APPROXIMATION

Now we will review the Wentzel-Kramers-Brillouin (WKB) approximation method specifically developed for those cases of quantum motion where the potential is a slowly varying function of x, and thus the motion is expected to be fairly classical.

We write the Schrödinger equation as

$$\frac{d^2\psi}{dx^2} + \frac{p^2}{\hbar^2}\psi = 0 \tag{8.13}$$

where p now represents the classical momentum at x:

$$p = \sqrt{2m(E - V(x))}$$

For a particle of energy E moving in a potential $V(x)$, the de Broglie wavelength is given as

$$\lambda = \frac{h}{p} = \frac{h}{\sqrt{2m(E - V(x))}}$$

If V varies slowly with x, it can be taken as a constant over a few de Broglie wavelengths at a time; then in a small region of space the wave function is a plane wave. And, in general, the wave function can be thought of as a plane wave whose wave number changes gradually from one region into another:

$$\begin{aligned}\psi(x) &= \phi(x)\exp\left[\pm i\int^x k(x)\,dx\right]\\ &= \phi(x)\exp\left[\pm(i/\hbar)\int^x p(x)\,dx\right]\end{aligned}$$

where ϕ is a slowly varying function. Substitute in the Schrödinger equation (8.13); this will give the differential equation satisfied by $\phi(x)$:

$$\frac{\hbar}{ip}\frac{d^2\phi}{dx^2} \pm \left(2\frac{d\phi}{dx} + \frac{1}{p}\frac{dp}{dx}\phi\right) = 0$$

We assume $\hbar/p$ to be small compared to the other dimensions of the problem, and since in addition ϕ is assumed to be slowly varying, the first term can be thrown away and we have

$$\frac{2}{\phi}\frac{d\phi}{dx} + \frac{1}{p}\frac{dp}{dx} = \frac{d}{dx}\ln(p\phi^2) = 0$$

This can be integrated giving

$$\phi = Cp^{-1/2}$$

where C is a constant of integration. Thus the WKB wave function is given as

$$\psi(x) = Cp^{-1/2} \exp\left[\pm (i/\hbar) \int^x p\, dx \right] \tag{8.14}$$

Note that if we substitute this form of ψ into the Schrödinger equation, we will get extra terms proportional to

$$(\hbar^2/p^3)\, d^2p/dx^2 \quad \text{and} \quad [(\hbar/p^2)\, dp/dx]^2$$

But the slow-varying condition on the potential can be written precisely as

$$|(\hbar^2/p^3)\, d^2p/dx^2| \ll 1 \quad \text{and} \quad |(\hbar/p^2)\, dp/dx| \ll 1 \tag{8.15}$$

It follows that in equation (8.14) we have a legitimate approximate solution of the Schrödinger equation when $V(x)$ varies slowly with x.

So far we are tacitly assuming $E > V(x)$, the classically allowed region. In the classically forbidden region, $E < V$, the wave number (and p) is imaginary. In analogy to equation (8.14), the solution in this region can be written as

$$\psi_\pm(x) = C_\pm |p|^{-1/2} \exp\left[\pm (1/\hbar) \int^x |p|\, dx \right] \tag{8.16}$$

where $C_\pm$ are constants. The wave function increases or decreases exponentially away from the classical "turning points" (see fig. 8.2 below) where $E = V$ and where none of the above solutions holds.

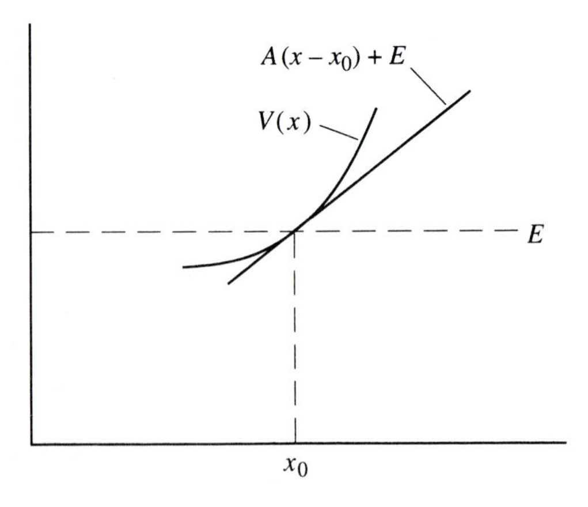

FIGURE 8.2
Near a turning point, a potential may be approximated by a linear potential.

The Turning Points

At the turning points, $E = V$ and $p = 0$, and the above method fails. Classically, the kinetic energy has to be greater than zero, and thus the classical motion turns about at these points; hence the name *turning points*. At these points, the conditions of equation (8.15) are violated, and in fact, the WKB solutions above cease to be valid within a few wavelengths of the turning points. However, we need the solution at these points in order to connect between the regions $E > V$ and $E < V$.

The trick to finding a solution near the turning point is to realize that for all slowly varying functions, a linear approximation to the actual variation of the potential is expected to hold for a short distance around any point. Thus the potential around a turning point x_0 can be taken as (see fig. 8.2)

$$V(x) = A(x - x_0) + E$$

The Schrödinger equation in the neighborhood of a turning point is then given as

$$\frac{d^2\psi}{dx^2} - \frac{2mA}{\hbar^2}(x - x_0)\psi = 0 \tag{8.17}$$

Now change the variable to

$$y = (2mA/\hbar^2)^{1/3}(x - x_0)$$

This gives

$$\frac{d^2\psi}{dy^2} - y\psi = 0 \tag{8.18}$$

The solution of this equation is given in terms of the Airy function $Ai(y)$ (some texts use Bessel functions of order 1/3, which are cousins of the Airy functions) introduced in chapter 4. With the help of the Airy function we will now find appropriate connection formulas.

Connection Formulas: The Case of Bound States

For bound states, we are interested in the Airy function that decays monotonically for $y > 0$ tending to zero as $y \to \infty$, and is oscillatory for $y < 0$, as shown in figure 8.3. We are interested presently in the asymptotic forms for large $|y|$:

$$Ai(y) \sim \frac{1}{2\sqrt{\pi}y^{1/4}} \exp\left(-\frac{2}{3} y^{3/2}\right) \qquad (y > 0) \tag{8.19}$$

$$\sim \frac{1}{\sqrt{\pi}(-y)^{1/4}} \sin\left[\frac{2}{3}(-y)^{3/2} + \frac{\pi}{4}\right] \qquad (y < 0) \tag{8.20}$$

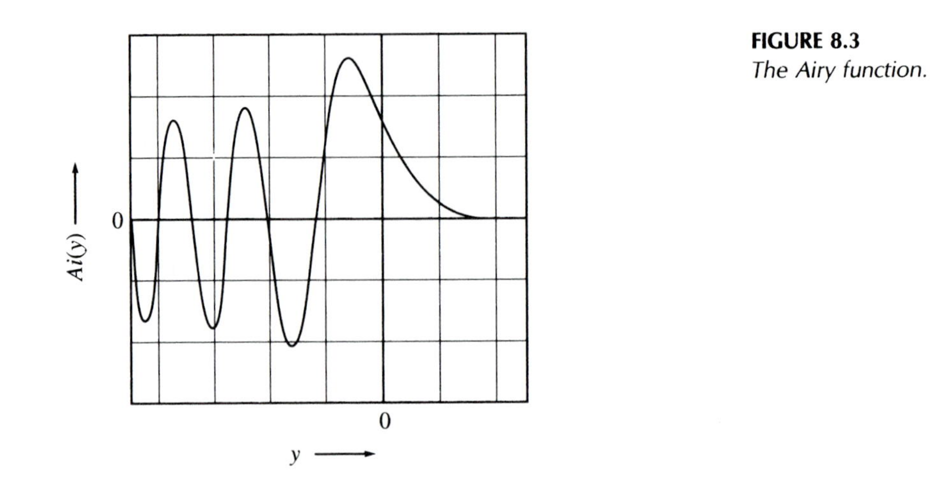

FIGURE 8.3
The Airy function.

Let's consider a particle in the potential well shown in figure 8.4 where we have indicated three regions. In region 1, the wave function must $\to 0$ as $x \to -\infty$. Thus it must be the positive exponent solution of equation (8.16):

$$\psi_1 = C_+ |p|^{-1/2} \exp\left[\frac{1}{\hbar}\int_{x_1}^{x} |p|\, dx\right]$$

On the other hand, in region 3 the wave function tends to zero as $x \to \infty$, and consequently, we have

$$\psi_3 = C_- |p|^{-1/2} \exp\left[-\frac{1}{\hbar}\int_{x_2}^{x} |p|\, dx\right] \tag{8.21}$$

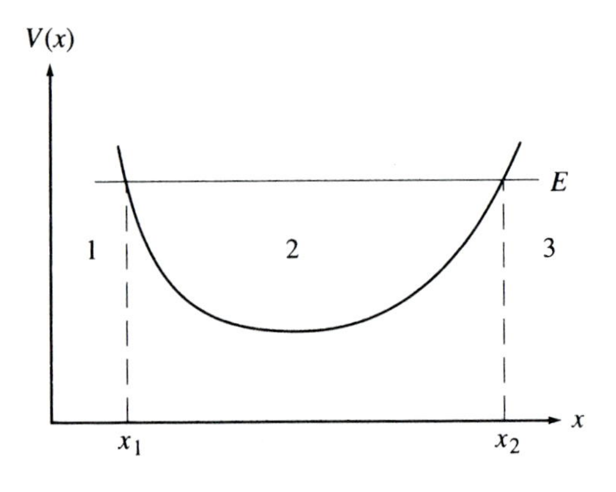

FIGURE 8.4
Potential well for the discussion of WKB treatment of the quantum bound-state problem.

These solutions have to connect to the oscillatory solution in the middle region 2, $x_1 < x < x_2$, a suitable linear combination of the two solutions in equation (8.14). But what linear combination? This is where we have to use the Airy functions, which go through the turning points smoothly and will tell us the correct way to extend ψ_1 and ψ_3 into region 2.

In the neighborhood of the turning point $x_0 = x_2$, we have

$$p^2 = 2m(E - V) = -2mA(x - x_2) = -(2mA\hbar)^{2/3}y$$

Therefore

$$-\frac{1}{\hbar}\int_{x_2}^{x} |p|\, dx = -\int_0^y \sqrt{y}\, dy = -\frac{2}{3}\, y^{3/2}$$

Thus away from the turning point, for $x > x_2$, the WKB solution (eq. [8.21]) agrees with the asymptotic form (eq. [8.19]) of the Airy solution. On the other hand, the integral in the exponent of the oscillatory functions in region 2 is given as

$$\frac{1}{\hbar}\int_{x}^{x_2} p\, dx = -\int_0^y \sqrt{-y}\, dy = \frac{2}{3}\,(-y)^{3/2}$$

Now compare this with the asymptotic expression for the Airy function on the left side of x_2, $x < x_2$, equation (8.20). The WKB oscillatory solution, equation (8.14), in region 2 will match the Airy solution, equation (8.20), only if we take a linear combination of the two exponentials to give us a sine function of the appropriate phase. In other words, the function that is represented for $x > x_2$ by the approximation

$$\psi \approx |p|^{-1/2}\exp\left[-\frac{1}{\hbar}\int_{x_2}^{x} |p|\, dx\right]$$

fits on the other side of $x_2 (x < x_2)$ to the approximation

$$\psi \approx 2p^{-1/2}\sin\left(\frac{1}{\hbar}\int_{x}^{x_2} p\, dx + \frac{\pi}{4}\right) \tag{8.22}$$

However, the story is not complete yet! We have to do the same thing for the turning point at $x = x_1$. Thus at x_1, we write the potential as

$$V(x) \approx E - B(x - x_1)$$

The Schrödinger equation in the vicinity of x_1 is then given by

$$\frac{d^2\psi}{dx^2} + \frac{2mB}{\hbar^2}(x - x_1)\psi = 0$$

But this, too, can be set in the form of equation (8.18) by setting

$$y = -\left(\frac{2mB}{\hbar^2}\right)^{1/3}(x - x_1)$$

The rest of the analysis is very similar, and the upshot is that in this case we find the oscillatory function in region 2 to be given by

$$\psi \approx 2p^{-1/2}\sin\left(\frac{1}{\hbar}\int_{x_1}^{x} p\,dx + \frac{\pi}{4}\right) \tag{8.23}$$

But the wave functions, equations (8.22) and (8.23), represent the same wave function in the same region; thus they must be identical within a constant C:

$$\sin\left(\frac{1}{\hbar}\int_{x_1}^{x} p\,dx + \frac{\pi}{4}\right) = C\sin\left(\frac{1}{\hbar}\int_{x}^{x_2} p\,dx + \frac{\pi}{4}\right)$$

Now

$$\int_{x_1}^{x} = \int_{x_1}^{x_2} - \int_{x}^{x_2}$$

Accordingly, we must have

$$\sin\left(\frac{1}{\hbar}\int_{x_1}^{x_2} p\,dx - \frac{1}{\hbar}\int_{x}^{x_2} p\,dx + \frac{\pi}{4}\right) = C\sin\left(\frac{1}{\hbar}\int_{x}^{x_2} p\,dx + \frac{\pi}{4}\right) \tag{8.24}$$

identically satisfied. This requires

$$\frac{1}{\hbar}\int_{x_1}^{x_2} p\,dx = \frac{1}{\hbar}\int_{x_1}^{x_2} \sqrt{2m(E - V(x))}\,dx = \left(n + \frac{1}{2}\right)\pi \tag{8.25}$$

where $n = 0,1,2,\ldots$ In order that the solutions connect smoothly at the turning points and satisfy the boundary conditions at $\pm\infty$, the allowed energies must satisfy the quantum condition, equation (8.25). Note that E appears in the integrand as well as in the limits of the integral since the turning points are those points at which the potential $V(x) = E$.

Additionally, the integral on the left-hand side in equation (8.25) measures the total change of phase of the WKB oscillatory wave function in region 2, between x_1 and x_2. It follows that there are altogether

$$\frac{(n + \frac{1}{2})\pi}{2\pi} = \frac{n}{2} + \frac{1}{4}$$

quasi wavelengths between x_1 and x_2. Hence n can be recognized as the number of nodes of the wave function.

Noting that the integration from x_1 to x_2 and back also takes the integrand through a complete cycle for classical motion, we can also write equation (8.25) in the form

$$\oint p\, dx = \left(n + \frac{1}{2}\right)h \tag{8.26}$$

For large n, when the classical approximation can be relied upon, the half integer can be ignored, and the quantum condition arrived at above is identical to the Sommerfeld-Wilson quantum condition.

The integral in equation (8.26) equals the area enclosed by the curve representing the classical motion in phase space (fig. 8.5) and is called the phase integral, denoted by J in classical mechanics. The area of phase space between two successive bound states is given by h. Hence the adage: Each quantum state occupies a volume h in phase space.

The constant C in equation (8.24) is also now fixed to be $(-1)^n$. In this way, except for normalization, we have arrived at the following WKB bound state wave functions for any slow-varying potential:

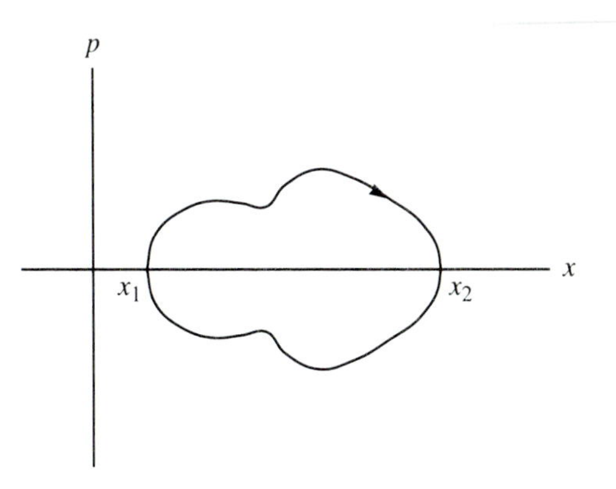

FIGURE 8.5
The periodic motion of a particle between two classical turning points, $x = x_1$ and $x = x_2$, is represented in the phase space.

$$\psi_{\text{WKB}} = (-1)^n |p|^{-1/2} \exp\left[-(1/\hbar)\int_x^{x_1} |p|\,dx\right] \qquad (x < x_1)$$

$$= (-1)^n 2p^{-1/2} \sin\left[1/\hbar \int_{x_1}^{x} p\,dx + \pi/4\right] \qquad (x_1 < x < x_2)$$

$$= |p|^{-1/2} \exp\left[-(1/\hbar)\int_{x_2}^{x} |p|\,dx\right] \qquad (x > x_2) \qquad (8.27)$$

A comment on the order of the connection procedure may be helpful. We have connected from region 3 into region 2 and again from region 1 into 2. But why not $1 \to 2 \to 3$, for example? Actually, we should, as we have, always make the connection in the direction of the decreasing (real) exponential. Because WKB is nothing but an approximation, there is a slight possibility that with any given solution, some of the other solution will be introduced. The rule "in from infinity" avoids this other solution being the growing (real) exponential (even small admixtures of which are dangerous).

Barrier Penetration

We will not treat the barrier penetration problem, because the formula one gets for the barrier transmission coefficient is the same that we finagled by generalizing the square barrier calculation (admittedly with some hand waving), equation (4.15); the only difference is that the WKB approximation enables us to properly identify the limits of integration in equation (4.15) as the classical turning points for the barrier. The reader can find the complete treatment in Powell and Crasemann, *Quantum Mechanics*.

■ **PROBLEM** Find the energy levels of a particle in the well $V(x) = V_0|x|$, $V_0 > 0$, using the WKB integral, equation (8.26).

SOLUTION Let's first consider the general case of $V(x) = V_0|x|^r$. The WKB integral gives upon substitution for the momentum p

$$\oint \sqrt{2m}\,\sqrt{E_n - V_0|x|^r}\,dx = \left(n + \frac{1}{2}\right)h$$

where we have written E_n instead of E to take the quantization of E into account. Since $E_n > 0$, we can write the integral above as

$$\sqrt{2mE_n}\,(E_n/V_0)^{1/r} \oint \sqrt{1 - |u|^r}\,du$$

where we have changed variables: $x = (E_n/V_0)^{1/r}u$. Thereupon we get the following expression for E_n:

$$E_n = \left[h\left(n+\frac{1}{2}\right)\right]^{2r/r+2}\left(\frac{V_0^{1/r}}{\sqrt{2m}\,I_r}\right)^{2r/r+2} \qquad \begin{array}{l} n = 0,1,2,\ldots \\ r > -2 \end{array}$$

where

$$I_r = \oint \sqrt{1-|u|^r}\,du$$

For $r = 1$, we have for the closed loop integral

$$I_1 = 2\int_{-1}^{1} \sqrt{1-|u|}\,du = 4\int_0^1 \sqrt{1-v}\,dv = \frac{8}{3}$$

Thereupon the quantized energy levels are given as

$$E_n = (n+\tfrac{1}{2})^{2/3}(V_0^2\hbar^2/m)^{1/3}(3\pi/4\sqrt{2})^{2/3}, \qquad n = 0,1,2,\ldots$$

■

PROBLEMS

1. Show that in the Heisenberg picture

$$\frac{d[A(t)]^2}{dt} = 2A(t)\frac{dA(t)}{dt} + \left[\frac{dA(t)}{dt}, A(t)\right]$$

2. Prove that

$$\frac{d[x(t)]^2}{dt} = 2x(t)\frac{dx(t)}{dt} + \frac{\hbar}{im}$$

3. Solve the Heisenberg equations of motion for x and p for the harmonic oscillator Hamiltonian directly (without using the a, $a^\dagger$ operators).
4. Use the Heisenberg equation of motion to solve for the time dependence of $x(t)$ given the Hamiltonian

$$H(t) = \frac{p^2(t)}{2m} + mgx(t)$$

5. Given the Hamiltonian

$$H = \frac{p^2}{2m} + \frac{1}{2}m(\omega_1^2 x^2 + \omega_2 x + \epsilon)$$

find and solve the equations describing the time dependence of $\langle x \rangle$ and $\langle p \rangle$.

6. Consider the quantum equation of motion of $\langle xp \rangle$. Show that for a stationary state $d\langle xp \rangle/dt = 0$. Hence show that

$$\langle p^2/m \rangle = \langle x dV/dx \rangle$$

7. Consider the Hamiltonian for the power law potential $V(x) = Cx^n$

$$H = \frac{p^2}{2m} + Cx^n$$

Now consider the equation of motion $\langle xp \rangle$ for this Hamiltonian and derive the quantum analog of the virial theorem:

$$\langle T \rangle = \frac{n}{2} \langle V \rangle$$

8. Solve the one-dimensional harmonic oscillator problem using the WKB method and compare with the solution of chapter 7.

9. Consider the applicability of the WKB method for the infinite potential well problem. Especially, work out the energy levels and compare with the exact quantum mechanical solution.

10. Find the WKB energy eigenvalues when $V(x) = Cx$ with $C > 0$ for $x > 0$ and $V(x) = \infty$ for $x < 0$. Compare with the exact solution given in chapter 4 in the limit of large E_n. (*Hint*: To find the zeroes of the Airy function for large negative argument use the asymptotic form, eq. [8.20].)

11. Determine the energy eigenvalues of a particle in the potential

$$V(x) = kx^4$$

using the WKB method.

ADDITIONAL PROBLEMS

A1. Estimate the energies of the ground and first excited states via the WKB method for the truncated harmonic oscillator potential:

$$\begin{aligned} V(x) &= \tfrac{1}{2}m\omega^2(x^2 - b^2), \quad && \text{for } |x| \leq b \\ &= 0, && \text{for } |x| \geq b \end{aligned}$$

Estimate the maximum value of b for which there is only one bound state.

A2. Use the WKB method to estimate the eigenvalues of the Hamiltonian for the displaced harmonic oscillator potential:

$$V(x) = \tfrac{1}{2}m\omega^2(x^2 + Cx).$$

Find an exact solution and compare with the WKB estimates.

REFERENCES

S. Gasiorowicz. *Quantum Physics.*
R. L. Liboff. *Quantum Mechanics.*
E. Merzbacher. *Quantum Mechanics.*
J. L. Powell and B. Crasemann. *Quantum Mechanics.*
D. Saxon. *Quantum Mechanics.*

9

Systems of Two Degrees of Freedom

So far, we have restricted ourselves to one-dimensional motion. The real world consists of three dimensions; to locate a system in real space takes three coordinates. It turns out, however, that it is useful to consider the motion of systems in two dimensions as a prelude to what comes with three dimensions. Most importantly, our treatment will lead to an introduction to angular momentum in quantum mechanics.

In actuality, real-world systems are also mostly *many-body* systems. And again, the consideration of the motion of two bodies in one dimension is a useful prelude to introduce the basic substance in the physics of many-body systems.

In short, in this chapter, we are going to talk about systems of two degrees of freedom, which includes the motion of one body in two dimensions, and the motion of two bodies in one dimension.

9.1 MOTION OF A SYSTEM IN TWO DIMENSIONS: FREE PARTICLES

Let's go over some basic stuff. We have two degrees of freedom; thus at least two measurements are needed to specify the state of a system. Accordingly, there must be at least two labels to specify the state. What are some of the appropriate labels? The answer to this question will lead to the answer to another—What is the form of the momentum operator in two dimensions? In some sense, the answer to the last question is intuitively obvious; however, going through the details of the answer reveals certain interesting points.

Consider a free particle. If we measure position, the eigenvalues x and y of the position operators $\hat{x}$ and $\hat{y}$, then the kets can be labeled by these eigenvalues as $|x,y\rangle$. They also form the base kets in two-dimensional position space. Similarly, if we measure momentum, the eigenvalues of the momentum operators $\hat{p}_x$ and $\hat{p}_y$ can be used to label the ket as $|p_x,p_y\rangle$ where p_x denotes the eigenvalue of $\hat{p}_x$, and so forth. These are momentum eigenkets. Can we start here and find the form of the momentum operators $\hat{p}_x$ and $\hat{p}_y$ in the position representation?

Let's first prove that the wave function of the free particle in position space, the bracket of the momentum eigenstate and the position base state, is given by

$$\psi(x,y) = \langle x,y|p_x,p_y\rangle = \frac{1}{2\pi\hbar}\, e^{i(p_x x+p_y y)/\hbar} \tag{9.1}$$

We can argue in the following way. We can always choose a coordinate system (x',y') such that the momentum of the particle lies along x'. In this coordinate system, the wave function $\langle x',y'|p,0\rangle$ is the probability amplitude of the event that the particle of momentum $\mathbf{p} = \mathbf{e}_{x'}p$ be found at x', where $\mathbf{e}$ denotes a unit vector. However, this is the same event that is described by the momentum wave function $\langle x'|p\rangle = C\exp(ipx'/\hbar)$ for one-dimensional motion (C is a constant). It follows that

$$\langle x',y'|p,0\rangle = C'e^{ipx'/\hbar} \tag{9.2}$$

where C' is a constant.

Now if we view this particle in the (x,y)-coordinate system (fig. 9.1), again the event described by the wave function $\langle x,y|p_x,p_y\rangle$ is the same as above. Thus

$$\langle x,y|p_x,p_y\rangle = \langle x',y'|p,0\rangle$$

All that remains to be done is to express px' in the exponent in equation (9.2) in terms of the (x,y)-coordinates. This is easily accomplished by realizing that px' is really the scalar product of two vectors $\mathbf{p}$ and $\mathbf{r}$:

$$px' = \mathbf{p}\cdot\mathbf{r} = p_x x + p_y y$$

Consequently, we get

$$\langle x,y|p_x,p_y\rangle = C'e^{i(p_x x+p_y y)/\hbar}$$

The constant C' has to be evaluated by the requirement of normalization in the two-dimensional (x,y) space. If we use the delta function normalization, we consider the scalar product $\langle p_x',p_y'|p_x,p_y\rangle$:

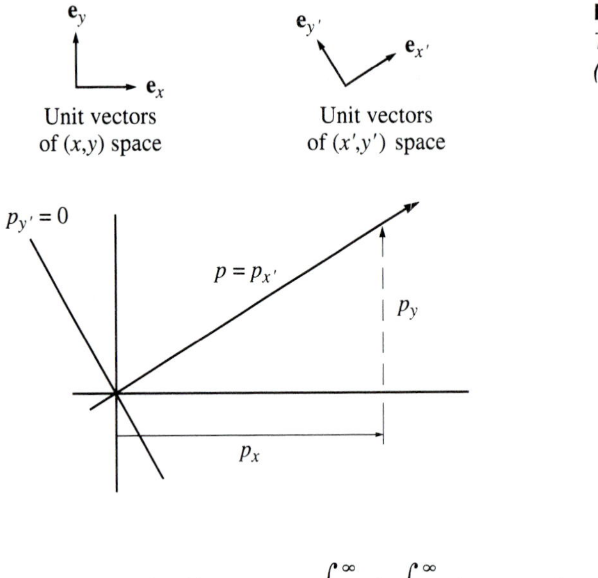

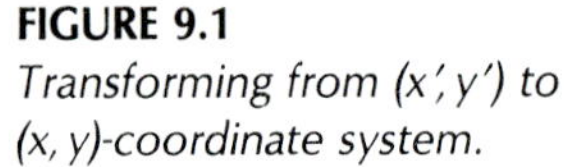

FIGURE 9.1
Transforming from (x', y') to (x, y)-coordinate system.

$$\langle p_x', p_y' | p_x, p_y \rangle = \int_{-\infty}^{\infty} dx \int_{-\infty}^{\infty} dy \, \langle p_x', p_y' | x, y \rangle \langle x, y | p_x, p_y \rangle$$
$$= (2\pi\hbar)^2 |C'|^2 \delta(p_x - p_x') \delta(p_y - p_y')$$

Thus C' must be equal to $1/2\pi\hbar$. Notice that we have used the ket-bra sum—as a two-dimensional integral in the (x, y)-space.

$$\int_{-\infty}^{\infty} \int_{-\infty}^{\infty} dx\, dy \, |x, y\rangle\langle x, y| = 1$$

In this way we see that equation (9.1) is the correct wave function to use for a free particle in two dimensions.

But $\psi(x, y)$ of equation (9.1) is the eigenfunction of the two operators $\hat{p}_x$ and $\hat{p}_y$ in position space belonging to the eigenvalues p_x and p_y. In order for this to be so, clearly we must have

$$\hat{p}_x = -i\hbar\partial/\partial x \qquad \hat{p}_y = -i\hbar\partial/\partial y \tag{9.3}$$

These are then the position-space representations of the momentum operators.

One element of caution is appropriate. From equation (9.1), we can see that

$$\psi(x, y) = \langle x, y | p_x, p_y \rangle = (1/2\pi\hbar)\exp[i(xp_x + yp_y)/\hbar]$$
$$= \langle x | p_x \rangle \langle y | p_y \rangle$$

However, it is very important to note that it is not always possible to achieve such a decomposition. Consider, for example, the description of the two-dimensional motion of a free particle in polar coordinates (ρ, ϕ) (fig. 9.2). Can we decompose $\langle \rho, \phi | p_x, p_y \rangle$? Since $x = \rho \cos \phi$, and $y = \rho \sin \phi$, we have

$$\begin{aligned} \langle \rho, \phi | p_x, p_y \rangle &= \int_{-\infty}^{\infty} \int_{-\infty}^{\infty} dx\, dy\, \langle \rho, \phi | x, y \rangle \langle x, y | p_x, p_y \rangle \\ &= \int_{-\infty}^{\infty} \int_{-\infty}^{\infty} dx\, dy\, \delta(x - \rho \cos \phi) \delta(y - \rho \sin \phi) \,. \\ &\quad \times (2\pi\hbar)^{-1} \exp[i(p_x x + p_y y)/\hbar] \end{aligned}$$

where we have used $\langle \rho, \phi | x, y \rangle = \delta(x - \rho \cos \phi) \delta(y - \rho \sin \phi)$. We get

$$\langle \rho, \phi | p_x, p_y \rangle = (2\pi\hbar)^{-1} \exp[i(p_x \cos \phi + p_y \sin \phi)\rho/\hbar]$$

And this cannot be decomposed as $\langle \rho | p_x \rangle \langle \phi | p_y \rangle$.

In general, then, the ket $|p_x, p_y\rangle$ *cannot* be decomposed as $|p_x\rangle |p_y\rangle$. We will denote the ket $|p_x, p_y\rangle$ by $|\mathbf{p}\rangle$, where $\mathbf{p}$ is the momentum vector in two di-

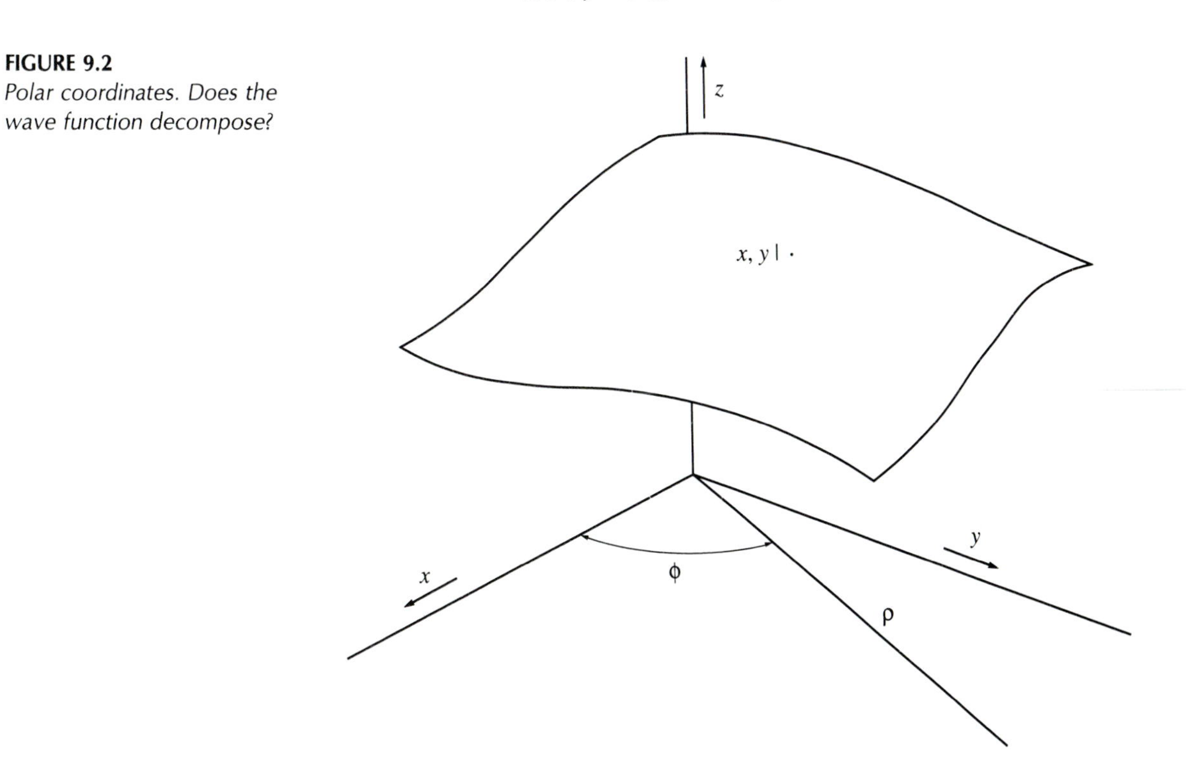

FIGURE 9.2
Polar coordinates. Does the wave function decompose?

mensions; and $|\mathbf{p}\rangle$ cannot be decomposed as $|p_x\rangle|p_y\rangle$. Similarly, the ket $|x, y\rangle$ will be denoted by $|\mathbf{r}\rangle$ ($\mathbf{r}$ is the position vector in two dimensions), which is not decomposable as $|x\rangle|y\rangle$. Only the bracket of $|\mathbf{p}\rangle$ with $\langle\mathbf{r}|$ may be decomposable in some coordinate systems.

Similarly, we introduce the notion of the vector operator $\mathbf{p}$ whose Cartesian components in two-dimensional position space are

$$p_x = -i\hbar\partial/\partial x \qquad p_y = -i\hbar\partial/\partial y$$

Thus, we will write

$$\mathbf{p} = p_x\mathbf{e}_x + p_y\mathbf{e}_y = -i\hbar\boldsymbol{\nabla} \tag{9.4}$$

where $\mathbf{e}_x$ and $\mathbf{e}_y$ are unit vectors and where we have introduced the del or grad operator $\boldsymbol{\nabla}$ in two dimensions:

$$\boldsymbol{\nabla} = \mathbf{e}_x\partial/\partial x + \mathbf{e}_y\partial/\partial y \tag{9.5}$$

Using these definitions, it is easy to generalize the commutation relation between $\hat{x}$ and $\hat{p}$, equation (3.5), to the two-dimensional case here. We have

$$[\hat{x}, \hat{p}_x] = [\hat{y}, \hat{p}_y] = i\hbar$$

Also note that $[\hat{x}, \hat{y}] = [\hat{p}_x, \hat{p}_y] = [\hat{x}, \hat{p}_y] = [\hat{y}, \hat{p}_x] = 0$.

The position representation of the Hamiltonian H in two dimensions can now be written as

$$H = \frac{\mathbf{p}^2}{2m} + V(\mathbf{r}) = -\hbar^2\boldsymbol{\nabla}^2/2m + V(\mathbf{r}) \tag{9.6}$$

where $\boldsymbol{\nabla}^2 = \partial^2/\partial x^2 + \partial^2/\partial y^2$. The Schrödinger equation is given as

$$H|\psi\rangle = E|\psi\rangle$$

Multiply by the base bra $\langle x, y|$ of the position representation from the left to obtain

$$\langle x, y|H|\psi\rangle = E\langle x, y|\psi\rangle$$

We now can use the position representation of the H operator, equation (9.6). Also put $\langle x, y|\psi\rangle = \psi(x, y) = \psi(\mathbf{r})$. Consequently, the Schrödinger equation in two dimensions is given as

$$i\hbar \frac{\partial \psi(\mathbf{r},t)}{\partial t} = H\psi(\mathbf{r},t) = \left[-\frac{\hbar^2}{2m} \nabla^2 + V(\mathbf{r}) \right] \psi(\mathbf{r},t) \tag{9.7}$$

for the time-dependent version; and

$$\nabla^2 \psi(\mathbf{r}) + \frac{2m}{\hbar^2} (E - V(\mathbf{r}))\psi(\mathbf{r}) = 0 \tag{9.8}$$

for the time-independent version.

We commented above that the minimum number of labels for the kets for two degrees of freedom is 2. If more labels are needed, that is a symptom of degeneracy. To see what kind of degeneracies arise, let's study the problem of the particle in a box in two dimensions.

9.2 PARTICLE IN A TWO-DIMENSIONAL SQUARE BOX

The potential $V(\mathbf{r}) = V(x,y)$ for a two-dimensional square box (fig. 9.3) is given as

$$\begin{aligned} V(x,y) &= 0 \qquad 0 \le x \le a \quad \text{and} \quad 0 \le y \le a \\ &= \infty \qquad x < 0,\, x > a \quad \text{or} \quad y < 0,\, y > a \end{aligned}$$

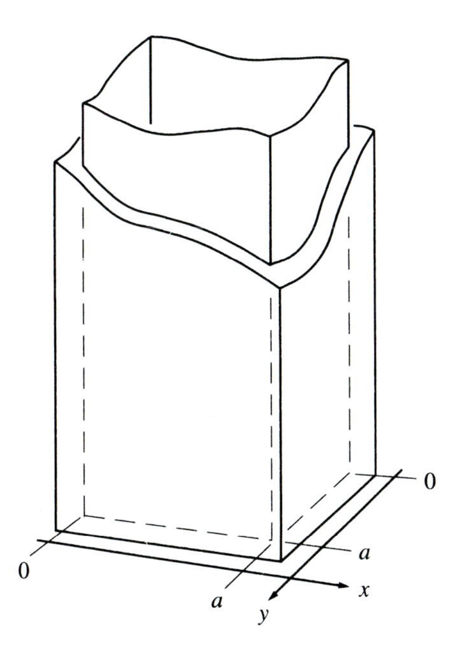

FIGURE 9.3
The two-dimensional square-box potential.

The Schrödinger equation inside the box is given as

$$\frac{\partial^2\psi(x,y)}{\partial x^2} + \frac{\partial^2\psi(x,y)}{\partial y^2} = -\frac{2mE}{\hbar^2}\,\psi(x,y) \tag{9.9}$$

Our job is to solve this equation with the boundary condition that the wave function vanishes at the walls of the box.

Since there are two variables, we will try the separation of variables technique. Put $\psi(x,y) = u(x)v(y)$. Upon insertion into equation (9.9) and dividing through by uv, we get

$$\frac{1}{u}\frac{d^2u}{dx^2} + \frac{1}{v}\frac{d^2v}{dy^2} = -\frac{2mE}{\hbar^2}$$

This equation can be satisfied only if each of the two terms on the left-hand side is separately a constant. That is,

$$\frac{d^2u}{dx^2} = -\frac{2mE^x}{\hbar^2}\,u$$

$$\frac{d^2v}{dy^2} = -\frac{2mE^y}{\hbar^2}\,v$$

Physically, you can think of E^x as the energy of the particle associated with its motion in the x-direction; E^y likewise has the significance of being the energy associated with the y-motion. And, of course, the total energy $E = E^x + E^y$. Now we have reduced the two-dimensional problem into two one-dimensional box problems with the same boundary conditions as we used in chapter 1. Therefore, we get for $\psi(x,y)$ and E

$$\psi_{n'n}(x,y) = u_{n'}(x)v_n(y) = \frac{2}{a}\sin\frac{n'\pi x}{a}\sin\frac{n\pi y}{a} \tag{9.10}$$

$$E_{n'n} = \frac{\pi^2\hbar^2}{2ma^2}(n'^2 + n^2) \tag{9.11}$$

Again we see the usual features as in the one-dimensional box: that the energy is quantized, that there is a zero-point energy, and all that. But there is a new feature, degeneracy. Now the allowed energy is the same when we interchange n' and n but the wave function is different

$$E_{nn'} = (\pi^2\hbar^2/2ma^2)(n'^2 + n^2)$$

$$\psi_{nn'} = (2/a)\sin(n\pi x/a)\sin(n'\pi y/a)$$

This is called a *symmetry* degeneracy, because one wave function can be changed into the other upon a reflection in the x-y plane or a rotation of the box by 90°.

Sometimes there arises a degeneracy even when there is no symmetry, when the box is not square. Suppose, for example, the x-length is a, and the y-length is $2a$. We put the origin in a corner as before. Now the allowed energies are

$$E_{n'n} = \frac{n'^2\hbar^2\pi^2}{2ma^2} + \frac{n^2\hbar^2\pi^2}{8ma^2}$$

We get the same energy eigenvalue for $n' = 2$, $n = 2$, and $n' = 1$, $n = 4$; this eigenvalue is doubly degenerate. This kind of degeneracy is called *accidental* degeneracy.

The Two-Dimensional Harmonic Oscillator

Now let's work out the eigenvalues and eigenfunctions for the problem of the two-dimensional harmonic oscillator in Cartesian coordinates. The general Hamiltonian is given as

$$H = \frac{p_x^2}{2m} + \frac{p_y^2}{2m} + \frac{1}{2}k_1x^2 + \frac{1}{2}k_2y^2 \tag{9.12}$$

In what follows we will assume that $k_1 = k_2 = k$—this case is called the *isotropic* harmonic oscillator.

The time-independent Schrödinger equation reads

$$\frac{\hbar^2}{2m}\left(\frac{\partial^2}{\partial x^2} + \frac{\partial^2}{\partial y^2}\right)u(x,y) + \left[E - \frac{1}{2}k(x^2+y^2)\right]u(x,y) = 0 \tag{9.13}$$

As in the case of the two-dimensional box, the Schrödinger equation is separable. Substitute $u(x,y) = u(x)v(y)$, divide through by $u(x)v(y)$, and rearrange terms in equation (9.13) to write it in the form

$$\frac{1}{u(x)}\left[\frac{\hbar^2}{2m}\frac{d^2}{dx^2} - \frac{1}{2}kx^2\right]u(x) + \frac{1}{v(y)}\left[\frac{\hbar^2}{2m}\frac{d^2}{dy^2} - \frac{1}{2}ky^2\right]v(y) = -E$$

The two terms on the left must now each separately be a constant. Call these constants E^x and E^y, as before. We have

$$\frac{d^2u(x)}{dx^2} + \frac{2m}{\hbar^2}\left(E^x - \frac{1}{2}kx^2\right)u(x) = 0$$

$$\frac{d^2v(y)}{dy^2} + \frac{2m}{\hbar^2}\left(E^y - \frac{1}{2}ky^2\right)v(y) = 0 \tag{9.14}$$

where $E^x + E^y = E$. We have gotten two one-dimensional oscillator equations in our hands for which we already know the solution! The eigenvalues of the two-dimensional oscillator are accordingly the sum of two one-dimensional oscillator eigenenergies and the eigenfunctions are the product of two one-dimensional eigenfunctions. Explicitly, the solution is

$$u_{n_1 n_2}(x, y) = u_{n_1}(x) u_{n_2}(y) \tag{9.15}$$

where u_{n_1} and u_{n_2} are one-dimensional oscillator eigenfunctions given by equation (7.18). And the eigenenergies are

$$E = (n_1 + n_2 + 1)\hbar\omega \tag{9.16}$$

with $\omega = (k/m)^{1/2}$.

Let's discuss degeneracy. There is no degeneracy for the ground state, $n_1 = 0$, $n_2 = 0$, and $E_0 = \hbar\omega$, but there is degeneracy in every other state. For the first excited state $E_1 = 2\hbar\omega$, but we have two combinations of n_1 and n_2 to achieve this, hence the degeneracy. The degenerate eigenfunctions corresponding to the energy $2\hbar\omega$ are given as

$$u_{01}(x, y) = u_0(x) u_1(y)$$

$$u_{10}(x, y) = u_1(x) u_0(y)$$

And as the excitation energy becomes greater, so does the degeneracy. For example, for the second excited state, $E_2 = 3\hbar\omega$, we have a threefold degeneracy corresponding to (1) $n_1 = 0$, $n_2 = 2$; (2) $n_1 = 1$, $n_2 = 1$; and (3) $n_1 = 2$, $n_2 = 0$. If we introduce a total quantum number $N = n_1 + n_2$, we can see that the energy depends only on N

$$E = (N + 1)\hbar\omega$$

and that there is an $(N + 1)$-fold degeneracy for each state.

9.3 PARTICLE CONFINED TO MOTION ON A RING: AN INTRODUCTION TO THE ANGULAR MOMENTUM OPERATOR

Consider a particle of mass m constrained to move on a circle of radius r that lies in the x-y plane. The Hamiltonian is given by just the kinetic energy; the potential energy by virtue of being a constant can be taken to be zero. Thus we have

$$H = p_x^2/2m + p_y^2/2m = (-\hbar^2/2m)(\partial^2/\partial x^2 + \partial^2/\partial y^2) \tag{9.17}$$

The Schrödinger equation for the problem is then given as

$$\left(\frac{\partial^2}{\partial x^2} + \frac{\partial^2}{\partial y^2}\right) \psi(x,y) + \frac{2mE}{\hbar^2} \psi(x,y) = 0 \tag{9.18}$$

But the problem has cylindrical symmetry; whenever there is a symmetry we should take advantage of it. In this case, we notice that in the cylindrical coordinates (ρ, ϕ), ρ is a constant; thus derivatives with respect to ρ vanish, and consequently in these coordinates, the Schrödinger equation will be reduced to one variable.

Coordinate Transformation

In cylindrical coordinates, $x = \rho \cos \phi$, $y = \rho \sin \phi$. We also have

$$\partial/\partial\rho = (\partial x/\partial\rho)\partial/\partial x + (\partial y/\partial\rho)\partial/\partial y$$

$$\partial/\partial\phi = (\partial x/\partial\phi)\partial/\partial x + (\partial y/\partial\phi)\partial/\partial y$$

Substituting for $\partial x/\partial\rho$ and the like, with a little manipulation we easily get

$$\partial^2/\partial\rho^2 + \frac{1}{\rho}\frac{\partial}{\partial\rho} + (1/\rho^2)\partial^2/\partial\phi^2 = \partial^2/\partial x^2 + \partial^2/\partial y^2$$

But in the present case ρ is a constant, hence $\partial/\partial\rho = \partial^2/\partial\rho^2 = 0$, and we get

$$\partial^2/\partial x^2 + \partial^2/\partial y^2 = (1/\rho^2)\partial^2/\partial\phi^2$$

In this way, the Schrödinger equation becomes

$$\frac{d^2\psi}{d\phi^2} + \frac{2IE}{\hbar^2}\psi = 0 \tag{9.19}$$

where I is the moment of inertia, $I = m\rho^2$. The solution of the Schrödinger equation (9.19) is standard:

$$\psi(\phi) = Ae^{im\phi} \tag{9.20}$$

where

$$m^2 = 2IE/\hbar^2 \tag{9.21}$$

and A is a constant to be evaluated from the normalization condition on $\psi(\phi)$.

The secret of discrete eigenvalues, as you know, is the boundary conditions. But in this problem, there is no wall, no range of the potential, not even any restriction on the wave function for large ϕ. However, there is the requirement that ψ be single-valued, which we will now invoke:

$$\psi(\phi + 2\pi) = \psi(\phi) \tag{9.22}$$

This condition is satisfied only if

$$e^{2\pi i m} = 1$$

In other words, $m = 0, \pm 1, \pm 2, \ldots$. Equation (9.21) now gives the allowed values of energy:

$$E = (\hbar^2/2I)m^2, \qquad m = 0, \pm 1, \pm 2, \ldots \tag{9.23}$$

The energy is quantized, as expected. The quantum number m is called the *azimuthal quantum number* (and also the *magnetic quantum number* for reasons that will be clear later; see chapter 13).

There is also a degeneracy. The energy depends on m^2; for any m, the two wave functions $\exp(im\phi)$ and $\exp(-im\phi)$ belong to the same energy eigenvalue. We can expect that there is an operator that commutes with the Hamiltonian that is responsible for this degeneracy. We will now show that the relevant operator is the *angular momentum* operator defined by

$$\hat{L}_z = \hat{x}\hat{p}_y - \hat{y}\hat{p}_x \tag{9.24}$$

where z is the axis perpendicular to the x-y plane of rotation. But $\hat{p}_x = -i\hbar\partial/\partial x$ and $\hat{p}_y = -i\hbar\partial/\partial y$. Therefore,

$$\hat{L}_z = -i\hbar(x\partial/\partial y - y\partial/\partial x) = -i\hbar\partial/\partial\phi \tag{9.25}$$

where the last step follows from our use of cylindrical coordinates. Now if L_z operates on the wave function, equation (9.20), we get

$$\hat{L}_z\psi(\phi) = -i\hbar(\partial/\partial\phi)Ae^{im\phi} = m\hbar Ae^{im\phi} \tag{9.26}$$

This is an eigenvalue-eigenfunction equation, and clearly the energy eigenfunction $A\exp(im\phi)$ is also an eigenfunction of the angular momentum operator $\hat{L}_z$ belonging to the eigenvalue $m\hbar$. Similarly, the eigenfunction $\exp(-im\phi)$ is an eigenfunction of $\hat{L}_z$ corresponding to the eigenvalue $-m\hbar$. The two wave functions correspond to equal and opposite values of the angular momentum, in other words, to an opposite sense of rotation.

Finally, let's normalize the eigenfunction, equation (9.20),

$$|A|^2 \int_0^{2\pi} d\phi\, e^{-im\phi}e^{im\phi} = 1 = 2\pi|A|^2$$

This gives $A = 1/(2\pi)^{1/2}$. In this way we find the normalized wave functions as

$$\psi_m(\phi) = (2\pi)^{-1/2}\exp(im\phi) \tag{9.27}$$

We note in passing that although we introduced the angular momentum through an operator relationship that is clearly borrowed from classical mechanics, the important difference between quantum and classical angular momentum is that the former is quantized. It can take on only specific values in contrast to the classical angular momentum, which can assume any value at all in a continuous fashion.

The Uncertainty Relation Between Angular Momentum and Rotation Angle

Since $L_z = -i\hbar\partial/\partial\phi$, the following commutation relation between L_z and ϕ is easily derived (cf. section 3.1):

$$[\phi, L_z] = i\hbar \tag{9.28}$$

It then follows from equation (6.23) that the following uncertainty relation is satisfied:

$$\Delta L_z \cdot \Delta\phi \geq \hbar/2 \tag{9.29}$$

However, although qualitatively correct and useful, this uncertainty relation enjoys only dubious popularity, since it is difficult to give physical meaning to uncertainties in ϕ that are greater than 2π.

9.4 TWO-PARTICLE SYSTEMS

Many real-world systems of interest to us are two-body systems, such as the hydrogen atom or the deuterium nucleus. If there is an interaction between the two particles, then we have to solve the Schrödinger equation in two variables even for one-dimensional motion, and this can be quite complicated. Fortunately, there is a quite versatile way to reduce the two-body problem to an equivalent one-body problem if the potential depends only on the separation of the two particles $x_1 - x_2$; that is, if $V(x_1, x_2) = V(x_1 - x_2)$.

The Hamiltonian is

$$H = \frac{p_1^2}{2m_1} + \frac{p_2^2}{2m_2} + V(x_1 - x_2) \tag{9.30}$$

The Schrödinger equation $H|\psi(1,2)\rangle = E|\psi(1,2)\rangle$ in the representation defined by the base kets $|x_1, x_2\rangle$ reads as

$$H\psi(x_1, x_2) = E\psi(x_1, x_2)$$

where $\langle x_1,x_2|\psi(1,2)\rangle = \psi(x_1,x_2)$. Now we can substitute $p_1 = -i\hbar\partial/\partial x_1$ and $p_2 = -i\hbar\partial/\partial x_2$:

$$\left(-\frac{\hbar^2}{2m_1}\frac{\partial^2}{\partial x_1^2} - \frac{\hbar^2}{2m_2}\frac{\partial^2}{\partial x_2^2}\right)\psi(x_1,x_2) + V(x_1-x_2)\psi(x_1,x_2) = E\psi(x_1,x_2) \quad (9.31)$$

Now introduce new variables x, X via the definitions

$$x = x_1 - x_2$$

$$X = \frac{m_1x_1 + m_2x_2}{m_1 + m_2} \quad (9.32)$$

You will recognize x as the relative coordinate and X as the center-of-mass coordinate of the two particles. Now introduce the *reduced mass* μ via the equation

$$\frac{1}{\mu} = \frac{1}{m_1} + \frac{1}{m_2} \quad (9.33)$$

Using the chain rules

$$\frac{\partial}{\partial x_1} = \frac{\partial}{\partial x}\frac{\partial x}{\partial x_1} + \frac{\partial}{\partial X}\frac{\partial X}{\partial x_1}$$

$$\frac{\partial}{\partial x_2} = \frac{\partial}{\partial x}\frac{\partial x}{\partial x_2} + \frac{\partial}{\partial X}\frac{\partial X}{\partial x_2}$$

after some algebra, the Schrödinger equation in this center-of-mass coordinate system is seen to be

$$\left(-\frac{\hbar^2}{2M}\frac{\partial^2}{\partial X^2} - \frac{\hbar^2}{2\mu}\frac{\partial^2}{\partial x^2} + V(x)\right)\psi(x,X) = E\psi(x,X) \quad (9.34)$$

where $M = m_1 + m_2$. Clearly the equation is separable. Write $\psi(x,X) = U(X)u(x)$. The equations for $U(X)$ and $u(x)$ are respectively given by

$$-\frac{\hbar^2}{2M}\frac{d^2U(X)}{dX^2} = E_{\rm cm}U(X) \quad (9.35)$$

$$-\frac{\hbar^2}{2\mu}\frac{d^2u(x)}{dx^2} + V(x)u(x) = E_{\rm rel}u(x) \quad (9.36)$$

where $E_{cm} + E_{rel} = E$. The first equation is a free-particle equation showing that the center of mass moves as a free particle in this case. The second equation is the desired one-body Schrödinger equation for the reduced mass.

If we call $|x, X\rangle$ the base kets of the Schrödinger representation for the center-of-mass coordinates x, X, then

$$\langle x, X | \psi(1,2)\rangle = u(x)U(X)$$

The mathematical convenience of this separability is obvious and has been stressed so far, but there is also a physical implication that goes quite deep. In this new representation, we can think of $u(x)$ and $U(X)$ being, respectively, the wave function of two new independent, noninteracting particles, one of reduced mass μ, the other of total mass M (i.e., $u(x) = \langle x|\psi(1_\mu)\rangle$ and $U(X) = \langle X|\psi(2_M)\rangle$). In comparison, in the original representation, $|x_1, x_2\rangle$, no such separation is possible. Physically, and this was first pointed out by Einstein, the two interacting particles become *correlated* and it is no longer possible to make a measurement on one without affecting the other; but this raises the old specter of action-at-a-distance (remember chapter 5?) so vividly that it is often referred to as the Einstein-Podolsky-Rosen (EPR) paradox. We will discuss the EPR paradox to the full extent it deserves in chapter 10.

Identical Particles and Exchange Degeneracy

In classical deterministic motion, we can distinguish between two identical billiard balls by following their individual deterministic trajectories. Alas, this is impossible in the quantum situation. In quantum mechanics, identical particles pose some radically new physics, because wave packets spread and overlap, and thus there is no way to keep track of each particle separately in a region where their wave packets overlap.

To examine how the new physics works, consider the case of two identical particles of mass m in a one-dimensional box with walls at $x = 0$ and $x = a$. Assume that the particles do not interact. Then the Hamiltonian of the system is given as

$$H = \frac{p_1^2}{2m} + \frac{p_2^2}{2m} + V(x_1) + V(x_2) \tag{9.37}$$

Here $V(x_1)$ and $V(x_2)$ are potentials relevant to a one-dimensional box; that is, the potential goes to zero inside the box and to infinity at the walls.

The Hamiltonian is the sum of two independent terms:

$$H = H(x_1) + H(x_2) \tag{9.38}$$

with

$$H(x_i) = \frac{p_i^2}{2m} + V(x_i) \tag{9.39}$$

where $i = 1,2$. The eigenvalues and eigenfunctions of $H(x_i)$ were worked out in chapter 1:

$$H(x_i)\psi_{n_i}(x_i) = E_{n_i}\psi_{n_i}(x_i)$$

where

$$E_n = \frac{n^2\pi^2\hbar^2}{2ma^2}$$

$$\psi_n = \sqrt{\frac{2}{a}}\sin\frac{n\pi x}{a}$$

Clearly, the solution of the two-body Schrödinger equation

$$H\psi(x_1,x_2) = E\psi(x_1,x_2)$$

partitions out as the product of two single-particle functions:

$$\psi_{n_1n_2}(x_1,x_2) = \psi_{n_1}(x_1)\psi_{n_2}(x_2) \tag{9.40}$$

And the eigenvalues are

$$\begin{aligned} E_{n_1n_2} &= E_{n_1} + E_{n_2} \\ &= \frac{n_1^2\pi^2\hbar^2}{2ma^2} + \frac{n_2^2\pi^2\hbar^2}{2ma^2} \end{aligned} \tag{9.41}$$

Now consider the eigenstate corresponding to $n_1 = 1$ and $n_2 = 2$. The eigenfunction is

$$\psi_{12}(x_1,x_2) = \frac{2}{a}\sin\frac{\pi x_1}{a}\sin\frac{2\pi x_2}{a}$$

with eigenenergy

$$E_{12} = 5\pi^2\hbar^2/2ma^2$$

But you can easily see that this eigenstate has the same energy as the one with eigenfunction

$$\psi_{12}(x_2,x_1) = \frac{2}{a}\sin\frac{\pi x_2}{a}\sin\frac{2\pi x_1}{a}$$

The eigenvalues are doubly degenerate. And this degeneracy exists for the two particles occupying any pair of states, n_1, n_2 ($n_1 \neq n_2$). What's the reason? From our previous experience, we should suspect a symmetry of some kind, some commuting operator lurking behind the scenes.

The Exchange Operator and Symmetric and Antisymmetric Wave Functions

We notice that the wave functions $\psi_{n_1 n_2}(x_1,x_2)$ and $\psi_{n_1 n_2}(x_2,x_1)$ above differ by an exchange of particles $1 \rightarrow 2$. Call the two-particle *exchange operator* P_{12}; then P_{12} operating on $\psi_{n_1 n_2}(x_1,x_2)$ converts it into $\psi_{n_1 n_2}(x_2,x_1)$:

$$P_{12}\psi_{n_1 n_2}(x_1,x_2) = \psi_{n_1 n_2}(x_2,x_1) \tag{9.42}$$

Also notice that the Hamiltonian H, equation (9.38), remains invariant under exchange of particles; in other words, $[H,P_{12}] = 0$. Thus P_{12} is the commuting operator we were looking for and is a constant of the motion.

We need to define simultaneous eigenfunctions of H and P_{12}. Since P_{12} operating twice on a wave function gives us back the original wave function, $P_{12}^2 = 1$, the eigenvalues of P_{12} are obviously ± 1. Thus there are two possibilities for the eigenfunctions. We can take the symmetric combination of $\psi(x_1,x_2)$ and $\psi(x_2,x_1)$:

$$\psi_{n_1 n_2}^{(s)} = \frac{1}{\sqrt{2}}\,[\psi_{n_1}(x_1)\psi_{n_2}(x_2) + \psi_{n_1}(x_2)\psi_{n_2}(x_1)] \tag{9.43}$$

which belongs to the eigenvalue $+1$ of P_{12} (the factor of $1/\sqrt{2}$ takes care of normalization); or we can take the antisymmetric combination

$$\psi_{n_1 n_2}^{(a)} = \frac{1}{\sqrt{2}}\,[\psi_{n_1}(x_1)\psi_{n_2}(x_2) - \psi_{n_1}(x_2)\psi_{n_2}(x_1)] \tag{9.44}$$

which belongs to the eigenvalue -1 of P_{12}, since it changes sign under the action of P_{12}.

What is most interesting in quantum mechanics—and indeed it is a law of the universe—is that symmetry and antisymmetry under particle exchange are innate characteristics of the particles and not anything that we can manipulate or arrange by suitable preparation or boundary conditions. The wave function of two electrons is always antisymmetric under exchange no matter how we prepare them, no matter how they spend their lives with what relationships and interactions. And there are other particles, such as the deuterium nuclei, that only exist in symmetric combination.

There is, in fact, a quantum number called spin, which is involved here. It is a kind of intrinsic angular momentum of a particle (we say "kind of" because, in truth, there is no known classical analog of spin), which we will introduce more formally later (see chapter 15). If the spin quantum number is a half integer for a particle (which is the case, for example, for electrons, protons, and neutrons), then the two-particle system must be described by antisymmetric wave functions. Such particles are called *fermions*. On the other hand, if the two-particle wave function exists only in the symmetric combination, the particles are of integer spin (such as deuteron above); such particles are called *bosons*.

In this way we see that in dealing with identical particles in quantum mechanics, we need to invoke a new degree of freedom, spin, for which no classical counterpart is known.

Notice in particular that the antisymmetric wave function, equation (9.44), vanishes for $x_1 = x_2$; thus the two particles, identical fermions in this case, cannot occupy the same place. Also, if $n_1 = n_2$, then $\psi^{(a)} = 0$; two identical fermions cannot occupy the same state. If one of them takes a given place or state to sit on, it excludes the other from that place or state. This is called the *Pauli exclusion principle*. A complete discussion of how the Pauli principle and antisymmetry operate needs the inclusion of the spin degree of freedom in the two-particle wave function, and we will take it up later. For bosons, instead of exclusion, we have *condensation*; bosons tend to condense together, all in the same ground state.

Finally, the symmetry or antisymmetry of the wave functions of identical particles, and especially their implications, such as the exclusion principle, should remind you of the global (nonlocal) nature of quantum systems. But does this mean that you need to worry about antisymmetrizing the wave function of an electron bound to a hydrogen atom on earth with one bound in an atom on Mars? Not really. If the wave functions of two identical particles never overlap appreciably, such as the electrons of two noninteracting H atoms right here on earth, we can assign independent regions for their respective Hamiltonians, and in effect can do without a two-particle antisymmetric formulation.

9.5 AN INTRODUCTION TO THE REAL-WORLD QUANTUM MECHANICS

As stated earlier, once you have crossed the hurdle of generalizing from one degree of freedom to two, further generalizations are straightforward. The vector notation for operators such as $\mathbf{p}$ and state vectors such as $|\mathbf{r}\rangle$ is easily generalized to three dimensions, precisely the reason for introducing such a notation. For the sake of completeness, let's write down the Schrödinger equation (in the complete time-dependent form) in the Cartesian coordinate representation:

$$i\hbar \frac{\partial \psi(\mathbf{r},t)}{\partial t} = -\frac{\hbar^2}{2m} \nabla^2 \psi(\mathbf{r},t) + V(\mathbf{r})\psi(\mathbf{r},t) \tag{9.45}$$

where the Laplacian operator ∇^2 is now given as

$$\nabla^2 = \frac{\partial^2}{\partial x^2} + \frac{\partial^2}{\partial y^2} + \frac{\partial^2}{\partial z^2} \tag{9.46}$$

Similarly, for N particles ψ becomes a function of all the $\mathbf{r}_i$, $i = 1, N$, $-\hbar^2/2m\nabla^2$ is replaced by the sum of all the $-\hbar^2/2m_i\nabla_i^2$, $i = 1, N$, and V is the sum of all the two-body interactions among the N particles:

$$i\hbar \frac{\partial\psi(\mathbf{r}_1,\mathbf{r}_2,\ldots)}{\partial t} = -\sum_i \frac{\hbar^2}{2m_i}\nabla_i^2\psi + \sum_{i>j}\sum_j V(\mathbf{r}_i,\mathbf{r}_j)\psi \tag{9.47}$$

But before we go on the adventurous path of using quantum mechanics to solve real-world problems, starting with the simple hydrogen atom all the way to complex systems such as solids, we will consider in chapter 10 some of the paradoxes that quantum mechanics poses. One of the paradoxes is the EPR paradox mentioned earlier; the other, Schrödinger's cat, goes to the heart of the problem we are faced with when we measure a quantum system. And both paradoxes raise the question, Is quantum mechanics a complete and universal theory?

PROBLEMS

1. Consider the problem of a two-dimensional harmonic oscillator in Cartesian coordinates. The Hamiltonian is given as

$$H = \frac{p_x^2}{2m} + \frac{p_y^2}{2m} + \frac{1}{2}k(x^2 + y^2)$$

(a) Write H and L_z in terms of the creation and annihilation operators in two dimensions, a_x, $a_x^\dagger$, a_y, $a_y^\dagger$. Show, using the commutation relation of the a and $a^\dagger$ operators, that L_z commutes with H. (b) Consider the two eigenstates $|u_{01}\rangle$ and $|u_{10}\rangle$ belonging to the first excited state of the oscillator ($N = 1$). Are these eigenstates also of L_z? Demonstrate your answer using the form of L_z you derived in part (a) in terms of $a, a^\dagger$ operators. Reconcile your answer with the conclusion reached in part (a).

2. Consider a particle confined to a circular box. Show that the Schrödinger equation is separable in cylindrical coordinates ρ, ϕ and find and solve the equations for the radial and the angular motion. (*Hint:* The radial equation is the Bessel equation and its solutions are called Bessel functions; see any book on special functions of mathematical physics.)

3. Calculate the rotational energy levels of a phonograph record (rotating freely in a plane) of radius 10 cm and mass 150 g. (*Hint:* This is the same problem as a particle rotating in a circle. Why?) To what value of m does 33 RPM correspond?
4. How long can an ordinary lead pencil be balanced on its point? Use the angular momentum-angle uncertainty relation and assume that the pencil is best balanced when the uncertainty in the kinetic energy of the pencil is of the same order as the uncertainty in its potential energy.
5. What is the reduced mass of an electron-proton system (as in the H atom)? What is the reduced mass of a neutron-proton system (as in the deuteron)? What is the reduced mass for a system of two identical particles?
6. Consider two coupled oscillators. The Hamiltonian is given as

$$H = \frac{p_1^2}{2m} + \frac{p_2^2}{2m} + \frac{1}{2}m\omega^2[x_1^2 + x_2^2 + 2\lambda(x_1 - x_2)^2]$$

   Separate the center of mass and relative motion and find the eigenfunctions and eigenvalues.
7. Prove that the exchange operator P_{12} is hermitian.
8. Separate the variables of the Schrödinger equation for a particle moving in a cubic box of dimension a; that is, $\psi(x,y,z) = u(x)v(y)w(z)$. Determine the energy eigenvalues and eigenfunctions. Discuss the degeneracies of the system.
9. Find the eigenfunctions and eigenvalues of the three-dimensional harmonic oscillator using the Cartesian coordinate system. Discuss the degeneracy of the states.

ADDITIONAL PROBLEMS

A1. Find the wave functions for the movement of a particle in an impenetrable box in three dimensions of length l_1, width l_2, and height l_3. (*Hint*: Try a product solution of terms such as that of eq. [1.19], one for each space coordinate.) Discuss the degeneracy of the first three wave functions for the case $l_1 = l_2 = l_3$.

A2. In connection with the difficulties of defining an uncertainty relation with the azimuthal angle ϕ, it has been pointed out (by W. Louisell, *Physics Letters*, 7, 60 (1963)) that $\sin\phi$ and $\cos\phi$ are better angle variables. Evaluate the commutators $[\sin\phi, L_z]$ and $[\cos\phi, L_z]$. Using these commutators, find the uncertainty relations between $\sin\phi$ and L_z and $\cos\phi$ and L_z.

A3. Suppose two identical quantum beads move on a circular thread. Assuming no interaction between the objects, solve the Schrödinger equation for eigenstates and eigenvalues. Is there any degeneracy?

A4. Show that the antisymmetric wave function of two identical particles can be written as a determinant.

A5. At $t = 0$, the wave function of a two particle system of unequal mass m_1 and m_2, respectively, in a one-dimensional impenetrable box is given as

$$\psi(x_1,x_2,0) = (1/\sqrt{5})\,[2u_1(x_1)u_3(x_2) + u_5(x_1)u_7(x_2)]$$

(a) Suppose we measure the energy of this system, what values will we find? with what probability?

(b) Let the measurement give the value E_{13}. What is the wave function of the system at time t in that case?

A6. What is the ground-state wave function for two identical particles in a one-dimensional box if the two particles are (a) fermions (b) bosons?

A7. Two identical particles, each of mass m, move in one dimension in the potential

$$V(x_1,x_2) = \tfrac{1}{2}A\,(x_1^2 + x_2^2) + \tfrac{1}{2}B(x_1 - x_2)^2$$

where A and B are positive constants and x_1 and x_2 denote the positions of the particles.

(a) Show that the Schrödinger equation is separable in the variables $x_1 + x_2$ and $x_1 - x_2$. Find the eigenvalues and the corresponding eigenfunctions.

(b) Discuss the symmetry of the eigenfunctions with respect to particle exchange.

REFERENCES

P. W. Atkins. *Molecular Quantum Mechanics.*
M. Chester. *Primer of Quantum Mechanics.*
C. Cohen-Tannoudji, B. Diu, and F. Laloe. *Quantum Mechanics.*
R. L. Liboff. *Quantum Mechanics.*

10

Quantum Paradoxes and the Copenhagen Interpretation

Before we begin our adventure in real-world quantum mechanics, let's take a break, a short detour into the question of the meaning and interpretation of quantum mechanics.

Why does quantum mechanics require an interpretation at all? Why does it take philosophy to understand it? Why can it not speak for itself? Let's count the reasons.

1. The state of a quantum system is determined by the Schrödinger equation, but the solution of the Schrödinger equation, the wave function, is not directly related to anything we observe. Thus the first question of interpretation is, What does the wave function represent? A single object? A group of similar events? An ensemble of objects? The square of the wave function determines probabilities, but how should we interpret the probabilities? This calls for interpretation.
2. Quantum objects are governed by the Heisenberg uncertainty principle—that it is impossible to simultaneously measure pairs of conjugate variables such as position and momentum. But is this purely an epistemological question or are there ontological aspects lurking behind it? (A quick reminder: Epistemology refers to how we know things; ontology refers to the being of things.) Again this is a question of interpretation and philosophy.
3. The paradox of wave-particle duality, that quantum objects have both wave and particle aspects, needs a resolution—which means interpretation and philosophy.
4. What is the ontological meaning of the coherent superposition?

5. Are discontinuity and quantum jumps truly fundamental aspects of the behavior of quantum systems? In particular, we have portrayed the collapse of a wave function or a coherent superposition in a measurement situation as a discontinuous event. But is collapse necessary?
6. The correspondence principle, due to Bohr, affirms that under certain conditions (for example, in the case of very closely spaced energy levels) quantum mechanical predictions reduce to those of classical mechanics. This guarantees that we can use classical mechanics for making predictions about macroobjects in most situations, but does it ensure the use of classicality for measurement apparatuses?
7. How should we interpret the meaning of quantum nonlocality? This has grave repercussions for philosophy.

In the preceding chapters, we have been implicitly assuming the standard interpretation, formally called the Copenhagen interpretation, developed principally by Bohr and Heisenberg. We have seen this interpretation deal quite adequately with items 1–3 above, but can it also deal successfully with the rest? This question has been debated for the past 60 years, but there is no consensus yet. There were two paradoxes raised in the thirties, one by Einstein and his collaborators, Podolsky and Rosen (the EPR paradox), and another by Schrödinger (the paradox of Schrödinger's cat), that cast grave doubts about the Copenhagen interpretation being the final answer to the meaning of quantum mechanics. The discussion of these paradoxes will constitute a major part of this chapter.

In chapter 5, we discussed the Bohr-Einstein debate. If that held your attention, the debate about this chapter's paradoxes is sure to mesmerize you. The debate is between what is called a realist position and its antirealist opposition. These are philosophical terms and let me give you a quick recapitulation of the relevant philosophies.

Realism. This philosophy holds that the fundamental elements of reality are independent of consciousness—this is the doctrine of *strong objectivity.* A tree in the forest is real even when it is not being perceived, the moon continues in its space-time orbit even when nobody is looking, and so forth. The doctrine of strong objectivity is further augmented by another doctrine—causal determinism—that the course of the universe is determined once for all by the initial conditions. There are many different subphilosophies within this basic realist view and I will mention only the one most often invoked in the discussion of quantum paradoxes: *Physical realism* considers matter (and its extensions, energy, and fields) to be the only fundamental reality. All else, including consciousness, are secondary phenomena and are ultimately reducible to matter. Furthermore, since the only reality is that defined by space-time, the doctrine of locality is held fundamental.

Idealism. This philosophy holds that the fundamental elements of reality must include the mind (or consciousness). Within this broad category I will mention one subcategory that will be important for our discussion in this chapter: *Dualism* considers mind (or consciousness) and body to be separate worlds, both having primary importance. (There is also *monistic idealism*, which considers consciousness as the primary phenomenon and matter secondary within it, but we will not use this philosophy in this chapter.)

Although diametrically opposite in their basic premises, both realists and idealists agree that it is possible to speak about reality; in fact, it is imperative since questions of the nature of reality are the most meaningful questions that face us. But in a third philosophy one refuses to talk about *reality*.

Logical positivism. This philosophy is summarized by Ludwig Wittgenstein's dictum, "Whereof one cannot speak, thereof one must remain silent." Or as John Wheeler puts it in the context of quantum mechanics, "no elementary phenomenon is a phenomenon until it is a registered [observed] phenomenon." (In the United States, this philosophy is sometimes called phenomenalism—one can only talk about phenomena and the relations among them. Phenomenalists define the task of physics as to give "definite prescriptions for *successfully* foretelling the results of observations.") Historically, logical positivism is important because it influenced both Bohr and Heisenberg, the proponents of the Copenhagen interpretation.

10.1 THE EPR PARADOX

Consider two noncommuting observables A and B: $[A,B] \neq 0$. Then for an eigenstate of A, according to quantum mechanics, B has no definite value, and only the probability of finding a particular value of B can be given. The mainstay of the argument of EPR is that if this is the case, then either (1) both A and B cannot have physical reality, or (2) quantum mechanics is incomplete. EPR then proceeds to prove that the first alternative leads to contradiction and attempts to conclude that we must accept the second alternative. The proof runs as follows.

Suppose two quantum objects come together, interact for a short time, and then stop interacting. After their interaction, the state of the two systems is given by $\psi(\alpha_1,\alpha_2)$ where α_1 and α_2 are the variables of the objects 1 and 2, respectively. Also suppose that A is an observable physical quantity pertaining to particle 1 with eigenstate $u_n(\alpha_1)$ belonging to the eigenvalue a_n; that is,

$$Au_n(\alpha_1) = a_n u_n(\alpha_1)$$

We now can expand $\psi(\alpha_1,\alpha_2)$ in terms of the complete set of eigenfunctions of A and write

$$\psi(\alpha_1,\alpha_2) = \sum_n \phi_n(\alpha_2)u_n(\alpha_1) \tag{10.1}$$

where $\phi_n(\alpha_2)$ are the coefficients of the expansion. If we measure A, obtaining the eigenvalue a_i for it, then we must conclude that as a result of measurement system 1 has been left in the state given by $u_i(\alpha_1)$. From equation (10.1), system 2 must be left in the state represented by $\phi_i(\alpha_2)$. In other words, the effect of the measurement is the reduction of the coherent superposition represented by equation (10.1) to the single term $\phi_i(\alpha_2)u_i(\alpha_1)$. All this you already know.

And again, instead of A, we can take another observable B with eigenfunctions $v_m(\alpha_1)$ with corresponding eigenvalues b_m, and we get

$$\psi(\alpha_1, \alpha_2) = \sum_m \psi_m(\alpha_2) v_m(\alpha_1) \tag{10.2}$$

where we denote the new expansion coefficients by $\psi_m(\alpha_2)$. If we measure B on particle 1 and find the eigenvalue b_j for its measured value, we similarly conclude that the coherent superposition, equation (10.2), has been reduced by the measurement to the single term $\psi_j(\alpha_2)v_j(\alpha_1)$.

Clearly, with two different measurements carried out on system 1, system 2 is left in states corresponding to two different wave functions. But, says EPR, since the two objects are not interacting, a measurement of system 1 cannot affect the *reality* of system 2. The conclusion of EPR is this. It is possible to assign two different wave functions (ϕ_i and ψ_j above) to the same reality. This is fine if the two wave functions are compatible; that is, they correspond to compatible (commuting) observables. But what if we can show that ϕ_i and ψ_j can correspond to noncommuting observables?

Consider. Let the objects 1 and 2 be particles in one-dimensional motion—the variables α_1 and α_2 are now just x_1 and x_2, the positions of the two particles. And suppose we have

$$\psi(x_1, x_2) = \int_{-\infty}^{\infty} dp\, e^{(i/\hbar)(x_1 - x_2 + x_0)p} \tag{10.3}$$

where x_0 is a constant. Choose the observable A to be the momentum of object 1; its eigenfunctions are

$$u_p(x_1) = \exp(ipx_1/\hbar)$$

We can write equation (10.3) in the form of a complete set expansion (remembering that for a continuous spectrum the Σ in eq. [10.1] must be replaced by an integral):

$$\psi(x_1, x_2) = \int_{-\infty}^{\infty} dp\, \phi_p(x_2) u_p(x_1)$$

with ϕ_p being given as

$$\phi_p(x_2) = e^{(i/\hbar)(x_0-x_2)p}$$

This is an eigenfunction of the momentum operator for particle 2, namely $-i\hbar\partial/\partial x_2$, belonging to the eigenvalue $-p$.

Next, let the observable B be the position of particle 1, x_1, with δ-function eigenfunctions

$$v_x(x_1) = \delta(x - x_1)$$

where x is the eigenvalue. We now can write equation (10.3) as

$$\psi(x_1,x_2) = \int_{-\infty}^{\infty} dx\, \psi_x(x_2)v_x(x_1)$$

where for $\psi_x(x_2)$ we now have

$$\psi_x(x_2) = \int_{-\infty}^{\infty} dp\, e^{(i/\hbar)(x-x_2+x_0)p} = 2\pi\hbar\delta(x_2 - x - x_0)$$

But this is the eigenfunction of the position operator x_2 of particle 2 with eigenvalue $x + x_0$.

In this way EPR showed that it is possible for ϕ_i and ψ_j above to correspond to eigenfunctions of two noncommuting operators, namely, the momentum and position operators of particle 2. If this is so, by measuring the momentum or the position of particle 1, we can predict with certainty, and without disturbing particle 2 (since it is not interacting with particle 1 during the measurement), the momentum or position of particle 2. According to EPR's criterion of reality, we can say that both the position and the momentum of particle 2 are then *real.* But this contradicts quantum mechanics; according to quantum mechanics, two physical quantities that correspond to noncommuting operators cannot simultaneously have physical reality, have definite measurable values. Behold! says EPR; the contradiction can be avoided only if we agree that the quantum mechanical description of nature is incomplete.

We cannot, EPR emphasizes, get out of the dilemma by insisting that since we are only able to predict either the momentum or the position of particle 2, but not both simultaneously, by making measurements on particle 1, maybe the momentum and position of particle 2 are not simultaneously real. This would be unreasonable, because it would imply that the reality of the position and momentum of particle 2 depends on the process of measurement on particle 1, at a distance, which should not have any disturbing effect on particle 2 (since there is no interaction between the two particles).

In classical physics, any quantity that can be predicted or measured forms an element of reality. Thus the position and momentum of an object would be elements of the physical reality under all circumstances. In quantum mechanics, one may be tempted to take a strict epistemological view that things become real only upon measurement (denying any reality of any ontological description before measurement; this philosophical position is called logical positivism—"no elementary phenomenon is a phenomenon until it is a registered phenomenon"); only when an object is in an eigenstate can the value of its measured attribute be predicted with certainty. Accordingly, only momentum or position can be regarded as an element of reality at a given time, never both. But what about an ontological description of quantum objects in between measurements when the normal Schrödinger evolution makes them into coherent superpositions? EPR is taking an intermediate position between the two extremes—classical ontology and quantum epistemology. For EPR, any physical quantity that can be predicted without disturbing the system in any way is an element of physical reality. Consequently, EPR's definition of reality is more restrictive than that of classical physics, yet EPR challenges the pure epistemological view of quantum mechanics on the ontological question. The pertinent issue is separability: Are quantum objects separate when they have no local interaction between them (as classical objects certainly are)?

Why is EPR's result considered a paradox? Philosophically speaking, the Einsteinian separability is part and parcel of the philosophy of physical realism, which Einstein defended throughout his later life. This is the philosophy that considers physical objects to be real independent of each other and of their measurement or observation (the doctrine of strong objectivity). But in quantum mechanics, as you have seen in chapter 5, the idea of the reality of physical objects independent of our measurement of them is difficult to uphold. Thus EPR's motive was to discredit quantum mechanics and reestablish physical realism as the undergirding philosophy of physics. EPR's result is a paradox from the point of view of physical realism; it seems to say that we have to choose between locality or separability and the completeness of quantum mechanics, and this is no choice at all since separability is imperative.

But is it? Indeed, the resolution of the EPR paradox is to realize an essential *inseparability* of quantum objects; measurement of one of two correlated particles affects its (EPR-)correlated partner (this was essentially Bohr's answer to EPR). When one object is collapsed in a state of momentum p, the other's wave function is collapsed also (in the state of momentum $-p$); and when one object is collapsed by position measurement at x, the other's wave function also collapses immediately to correspond to the position $x + x_0$. The collapse is nonlocal, just as the correlation is nonlocal. EPR-correlated objects have a nonlocal ontological connection (inseparability) with a signalless instantaneous influence upon each other, as hard as it is to believe from the point of view of physical realism and the locality principle. Separability is the result of collapse; only after collapse are there independent objects. Thus EPR forces us to admit that quantum ontology, if one attempts to give an ontological description of quantum objects such as Heisenberg's potentia that we have adapted, must be a nonlocal ontology. In other words, Heisenberg's potentia defines a nonlocal domain of reality outside local space-time, outside of the jurisdiction of Einsteinian speed limits.

The Copenhagen Interpretation

Neither Bohr nor Heisenberg ever laid down the Copenhagen interpretation in detail. Thus what comprises the Copenhagen interpretation varies somewhat from author to author. For future reference, we will consider the following five tenets to comprise the Copenhagen interpretation of quantum mechanics:

1. The state of a quantum system is determined by the solution of the Schrödinger equation, the wave function. The square of the wave function determines the probability of finding a certain result, the average of a group of similar events.
2. Quantum objects are governed by the Heisenberg uncertainty principle—that it is impossible to simultaneously measure pairs of conjugate variables such as position and momentum.
3. The complementarity principle of Bohr asserts that quantum objects have complementary wave and particle aspects, and only one of these aspects can be measured with any given experimental arrangement.
4. Discontinuity and quantum jumps are a fundamental aspect of the behavior of quantum systems. For example, a measurement leads to a discontinuous collapse of the wave function of a system from a coherent superposition to an eigenstate of the observable being measured. The coherent superposition is a state of multifaceted possibilities in potentia; the collapse is the actualization of one of these possibilities. And the collapse is nonlocal.
5. Inseparability—quantum systems cannot be unambiguously separated from their measurement process. Furthermore, for EPR-correlated objects, there is an essential inseparability between the objects so that the measurement of one also collapses the wave function of the other.

Heisenberg also said that the wave function represents not the real system but our knowledge of the system. Thus, according to Heisenberg, the collapse of the wave function is not a real physical event, but represents a change in our knowledge of the system as a result of our measurement. This is a little vague, and will not be considered as part of the standard Copenhagen interpretation. However, Heisenberg's idea may very well be visionary, and we will return to it in chapter 24.

Hidden Variables

It is the inseparability aspect of the Copenhagen interpretation that causes the most consternation for the realist. Ultimately, all measuring apparatuses refer to a subject doing the measurement; thus in the Copenhagen interpretation, a mixing of subjects and objects is threatened and this compromises the strong objectivity of physical realism. Moreover, nonlocal collapse is unpalatable to realists (that included Einstein, as you have seen), and we must consider the possibility of alternative *realistic* interpretations of quantum mechanics. One such interpretation is based on the idea of *hidden variables*—hidden unknown parameters that should provide a realistic ontological description of quantum objects, trajectories and all.

Can such a hidden-variable theory be constructed? The answer is yes, according to Bohm. We will not go into Bohm's construction here, but an important twist has come from the work of Bell, who has shown that hidden-variable theories, in order to be compatible with quantum mechanics, must themselves be nonlocal. We will take up Bell's work in chapter 17; however, the full import of the EPR work should now be clear. Any ontological description of quantum objects, with or without hidden variables, will have to include nonlocality. The bell tolls clearly—local hidden variables cannot save physical realism.

Bell's work has also led to the experimental verification of EPR correlation of quantum objects (see chapter 17).

10.2 THE PARADOX OF SCHRÖDINGER'S CAT

A fundamental aspect of the postulates of quantum mechanics is that the state of a system is often given by a coherent superposition. However, we never see a coherent superposition! The reduction postulate says that the effect of a quantum measurement is to reduce the state vector of a system from a coherent superposition to an eigenstate. Thus whenever we look at a system, we look at one of its eigenstates. Yet when we are not looking, we admit that the state of the system can develop into a coherent superposition. What is the ontological meaning of a coherent superposition? What constitutes "looking" that it can do the magic of reduction of the state vector? These questions were raised in a very clever manner by the paradox of Schrödinger's cat.

Suppose we put a cat in a cage with a radioactive atom, a Geiger counter, a hammer, and a poison bottle; further suppose that the atom in the cage has a half-life of 1 hour, a fifty-fifty chance of decaying within the hour. If the atom decays, the Geiger counter will tick; the triggering of the counter will activate the hammer, which will break the poison bottle, which will kill the cat. If the atom doesn't decay, none of the above things happens, and the cat will be alive. Now the question, What is the state of the cat after the hour?

If you think classically, you will make a mental analogy of the situation with a coin that somebody has flipped, but which is hidden under the flipper's palm, and you don't know if it is heads or tails. But of course, you do know it is either heads or tails, the cat is either dead or alive with a 50% chance for each outcome, you just don't know which.

But quantum mechanics forces you to think differently. How does quantum mechanics describe the state of the cat after the hour? Literally, as a half-alive and half-dead cat, as a coherent superposition (fig. 10.1),

$$|\text{Cat}\rangle = (1/\sqrt{2})\,[|\text{live cat}\rangle + e^{i\theta}\,|\text{dead cat}\rangle] \tag{10.4}$$

where θ is an arbitrary phase. What is the meaning of such a diabolical dichotomy?

Of course, nobody has actually *seen* a coherent superposition. Indeed, if we make an observation, the cat is found either alive or dead. And then the question arises, What's so special about our making an observation that resolves the dichotomy?

The Copenhagen interpretation explains this paradox up to a point. For a start, we can use the complementarity principle. The coherent superposition is an abstraction; as an abstraction, the cat is able to exist as both live and dead. This is a complementary description, complementary to the dead or alive description that we give when we do see the cat. Moreover, following Heisenberg, we can say that the state, equation (10.4), describes a cat in potentia, in the realm of possibilities; only observation manifests the alive or dead actuality of the cat.

But even so, further paradoxical questions remain: Who or what determines the outcome when an observation is made, when the cage of the cat is opened? The measuring apparatus for the atom, such as the Geiger counter? the cat? or is it the human observer? As Einstein would have asked, Who plays dice with the fate of the cat? When, at what stage of measurement, does the wave

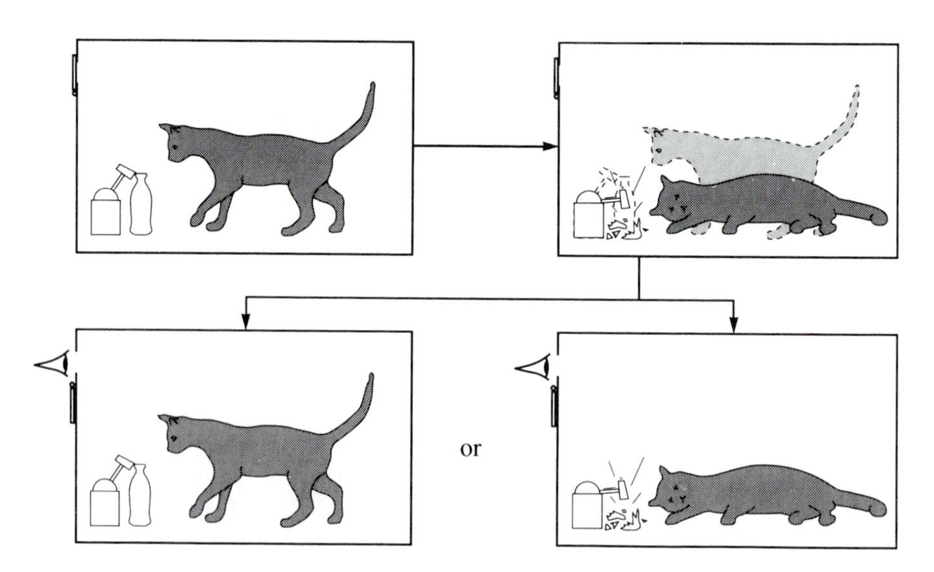

FIGURE 10.1
The paradox of Schrödinger's cat. After an hour, the cat's state becomes a coherent superposition (upper right). Observation collapses either a live cat (shown on the left) or a dead cat (on the right).

function collapse; that is, what constitutes an observation? These questions will be taken up a little later.

The Ensemble or Statistical Interpretation

One way out of the predicament you already know—it is to say that the mathematics of quantum mechanics, the prediction of coherent superpositions, as above, must not be taken literally. Instead, we can opt for the ensemble or statistical interpretation (see also chapter 5) that quantum mechanics only makes predictions about experiments involving ensembles.

Technically, we can distinguish between what is called a *pure state* and a *mixture*. A pure state of an ensemble occurs when every element of an ensemble of objects is in the same coherent superposition. Pure state is the quantum mechanical description of an ensemble, according to the standard interpretation, which also says that it takes a collapse, a reduction of the state vector, to discontinuously transform a pure state into a mixture of objects in various eigenstates. The actual composition of the mixture, upon measurement, depends on the statistical weights of the various eigenstates in the original coherent superposition.

Now here is what the statistical interpretation says. Suppose an ensemble is always a mixture; quantum mechanics just determines the probabilities with which the various eigenstates occur in the mixture. The advantage of such an assumption is obvious: There is no longer any need for collapse. The disadvantages should be equally obvious: Even the double-slit experiment is difficult to explain without the coherent superposition.

Central to the statistical interpretation is the idea that for a single object, an electron or a cat, the quantum theory just does not apply. On the other hand, isn't this tantamount to giving up on finding a physical theory for the description of a single object or single events? But single events do occur, and now even single electrons have been isolated (in the so-called Penning trap). So a resolution for the paradox of quantum measurement of a single quantum process, as raised in the example of Schrödinger's cat, does need to be worked out. Hence the implication of the statistical interpretation is that quantum mechanics must be supplemented by some sort of hidden-variable theory for the single object.

The Many-Worlds Interpretation

Since collapse is so problematic, can we find an interpretation of quantum mechanics that does away with the collapse while retaining quantum mechanics as a single-particle theory? An affirmative answer has been given by Hugh Everett. In his interpretation, both possibilities, live cat and dead cat, occur, but in different realities, parallel universes. For every live cat we find in the cage, prototypes of us in a parallel universe open a prototype cage only to discover a prototype cat that is dead. The dichotomy of the cat's state upon observation forces the universe itself to split into parallel branches that do not interact! Thus there is no way to verify the existence of these parallel universes.

From the philosophical standpoint, does this many-worlds interpretation save physical realism? Not likely. When we carefully examine many-worlds theory, we find that observation still plays a subtle but decisive role. For example, how does one define when a branching of the universe occurs? If this is connected with measurement, doesn't the definition of measurement bring to the fore the role of the observer that realists try to avoid? Moreover, try to analyze the EPR paradox with the many-worlds interpretation (see problem 2 at the end of the chapter); you will find that nonlocality still lingers on.

Nevertheless, the many-worlds theory is an interesting idea, although clearly science fictiony in flavor (and some science fiction writers, notably Philip K. Dick, make good use of it. You may read Dick's novel *The Man in the High Castle*, which is about a many-world reality). Unfortunately, it is a costly idea, too many universes, too many unverifiable entities for most people's taste; it violates the philosophical principle of parsimony (sometimes referred to as Occam's razor) very badly.

10.3 WHEN IS A MEASUREMENT?

Let's return to the question, What defines a measurement? Put slightly differently, When can we say that a quantum measurement is completed? This is a question with a long history.

In order to elucidate the uncertainty principle, its discoverer Werner Heisenberg formulated a thought experiment, the Heisenberg microscope described in chapter 5, that Bohr clarified further. Let's recapitulate the argument if only to make a point. Suppose a particle is at rest in the target plane of a microscope, and we are analyzing its observation in terms of classical physics. To observe the target particle, we focus (with the help of the microscope) another particle that is deflected by the target particle onto a photographic emulsion plate where it leaves a track. From the track, and knowing how the microscope works, according to classical physics, we can determine both the position of the target particle and the momentum imparted to it at the moment of deflection. The intermediary, the specific experimental conditions, drop out from the final result.

But all this changes in quantum mechanics. If the target particle is an atom, and we are looking at it through an electron microscope in which an electron is deflected from the atom onto a photographic plate (fig. 10.2), the following four considerations enter:

1. The deflected electron must be described as both a wave (while it is traveling from the object O to the image P) and a particle (at arrival at P and while leaving the track T).
2. Because of this wave aspect of the electron, the image point P tells us only the probability distribution of the position of the object O. In other words, the position is determined only within a certain uncertainty Δx.
3. Similarly, argued Heisenberg, the direction of the track T gives us only the probability distribution of the momentum of O, and thus determines

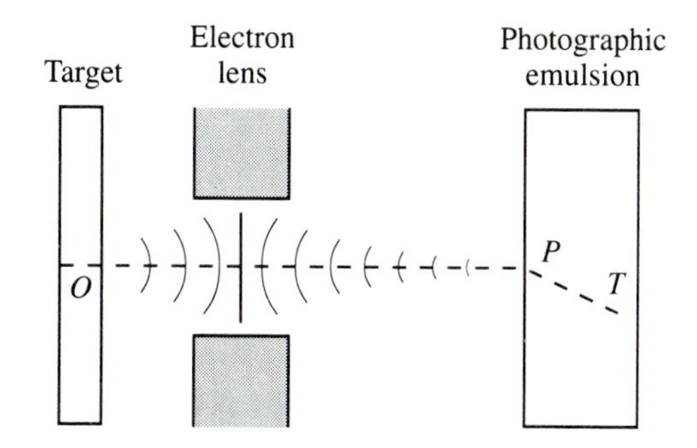

FIGURE 10.2
A second look at the Heisenberg-Bohr microscope. (See text for explanation.) (Reprinted with permission from J. A. Schumacher.)

the momentum only within an uncertainty Δp. As we saw in chapter 5, Heisenberg was able to show that the product of the two uncertainties is given as

$$\Delta x \cdot \Delta p \geq h$$

4. In a more detailed mathematical account, according to quantum mechanics, Bohr pointed out that it is impossible to specify the wave function of the observed atom separately from the electron used to see it. And in truth, said Bohr, the wave function of the electron cannot be disentangled from that of the photographic emulsion. In other words, the description of the experimental conditions does not drop out as a mere intermediary link of inference as in the classical case, but remains inseparable from the description of what is observed.

And more. We cannot draw the line in this chain with unambiguity, observed Bohr. But in spite of the ambiguity in drawing the line, Bohr also felt that we *must* draw the line because of the "indispensable use of classical concepts in the interpretation of all proper measurements." The experimental arrangement, said Bohr reluctantly, must be described in totally classical terms; the dichotomy of quantum waves must be assumed to terminate with the measuring apparatus. But the point can be made that all actual experiments have a second Heisenberg microscope built into them: The process of *seeing* the emulsion track involves the same kind of consideration that led Heisenberg to the uncertainty principle (fig. 10.3). Can we ignore the quantum mechanics of our own seeing? And if not, isn't our mind-brain-consciousness inexorably connected with the measurement process?

When you think about it, it becomes clear that Bohr was curing one dichotomy, that of the cat, by creating another—that the world is divided into quantum and classical; some objects obey quantum mechanics and some classical.

Returning to Schrödinger's cat, according to Bohr, we cannot separate the wave function of the atom from the rest of the environment in the cat's cage

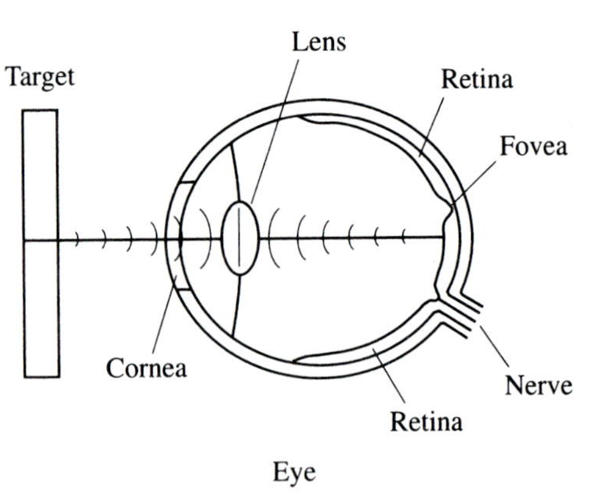

FIGURE 10.3
The mechanics of seeing. Another Heisenberg microscope in operation? (Reprinted with permission from J. A. Schumacher.)

(the various measuring devices for the atom's decay, such as the Geiger counter, the poison bottle, and even the cat), and the line we draw between the microworld and the macroworld is quite arbitrary. Unfortunately, Bohr said that it is necessary to go along with the idea that the observation by a machine, a measuring apparatus, resolves the dichotomy of a quantum wave function, if only for the sake of communication.

But one can argue that any macrobody (the cat or any observing machine) is ultimately a quantum object; there is no such thing as a classical body unless we are willing to admit a vicious quantum/classical dichotomy in physics. It is true that a macrobody's behavior can be predicted in most situations from the rules of classical mechanics (quantum mechanics gives the same mathematical prediction as classical mechanics on such occasions—this is the correspondence principle, which Bohr himself pioneered); this is the reason that we often loosely refer to macrobodies as classical. But the measurement process is not one of these occasions; the correspondence principle does not apply here. Bohr knew this, of course; in his debate with Einstein, he invoked quantum mechanics for describing macrobodies of measurement in order to refute the acute objections that Einstein raised to probability waves and the uncertainty principle (see chapter 5).

Thus, it is difficult, if not impossible, to deny that all objects ultimately obey quantum uncertainty and acquire quantum dichotomy. And then, as von Neumann first argued, if a chain of material machines measures a dichotomic quantum object such as Schrödinger's cat in the coherent superposition of live and dead cats, they all in turn pick up the dichotomy of the object, ad infinitum (fig. 10.4). How do we get out of such a logjam? The answer is startling. By jumping out of the system!

We *know* that an observation by a conscious observer ends the dichotomy. Thus, said von Neumann—and later other authors such as London and Bauers, and Wigner supported his idea—it is our consciousness, acting from outside of the system, that collapses the quantum wave function.

FIGURE 10.4
Von Neumann chain illustrated. If even our mind-brain catches the dichotomy of the cat, how does the chain terminate?

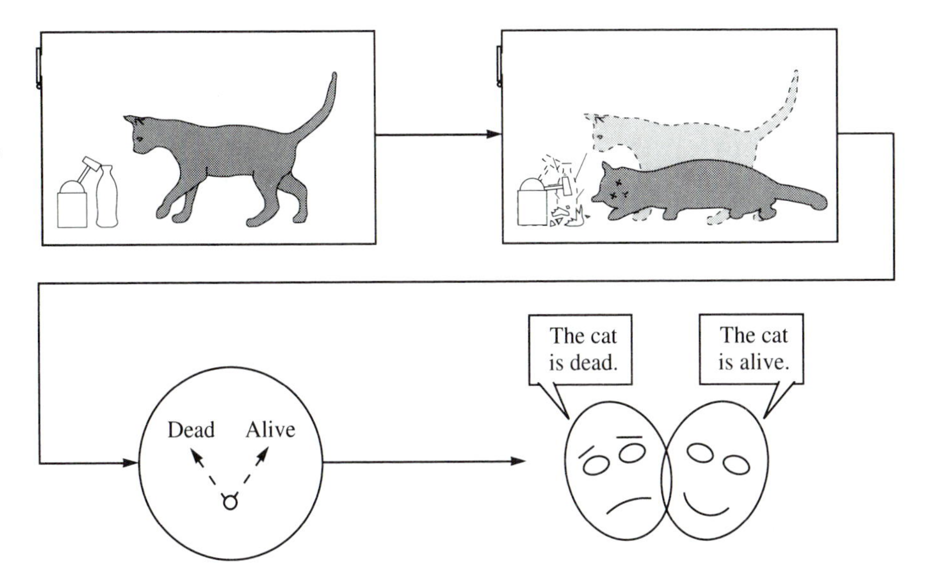

Can our consciousness be outside the system, the physical reality? According to the philosophy of physical realism, all things are made up of elementary atoms and are part of the physical reality; even consciousness, according to this philosophy, is nothing but an epiphenomenon of matter. Thus in postulating consciousness as the agent of collapse, von Neumann is stepping outside the philosophy of physical realism into idealism, which holds that consciousness is independent of matter. Subsequently, in Wigner's writing a dualistic view is promulgated—consciousness (mind) and body form two different independent realities. Consequently, the von Neumann-Wigner resolution of the quantum measurement paradox presents its own series of difficulties (see problems 3 and 4 at the end of the chapter). Nevertheless, we should be patient before we turn in a final verdict. After all, do we really understand consciousness? Maybe when we do, it would not surprise us that we need to invoke it to resolve the paradoxes of quantum mechanics. These matters will be taken up again in chapter 24.

Macrorealism

Let us now focus on another line of thought to answer the question, Where do we draw the line in quantum measurement? One answer is squarely to propose a classical/quantum dichotomy, a real division of the world into two categories of objects, some obeying quantum mechanics, some classical. The argument (sometimes called macrorealism) goes as follows. It is a fact that we do not see quantum objects without amplification and without the intermediary of some macroapparatus. This could be a clue of some sort suggesting that the laws of the macrobodies are different than the quantum laws. You say, no, this cannot be the case, because then, as the Bohr-Einstein debate showed, we can use the classical nature of macromeasuring apparatuses to refute the uncertainty principle. But suppose there is a complementarity principle operating here. Suppose that sometimes the macroapparatus does take on quantum dichotomy, but other times it doesn't. The former occurs in the case of the diaphragm of the Bohr-Einstein debate (see chapter 5), but the latter is true for a measuring apparatus. The idea is interesting because it has inspired an important experimental program named SQUID (an acronym for superconducting quantum interference device). We shall have more to say about SQUID in chapter 14.

Outlook

I hope you have begun to see why sometimes a comment such as "nobody understands quantum mechanics" is made even by serious physicists. I hope you also see why it is important to delve into the meaning of quantum mechanics, because here we have an opportunity to study within science some basic philosophical questions about reality. What I have given here is only a modicum of all the interpretive work that has been carried out in the last half of the century, and you are strongly encouraged to check out some of the references given at the end of the chapter for further details and ideas. Happy trails!

PROBLEMS

The problems below are mostly "think" problems, somewhat open-ended at this point in our understanding of quantum mechanics.

1. Show that the interacting two-particle system of section 9.3 in the last chapter also suffers from EPR correlations. Only qualitative consideration is necessary.
2. Analyze the EPR paradox from a many-worlds point of view.
3. One criticism of the von Neumann resolution to the paradox of Schrödinger's cat is to ask, What do we do to the cat, what power does our tiny little peek have that it can resolve the cat's alive or dead dichotomy? Can you think of any way to answer this question?
4. The paradox of Wigner's friend: Suppose instead of observing the cat himself, Wigner sends his friend to do so. When does the cat's wave function collapse? When Wigner asks his friend or when his friend makes the observation of the cat? Criticize both answers.
5. Critique the following cosmological criticism of the von Neumann-Wigner interpretation: If consciousness is needed to collapse dichotomic wave functions, then how did the universe evolve during all the time that no conscious being was around? Surely, thanks to radioactivity and such, dichotomic states are created in nature quite frequently (as in mutation).
6. It has been argued (B. d'Espagnat, *In Search of Reality*) that in quantum mechanics, the criterion of strong objectivity (observer independence) must be replaced by the criterion of *weak objectivity* (observer invariance—that is, independence of the result of an observation from a *particular* observer). Evaluate this argument.

ADDITIONAL PROBLEM

A1. Analyze the double-slit experiment from a many-worlds point of view. Does the world split when an electron passes through the double slit? Does the world split when the electron falls on the fluorescent plate afterward? How does the electron know the difference, if any, between a situation which is an ordinary interaction and a measurement situation? In view of any difficulty, what is your opinion regarding the veracity of the many-worlds interpretation? (You may read E. J. Squares, *European Journal of Physics*, *8*, 171 (1987).)

REFERENCES

D. Bohm. *Quantum Theory.*

D. Bohm. *Wholeness and Implicate Order.*

B. d'Espagnat. *In Search of Reality.*

J. A. Schumacher, in *Fundamental Questions in Quantum Mechanics*, L. M. Roth and A. Inomata (eds.).

J. A. Wheeler and W. Zurek (eds.). *Quantum Theory and Measurement.* This collection has most of the original papers discussed in this chapter (e.g., the work of EPR, Schrödinger, Everett, and von Neumann).

E. P. Wigner. *Symmetries and Reflections.*

11

Angular Momentum

In the quantum mechanics of motion in three dimensions, the concept of angular momentum plays a crucial role. This should not be surprising, since this is the case in classical mechanics as well, what with the treatment of orbits, gyroscopes, and such. However, there is a difference between the classical and quantum introductions to angular momentum. In classical mechanics, we simply relate angular momentum to torque—it is a quantity that is changed by a torque. But that is not so useful in quantum mechanics. Instead, we find a classical Hamiltonian that depends only on angular momentum. Once we have adapted this Hamiltonian for quantum mechanics, angular momentum makes its entry into the theory.

Let's see how this works. In chapter 9, we introduced the angular momentum operator L_z. We saw how it comes in naturally when we discuss the Hamiltonian of the angular motion of a particle in a ring. Let's now consider a particle constrained to move on the surface of a sphere; the particle can go anywhere except that it must remain at a fixed distance R from some center (fig. 11.1). The constraint is expressed mathematically by saying that if $\mathbf{r}$ is the position vector of the particle, then

$$\mathbf{r}\cdot\mathbf{r} = r^2 = R^2$$

Thus, in these polar coordinates, the position space is reduced to the two angular coordinates (θ, ϕ), and r is no longer a variable.

There is also a constraint on the momentum of a particle moving on a

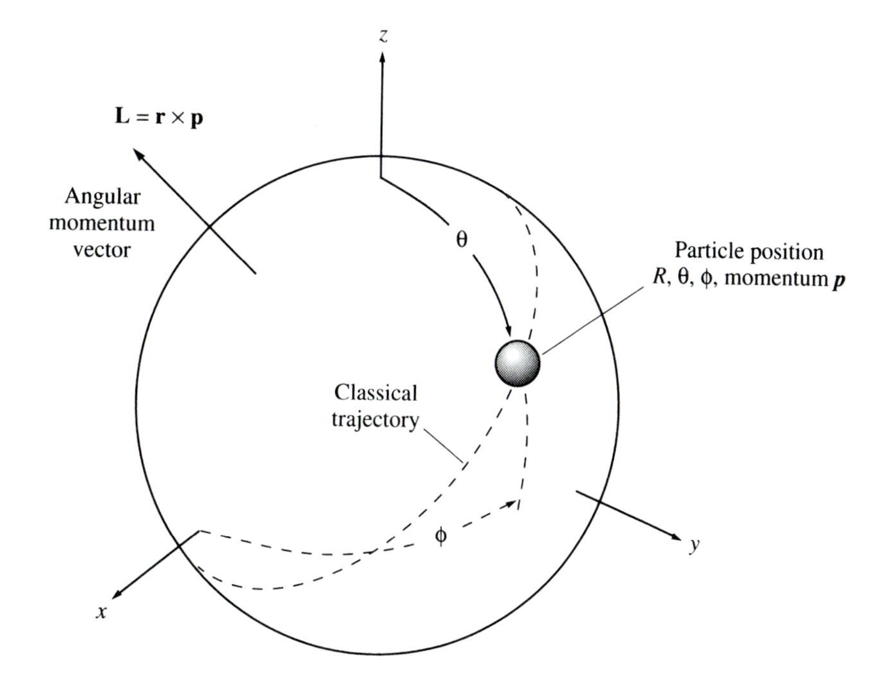

FIGURE 11.1
A particle constrained to move on a spherical shell at a fixed distance R from the origin.

sphere. The momentum vector $\mathbf{p}$ must always be perpendicular to the position vector $\mathbf{r}$; that is,

$$\mathbf{p} \cdot \mathbf{r} = 0$$

This means that only two of the momentum components are independent.

The classical particle constrained to move on a sphere also has an angular momentum vector given by

$$\mathbf{L} = \mathbf{r} \times \mathbf{p} \tag{11.1}$$

In view of the constraints above, the square of the angular momentum is given as

$$\begin{aligned} \mathbf{L}^2 &= (\mathbf{r} \times \mathbf{p}) \cdot (\mathbf{r} \times \mathbf{p}) \\ &= (\mathbf{r} \cdot \mathbf{r})(\mathbf{p} \cdot \mathbf{p}) - (\mathbf{r} \cdot \mathbf{p})(\mathbf{p} \cdot \mathbf{r}) = r^2 p^2 = R^2 p^2 \end{aligned} \tag{11.2}$$

There is no potential energy in the problem, only kinetic energy. The Hamiltonian H for the motion is given as

$$H = p^2/2m = L^2/2mR^2 = L^2/2I \tag{11.3}$$

where I is the moment of inertia and where we have used equation (11.2).

Observe that the Hamiltonian is purely dependent on the angular momentum, exactly the situation we were seeking. The adaptation of this Hamiltonian to quantum mechanics is straightforward: Regard angular momentum as an operator (of course, for large systems, this will give us back classical angular momentum). Solving the quantum mechanical motion of a particle on a sphere involves finding the eigenvalues and eigenfunctions of the angular momentum operator. This is the concern of this chapter.

A few more words about motivation are helpful. The same Hamiltonian, equation (11.3), applies to the rigid rotator, a rigid body that rotates about only two of its principal axes but not the third and for which the principal moments of inertia are the same. The motion of many quantum systems can be treated, in part, as that of a rigid rotator. In chapter 12, we will see also that the quantum mechanical solution of the three-dimensional motion of a particle under a central force (a force that depends only on the magnitude of **r**, not the direction) involves the angular momentum eigenstates.

We will work out the angular momentum eigenvalue problem in two different ways, first, by using the coordinate representation, and second, by the representation-independent operator method. Both methods, as we will see, have virtues.

11.1 THE ANGULAR MOMENTUM OPERATOR AND ITS EIGENVALUE PROBLEM IN POSITION REPRESENTATION

The classical definition of angular momentum $\mathbf{L} = \mathbf{r} \times \mathbf{p}$ gives for the components

$$L_x = yp_z - zp_y$$

$$L_y = zp_x - xp_z$$

$$L_z = xp_y - yp_x \tag{11.4}$$

We will adopt the same definitions for the components of the angular momentum operators, except now p_i are the operators $-i\hbar(\partial/\partial x_i)(\mathbf{p} = -i\hbar\boldsymbol{\nabla})$. Notice that we do not face any difficulty here from products of operators that fail to commute; all the operator products above are those of commuting operators.

Since a coordinate operator and its momentum conjugate do not commute, we should expect that the components of the angular momentum operator likewise do not commute with one another. Our first task is to derive the commutation relations of these operators.

To this end, we evaluate the commutation relations of the components of **L** with those of **r** and **p**:

$$[L_x,z] = [yp_z - zp_y,z] = y[p_z,z] = -i\hbar y$$

$$[L_x,p_z] = [yp_z - zp_y,p_z] = -[z,p_z]p_y = -i\hbar p_y$$

$$[L_x,x] = [L_x,p_x] = 0 \tag{11.5}$$

And similarly for the rest of the components. Now use these commutation relations to find those between the components of **L**. For example, we have

$$\begin{aligned}[L_x,L_y] &= [L_x,zp_x - xp_z] = [L_x,z]p_x - x[L_x,p_z] \\ &= -i\hbar yp_x + i\hbar xp_y = i\hbar L_z \end{aligned} \tag{11.6a}$$

We can write down the other commutation relations by cyclic permutation $(x \to y \to z \to x)$ of this result:

$$[L_y,L_z] = i\hbar L_x; \qquad [L_z,L_x] = i\hbar L_y \tag{11.6b}$$

These relationships are summarized by the easy-to-memorize expression

$$\mathbf{L} \times \mathbf{L} = i\hbar\mathbf{L} \tag{11.7}$$

a relationship that would be absurd for ordinary vectors!

What does the existence of the commutation relations, equations (11.6a,b), mean? It means that in general we cannot simultaneously assign eigenvalues to all three components of angular momentum. Let's look at this from the point of view of the generalized uncertainty relation, equation (6.23); we have

$$\Delta L_x \Delta L_y \geq (\hbar/2)|\langle L_z \rangle|, \qquad x,y,z \text{ cyclic}$$

A necessary condition for the simultaneous determination of all three components of angular momentum is

$$\langle L_x \rangle = \langle L_y \rangle = \langle L_z \rangle = 0$$

Or, in other words, $\langle \mathbf{L} \rangle = 0$. However, we also have to recognize that

$$(\Delta L_i)^2 = \langle L_i^2 \rangle - \langle L_i \rangle^2, \qquad i = x,y,z$$

Consequently, $\langle \Delta \mathbf{L} \rangle = 0$ only if we have in addition to $\langle \mathbf{L} \rangle = 0$,

$$\langle L_x^2 \rangle = \langle L_y^2 \rangle = \langle L_z^2 \rangle = 0$$

In other words, the necessary and sufficient condition then is that the state in question be an eigenstate of each component of **L** with eigenvalue 0. Accordingly, a state $|\psi\rangle$ is a simultaneous eigenstate of all three components of angular momentum only if

$$\mathbf{L}|\psi\rangle = 0 \tag{11.8}$$

For any other $|\psi\rangle$, a measurement of any component of angular momentum introduces an uncontrollable uncertainty of the order of $\hbar$ in our knowledge of the other two components. A measurement of L_z produces an eigenstate of L_z, which cannot be an eigenstate of L_x or L_y. If we now measure L_x, we will have an eigenstate of L_x, and all knowledge that we gathered about L_z from the previous measurement will be for naught; the value of L_z will become quite uncertain again.

Now consider the operator corresponding to the square of the angular momentum, defined as

$$\mathbf{L}^2 = L_x^2 + L_y^2 + L_z^2 \tag{11.9}$$

and let's compute the commutators of $\mathbf{L}^2$ with the components L_i.

$$\begin{aligned}
[\mathbf{L}^2, L_z] &= [L_x^2 + L_y^2 + L_z^2, L_z] = [L_x^2 + L_y^2, L_z] \\
&= L_x[L_x, L_z] + [L_x, L_z]L_x + L_y[L_y, L_z] + [L_y, L_z]L_y \\
&= -i\hbar L_x L_y - i\hbar L_y L_x + i\hbar L_y L_x + i\hbar L_x L_y = 0
\end{aligned}$$

In this way, we can see that

$$[\mathbf{L}^2, L_i] = 0, \qquad i = x, y, z \tag{11.10}$$

Since $\mathbf{L}^2$ commutes with all the components of angular momentum, we can find simultaneous eigenstates of $\mathbf{L}^2$ and any one component of $\mathbf{L}$. In other words, out of the four operators,

$$\mathbf{L}^2,\ L_x,\ L_y, \text{ and } L_z$$

the eigenvalues of only two can be used to label the angular momentum eigenstates.

You can understand this last result from another point of view. Remember that the angular momentum eigenvalue problem is the same as the eigenvalue problem of a particle constrained to move on the surface of a sphere. But for the latter problem, there are only two dimensions, θ and ϕ, and only these two quantities can take up measurable values. Because of this the dimensionality of space defined by the angular momentum operators must also be two dimensional; only two of the four angular momentum operators above take on measurable values.

The Eigenvalue Problem

It is customary to choose $\mathbf{L}^2$ and L_z as the two operators whose eigenvalues label the angular momentum eigenstates. Denoting these eigenvalues by λ and m, respectively, and the normalized eigenstates as $|\lambda m\rangle$, we can write the eigenvalue equations as

$$\mathbf{L}^2|\lambda m\rangle = \lambda|\lambda m\rangle$$

$$L_z|\lambda m\rangle = m\hbar|\lambda m\rangle \tag{11.11}$$

For our problem, the position representation is the (θ,ϕ)-representation and our job is to solve for the eigenfunctions

$$\langle\theta,\phi|\lambda m\rangle = \psi_{\lambda m}(\theta,\phi)$$

We have

$$\langle\theta,\phi|\mathbf{L}^2|\lambda m\rangle = \lambda\langle\theta,\phi|\lambda m\rangle$$

$$\langle\theta,\phi|L_z|\lambda m\rangle = m\hbar\langle\theta,\phi|\lambda m\rangle \tag{11.12}$$

To execute the operations on the left-hand side of equation (11.12), however, we need to express the operators $\mathbf{L}^2$ and L_z in the (θ,ϕ)-representation.

Coordinate Transformation

The relations of rectangular and spherical polar coordinates are given by (see fig. 11.2)

$$x = r\sin\theta\cos\phi$$

$$y = r\sin\theta\sin\phi$$

$$z = r\cos\theta \tag{11.13}$$

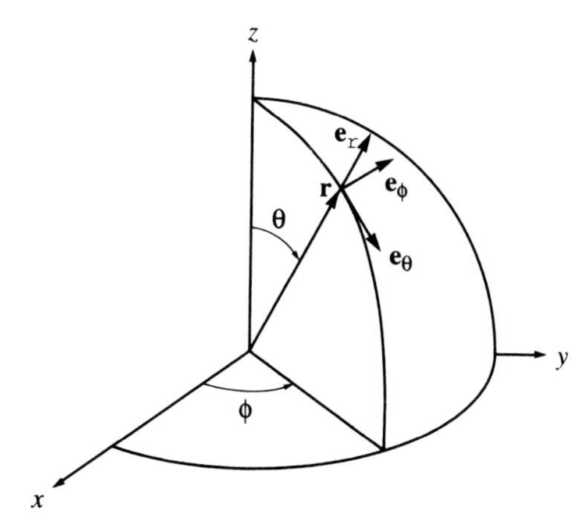

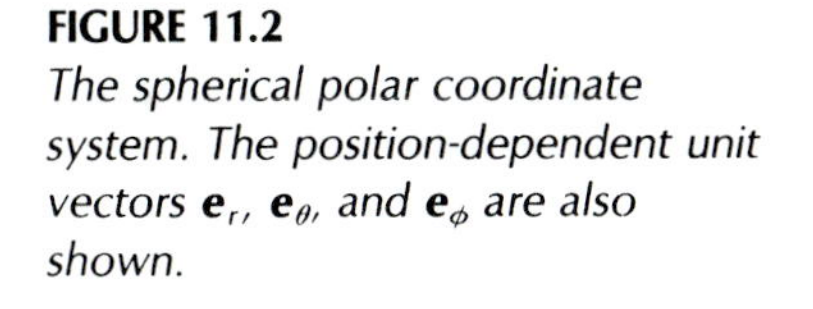

FIGURE 11.2
The spherical polar coordinate system. The position-dependent unit vectors $\mathbf{e}_r$, $\mathbf{e}_\theta$, and $\mathbf{e}_\phi$ are also shown.

The orthonormal basis vectors $\mathbf{e}_r$, $\mathbf{e}_\theta$, and $\mathbf{e}_\phi$ of the spherical coordinates satisfy the relations

$$\mathbf{e}_r \cdot \mathbf{e}_\theta = \mathbf{e}_r \cdot \mathbf{e}_\phi = \mathbf{e}_\phi \cdot \mathbf{e}_\theta = 0$$

$$\mathbf{e}_r \cdot \mathbf{e}_r = \mathbf{e}_\theta \cdot \mathbf{e}_\theta = \mathbf{e}_\phi \cdot \mathbf{e}_\phi = 1$$

$$\mathbf{e}_r \times \mathbf{e}_\theta = \mathbf{e}_\phi; \qquad \mathbf{e}_r \times \mathbf{e}_\phi = -\mathbf{e}_\theta; \qquad \mathbf{e}_\theta \times \mathbf{e}_\phi = \mathbf{e}_r \tag{11.14}$$

These unit vectors can be written in terms of the rectangular unit vectors $\mathbf{e}_x$, $\mathbf{e}_y$, and $\mathbf{e}_z$ (consult fig. 11.2):

$$\mathbf{e}_r = \mathbf{e}_x \sin\theta\cos\phi + \mathbf{e}_y \sin\theta\sin\phi + \mathbf{e}_z \cos\theta$$

$$\mathbf{e}_\theta = \mathbf{e}_x \cos\theta\cos\phi + \mathbf{e}_y \cos\theta\sin\phi - \mathbf{e}_z \sin\theta$$

$$\mathbf{e}_\phi = -\mathbf{e}_x \sin\phi + \mathbf{e}_y \cos\phi \tag{11.15}$$

The inverse relations are also useful:

$$\mathbf{e}_x = \mathbf{e}_r \sin\theta\cos\phi + \mathbf{e}_\theta \cos\theta\cos\phi - \mathbf{e}_\phi \sin\phi$$

$$\mathbf{e}_y = \mathbf{e}_r \sin\theta\sin\phi + \mathbf{e}_\theta \cos\theta\sin\phi + \mathbf{e}_\phi \cos\phi$$

$$\mathbf{e}_z = \mathbf{e}_r \cos\theta - \mathbf{e}_\theta \sin\theta \tag{11.16}$$

We are now ready to express the gradient operator $\boldsymbol{\nabla}$ in spherical coordinates

$$\boldsymbol{\nabla} = \mathbf{e}_x \partial/\partial x + \mathbf{e}_y \partial/\partial y + \mathbf{e}_z \partial/\partial z$$

We substitute for $\mathbf{e}_x$, $\mathbf{e}_y$, and $\mathbf{e}_z$ from equation (11.16) and collect terms. This gives

$$\begin{aligned}\boldsymbol{\nabla} = \mathbf{e}_r&[\sin\theta\cos\phi(\partial/\partial x) + \sin\theta\sin\phi(\partial/\partial y) + \cos\theta(\partial/\partial z)] \\ &+ \mathbf{e}_\theta[\cos\theta\cos\phi(\partial/\partial x) + \cos\theta\sin\phi(\partial/\partial y) - \sin\theta\partial/\partial z] \\ &+ \mathbf{e}_\phi[-\sin\phi(\partial/\partial x) + \cos\phi\partial/\partial y]\end{aligned}$$

The quantities within the square brackets are easily recognized as $\partial/\partial r$, and so forth, via the chain rule; for example,

$$\begin{aligned}\partial/\partial r &= (\partial x/\partial r)(\partial/\partial x) + (\partial y/\partial r)(\partial/\partial y) + (\partial z/\partial r)(\partial/\partial z) \\ &= \sin\theta\cos\phi(\partial/\partial x) + \sin\theta\sin\phi(\partial/\partial y) + \cos\theta(\partial/\partial z)\end{aligned}$$

where we have used the definitions of equation (11.13). In this way we get

$$\nabla = \mathbf{e}_r \frac{\partial}{\partial r} + \mathbf{e}_\theta \frac{1}{r} \frac{\partial}{\partial \theta} + \mathbf{e}_\phi \frac{1}{r \sin\theta} \frac{\partial}{\partial \phi} \tag{11.17}$$

a very important expression. The angular momentum operator in the (θ, ϕ)-representation can now be obtained easily:

$$\begin{aligned}
\mathbf{L} &= \mathbf{r} \times \mathbf{p} = r[\mathbf{e}_r \times (-i\hbar\nabla)] \\
&= -i\hbar r \mathbf{e}_r \times \left[\mathbf{e}_r \frac{\partial}{\partial r} + \mathbf{e}_\theta \frac{1}{r} \frac{\partial}{\partial \theta} + \mathbf{e}_\phi \frac{1}{r \sin\theta} \frac{\partial}{\partial \phi}\right] \\
&= -i\hbar r \left[\mathbf{e}_\phi \frac{1}{r} \frac{\partial}{\partial \theta} - \mathbf{e}_\theta \frac{1}{r \sin\theta} \frac{\partial}{\partial \phi}\right] \\
&= -i\hbar \left[\mathbf{e}_\phi \frac{\partial}{\partial \theta} - \mathbf{e}_\theta \frac{1}{\sin\theta} \frac{\partial}{\partial \phi}\right]
\end{aligned} \tag{11.18}$$

where we have used equation (11.14). From equation (11.18) we can obtain L_z in the (θ, ϕ)-representation as

$$\begin{aligned}
L_z &= \mathbf{e}_z \cdot \mathbf{L} = (\mathbf{e}_r \cos\theta - \mathbf{e}_\theta \sin\theta) \cdot (-i\hbar) \left(\mathbf{e}_\phi \frac{\partial}{\partial \theta} - \mathbf{e}_\theta \frac{1}{\sin\theta} \frac{\partial}{\partial \phi}\right) \\
&= -i\hbar \frac{\partial}{\partial \phi}
\end{aligned} \tag{11.19}$$

This is the same as equation (9.25), as it must be. The evaluation of $\mathbf{L}^2$ is only slightly more complicated.

$$\mathbf{L}^2 = \mathbf{L} \cdot \mathbf{L} = -\hbar^2 \left(\mathbf{e}_\phi \frac{\partial}{\partial \theta} - \mathbf{e}_\theta \frac{1}{\sin\theta} \frac{\partial}{\partial \phi}\right) \cdot \left(\mathbf{e}_\phi \frac{\partial}{\partial \theta} - \mathbf{e}_\theta \frac{1}{\sin\theta} \frac{\partial}{\partial \phi}\right)$$

The derivatives of the unit vectors are easily obtained from equation (11.15):

$$\frac{\partial}{\partial \theta} \mathbf{e}_\theta = -\mathbf{e}_r, \qquad \frac{\partial}{\partial \phi} \mathbf{e}_\theta = \mathbf{e}_\phi \cos\theta$$

$$\frac{\partial}{\partial \theta} \mathbf{e}_\phi = 0, \qquad \frac{\partial}{\partial \phi} \mathbf{e}_\phi = -(\mathbf{e}_r \sin\theta + \mathbf{e}_\theta \cos\theta)$$

Substituting in the equation for $\mathbf{L}^2$ above, we obtain after some algebra

$$\mathbf{L}^2 = -\hbar^2\left[\frac{1}{\sin\theta}\frac{\partial}{\partial\theta}\left(\sin\theta\frac{\partial}{\partial\theta}\right) + \frac{1}{\sin^2\theta}\frac{\partial^2}{\partial\phi^2}\right] \tag{11.20}$$

Now substitute the expression for L_z in the (θ,ϕ)-representation, equation (11.19), into the eigenvalue equation for L_z, equation (11.12). This gives

$$\langle\theta,\phi|L_z|\lambda m\rangle = -i\hbar\frac{\partial}{\partial\phi}\langle\theta,\phi|\lambda m\rangle = m\hbar\langle\theta,\phi|\lambda m\rangle \tag{11.21}$$

The bracket $\langle\theta,\phi|\lambda m\rangle$ is the probability amplitude for finding the particle at the physical position (θ,ϕ). Consequently, the bracket must be continuous, square integrable, and single-valued in (θ,ϕ)-space.

As you recall from chapter 9, it is the condition of single valuedness that is crucial. On the surface of a sphere, the point $(\theta,\phi+2\pi)$ refers to the same point as (θ,ϕ). Accordingly, we must insist that

$$\langle\theta,\phi+2\pi|\lambda m\rangle = \langle\theta,\phi|\lambda m\rangle \tag{11.22}$$

The solution of the differential equation (11.21) is given as

$$\langle\theta,\phi|\lambda m\rangle = \exp(im\phi)f(\theta) \tag{11.23}$$

where $f(\theta)$ is a function of θ alone. Now the single valuedness condition, equation (11.22), can be satisfied only if m is an integer (positive or negative) or zero. We have

$$L_z|\lambda m\rangle = m\hbar|\lambda m\rangle \tag{11.24}$$

and

$$\langle\theta,\phi|\lambda m\rangle = \exp(im\phi)f(\theta) \tag{11.25}$$

To obtain the function $f(\theta)$, we must resort to the eigenvalue equation for $\mathbf{L}^2$, equation (11.12), with help from equations (11.20), (11.24), and (11.25). We have

$$-\frac{1}{\sin\theta}\frac{d}{d\theta}\sin\theta\frac{d}{d\theta}f(\theta) + \frac{m^2}{\sin^2\theta}f(\theta) = \frac{\lambda}{\hbar^2}f(\theta) \tag{11.26}$$

To solve this equation, we need, first of all, a change of variables

$$\zeta = \cos\theta \tag{11.27}$$

Call the new eigenfunction $P(\zeta) = f(\theta)$. The equation satisfied by $P(\zeta)$ is easily seen to be

$$\frac{d}{d\zeta}(1-\zeta^2)\frac{d}{d\zeta}P(\zeta) + \left(\frac{\lambda}{\hbar^2} - \frac{m^2}{1-\zeta^2}\right)P(\zeta) = 0 \tag{11.28}$$

Since θ runs from 0 to π, the domain of ζ is from -1 to 1; our task is to find solutions that are continuous, square integrable, and single valued throughout this region of ζ.

Actually, equation (11.28) is a well-known equation of mathematical physics, but its solution is complicated by the presence of the singularities (blow-up) for $\zeta = \pm 1$. We will, however, solve the equation for the special case $m = 0$ by the same power series technique that we used for the harmonic oscillator. This is instructive since the quantization of the eigenvalues λ will follow from the boundary conditions imposed on $P(\zeta)$. For $m = 0$, we have

$$\frac{d}{d\zeta}(1-\zeta^2)\frac{d}{d\zeta}P(\zeta) + \frac{\lambda}{\hbar^2}P(\zeta) = 0 \tag{11.29}$$

This is known as the *Legendre equation*. Try the polynomial solution

$$P(\zeta) = \sum_n c_n \zeta^n \tag{11.30}$$

Substitute in equation (11.29), and follow exactly the same procedure as in the case of the harmonic oscillator; your patience will reward you with the recursion relation

$$c_{n+2} = \frac{n(n+1) - (\lambda/\hbar^2)}{(n+1)(n+2)}\,c_n \tag{11.31}$$

for the coefficients c_n. Note that

$$\frac{c_{n+2}}{c_n} \to 1 \text{ as } n \to \infty$$

and argue that the polynomial will diverge for $\zeta = 1$ unless the series terminates for some $n = l$. The condition of termination is the quantization condition for the eigenvalues λ of $\mathbf{L}^2$:

$$\lambda = l(l+1)\hbar^2 \tag{11.32}$$

where l is zero or a *positive* integer.

We will denote the polynomial of the lth order by $P_l(\zeta)$, which is called the Legendre polynomial and gives us the solution of the θ-equation for $m = 0$, equation (11.29). The conventional form of the polynomial solution $P_l(\zeta)$ is given as

$$P_l(\zeta) = \frac{1}{2^l \cdot l!} \frac{d^l}{d\zeta^l} (\zeta^2 - 1)^l \tag{11.33}$$

We will tabulate the first few Legendre polynomials:

$$P_0(\zeta) = 1$$

$$P_1(\zeta) = \zeta$$

$$P_2(\zeta) = \frac{1}{2} (3\zeta^2 - 1) \tag{11.34}$$

As required for any eigenfunction of a hermitian operator, the Legendre polynomials form an orthogonal set:

$$\int_{-1}^{1} d\zeta \, P_l(\zeta) P_{l'}(\zeta) = 0, \qquad l \neq l'$$

No complex conjugation is needed since the Legendre polynomials are real functions. Furthermore, the normalization integral of the P_l's is easily evaluated by l-fold integration by parts (check it out!):

$$\int_{-1}^{1} d\zeta \, [P_l(\zeta)]^2 = \frac{2}{2l + 1}$$

All this is for $m = 0$. For $m \neq 0$, we define

$$\begin{aligned} P_l^m(\zeta) &= (1 - \zeta^2)^{m/2} \frac{d^m}{d\zeta^m} P_l(\zeta) \\ &= \frac{1}{2^l \cdot l!} (1 - \zeta^2)^{m/2} \frac{d^{l+m}}{d\zeta^{l+m}} (\zeta^2 - 1)^l \end{aligned} \tag{11.35}$$

for *positive* m ($\leq l$), and you can check by direct substitution that $P_l^m(\zeta)$ is a solution of equation (11.28) with $\lambda = l(l + 1)\hbar^2$. The functions $P_l^m(\zeta)$ are called *associated Legendre functions* (alas, they are no longer polynomials; they have lost their simple polynomial nature by having the factor $[1 - \zeta^2]^{1/2}$).

An alternative form of $P_l^m(\zeta)$ is also used.

$$P_l^m(\zeta) = \frac{(-1)^m}{2^l} \frac{(l + m)!}{l!\,(l - m)!} (1 - \zeta^2)^{-m/2} \frac{d^{l-m}}{d\zeta^{l-m}} (\zeta^2 - 1)^l \tag{11.36}$$

The associated Legendre functions satisfy an orthogonality relationship, but only for the same m (why?):

$$\int_{-1}^{1} d\zeta\, P_l^m(\zeta) P_{l'}^m(\zeta) = 0, \qquad l \neq l' \tag{11.37}$$

The normalization integral for $P_l^m(\zeta)$ is tedious, but the diligent reader can work it out:

$$\int_{-1}^{1} d\zeta\, [P_l^m]^2 = \frac{2}{2l+1} \frac{(l+m)!}{(l-m)!} \tag{11.38}$$

Now notice that equation (11.28), satisfied by P_l^m, remains unchanged by the shuffle $m \to -m$; this combined with the fact that P_l^m is the only physically admissible solution of our θ-equation, equation (11.26), makes it clear that the same associated Legendre function must be used for a given absolute value of m.

To summarize, if we denote the angular momentum eigenstates as $|lm\rangle$ using the quantum numbers l and m, the eigenvalues of L_z are $m\hbar$

$$L_z|lm\rangle = m\hbar|lm\rangle \qquad \text{with } m = 0, \pm 1, \pm 2, \ldots$$

The eigenvalues of $\mathbf{L}^2$ are $l(l+1)\hbar^2$

$$\mathbf{L}^2|lm\rangle = l(l+1)\hbar^2|lm\rangle \qquad \text{with } l = 0, 1, 2, \ldots$$

Remember that the eigenvalue of $\mathbf{L}$, if we can talk of such, is then $[l(l+1)]^{1/2}$, not l, although we loosely refer to l as the angular momentum of the state.

Furthermore, from the structure of equation (11.35), it is clear that

$$l \geq |m|$$

otherwise $P_l^m(\zeta)$ would be trivially equal to 0. Explicitly, we have

$$m = -l, -l+1, -l+2, \ldots, l-1, l$$

The Vector Model

What are some of the surprises about the quantum mechanical angular momentum compared to the classical? Classically, for the same magnitude of the angular momentum, there are an infinite number of states obtained by changing the direction of the angular momentum vector. But quantum mechanically, there are only a finite number of states, a quantized number. Moreover, in quantum mechanics, we do not describe a state by specifying the direction of the angular momentum vector, but instead by giving the component of the angular momentum along some direction (which we conveniently choose as the z-axis).

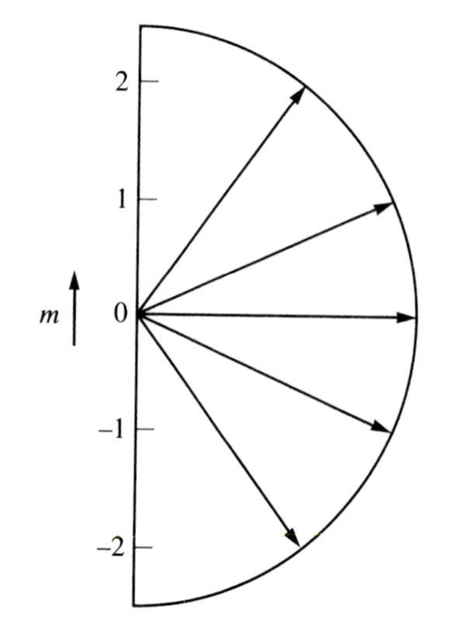

FIGURE 11.3
Vector model for angular momentum is illustrated for $l = 2$. The radius of the circle is $[2(2 + 1)]^{1/2}$. The multiplicity of states is 5.

Nevertheless, there is a useful pictorial way to convey the quantum mechanical results—this is the so-called vector model for angular momentum. In this model, we represent the angular momentum by a vector of length $[l(l+1)]^{1/2}$. We also realize that the component of $\mathbf{L}$ along the z-axis is quantized, which means that the vector $\mathbf{L}$ can make only certain quantized angles with the z-axis. A measurement of L_z will yield only the $2l+1$ quantized values. Finally, the maximum value of L_z is l, which is less than the magnitude of the vector (for $l \neq 0$) $\mathbf{L}$. This also means that unless $l = 0$, there is always a nonzero value of L_x and L_y; however, these values are not quantized. We visualize the angular momentum vector sweeping out in all possible directions in the x-y plane, no holds barred. Figure 11.3 shows the vector picture of angular momentum for $l = 2$.

Spherical Harmonics

The complete eigenfunction of $\mathbf{L}^2$ and L_z, $\langle \theta, \phi | lm \rangle$, is given as

$$\langle \theta, \phi | lm \rangle = N_l^m P_l^m(\cos\theta)\exp(im\phi) = Y_{lm}(\theta, \phi)$$

where the functions Y_{lm} are called *spherical harmonics* and N_l^m is a normalization constant such that Y_{lm} is normalized with respect to an integration over the entire solid angle:

$$\int_0^{2\pi} d\phi \int_0^{\pi} \sin\theta \, d\theta \, Y_{lm}^*(\theta, \phi) Y_{lm}(\theta, \phi) = 1 \tag{11.39}$$

The normalization of Y_{lm} follows from the fact that the eigenstates $|lm\rangle$ are normalized and that the basic states $|\theta\phi\rangle$ form a complete set in the total angular space:

$$1 = \langle lm | lm \rangle = \int_0^{2\pi} d\phi \int_0^{\pi} \sin\theta \, d\theta \langle lm | \theta\phi \rangle \langle \theta\phi | lm \rangle$$

Since $\langle lm | \theta\phi \rangle = Y_{lm}^*$, this is the same equation as equation (11.39). We will write down for $m \geq 0$, the following final neat normalized expression for the spherical harmonics with the established phase convention:

$$Y_{lm}(\theta, \phi) = \left[\frac{2l+1}{4\pi} \frac{(l-m)!}{(l+m)!} \right]^{1/2} (-1)^m e^{im\phi} P_l^m(\cos\theta) \tag{11.40}$$

The spherical harmonics for negative m (but in keeping with $-l \leq m \leq l$) are defined by

$$Y_{l-m} = (-1)^m Y_{lm}^* \tag{11.41}$$

In the coordinate representation, the spherical harmonics represent the normalized simultaneous eigenfunctions of the coordinate representations of the two operators $\mathbf{L}^2$ and L_z:

$$L_z Y_{lm} = -i\hbar \partial Y_{lm}/\partial\phi = m\hbar Y_{lm}$$

$$\mathbf{L}^2 Y_{lm} = l(l+1)\hbar^2 Y_{lm} \tag{11.42}$$

The following is a list of the first few spherical harmonics:

$$Y_{00} = \left(\frac{1}{4\pi} \right)^{1/2}$$

$$Y_{10} = \left(\frac{3}{4\pi} \right)^{1/2} \cos\theta = \left(\frac{3}{4\pi} \right)^{1/2} \frac{z}{r}$$

$$Y_{1\pm1} = \mp \left(\frac{3}{8\pi} \right)^{1/2} e^{\pm i\phi} \sin\theta = \mp \left(\frac{3}{8\pi} \right)^{1/2} \frac{x \pm iy}{r}$$

$$Y_{20} = \left(\frac{5}{16\pi} \right)^{1/2} (3\cos^2\theta - 1) = \left(\frac{5}{16\pi} \right)^{1/2} \frac{3z^2 - r^2}{r^2}$$

$$Y_{2\pm1} = \mp \left(\frac{15}{8\pi} \right)^{1/2} e^{\pm i\phi} \cos\theta \sin\theta = \mp \left(\frac{15}{8\pi} \right)^{1/2} \frac{(x \pm iy)z}{r^2}$$

$$Y_{2\pm2} = \left(\frac{15}{32\pi} \right)^{1/2} e^{\pm 2i\phi} \sin^2\theta = \left(\frac{15}{32\pi} \right)^{1/2} \frac{(x \pm iy)^2}{r^2} \tag{11.43}$$

Parity

Note also that under space reflection, the operation of the parity operator, both the momentum operator **p** and the position operator **r** change sign, leaving the angular momentum $\mathbf{L} = \mathbf{r} \times \mathbf{p}$ invariant. This means that the eigenfunctions of angular momentum should also be simultaneous eigenfunctions of the parity operator P. To demonstrate this, note that under the parity operation $\mathbf{r} \rightarrow -\mathbf{r}$. In spherical coordinates, this translates into (fig. 11.4):

$$r \rightarrow r$$

$$\theta \rightarrow \pi - \theta$$

$$\phi \rightarrow \pi + \phi$$

Now use the explicit form of the spherical harmonics Y_{lm} given by equation (11.40) to see (with the help of eq. [11.35]) that

$$Y_{lm}(\pi - \theta, \phi + \pi) = (-1)^l Y_{lm}(\theta, \phi) \tag{11.44}$$

It follows that

$$PY_{lm}(\theta, \phi) = (-1)^l Y_{lm}(\theta, \phi) \tag{11.45}$$

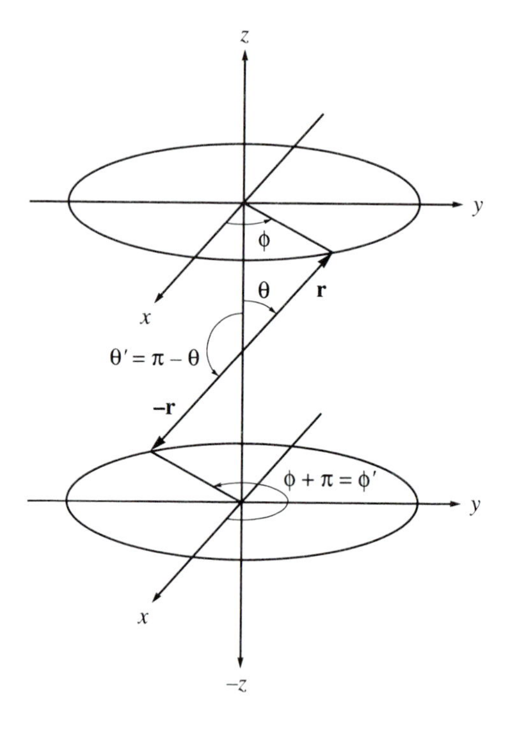

FIGURE 11.4
Space reflection or parity. How the angles of **r** *and* −**r** *are related while their magnitudes remain the same. (Adapted from Anton Capri,* Nonrelativistic Quantum Mechanics, *© 1985, by Addison-Wesley Publishing Company, Inc., Reading, Massachusetts. Figure 10-2, page 293. Reprinted with permission of the publisher.)*

which shows that the spherical harmonics are indeed eigenfunctions of parity; the parity is even for even l and the parity is odd for odd l.

The solution of the angular momentum eigenvalue problem also solves the eigenvalue problem for the particle moving on a spherical surface as well as that of the rigid rotator. Of these the rigid rotator deserves some discussion since there are physical systems such as molecules part of whose Hamiltonian can be modeled after a rigid rotator.

The Rigid Rotator

The simplest diatomic molecule, two atoms of mass m separated by a distance R rotating as a whole about their midpoint fixed in space, can be thought of as a rigid rotator of moment of inertia

$$I = mR^2/2$$

The Hamiltonian for the rotational motion is given by

$$H = \mathbf{L}^2/2I$$

The eigenvalues of this Hamiltonian are given as

$$E_l = (\hbar^2/2I)l(l + 1)$$

The energy is quantized, of course, and in a very regular pattern, almost as spectacular as the equispaced pattern of the oscillator (fig. 11.5). Experimentally,

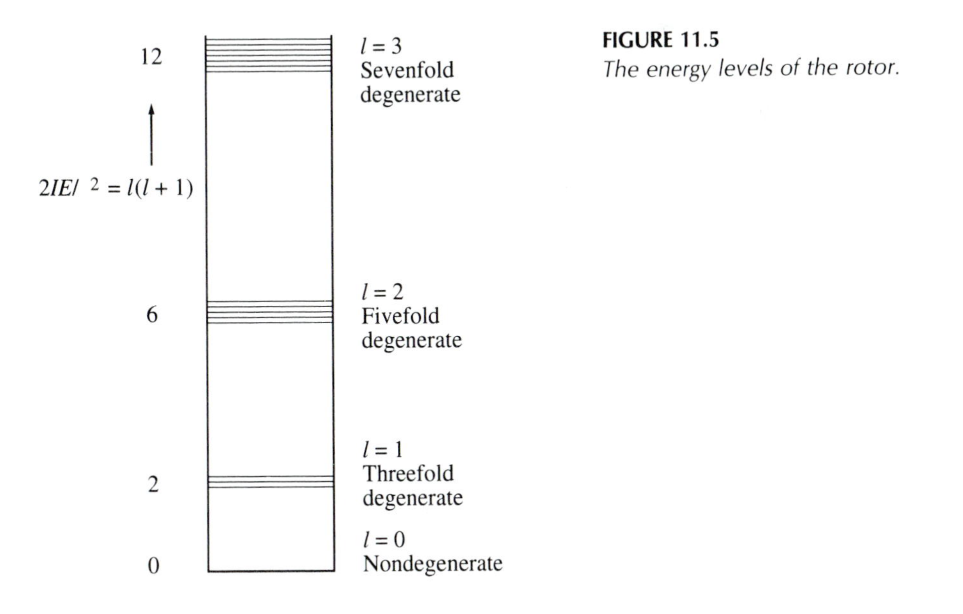

FIGURE 11.5
The energy levels of the rotor.

bands of such spectra with a clear rotational signature have been found for diatomic molecules, and even atomic nuclei.

Note, furthermore, that the energy is independent of m, the energy is $(2l + 1)$-fold degenerate; that is, for any eigenvalue of energy E_l, there are $2l + 1$ eigenfunctions

$$Y_{ll}, Y_{ll-1}, \ldots, Y_{l-l}$$

11.2 THE OPERATOR METHOD FOR ANGULAR MOMENTUM

Let's now investigate the eigenvalue problem via the quantum mechanical operator algebra without using any specific representation. Since the angular momentum commutation relations, being operator relations, are independent of any particular representation, they can be used as the defining relationships for a triplet of operators J_x, J_y, and J_z. They will be called angular momentum operators, but they are not necessarily restricted to the triplet L_x, L_y, and L_z defined through the classical relations, equation (11.4).

From our definition above, we have

$$[J_x, J_y] = i\hbar J_z, \qquad xyz \text{ cyclic} \tag{11.46}$$

Define

$$J^2 = J_x^2 + J_y^2 + J_z^2 \tag{11.47}$$

Then, as before,

$$[J^2, J_i] = 0, \qquad i = x, y, \text{ or } z \tag{11.48}$$

Let's choose the eigenstates to be simultaneous eigenstates of J^2 and J_z whose eigenvalues, denoted by λ_J and $m\hbar$, respectively, we will use to label these states:

$$\begin{aligned} \mathbf{J}^2|\lambda_J m\rangle &= \lambda_J|\lambda_J m\rangle \\ J_z|\lambda_J m\rangle &= m\hbar|\lambda_J m\rangle \end{aligned} \tag{11.49}$$

Now define the nonhermitian operators

$$\begin{aligned} J_+ &= J_x + iJ_y \\ J_- &= J_x - iJ_y \end{aligned} \tag{11.50}$$

The commutation relations of these operators with one another and with J_z are easily evaluated:

$$[J_z, J_+] = \hbar J_+; \qquad [J_z, J_-] = -\hbar J_-$$
$$[J_+, J_-] = 2\hbar J_z \tag{11.51}$$

Additionally, take note that

$$J_+J_- = (J_x + iJ_y)(J_x - iJ_y) = J_x^2 + J_y^2 + i(J_yJ_x - J_xJ_y)$$
$$= \mathbf{J}^2 - J_z^2 + \hbar J_z \tag{11.52}$$

Similarly,

$$J_-J_+ = \mathbf{J}^2 - J_z^2 - \hbar J_z \tag{11.53}$$

The introduction of the $J_\pm$ operators should not surprise you; surely you see their similarity with operators we introduced for the operator treatment of the harmonic oscillator. Now we will show that J_+ and J_- are indeed raising and lowering operators analogous to the creation and annihilation operators for the case of the harmonic oscillator.

From the commutation relations, equation (11.51), we have

$$J_zJ_+ = J_+(J_z + \hbar)$$

Accordingly,

$$J_zJ_+|\lambda_J m\rangle = J_+(J_z + \hbar)|\lambda_J m\rangle = (m + 1)\hbar J_+|\lambda_J m\rangle$$

In this way you can see that $J_+|\lambda_J m\rangle$ is an eigenstate of J_z belonging to the eigenvalue $(m + 1)\hbar$. Hence J_+ is a raising operator; it raises the eigenvalue of J_z by one unit of $\hbar$. Notice that the operation by J_+ has no effect on λ, the eigenvalue of $\mathbf{J}^2$. This is because $[\mathbf{J}^2, J_+] = 0$. In exactly the same way

$$J_zJ_-|\lambda_J m\rangle = (m - 1)\hbar J_-|\lambda_J m\rangle$$

Hence J_- is the lowering operator. When acting upon an angular momentum eigenstate, it lowers the eigenvalue of J_z by $\hbar$, but has no effect on the eigenvalue of J^2.

Let's sum up these results by writing

$$J_+|\lambda_J m\rangle = c_{\lambda_J m}|\lambda_J m + 1\rangle$$
$$J_-|\lambda_J m\rangle = d_{\lambda_J m}|\lambda_J m - 1\rangle \tag{11.54}$$

where c and d are quantities to be determined.

But that can wait. First, let's determine the eigenvalues $m\hbar$ and λ_J. To this end, we first show that m is bounded. That is easy enough; we simply argue that the value of the component of a vector cannot be greater than the value of the vector itself, $(m\hbar)^2 \leq \lambda_J$. Therefore, for fixed λ_J, the value of m is bounded; call these bounds $m_{\max}$ and $m_{\min}$.

Another way of seeing the same result is to recognize that the expectation value of the square of a hermitian operator cannot be negative. This means that the expectation value of $J_x^2 + J_y^2$ in the state $|\lambda_J m\rangle$ is nonnegative. Consequently,

$$0 \leq \langle \lambda_J m | J_x^2 + J_y^2 | \lambda_J m \rangle = \langle \lambda_J m | \mathbf{J}^2 - J_z^2 | \lambda_J m \rangle = \lambda_J - (m\hbar)^2$$

Second, what is $m_{\max}$? Since the eigenvalues mh can be increased or decreased by the raising and lowering operators in discrete steps of one unit of $\hbar$, m must be discrete forming a ladder of unit steps; then $m_{\max}$ is the highest rung of the ladder. Even the magic of the raising operator cannot take us any higher

$$J_+ | \lambda_J m_{\max} \rangle = 0$$

Operate on this by J_-; then we get

$$J_- J_+ | \lambda_J m_{\max} \rangle = 0$$

Now use equation (11.53). We have

$$(\mathbf{J}^2 - J_z^2 - \hbar J_z) | \lambda_J m_{\max} \rangle = 0$$

This gives

$$\lambda_J - m_{\max}(m_{\max} + 1)\hbar^2 = 0 \tag{11.55}$$

Similarly, there is an $m_{\min}$, a bottom rung of the eigenvalue ladder of J_z, such that

$$J_- | \lambda_J m_{\min} \rangle = 0$$

Act on this by J_+, and realize that $J_+ J_- = \mathbf{J}^2 - J_z(J_z - \hbar)$. We get

$$\lambda_J - m_{\min}(m_{\min} - 1)\hbar^2 = 0 \tag{11.56}$$

There are two solutions to these two equations, equations (11.55) and (11.56). One solution is

$$m_{\min} = m_{\max} + 1$$

which is obviously unacceptable since it violates our very definitions of $m_{\max}$ and $m_{\min}$. The other solution, the right one, is

$$m_{\max} = -m_{\min} = j \quad \text{(say)} \tag{11.57}$$

Equation (11.57) defines the quantum number j. Substituting equation (11.57) in either equation (11.55) or (11.56), we get

$$\lambda_J = j(j+1)\hbar^2 \tag{11.58}$$

This compares well with what we found for the eigenvalue of $\mathbf{L}^2$ in section 11.1. However, it is too soon to identify **J** with **L**. We have a surprise of sorts in our hands.

Ask yourself, Can j take arbitrary values? Let's try to prove that it cannot. To this effect, let's note that we can generate the highest state of the ladder of J_z-eigenstates by operating repeatedly on the bottommost state $|\lambda_J m_{\min}\rangle$ with J_+. But each time J_+ operates, the value of J_z just goes up by one unit of $\hbar$. It follows that the difference in value between $m_{\max}$ and $m_{\min}$ must be either a positive integer or 0. But

$$m_{\max} - m_{\min} = j - (-j) = 2j$$

So the value of j cannot be arbitrary; $2j$ must be a positive integer or 0. Thus j can only be an integer, zero, or a half-integer.

It is this last conclusion that is a little unexpected. If $\mathbf{J}^2$ were identical with $\mathbf{L}^2$, we would have gotten j to be an integer or 0, the same as l. Obviously, $\mathbf{L}^2$ can be regarded as a special case of $\mathbf{J}^2$, since what we have done with the algebra of the J_i operators holds exactly for the L_i operators as well. But the J_i algebra is more general since it permits half-integer j-values.

There is a saying in physics that people sometimes use, "Everything that is not forbidden must be compulsory." Since half-integer j-values are not forbidden, are they compulsory? In other words, is there a new kind of angular momentum (for which obviously no classical analog exists) connected with these half-integer j-values? The answer is yes; it is called *spin* angular momentum, which we will treat in more detail in later chapters. However, in anticipation, let's develop the following nomenclature. We will call **L** the *orbital* angular momentum, the angular momentum directly borrowed from classical mechanics. It is directly related to the orbital motion of the system. Then we have spin for which we will reserve the notation **S**. Finally, the notation **J** will refer to both kinds of angular momentum, **L**, **S**, or their sum.

Finally, let's evaluate the quantities c and d in equation (11.54). For this purpose, let's calculate the expectation value of J_-J_+ in the state $|\lambda_J m\rangle$, which we will now denote as simply $|jm\rangle$; since $J_- = J_+^\dagger$, we get

$$\langle jm|J_-J_+|jm\rangle = \langle (J_+)jm|J_+|jm\rangle = |c_{jm}|^2 \tag{11.59}$$

where we have replaced λ_J by j in the notation for the coefficient c and we have used the normalization of the states: $\langle jm+1|jm+1\rangle = 1$. But equation (11.59) is equivalent to

$$\langle jm|\mathbf{J}^2 - J_z^2 - \hbar J_z|jm\rangle = |c_{jm}|^2$$

The expectation value on the left-hand side can be evaluated and this gives us the desired expression for c_{jm}:

$$c_{jm} = \hbar[j(j+1) - m(m+1)]^{1/2} \tag{11.60}$$

Similarly, by considering the expectation of J_+J_- in the state $|jm\rangle$, it is easy to evaluate d_{jm}:

$$d_{jm} = \hbar[j(j+1) - m(m-1)]^{1/2} \tag{11.61}$$

Now we can summarize the effect of raising and lowering operators on angular momentum eigenstates:

$$J_+|jm\rangle = \hbar\sqrt{j(j+1) - m(m+1)}\,|jm+1\rangle$$

$$J_-|jm\rangle = \hbar\sqrt{j(j+1) - m(m-1)}\,|jm-1\rangle \tag{11.62}$$

Use of Raising and Lowering Operators to Generate Spherical Harmonics

There is a nice application of the raising and lowering operators for the case of the orbital angular momentum, when **J** is **L**. They can be used to generate the spherical harmonics algebraically without having to solve the differential equation that we did in section 11.1. Let's sketch the procedure.

For any given l, the highest value m can attain is l. Therefore the state $|ll\rangle$ has the property that L_+ acting on it must give zero. We now can implement the resulting equation in $(\theta\phi)$-space; we will need the representation of the L_+ operator in spherical coordinates. This is found to be (using eq. [11.18] and taking the appropriate scalar products and linear combinations):

$$L_+ = \hbar e^{i\phi}\left[\frac{\partial}{\partial\theta} + i\cot\theta\,\frac{\partial}{\partial\phi}\right] \tag{11.63}$$

In this way we get

$$0 = \langle\theta\phi|L_+|ll\rangle = \hbar e^{i\phi}\left[\frac{\partial}{\partial\theta} + i\cot\theta\,\frac{\partial}{\partial\phi}\right]\langle\theta\phi|ll\rangle$$

This is a differential equation for the wave function $\langle\theta\phi|ll\rangle$. Use the known ϕ-dependence, $\exp(il\phi)$. Now do the θ-integration. We have

$$\langle\theta\phi|ll\rangle = (\text{normalization constant})\exp(il\phi)\sin^l\theta$$

which is indeed the form of $Y_{ll}(\theta)$; see equations (11.40) and (11.36).

The wave functions for $m = l - 1$ can be obtained by using the lowering operator L_- on $\langle\theta\phi|ll\rangle$ and using the (θ,ϕ)-representation of L_-. The procedure can be repeated to generate all the rest of the spherical harmonics. The details will be left as a problem.

PROBLEMS

1. If O is an operator such that

$$[O, L_x] = [O, L_y] = 0$$

calculate $[O, L_z]$ and $[O, \mathbf{L}^2]$.

2. Determine the rotational energy levels of HCl. Assume the interatomic distance to be a constant 1.27×10^{-8} cm.
3. A particle is in an angular momentum state of $l = 2$, $m = 1$. What is the probability of finding it at the position $\theta = \pi/4$ and $\phi = \pi/4$ within $d\theta =$ 0.01 radians and $d\phi$ the same?
4. The angular position measurement on a particle turns up with $(\theta,\phi) = (\pi/6, \pi/4)$ in $d\theta = 0.01$ radians and $d\phi$ the same. What is the probability of finding $l = 1$ if an instantaneous subsequent measurement of $\mathbf{L}^2$ is made on the particle?
5. A particle moving in a potential is described by the wave packet

$$\psi(x,y,z) = (xy + yz + zx)\exp[-\alpha(x^2 + y^2 + z^2)]$$

What is the probability that a measurement of $\mathbf{L}^2$ and L_z yields the results $6\hbar^2$ and $\hbar$, respectively?

6. Assume that a particle is in an eigenstate $|lm\rangle$. Show that

$$\langle L_x\rangle = \langle L_y\rangle = 0$$

Also show that

$$\langle L_x^2\rangle = \langle L_y^2\rangle = \frac{1}{2}\,[l(l+1)\hbar^2 - m^2\hbar^2]$$

7. A system is in the following coherent superposition of angular momentum eigenstates $|lm\rangle$:

$$|\psi\rangle = \alpha|11\rangle + \beta|10\rangle + \gamma|1-1\rangle$$

Calculate the expectation value of L_x in this state.

8. Generate the spherical harmonics Y_{ll-1} and Y_{ll-2} from the procedure described in the last subsection of the chapter using the operator L_-. From your calculation you will be able to infer that the general form of Y_{lm} must be

$$Y_{lm} = (\text{constant})\,\frac{\exp(im\phi)}{(\sin\theta)^m}\left(\frac{d}{d(\cos\theta)}\right)^{l-m}[(1-\cos^2\theta)^l]$$

9. Justify the alternative definition of the associated Legendre functions given in the text $(m \geq 0)$:

$$P_l^m(\zeta) = \frac{(-1)^m}{2^l}\,\frac{(l+m)!}{l!\,(l-m)!}\,(1-\zeta^2)^{-m/2}\,\frac{d^{l-m}}{d\zeta^{l-m}}\,(\zeta^2-1)^l$$

10. Discuss rotations in four spatial dimensions following the same procedure as outlined in the chapter. First generalize $\mathbf{L}$ to

$$L_{ij} = -i\hbar(x_i\partial/\partial x_j - x_j\partial/\partial x_i)$$

with $i,j = 1,2,3,4$. Next introduce the quantities

$$(J_1, J_2, J_3) = (L_{23}, L_{31}, L_{12})$$

and

$$(K_1, K_2, K_3) = (L_{14}, L_{24}, L_{34})$$

Find the commutation relations of all six operators J_i and K_i $(i = 1,2,3)$ among one another. Then show that the operators

$$\mathbf{J}^{(+)} = (\mathbf{J} + \mathbf{K})/2$$

and

$$\mathbf{J}^{(-)} = (\mathbf{J} - \mathbf{K})/2$$

are angular momentum operators (that is, their components obey angular momentum commutation relations). Also show that $\mathbf{J}^{(+)}$ and $\mathbf{J}^{(-)}$ commute. Finally, what quantum numbers would you use to label the eigenstates?

ADDITIONAL PROBLEMS

A1. Find the commutators $[L^2,x_i]$, $i = 1,2,3$.

A2. Suppose we write the angular momentum operator as

$$L_i = \Sigma_{jk}\epsilon_{ijk}x_j p_k$$

Define ϵ_{ijk}. Using this expression for L_i, show that

$$[L_i,x_j] = i\hbar\Sigma_k\epsilon_{ijk}x_k$$

A3. Show that the results $[L^2,x_i]$ of problem A1 can be written as

$$[\mathbf{L}^2,x_k] = (-i\hbar)\Sigma_{ij}\epsilon_{ijk}(L_ix_j + x_jL_i)$$

A4. Show that

$$[\mathbf{L}^2,\ [\mathbf{L}^2,\mathbf{x}]] = 2\hbar^2(\mathbf{x}L^2 + L^2\mathbf{x})$$

A5. Carry out an alternative derivation of equation (11.20) for the operator for L^2 in the following way: Start with

$$L^2 = (\mathbf{r} \times \mathbf{p})\cdot(\mathbf{r} \times \mathbf{p})$$

and note that since we are dealing with angular momentum, the radial component of the momentum operator p_r is inoperative, and only the component in the plane perpendicular to the radial direction $p_\perp$ need be considered. Now use the standard expression

$$(\mathbf{a} \times \mathbf{b})\cdot(\mathbf{c} \times \mathbf{d}) = (\mathbf{a}\cdot\mathbf{c})(\mathbf{b}\cdot\mathbf{d}) - (\mathbf{a}\cdot\mathbf{d})(\mathbf{b}\cdot\mathbf{c})$$

and make the usual conversion from classical c-numbers to quantum operator for $p_\perp$.

A6. A rigid rotator is found to be in the state

$$\psi(\theta,\phi) = (15/8\pi)^{1/2}\cos\theta\sin\theta\cos\phi$$

(a) What are the possible values of L_z that measurement will give and with what probability? (b) Determine the value of $\langle L_x \rangle$ for this state.

A7. Consider the Hamiltonian of two uncoupled, identical rigid rotators:

$$H = (1/2I)[L^2(1) + L^2(2)]$$

where I is the moment of inertia. Write down expressions for the energy eigenvalues and for the symmetric and antisymmetric eigenfunctions. Discuss the degeneracy of these eigenfunctions.

REFERENCES

M. Chester. *Primer of Quantum Mechanics.*
Fayyazuddin and Riazuddin. *Quantum Mechanics.*
S. Gasiorowicz. *Quantum Physics.*
E. Merzbacher. *Quantum Mechanics.*
R. L. White. *Basic Quantum Mechanics.*

12

Motion in Central Potential

We are now in a position to treat motion in three dimensions, at least an important class of three-dimensional motion—motion under a central potential, a potential that depends only on the distance r from a center. We can use the angular momentum eigenstates of chapter 11 to deal with the angular part of the motion, and all that remains is to treat the radial part. How so?

The key point is the conservation of angular momentum. When is angular momentum conserved? If you think classically, you will say, "when there is no torque," which is true for central potentials. But torque thinking is not useful in quantum mechanics, so here we approach things differently. When is angular momentum conserved? In quantum mechanics, the answer is, when there is rotational symmetry, when the Hamiltonian of the system is invariant under rotations. Of course, such an important assertion needs a proof, and one of the first tasks in this chapter is to construct such a proof.

Clearly, with angular momentum conservation, the central-potential problem becomes one of finding the simultaneous eigenstates of the Hamiltonian and the angular momentum. The knowledge of the eigenstates of the latter is then used to simplify the simultaneous eigenstate problem.

Central-potential problems are important in classical mechanics, and Kepler motion is the showpiece example. The same is true for quantum mechanics. The major, and historically the most important, central-force motion in quantum mechanics is the Kepler motion found in the hydrogen atom, but this will be treated in chapter 13. Instead, in this chapter we will consider a few pedagogical examples of quantum mechanical motion under central potential (consider them as practice problems) with only an occasional glimpse at real-world applications.

The classical Kepler problem is a two-body problem. Many quantum central-potential problems in the real world are also two-body problems. But we know how to solve only one-body Schrödinger equations exactly. Therefore, before all else, let's go through the machinations of how to convert the two-body problem into a doable one-body problem following the basic idea already developed in section 9.4.

12.1 REDUCTION OF TWO-BODY TO ONE-BODY PROBLEM

The Hamiltonian for two-body central force motion in three dimensions is given as

$$H = \frac{\mathbf{p}_1^2}{2m_1} + \frac{\mathbf{p}_2^2}{2m_2} + V(|\mathbf{r}_1 - \mathbf{r}_2|) \tag{12.1}$$

which is a six-dimensional Hamiltonian. As in chapter 9, the trick of the reduction of the two-body problem to one-body is to replace the two initial bodies (m_1, m_2) by two new bodies (M, μ) where

$$M = m_1 + m_2$$

and

$$\frac{1}{\mu} = \frac{1}{m_1} + \frac{1}{m_2}$$

The massive particle moves free of any potential. The light particle of reduced mass μ moves within the central potential of interest. Mathematically, what we do is a coordinate transformation in three-dimensional space taking us to the center-of-mass reference frame (fig. 12.1) defined by

$$M\mathbf{R} = m_1\mathbf{r}_1 + m_2\mathbf{r}_2; \qquad \mathbf{r} = \mathbf{r}_1 - \mathbf{r}_2$$

$$\mathbf{P} = \mathbf{p}_1 + \mathbf{p}_2; \qquad \mathbf{p}/\mu = (\mathbf{p}_1/m_1) - (\mathbf{p}_2/m_2)$$

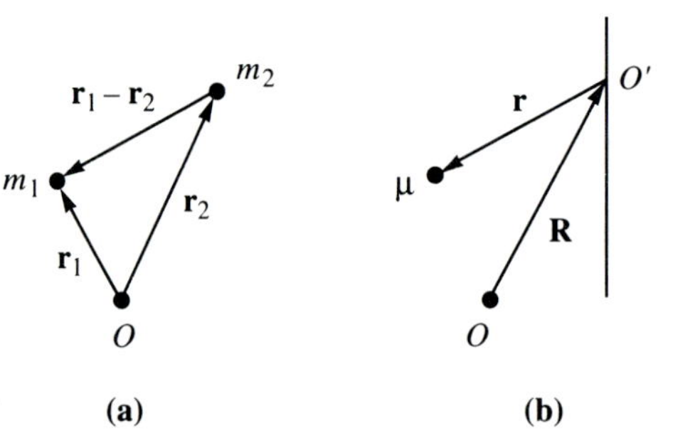

FIGURE 12.1
Transformation to the center-of-mass coordinate system illustrated. (a) Coordinate system with origin O fixed in space; (b) coordinate system with origin at the center of mass (O′).

In terms of these new variables, the Hamiltonian, equation (12.1), is easily seen to be

$$H = \frac{\mathbf{P}^2}{2M} + \frac{\mathbf{p}^2}{2\mu} + V(|\mathbf{r}|) \tag{12.2}$$

This is the Hamiltonian of two independently moving particles, and therefore the reduction has been achieved. As you know, whenever the Hamiltonian is the sum of two independent parts, the eigenvalue problem can be separated. The Hamiltonian of the one-body Schrödinger equation for the particle μ (eq. [12.3] below) easily follows via the same procedure as in section 9.4 (note that $\mathbf{p} = -i\hbar\nabla_{\mathbf{r}}$). We will now solve this Hamiltonian by finding its important symmetry—symmetry under rotation.

12.2 ROTATIONAL SYMMETRY AND SEPARATION OF VARIABLES

The Hamiltonian for the central-potential problem (two-body or one-body) is given as

$$H = \frac{\mathbf{p}^2}{2\mu} + V(|\mathbf{r}|) = -\frac{\hbar^2}{2\mu}\nabla^2 + V(r) \tag{12.3}$$

where for the two-body problem reduced to one-body, μ has to be interpreted as the reduced mass.

Let's consider the symmetry of this Hamiltonian under rotation of the coordinate system. Suppose we rotate the coordinate system about the z-axis by an angle θ. The rotated coordinates, call them x', y', and z', are related to the original coordinates x, y, and z via the relations

$$\begin{aligned} x' &= x\cos\theta - y\sin\theta \\ y' &= x\sin\theta + y\cos\theta \\ z' &= z \end{aligned} \tag{12.4}$$

The inverse relations are

$$\begin{aligned} x &= x'\cos\theta + y'\sin\theta \\ y &= -x'\sin\theta + y'\cos\theta \\ z &= z' \end{aligned} \tag{12.5}$$

To prove the invariance of H under the rotation above, we first note that

$$r' = (x'^2 + y'^2 + z'^2)^{1/2} = (x^2 + y^2 + z^2)^{1/2} = r$$

Thus the potential term does not change with rotation. For the kinetic energy term we note

$$\begin{aligned}
\nabla'^2 &= [(\partial/\partial x')^2 + (\partial/\partial y')^2 + (\partial/\partial z')^2] \\
&= [(\partial x/\partial x')(\partial/\partial x) + (\partial y/\partial x')(\partial/\partial y)]^2 \\
&\quad + [(\partial x/\partial y')(\partial/\partial x) + (\partial y/\partial y')(\partial/\partial y)]^2 + (\partial/\partial z)^2 \\
&= [\cos\theta(\partial/\partial x) - \sin\theta(\partial/\partial y)]^2 \\
&\quad + [\sin\theta(\partial/\partial x) + \cos\theta(\partial/\partial y)]^2 + (\partial/\partial z)^2 \\
&= \nabla^2
\end{aligned}$$

Accordingly, the Hamiltonian, kinetic plus potential, remains unchanged under rotation. If there is an invariance, there should be a conservation law, and we will see that there is an operator connected with rotation that commutes with the rotationally invariant Hamiltonian.

To find this operator, consider an infinitesimal rotation about the z-axis, that is, put $\cos\theta = 1$ and $\sin\theta = \theta$ in equation (12.4):

$$\begin{aligned}
x' &= x - \theta y \\
y' &= y + \theta x \\
z' &= z
\end{aligned} \tag{12.6}$$

Let $\psi(x, y, z)$ denote an eigenfunction of H

$$H\psi(x, y, z) = E\psi(x, y, z) \tag{12.7}$$

Now we impose the condition that

$$H\psi(x', y', z') = E\psi(x', y', z')$$

Or, substituting for x', y', and z' from equation (12.6), we have

$$H\psi(x - \theta y, y + \theta x, z) = E\psi(x - \theta y, y + \theta x, z)$$

If we Taylor-expand ψ in the above equation, keeping terms only up to the first order in θ, and then subtract equation (12.7) from it, we get

$$H\left(x\frac{\partial}{\partial y} - y\frac{\partial}{\partial x}\right)\psi(x,y,z) = E\left(x\frac{\partial}{\partial y} - y\frac{\partial}{\partial x}\right)\psi(x,y,z)$$

$$= \left(x\frac{\partial}{\partial y} - y\frac{\partial}{\partial x}\right)H\psi(x,y,z)$$

But

$$x\frac{\partial}{\partial y} - y\frac{\partial}{\partial x} = \frac{i}{\hbar}L_z$$

where L_z, you recognize, is the z-component of angular momentum. In this way, we see that L_z commutes with H, $[L_z, H] = 0$, as far as operation on ψ is concerned. However, the functions $\psi(x,y,z)$, being eigenfunctions of H, form a complete set and any wave function can be expanded in terms of them. It follows that the commutation relation

$$[H, L_z] = 0 \tag{12.8}$$

holds true in general.

Similarly, by considering the invariance of the Schrödinger equation under infinitesimal rotations about the x- and y-axes, we can easily show that

$$[H, L_x] = [H, L_y] = 0 \tag{12.9}$$

Since all three components of angular momentum commute with H, we conclude that the angular momentum is the conserved quantity we have been looking for.

Note also that for infinitesimal rotation about the z-axis

$$\psi(x', y', z) = (1 + iL_z\theta/\hbar)\psi(x,y,z) \tag{12.10}$$

Consequently, L_z is recognized as the generator of infinitesimal rotation about the z-axis (in the same spirit that we recognize the Hamiltonian H as the generator of translations in time; see equation [6.26]).

Coming back to the treatment of the central-potential Hamiltonian, how do we take advantage of the conserved quantities, namely, L_x, L_y, and L_z discovered above? There is a slight problem; these operators do not form a commuting set and therefore cannot *all* have simultaneous eigenvalues with H, and cannot *all* be used as labels for the eigenstates of H. There is, however, the operator $\mathbf{L}^2$, which commutes with all the L_i ($i = x, y, z$) and with H. Accordingly, we seek simultaneous eigenvalues of H, $\mathbf{L}^2$, and L_z (the last one by convention). Note that since we are going to calculate eigenstates in three dimensions, we must have three labels for them; the above three eigenvalues provide these labels.

The Hamiltonian in Spherical Coordinates

For the Hamiltonian, equation (12.3), the potential energy term is already in spherical coordinates; thus the job is to express the kinetic energy in terms of r, θ, and ϕ. To this end, we will use equation (11.17) for the ∇ operator to evaluate ∇^2:

$$\nabla^2 = \left(\mathbf{e}_r \frac{\partial}{\partial r} + \mathbf{e}_\theta \frac{1}{r}\frac{\partial}{\partial \theta} + \mathbf{e}_\phi \frac{1}{r\sin\theta}\frac{\partial}{\partial \phi}\right) \cdot \left(\mathbf{e}_r \frac{\partial}{\partial r} + \mathbf{e}_\theta \frac{1}{r}\frac{\partial}{\partial \theta} + \mathbf{e}_\phi \frac{1}{r\sin\theta}\frac{\partial}{\partial \phi}\right)$$

We can perform the required differentiations easily by using equation (11.15). The rest is algebra. In this way we obtain the very useful equation for the Laplacian ∇^2 in spherical coordinates:

$$\nabla^2 = \frac{1}{r^2}\frac{\partial}{\partial r}\left(r^2 \frac{\partial}{\partial r}\right) + \frac{1}{r^2 \sin\theta}\frac{\partial}{\partial \theta}\left(\sin\theta \frac{\partial}{\partial \theta}\right) + \frac{1}{r^2\sin^2\theta}\frac{\partial^2}{\partial \phi^2} \tag{12.11}$$

Now notice that the (θ, ϕ)-part of the operator above is but $-\mathbf{L}^2/r^2\hbar^2$ since $\mathbf{L}^2$ is given by equation (11.20). Consequently, we have

$$\mathbf{p}^2 = -\hbar^2\nabla^2 = -\hbar^2\left[\frac{1}{r^2}\frac{\partial}{\partial r}\left(r^2\frac{\partial}{\partial r}\right) - \frac{\mathbf{L}^2}{\hbar^2 r^2}\right] \tag{12.12}$$

Finally, the central-field Hamiltonian in spherical coordinates can be written down:

$$H = -\frac{\hbar^2}{2\mu}\left[\frac{1}{r^2}\frac{\partial}{\partial r}\left(r^2\frac{\partial}{\partial r}\right) - \frac{\mathbf{L}^2}{\hbar^2 r^2}\right] + V(r) \tag{12.13}$$

The "Radial Momentum" Operator

There is a shortcut to equation (12.12), but it's tricky. We can try to use the definition $\mathbf{L} = \mathbf{r} \times \mathbf{p}$ and take its scalar product with itself:

$$\mathbf{L}^2 = (\mathbf{r} \times \mathbf{p}) \cdot (\mathbf{r} \times \mathbf{p}) = \mathbf{r}^2\mathbf{p}^2 - (\mathbf{r}\cdot\mathbf{p})^2 \tag{12.14}$$

but this is not kosher in quantum mechanics because we are dealing with products of noncommuting operators. Actually, the first term is okay, but is there a way to symmetrize the difficult second term on the right-hand side that works?

We may start by defining the operator

$$p_r = -i\hbar\partial/\partial r \tag{12.15}$$

You may call this operator the "radial momentum," but beware! It is not a hermitian operator, and therefore not measurable and not an observable.

Clearly,

$$\mathbf{r}\cdot\mathbf{p} = -i\hbar\left(x\frac{\partial}{\partial x} + y\frac{\partial}{\partial y} + z\frac{\partial}{\partial z}\right) = -i\hbar r\frac{\partial}{\partial r} = rp_r \tag{12.16}$$

It turns out that the proper symmetric way to write the operator product $(\mathbf{r}\cdot\mathbf{p})^2$ is

$$(\mathbf{r}\cdot\mathbf{p})^2 = p_r r^2 p_r \tag{12.17}$$

Let's check this out. Using equation (12.17), let's assume that the proper way to symmetrize $\mathbf{L}^2$ in equation (12.14) is

$$\mathbf{L}^2 = \mathbf{r}^2\mathbf{p}^2 - p_r r^2 p_r \tag{12.18}$$

This gives for $\mathbf{p}^2$

$$\mathbf{p}^2 = (1/r^2)(p_r r^2 p_r + \mathbf{L}^2)$$

and this is the same as equation (12.12). The advantage of equation (12.18) is the ease with which you can memorize it. It gives you an alternative way to memorize the Laplacian in spherical coordinates.

The Radial Equation

Now let's label the states as $|nlm\rangle$, as planned, n being a quantum number related to the energy eigenvalue, and you already know what l and m signify. We are interested only in the bound-state problem right now, thus no generality is lost by assuming discrete energy levels. The eigenvalue equation for the Hamiltonian

$$H|nlm\rangle = E|nlm\rangle$$

becomes for spherical coordinate representation

$$\frac{2\mu}{\hbar^2}\langle r\theta\phi|H|nlm\rangle = \langle r\theta\phi|\frac{L^2}{r^2\hbar^2} - \frac{1}{r^2}\frac{\partial}{\partial r}\left(r^2\frac{\partial}{\partial r}\right) + \frac{2\mu V(r)}{\hbar^2}|nlm\rangle$$

$$= \frac{2\mu E}{\hbar^2}\langle r\theta\phi|nlm\rangle$$

Since the Hamiltonian contains additive terms that act on separate subspaces, the angular and radial spaces, respectively, the Schrödinger equation for the eigenfunction

$$\langle r\theta\phi | nlm \rangle = \psi_{nlm}(\mathbf{r})$$

clearly is separable in the angular and radial parts, the angular part being the eigenfunction of $\mathbf{L}^2$. Accordingly, we factorize ψ_{nlm} in the form

$$\psi_{nlm}(\mathbf{r}) = R_{nl}(r) Y_{lm}(\theta\phi)$$

where the quantum number n (as the quantum number related to energy) will emerge from the solution of the radial equation. And since

$$\mathbf{L}^2 Y_{lm}(\theta\phi) = l(l+1)\hbar^2 Y_{lm}(\theta\phi)$$

we get for the radial equation

$$\left[\frac{1}{r^2}\frac{d}{dr}\left(r^2 \frac{d}{dr}\right) + \frac{2\mu}{\hbar^2}\left(E - V(r) - \frac{\hbar^2 l(l+1)}{2\mu r^2}\right)\right] R_{nl}(r) = 0 \tag{12.19}$$

Notice that the radial equation has explicit reference to l; thus it is that we have used the label l (as well as n) for the radial function R.

12.3 SOLUTION OF THE RADIAL EQUATION FOR BOUND STATES

In this section we will first study the radial equation (12.19) in a general way, and second, solve it for the simplest situation you can imagine. We will assume that $rV(r)$ is nonsingular. And also that $r^2 V(r) \to 0$, as $r \to 0$.

Consider normalization—the total wave function $\psi_{nlm}(\mathbf{r})$ must be normalized:

$$\int |\psi_{nlm}(r, \theta, \phi)|^2 \, d^3r = 1$$

Substitute for $\psi_{nlm} = R_{nl}(r) Y_{lm}(\theta, \phi)$, and realize that the spherical harmonics are normalized functions integrated over the solid angle of $\mathbf{r}$. This gives

$$\int_0^\infty r^2 \, dr \, |R_{nl}(r)|^2 = 1 \tag{12.20}$$

We will be interested for now in finding only the bound-state solutions ($E < 0$) of the radial equation and take up the continuum solutions much later (see chapter 23). The reason is partly historical. Historically, atomic physics dominated the early development of quantum mechanics, and the study of bound states tells us much that there is to know about atoms. The continuum solutions enter for the treatment of scattering problems. Scattering is a useful, in fact essential, probe for nuclei and elementary particles but until recently has played a relatively minor role in atomic physics.

For bound states, the system remains localized and the probability of finding it at very large distances from the center of the field must go to 0. Thus we have to impose the following boundary condition on R_{nl}:

$$rR_{nl}(r) \to 0 \quad \text{as} \quad r \to \infty \tag{12.21a}$$

Note that this boundary condition also ensures square integrability, the normalization condition, equation (12.20). There is also a requirement for $r \to 0$

$$rR_{nl}(r) \to 0 \quad \text{as} \quad r \to 0 \tag{12.21b}$$

Why? The reason is that otherwise we would have a finite probability for finding the particle in a sphere of arbitrarily small radius about the origin.

Since both of the above boundary conditions refer to $rR_{nl}(r)$, we may as well rewrite the radial equation for the function

$$u_{nl}(r) = rR_{nl}(r) \tag{12.22}$$

Since

$$\left[\frac{1}{r^2}\frac{d}{dr}\left(r^2\frac{d}{dr}\right)\right]\frac{u_{nl}}{r} = \left[\frac{d^2}{dr^2} + \frac{2}{r}\frac{d}{dr}\right]\frac{u_{nl}}{r} = \frac{1}{r}\frac{d^2u_{nl}}{dr^2}$$

the radial equation actually is simpler in terms of $u_{nl}(r)$ and is given as

$$\frac{d^2u_{nl}}{dr^2} + \frac{2\mu}{\hbar^2}\left[E_{nl} - V(r) - \frac{\hbar^2 l(l+1)}{2\mu r^2}\right]u_{nl}(r) = 0 \tag{12.23}$$

This equation has to be solved under the boundary conditions

$$u_{nl}(r) \to 0 \quad \text{as} \quad r \to 0$$

$$u_{nl}(r) \to 0 \quad \text{as} \quad r \to \infty \tag{12.24}$$

You can see the similarity of the radial Schrödinger problem expressed in this fashion with the one-dimensional Schrödinger equation. Of course, there is the boundary condition at the origin (in the one-dimensional case this can be simulated by erecting an infinite potential barrier at the origin); and there is also something extra in the form of a repulsive potential energy $\hbar^2 l(l+1)/2\mu r^2$. This is called the *centrifugal barrier.*

The origin of the centrifugal term can be understood in the following manner using classical correspondence. For a particle of mass μ moving in a circular orbit of radius r, classically there is a centrifugal force directed radially outward (fig. 12.2). The magnitude of the force is $\mu v^2/r = L^2/\mu r^3$, where $L = \mu v r$ for a circular orbit. The potential corresponding to such a force is $L^2/2\mu r^2$ (since $F = -\partial V/\partial r$). In quantum mechanics we must replace L^2 by its eigenvalue $l(l+1)\hbar^2$, hence we obtain the quantum mechanical expression for the centrifugal potential.

Figure 12.3 shows a typical attractive central potential $V(r)$, the centrifugal potential, and the *effective* potential, the sum total of the two:

$$V_{\text{eff}}(r) = V(r) + \hbar^2 l(l+1)/2\mu r^2 \tag{12.25}$$

You can see that the effect of the centrifugal term is to reduce the effective depth of the well and to introduce a repulsive potential (a barrier) at a short range.

Example: Particle in a Spherical Box

Now a useful practice problem—the case of a quantum object trapped in a spherical box. A spherical box is defined as a potential problem as follows:

$$V(r) = 0, \qquad \text{for } r < a$$

$$V(r) = \infty, \qquad \text{for } r > a$$

The radial equation (12.19) inside the box becomes

$$\left[\frac{d^2}{dr^2} + \frac{2}{r}\frac{d}{dr} - \frac{l(l+1)}{r^2}\right]R(r) + k^2 R(r) = 0 \tag{12.26}$$

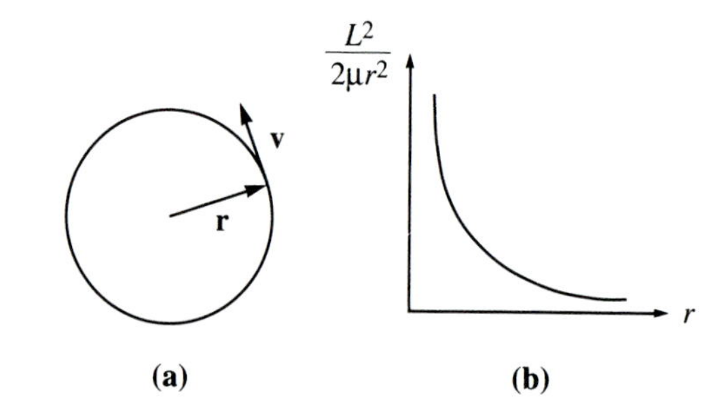

FIGURE 12.2
(a) A particle moving in a circular orbit of angular momentum l; (b) the centrifugal potential.

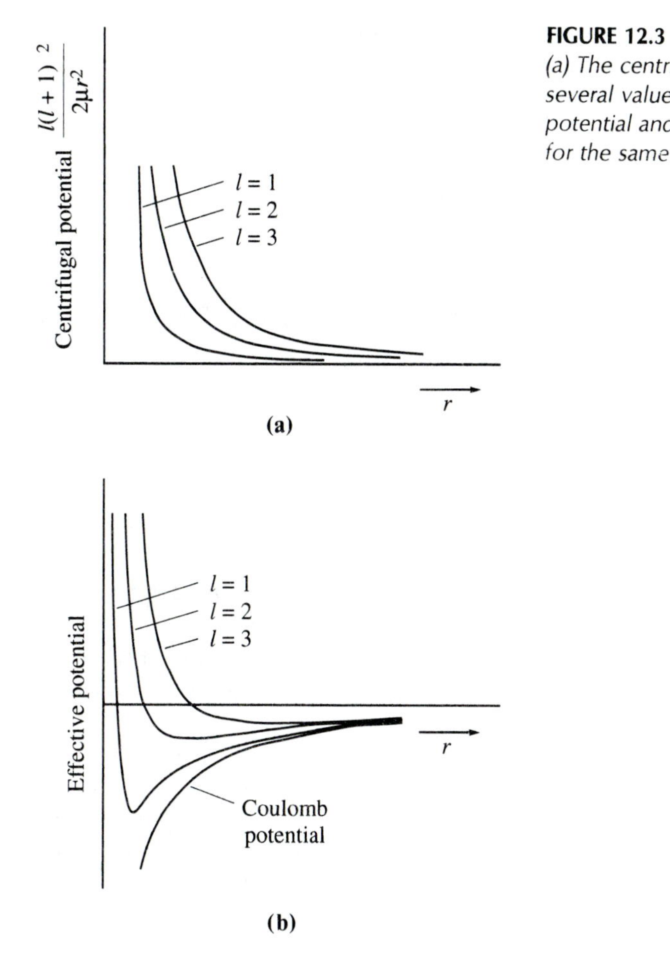

FIGURE 12.3
(a) The centrifugal potential for several values of l; (b) the Coulomb potential and the effective potential for the same values of l.

where we have introduced $k^2 = 2\mu E/\hbar^2$. Make a change of variables to the dimensionless $\rho = kr$. The radial equation now reads

$$\frac{d^2R}{d\rho^2} + \frac{2}{\rho}\frac{dR}{d\rho} + \left[1 - \frac{l(l+1)}{\rho^2}\right]R = 0 \tag{12.27}$$

The solutions of this equation are well-known simple functions. We are interested only in the solution that remains regular at the origin. The regular solution of the equation is called the *spherical Bessel function* and is given as

$$j_l(\rho) = \left(\frac{\pi}{2\rho}\right)^{1/2} J_{l+1/2}(\rho)$$

where $J_{l+1/2}(\rho)$ is an ordinary Bessel function of half-odd integer order. The first few functions are

$$j_0(\rho) = \sin\rho/\rho$$

$$j_1(\rho) = (\sin\rho/\rho^2) - \cos\rho/\rho$$

$$j_2(\rho) = [(3/\rho^3) - 1/\rho]\sin\rho - (3/\rho^2)\cos\rho \tag{12.28}$$

In this way we see that the solution of the radial equation (12.27), subject to the boundary conditions in equation (12.21), can be written as

$$R(r) = Cj_l(kr) \tag{12.29}$$

where C is a normalization constant. Since the wave function must vanish at the wall of the box, $r = a$, we must further impose the condition

$$j_l(ka) = 0 \tag{12.30}$$

The eigenvalues are therefore given by the zeroes of the spherical Bessel function. Below is a list of the first few zeroes for each of the first few l's:

$l = 0$	1	2	3	4
3.14	4.49	5.76	6.99	8.18
6.28	7.73	9.10	10.42	
9.42				

Figure 12.4 shows the eigenvalue spectrum, where for the sake of historical idiosyncrasy, the so-called spectroscopic notation (the notation originated in atomic spectroscopy) has been used for denoting states of different l (you may as well memorize this notation, because we will often use it; note also that some authors use lowercase letters):

$l = 0$ S (sharp)

$l = 1$ P (principal)

$l = 2$ D (diffuse)

$l = 3$ F (fine)

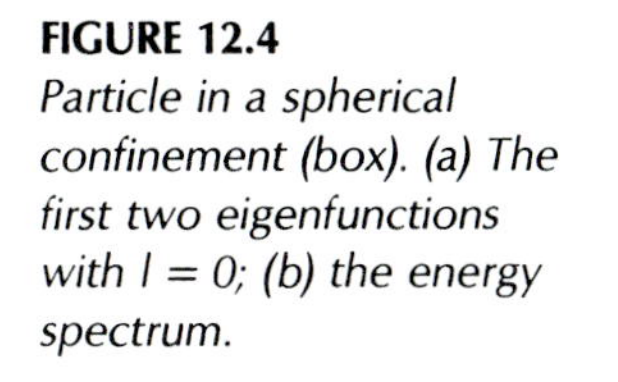

FIGURE 12.4
Particle in a spherical confinement (box). (a) The first two eigenfunctions with $l = 0$; (b) the energy spectrum.

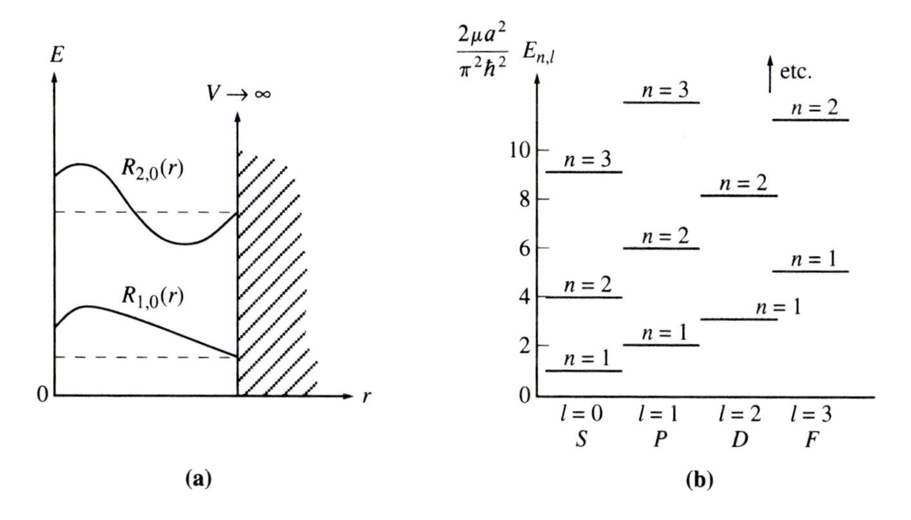

For each l, the first zero of j_l will be given the quantum number $n = 1$, the second $n = 2$, and so forth. In this way the order in which the energy levels occur is found to be (fig. 12.4):

$$1S,\ 1P,\ 1D,\ 2S,\ 1F,\ 2P,\ 1G,\ 2D,\ \text{etc.}$$

A notable feature is the ordering "theorem" that the solutions reveal: If two solutions have the same number of radial nodes, the one with higher l has the higher energy.

12.4 THE (SPHERICAL) SQUARE-WELL POTENTIAL

Of special importance is the attractive (spherical) square-well potential (fig. 12.5):

$$V(r) = -V_0 \quad \text{for } r < a$$
$$= 0 \quad \text{for } r > a \tag{12.31}$$

Admittedly, this potential has abrupt behavior and is not expected to occur in nature; yet it has been applied quite successfully in the two-nucleon problem, for example, for the case of the ground state of the deuteron, which we will consider below.

The Case of the S-State, $l = 0$

Let's consider the simpler S-state case first, and to this purpose, consider the radial equation in terms of $u = rR$. For bound states, we put $E = -|E|$ and obtain for $r < a$

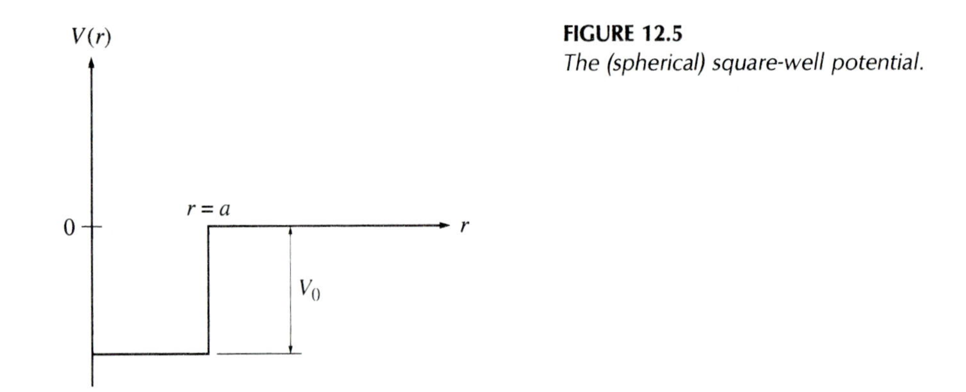

FIGURE 12.5
The (spherical) square-well potential.

$$\left[\frac{d^2}{dr^2} + \frac{2\mu}{\hbar^2}(V_0 - |E|)\right]u(r) = 0 \qquad (12.32a)$$

For $r > a$, we have

$$\left[\frac{d^2}{dr^2} - \frac{2\mu}{\hbar^2}|E|\right]u(r) = 0 \qquad (12.32b)$$

It is convenient to define the inside and outside wave numbers denoted by k_{in} and k_B, respectively:

$$2\mu(V_0 - |E|)/\hbar^2 = k_{\text{in}}^2$$

$$2\mu|E|/\hbar^2 = k_B^2 \qquad (12.33)$$

The radial equations (12.32a,b) now become

$$\frac{d^2u(r)}{dr^2} + k_{\text{in}}^2 u(r) = 0 \qquad (12.34a)$$

for $r < a$ and

$$\frac{d^2u(r)}{dr^2} - k_B^2 u(r) = 0 \qquad (12.34b)$$

for $r > a$.

Notice that these equations are rather similar to the one-dimensional case treated in chapter 4, and the procedure to obtain the eigenvalue condition is also the same. We take the appropriate inside and outside solutions and match them at $r = a$. The most general solution for equation (12.34a), $r < a$, is given as

$$u(r) = A \sin k_{\text{in}} r + B \cos k_{\text{in}} r$$

where A and B are constants. But remember, $u(r)$ must go to 0 at $r = 0$. As a result, we must discard the (irregular) cosine solution for the inside wave function and set $B = 0$. Therefore,

$$u(r) = A \sin k_{\text{in}} r \tag{12.35}$$

for $r < a$.

For $r > a$, equation (12.34b) has both decaying and growing exponential solutions; again we must invoke the boundary condition for a bound state, that $u \to 0$ as $r \to \infty$, to discard the growing exponential. Hence the solution for $r > a$ is given as

$$u(r) = C \exp(-k_B r) \tag{12.36}$$

The matching at $r = a$ of these two solutions is best carried out by equating the logarithmic derivatives (this gets rid of the two constants A and C in which we have not much interest at the moment, since we are concerned only with the eigenvalues); we put

$$\left.\frac{1}{u}\frac{du}{dr}\right|_{r=a_-} = \left.\frac{1}{u}\frac{du}{dr}\right|_{r=a_+}$$

This gives

$$k_{\text{in}} \cot k_{\text{in}} a = -k_B \tag{12.37}$$

which is the desired equation for the eigenvalues $-|E|$ of the problem

$$|E| = \hbar^2 k_B^2 / 2\mu \tag{12.38}$$

Application to Deuteron

Let's now apply all this to the deuteron nucleus, the bound state of a neutron and a proton held together by the nuclear potential; the binding energy of deuteron is known, $|E| = 2.226$ MeV, and therefore k_B is known from equation (12.38). We now can use equation (12.37), assuming that the ground state of deuteron is an S-state, to determine the parameters of the square-well potential that fit the deuteron binding energy. For a given energy eigenvalue, equation (12.37) is just a relationship between the range and depth parameters of the potential—the range-depth relationship previously discussed in chapter 4.

Since the binding energy of the deuteron is rather small compared to the depth of the potential V_0 (we will check this out shortly), it is instructive to choose it to be zero for an approximate estimate of the range-depth relation. If $k_B \approx 0$, we have

$$\cot k_{\text{in}} a = 0$$

which gives

$$k_{\text{in}} a = \pi/2 \tag{12.39}$$

Squaring and substituting for k_{in}^2 (remember, $|E| \approx 0$), we get

$$\frac{2\mu V_0 a^2}{\hbar^2} = \frac{\pi^2}{4}$$

Therefore, with the assumption of negligible binding energy, the range-depth relationship is given as

$$V_0 a^2 = \frac{\pi^2 \hbar^2}{8\mu} = \frac{\pi^2 \hbar^2}{4m} \tag{12.40}$$

since the reduced mass $\mu = m/2$, m being the mass of a nucleon (assuming the neutron and proton masses to be equal). The range of nuclear potential is approximately equal to the Compton wavelength of the pion which, to a first approximation, mediates the interaction between nucleons; that is, $a \approx \hbar/m_\pi c = 1.4$ F, where 1 F (fermi) $= 10^{-13}$ cm. This gives

$$V_0 = \frac{\pi^2 \hbar^2}{4ma^2} = \frac{\pi^2 (\hbar c)^2}{4mc^2 a^2} \approx 52 \text{ MeV}$$

where we have substituted $\hbar c = 200$ MeV-F, and $mc^2 \approx 1000$ MeV. We find that our initial assumption $|E| \ll V_0$ is justified.

The exact range-depth relationship can also be obtained easily. We rewrite equation (12.37) in the form

$$k_{\text{in}} a = \arctan(-k_{\text{in}}/k_B) \tag{12.41}$$

This equation has many solutions; however, only the smallest value of the energy eigenvalue is consistent with the interpretation that we are dealing with the ground state of the deuteron. We conclude that the arctangent lies in the second quadrant; note that this also means that $k_{in}a$ also lies in the second quadrant, that is,

$$\pi/2 \leq k_{in}a \leq \pi$$

The range-depth relation for the square-well potential is plotted in figure 12.6.

From the relationship of the binding energy and the exterior wave number k_B, it is clear that the binding energy determines the exponential fall-off of the deuteron wave function. Moreover, $1/k_B$ is seen to give the measure of the spatial extent of the deuteron—the region in which there is a significant probability for finding it. From the value of the observed binding energy the spatial extent is seen to be 4.31×10^{-13} cm, which is considerably larger than the range of the potential. It follows that the wave function is relatively insensitive to the shape of the potential, and this justifies the square-well treatment.

We assumed in our treatment that the range of the nuclear potential can be taken to be the Compton wavelength of the pion. A justification can be attempted as follows using the time-energy uncertainty relation. You can violate the principle of conservation of energy so long as you do it within the time that the uncertainty relation permits; it is a sort of borrowing with uncertainty as

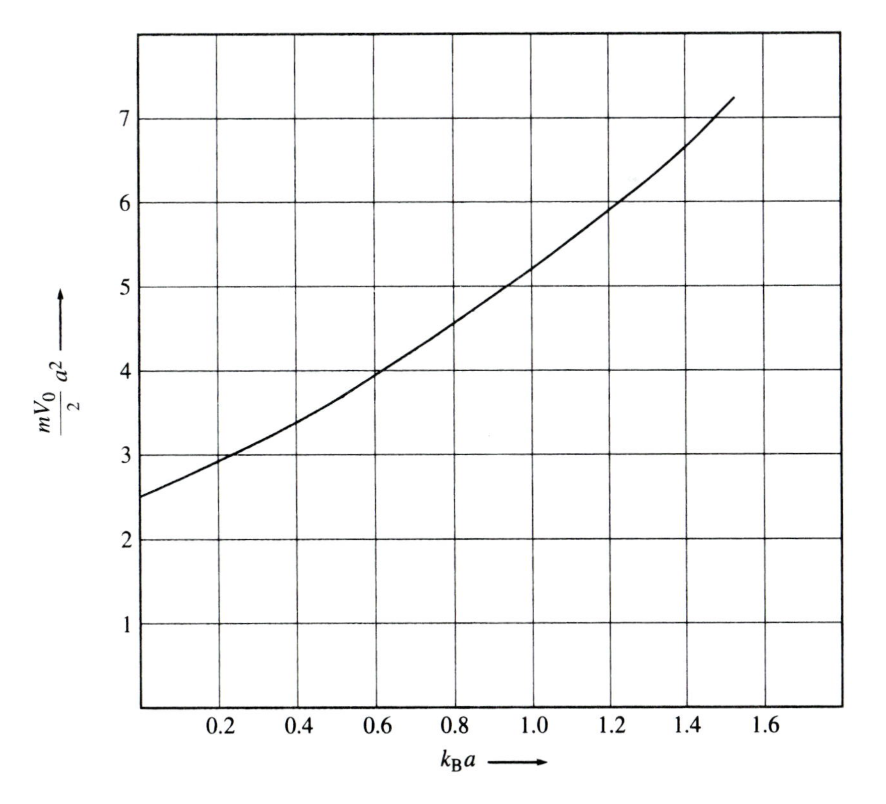

FIGURE 12.6
The range-depth relation for the square-well potential is exhibited by plotting $mV_0a^2/\hbar^2$ as a function of k_Ba. See text for notation.

your collateral. Therefore, a nucleon can cough up a pion spontaneously, violating the energy conservation by an amount $m_\pi c^2$, provided its interacting partner catches it within the time

$$\Delta t \approx \hbar/\Delta E = \hbar/m_\pi c^2$$

During this time the pion can travel at most the distance

$$c \times (\hbar/m_\pi c^2) = \hbar/m_\pi c$$

which is therefore a reasonable measure of how long the range of the interaction can be.

Finally, how good is our assumption that the ground state of the deuteron is an S-state? This is an easily discernible experimental question, because if the deuteron is an S-state, its charge distribution should be entirely spherically symmetric. It turns out that there is a small amount of deviation from spherical symmetry, an amount that suggests only about a 5% admixture of the D-wave in the ground state. (Why can't the P-wave be mixed in?) Such an admixture could not come about if there were only central potentials; but nature does not live by central potentials alone.

Solution for Arbitrary l

For $l \neq 0$, it is more convenient to use the radial equation for R, and make a change of variables. The radial equation in terms of R for the square-well potential is given as

$$\frac{d^2R}{dr^2} + \frac{2}{r}\frac{dR}{dr} + \frac{2\mu}{\hbar^2}(V_0 - |E|)R - \frac{l(l+1)}{r^2}R = 0 \qquad r < a$$

$$\frac{d^2R}{dr^2} + \frac{2}{r}\frac{dR}{dr} - \frac{2\mu}{\hbar^2}|E|R - \frac{l(l+1)}{r^2}R = 0 \qquad r > a \tag{12.42}$$

For the interior solution, we change variables from r to $\rho = k_{\text{in}} r$, k_{in} given by equation (12.33). The wave equation then becomes (for $r < a$)

$$\frac{d^2R_l}{d\rho^2} + \frac{2}{\rho}\frac{dR_l}{d\rho} + \left[1 - \frac{l(l+1)}{\rho^2}\right]R_l = 0 \tag{12.43}$$

This is the same equation as (12.27), and the solution regular at the origin that we are interested in is given by the spherical Bessel function $j_l(\rho)$, as before:

$$R_l(r) = A_l j_l(k_{\text{in}} r) \tag{12.44}$$

For $r > a$, we make the substitution, $r \to \rho = ik_B r$ (k_B given by eq. [12.33]), which again leads to the spherical Bessel equation. However, the domain of ρ

no longer includes $\rho = 0$, and therefore, not only the solution $j_l(\rho)$ regular at the origin is permitted, but also the so-called spherical Neumann function $n_l(\rho)$, which is irregular at the origin, is permitted. The spherical Neumann function n_l is given as

$$n_l(\rho) = (-1)^{l+1}(\pi/2\rho)^{1/2}J_{-l-1/2}(\rho) \tag{12.45}$$

The first few spherical Neumann functions are

$$\begin{aligned} n_0(\rho) &= -\cos\rho/\rho \\ n_1(\rho) &= -(\cos\rho/\rho^2) - \sin\rho/\rho \\ n_2(\rho) &= -[(3/\rho^3) - (1/\rho)]\cos\rho - (3/\rho^2)\sin\rho \end{aligned} \tag{12.46}$$

The general exterior solution is a linear combination of j_l and n_l; the particular linear combination to be chosen is determined by the usual asymptotic boundary condition for a bound state

$$R \sim \exp(-k_B r)$$

Define the functions

$$\begin{aligned} h_l^{(1)}(\rho) &= j_l(\rho) + in_l(\rho) \\ h_l^{(2)}(\rho) &= j_l(\rho) - in_l(\rho) \end{aligned} \tag{12.47}$$

These are called the spherical Hankel functions of the first and second kind, respectively. We are particularly interested in the asymptotic forms. To this end we note

$$\begin{aligned} j_l(\rho) &\underset{\rho\to\infty}{\longrightarrow} \rho^{-1}\cos[\rho - (l+1)\pi/2] \\ n_l(\rho) &\underset{\rho\to\infty}{\longrightarrow} \rho^{-1}\sin[\rho - (l+1)\pi/2] \end{aligned} \tag{12.48}$$

It is clear that only $h_l^{(1)}(\rho)$ has the correct asymptotic behavior, equation (12.36), for a bound state. This suggests that for the exterior solution, $r > a$, we choose

$$\begin{aligned} R(r) &= Bh_l^{(1)}(ik_B r) \\ &= B[j_l(ik_B r) + in_l(ik_B r)] \end{aligned} \tag{12.49}$$

The interior and exterior solutions must match at $r = a$. Equating their log-derivatives, we obtain

$$k_{\text{in}}\left[\frac{dj_l(\rho)/d\rho}{j_l(\rho)}\right]_{\rho=k_{\text{in}}a} = ik_B\left[\frac{dh_l^{(1)}(\rho)/d\rho}{h_l^{(1)}(\rho)}\right]_{\rho=k_B a} \tag{12.50}$$

Admittedly, this is a very complicated transcendental equation, but for any value of l, it does give us an eigenvalue equation or a range-depth relationship, whichever is desired.

12.5 THE CASE OF THE LINEAR POTENTIAL FOR $l = 0$

Finally, we will solve the radial equation for a potential with an explicit r-dependence—the linear potential (fig. 12.7). We previously discussed the one-dimensional linear potential on two different occasions, in chapter 4 and again in chapter 8. The Schrödinger equation for the linear potential in the three-dimensional case for the S-state ($l = 0$) is very similar to the one-dimensional case and its solutions are found in terms of the Airy functions now familiar to the reader. We include it here because the potential has important use as the interaction between quarks, the elementary particles of which strongly interacting particles are supposedly made. For example, a nucleon is thought of as a "bag" of three quarks, and mesons as bags of quark-antiquark pairs. The linear potential, being proportional to r, goes to zero as $r \to 0$, and becomes large as r becomes large, mimicking what we know about quarks. Quarks seem to behave as free particles as $r \to 0$; on the other hand, for large r, it takes too much energy to separate a quark from its bag. Consequently, the three-dimensional linear potential gives us a good model for quark confinement—why we never see quarks free in the daylight.

We will write the linear potential as

$$V(r) = Cr \tag{12.51}$$

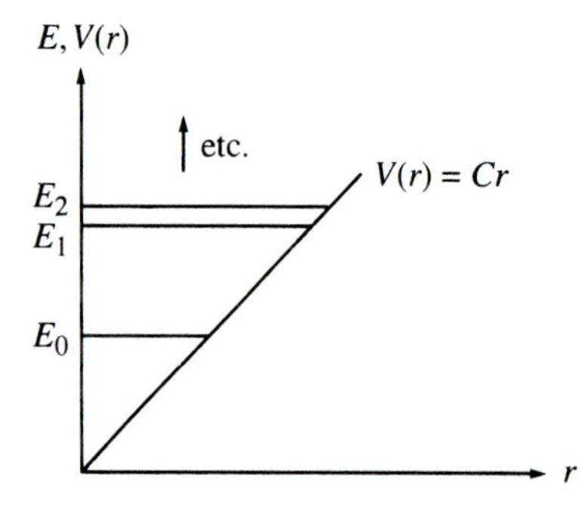

FIGURE 12.7
The linear potential Cr and some of the energy levels for motion in such a potential.

Then the radial Schrödinger equation (12.23) for $l = 0$ and for this potential becomes (notice we are using the equation for $u = rR(r)$; it is always more convenient for $l = 0$)

$$\left[\frac{d^2}{dr^2} + \frac{2\mu}{\hbar^2}(E_n - Cr)\right]u_n(r) = 0 \tag{12.52}$$

where the subscript l being fixed as zero is no longer needed. The boundary conditions on the wave function u, if you recall, are

$$u_n(r) \to 0 \quad \text{as} \quad r \to 0 \quad \text{and} \quad r \to \infty \tag{12.24}$$

We now make a change of variables. Define

$$x = (2\mu C/\hbar^2)^{1/3}(r - E/C)$$

so that

$$r = (2\mu C/\hbar^2)^{-1/3}x + E/C$$

Consequently x lies in the range $-E(2\mu/\hbar^2C^2)^{1/3} \le x \le \infty$. The radial equation in terms of x reads

$$\left[\frac{d^2}{dx^2} - x\right]u_n(x) = 0 \tag{12.53}$$

And the boundary conditions, equation (12.24), now give

$$u(x = -E(2\mu/\hbar^2C^2)^{1/3}) = 0 \quad \text{and} \quad u(x \to \infty) = 0 \tag{12.54}$$

The solutions of equation (12.53), you already know; they are the Airy functions $Ai(x)$ that conveniently go to zero for $x \to \infty$ (see fig. 8.3). Clearly, the eigenvalues of the bound-state energy are given from the zeroes of $Ai(x)$, which has to satisfy the boundary conditions of equation (12.54). Calling these zeroes a_n, the eigenvalues E_{n0} (the extra subscript stands for l, which is 0) are given by

$$E_{n0} = -(\hbar^2C^2/2\mu)^{1/3}a_n \tag{12.55}$$

The numerical values of the first few a_n's have already been given in chapter 4. Figure 12.8 displays the wave functions $u_n(r)$ for the first few n's.

Let's now determine the constant C of the linear potential by looking at the data for some suitable quark bag. It is believed that the J/ψ mesons are made

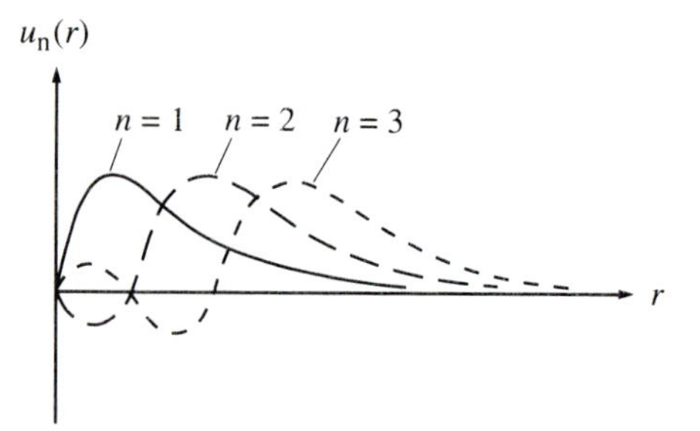

FIGURE 12.8
The radial eigenfunctions of the linear potential $rR_{nl=0}$ are shown for the first three n's.

up of quarks and antiquarks named charm, and for these particles the excitation spectrum is quite well known. In particular the ground-state ($n = 1, l = 0$) energy (the mass) is given as

$$E_{10} = 3.097 \text{ GeV}$$

where 1 GeV $= 10^9$ eV. The excitation energy of the $n = 2$, $l = 0$ state is also known:

$$E_{20} = 3.685 \text{ GeV}$$

Since quarks and antiquarks have the same mass, the reduced mass μ of the system is $(1/2)m_c$, where m_c is the mass of a charmed quark. Therefore, the total energy of $l = 0$ states including the mass energy is given as

$$E_{n0} = 2m_c c^2 - [\hbar^2 C^2/m_c]^{1/3} a_n$$

where we have used equation (12.55). We also have for the zeroes of Airy functions

$$a_1 = -2.338$$

$$a_2 = -4.088$$

Since we have two unknowns and two pieces of data, we can easily solve for the unknowns and obtain

$$m_c c^2 = 1.16 \text{ GeV} \qquad (\hbar^2 C^2/m_c) = 0.0379 \text{ GeV}^3$$

Finally, writing

$$\hbar^2 C^2/m_c = (\hbar c)^2 C^2/m_c c^2$$

and substituting $m_c c^2 = 1.16$ GeV, and $\hbar c = 0.2$ GeV-F, we get

$$C \approx 1 \text{ GeV/F}$$

The quark-antiquark interaction is therefore ~30 times stronger than the nucleon-nucleon interaction at the distance of 1.4 F, which is roughly the range of nuclear force.

PROBLEMS

1. Following the same procedure as the case of infinitesimal rotation, show that the momentum operator p_x is the generator of infinitesimal translation along the x-axis.
2. Show that the operator corresponding to a finite rotation θ of the coordinate system about an axis $\mathbf{e}_n$ is given as

$$R_\theta = \exp(i\theta \mathbf{e}_n \cdot \mathbf{L}/\hbar)$$

   Hence show that finite rotations about different coordinate axes do not commute.
3. Show that the commutator of the operator p_r $(= -i\hbar\partial/\partial r)$ with r is given as

$$[r, p_r] = i\hbar$$

   Hence show that

$$\mathbf{p}\cdot\mathbf{r} = \mathbf{r}\cdot\mathbf{p} - 3i\hbar \neq p_r r$$

4. Prove the following equivalence:

$$p_r r^2 p_r = r p_r^2 r$$

   (*Hint:* Use the commutator in problem 3.)
5. Find the expression for ∇^2 in cylindrical coordinates ρ, ϕ, and z. Then show that

$$\rho^2 \mathbf{p}^2 = (\rho p_\rho)^2 + L_z^2 + \rho^2 p_z^2$$

   where $p_\rho = -i\hbar\partial/\partial\rho$.
6. A neutron is bound in a square-well potential of range 1.4 fm and depth 400 MeV. Find the binding energy of all the possible S-wave bound states.

7. Normalize the deuteron ground-state (S-wave) wave function obtained with the square-well potential.
8. For the ground state of the deuteron obtained with the square-well potential, find and plot the radial probability density $P(r)$ $[= r^2(R)^2]$ of finding the neutron at a distance r from the proton.
9. Consider a potential of arbitrary shape; however, it is given that the potential vanishes for $r \geq a$. Furthermore, for the S-state ($l = 0$), the log-derivative of the radial function inside the potential at $r = a$ is given as

$$\frac{1}{R}\frac{dR(r)}{dr}\bigg|_{r=a} = f_0(E)$$

Determine $f_0(E_B)$ where E_B is the energy at which the potential has a bound state.
10. Solve the radial Schrödinger equation for $l = 0$ for the attractive exponential potential

$$V(r) = -V_0 \exp(-2\beta r)$$

(*Hint:* Change variables to $\rho = \exp(-\beta r)$ to obtain the Bessel equation.) Also pay attention to the boundary conditions.

ADDITIONAL PROBLEMS

A1. A particle moves in a potential

$$V(x) = V_0 \sin(2\pi x/a)$$

which is invariant under the translation $x \to x + na$, where n is an integer. Is momentum conserved? Why or why not?

A2. Estimate the average value of r for the ground-state square-well-model wave function of the deuteron.

A3. In chapter 4, you saw that the one-dimensional square-well potential has a bound state whenever $V_0 a^2 > 0$. In this chapter, you have seen that the three-dimensional square well gives a bound state only if $V_0 a^2 > \pi^2\hbar^2/8\mu$. Can you tell what the condition is for a bound state for the two-dimensional circularly symmetric square-well potential?

A4. Solve the Schrödinger equation for the cylindrical (impenetrable) box problem. Take the box to be of radius a and height b. Use the expression for ∇^2 from problem 5 above.

A5. Consider the three-dimensional δ-shell potential

$$V(r) = -\lambda\delta(r - a),\ \lambda > 0.$$

(a) Determine the eigenvalue equation for bound states (*Hint*: $j_l(x)n_l'(x) - j_l'(x)n_l(x) = 1/x^2$).

(b) How large must the value of λ be in order for a bound state of angular momentum l to exist?

A6. A particle of mass m moves in a central potential given by

$$V(r) = -V_0 \exp[-r/a]$$

where V_0 is a constant > 0. Find the expectation value of the energy of the particle if its wave function is given by

$$\psi(r) = A \exp[-r/b]$$

in terms of the parameters a and b.

REFERENCES

M. Chester. *Primer of Quantum Mechanics.*
A. Das and A. C. Melissinos. *Quantum Mechanics: A Modern Introduction.*
S. Gasiorowicz. *Quantum Physics.*
L. I. Schiff. *Quantum Mechanics.*

13

The Hydrogen Atom

When quantum mechanics was discovered back in the nineteen-twenties, one of its first (and also one of the most important) applications was to the understanding of the hydrogen atom and hydrogenlike atoms (atoms with one valence electron). This is the Kepler problem of quantum mechanics—motion of an electron in the $1/r$ Coulomb potential of the nucleus, and it is exactly solvable.

Strictly speaking, of course, we also have the nuclei to worry about. However, we already know how to convert the two-body electron-nucleus problem into a one-body problem for a particle of suitable reduced mass μ (the reduced mass will be slightly different for different hydrogenlike atoms) moving in the Coulomb field

$$V(r) = -Ze^2/r$$

where Ze is the charge of the appropriate nucleus. Note also that this Coulomb field does not describe the situation exactly. There is the finite charge distribution of the nucleus, and there are also the screening effects of the other electrons (for $Z > 1$), but we will ignore such effects.

We can see that the problem of hydrogenlike atoms is a special case of the central-potential problem; therefore, the machinery developed in chapter 12 will be our starting point. The general solution of the time-independent Schrödinger equation for the central-potential problem is given as

$$\psi(r,\theta,\phi) = \sum_{nlm} C_{nlm} R_n(r) Y_{lm}(\theta,\phi)$$

where the radial function $R(r)$ has been given the subscript n in anticipation and will have to be determined from the solution of the radial equation, which is our main task in this chapter.

13.1 SOLUTION OF THE RADIAL EQUATION FOR THE COULOMB POTENTIAL: THE EIGENVALUES FOR $E < 0$

The radial equation for the Coulomb potential is given by

$$\left(\frac{d^2}{dr^2} + \frac{2}{r}\frac{d}{dr}\right)R + \frac{2\mu}{\hbar^2}\left(E + \frac{Ze^2}{r} - \frac{\hbar^2 l(l+1)}{2\mu r^2}\right)R = 0 \tag{13.1}$$

We will consider only bound states, $E < 0$; so we put $E = -|E|$ in equation (13.1). It is convenient to rescale distances and energies to define dimensionless variables; therefore, we introduce the variables

$$\rho = \left(\frac{8\mu|E|}{\hbar^2}\right)^{1/2} r \tag{13.2}$$

$$\lambda = \frac{Ze^2}{\hbar}\left(\frac{\mu}{2|E|}\right)^{1/2} \tag{13.3}$$

Upon substitution, equation (13.1) becomes

$$\frac{d^2R}{d\rho^2} + \frac{2}{\rho}\frac{dR}{d\rho} - \frac{l(l+1)}{\rho^2}R + \left(\frac{\lambda}{\rho} - \frac{1}{4}\right)R = 0 \tag{13.4}$$

This equation is isomorphic with one of the famous equations of mathematical physics; the solutions that we desire, single-valued, continuous, and square-integrable, are given in terms of special functions called associated Laguerre polynomials. For the fun of it, we will derive these solutions using the (by now familiar) technique of power-series expansion.

First, we look at the asymptotic equation, large ρ. We have

$$\frac{d^2R}{d\rho^2} - \frac{1}{4}R = 0 \tag{13.5}$$

The permitted solution for bound states is $R \sim \exp(-\rho/2)$. Therefore, for the general solution of equation (13.4), let's try

$$R = \exp(-\rho/2)G(\rho) \tag{13.6}$$

Substitution in equation (13.4) yields the equation for $G(\rho)$:

$$\frac{d^2G}{d\rho^2} - \left(1 - \frac{2}{\rho}\right)\frac{dG}{d\rho} + \left[\frac{\lambda - 1}{\rho} - \frac{l(l+1)}{\rho^2}\right]G = 0 \tag{13.7}$$

There is a slight problem here; equation (13.7) above is singular at $\rho = 0$. To remove the singularity, let's set

$$G(\rho) = \rho^s \Sigma_\nu a_\nu \rho^\nu = \rho^s H(\rho) \tag{13.8}$$

Equation (13.8) defines $H(\rho)$. Upon substitution, we obtain the following equation for H:

$$\rho^2 \frac{d^2H}{d\rho^2} + \rho[2(s+1) - \rho] \frac{dH}{d\rho} + [\rho(\lambda - s - 1) + s(s+1) - l(l+1)]H = 0$$

If now ρ is put equal to 0, it follows from the form of $H(\rho)$ that we must have

$$s(s+1) - l(l+1) = 0$$

There are two solutions to the equation above:

$$s = l \quad \text{and} \quad s = -l - 1$$

However, the radial wave function R must be finite at the origin. This rules out the second of the above solutions, and we must set $s = l$. The equation for H now becomes

$$\frac{d^2H}{d\rho^2} + \left(\frac{2l+2}{\rho} - 1\right)\frac{dH}{d\rho} + \frac{\lambda - l - 1}{\rho} H = 0 \tag{13.9}$$

Now substitute the power-series expansion for H:

$$H(\rho) = \Sigma_\nu a_\nu \rho^\nu$$

This gives

$$\sum_{\nu=0}^{\infty} \left[\nu(\nu-1)a_\nu \rho^{\nu-2} + \nu a_\nu \rho^{\nu-1}\left(\frac{2l+2}{\rho} - 1\right) + (\lambda - l - 1)a_\nu \rho^{\nu-1}\right] = 0$$

which is the same as

$$\sum_{\nu=0}^{\infty} \{(\nu+1)[\nu + 2l + 2]a_{\nu+1} + (\lambda - 1 - l - \nu)a_\nu\}\rho^{\nu-1} = 0$$

For this equation to hold, the coefficient of each term must vanish. Consequently, we get the recursion relation

$$\frac{a_{\nu+1}}{a_\nu} = \frac{\nu + l + 1 - \lambda}{(\nu+1)(\nu + 2l + 2)} \tag{13.10}$$

For large ν, we have

$$\frac{a_{\nu+1}}{a_\nu} \sim \frac{1}{\nu}$$

a ratio characteristic of a series that is not well behaved at infinity. In fact, the ratio is the same as that of the series for $\exp(\rho)$. Thus for a fixed value of l, we have

$$\begin{aligned} R(\rho) &= H(\rho)\rho^l \exp(-\rho/2) \\ &\rightarrow \exp(\rho/2) \rightarrow \infty \end{aligned}$$

as $\rho \rightarrow \infty$. This is the same situation that we found for the harmonic oscillator, and the reasoning leading to our rescue is the same: The series has to terminate, giving us the quantum condition. Explicitly, for a given value of l, for some $\nu = n_r$, the series terminates if

$$\lambda = n_r + l + 1 \tag{13.11}$$

This, then, is the condition for the quantized energy levels in the hydrogen atom. It is customary to introduce the *principal quantum number n* to denote λ:

$$n \equiv \lambda = n_r + l + 1$$

We find that n is a positive integer. Moreover, since $n_r \geq 0$, $n \geq l + 1$, or

$$l \leq n - 1$$

Finally, from equation (13.3)

$$E_n = -\frac{\mu Z^2 e^4}{2\hbar^2 n^2} \tag{13.12}$$

where we have put the appropriate subscript on E. We find that we have recovered the Bohr formula (but not the entire Bohr framework; Bohr could not quite anticipate the correct assignment of angular momentum to the energy levels).

Numerical Estimates

Since the hydrogenic results are quite frequently used in atomic calculations, it is useful to consider certain shortcuts to numerical estimates in connection with the hydrogen atom.

Consider first the numerical calculation of the Bohr radius a_0

$$a_0 = \hbar^2/\mu e^2$$

Since the proton is about 1800 times heavier than the electron, the reduced mass $\mu \approx m$, the mass of the electron. To calculate a_0, we now proceed as follows:

$$a_0 \approx \frac{\hbar^2}{me^2} = \frac{\hbar c}{mc^2}\frac{1}{\alpha}$$

where α is the *fine structure constant* given by

$$\alpha = e^2/\hbar c = 1/137$$

Also, $\hbar c \approx 2$ keV-Å, with 1 Å $= 10^{-8}$ cm; and $mc^2 \approx 0.5$ MeV. Combining all this, we get

$$a_0 \approx (2 \times 10^3 \times 137/0.5 \times 10^6) = 0.55 \text{ Å}$$

which compares pretty favorably with the accurate calculation of 0.53 Å.

Consider next the estimate of the energy levels. It is customary to write the Bohr formula, equation (13.12), in the form (for $Z = 1$)

$$E_n = -Ry/n^2$$

Let's now estimate Ry (which stands for Rydberg) with our newfound tricks:

$$Ry = \frac{me^4}{2\hbar^2} = \frac{mc^2}{2}(e^2/\hbar c)^2 \approx \frac{0.5 \times 10^6}{2}(1/137)^2 \text{ eV} \approx 13.3 \text{ eV}$$

which compares quite favorably with the more exact calculation of

$$Ry = 13.6 \text{ eV}$$

Two other electronic length scales are also useful in making numerical estimates. One of them is the electron Compton wavelength $\lambda_c = \hbar/mc$. We can calculate it as follows:

$$\lambda_c = \alpha a_0 \approx 0.5/137 \text{ Å} \approx 4 \times 10^{-3} \text{ Å}$$

Incidentally, the electron Compton wavelength represents the lower limit of how well the electron can be localized. You can see it in this way. To locate an electron to the accuracy of Δx, we require a photon of momentum

$$\Delta p \approx \hbar/\Delta x$$

Since the photon is massless, the momentum uncertainty corresponds to an energy uncertainty $\Delta E \sim \hbar c/\Delta x$. But if the energy uncertainty exceeds $2mc^2$, an electron-positron pair will be produced, and instead of one particle we will have three. Consequently,

$$\Delta E \leq 2mc^2$$

which gives

$$\Delta x \geq \hbar/2mc \sim \hbar/mc$$

Notice also that since the quantum "size" of the electron λ_c is 1/137 times smaller than the Bohr radius, we are justified in treating the electron as a point charge, as we have.

Finally, there is another length, the "classical electron radius," r_e, which is sometimes useful in numerical estimates:

$$r_e = e^2/mc^2 = (\hbar/mc)(e^2/\hbar c) = \lambda_c \cdot \alpha \approx 3 \times 10^{-5} \text{ Å}$$

Comments on the Eigenvalues

The following comments on the Coulomb potential-energy spectrum are now in order:

1. Unlike the square-well potential, the Coulomb potential, for any finite Z, gives an infinite number of bound states, starting at $\mu Z^2 e^4/2\hbar^2$ and ending all the way with 0. This is due to the slow falling off of the Coulomb potential for large r (as opposed to the sharp decline of the square-well potential).
2. We have an interesting degeneracy in the spectrum, the *l-degeneracy.* The energy does not depend on l, but only on n; yet for a fixed n, the possible l-values are $l = 0, 1, 2, \ldots, n-1$. Initially, it was thought that this must be in the class of accidental degeneracies, but it was found that the l-degeneracy is due to a not-so-obvious symmetry of the Coulomb potential Hamiltonian.†
3. In addition to the l-degeneracy, there is also the m-degeneracy, the result of spherical symmetry. For each l, m goes from $-l$ to $+l$, giving us $2l+1$ degenerate levels. For any n, the total degeneracy, then, is

$$\sum_{l=0}^{n-1} (2l+1) = n^2 \tag{13.13}$$

And if we take into account the two-valuedness called spin, to which we have alluded before, the total degeneracy is $2n^2$.

4. The energy spectrum for the H atom ($Z = 1$, $\mu \approx m$, the electron mass) given from equation (13.12) is shown in figure 13.1. The success of this

†For readers familiar with group theory, this is the O_4-symmetry, symmetry under four-dimensional rotations, in contrast to ordinary three-dimensional rotations that belong to the group O_3.

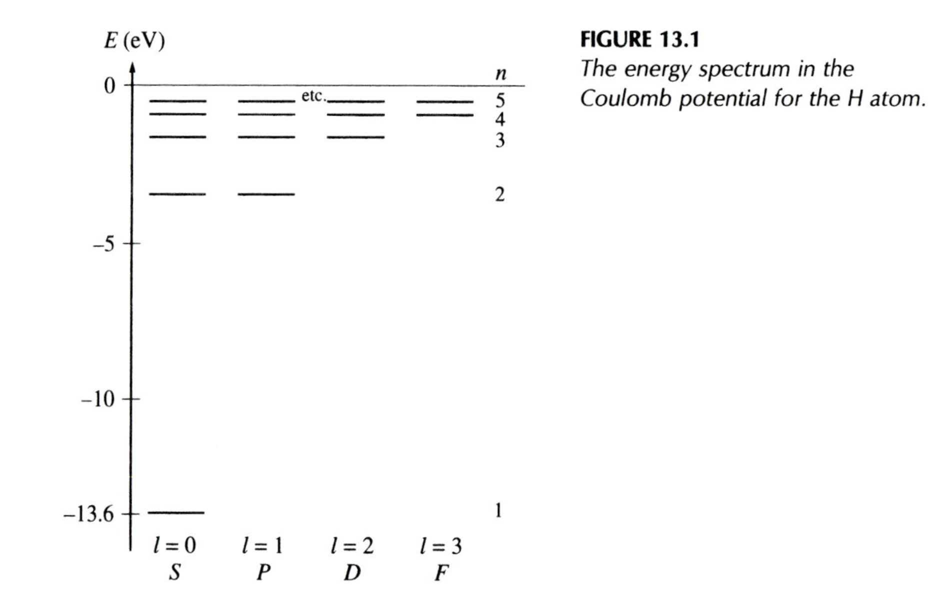

FIGURE 13.1
The energy spectrum in the Coulomb potential for the H atom.

energy calculation in explaining the observed line spectra is well known. However, when we compare this with very accurate experimental data, we find some discrepancies because in the real hydrogen atom there are other interactions that we have neglected here. Some of these interactions will be treated later when we develop approximate methods for calculating their effects.

5. The quantum number n_r is sometimes called the *radial quantum number*; it denotes the number of radial nodes of the wave function (see below). However, the eigenfunctions can be completely specified in terms of n and l; therefore, n_r is redundant for the description of the states.

13.2 HYDROGENIC WAVE FUNCTIONS

Let us return to the solution of the radial equation. Substituting $\lambda = n$ in the recursion relation (13.10), we get

$$a_{\nu+1} = \frac{\nu + l + 1 - n}{(\nu + 1)(\nu + 2l + 2)} a_\nu$$

In this way, we obtain

$$a_{\nu+1} = (-1)^{\nu+1} \frac{n - (\nu + l + 1)}{(\nu + 1)(\nu + 2l + 2)} \cdot \frac{n - (\nu + l)}{\nu(\nu + 2l + 1)} \cdots \frac{n - (l + 1)}{1 \cdot (2l + 2)} a_0 \tag{13.14}$$

This determines for us the power series expansion of $H(\rho)$. However, the $H(\rho)$ we obtain in this way is closely related to functions, already tabulated in the mathematical literature, called the *associated Laguerre polynomials*, denoted by $L(\rho)$; it is these functions that we will use for $H(\rho)$.

Associated Laguerre Polynomials

First, let's introduce the Laguerre polynomials via the following representation:

$$L_q(x) = e^x \frac{d^q}{dx^q}(x^q e^{-x}) \tag{13.15}$$

Next we define the associated Laguerre polynomials according to the equation

$$L_q^p(x) = \frac{d^p}{dx^p} L_q(x) \tag{13.16}$$

A first few of the polynomials are:

$$L_0(x) = 1$$

$$L_1(x) = 1 - x \qquad L_1^1(x) = -1$$

$$L_2(x) = 2 - 4x + x^2 \qquad L_2^1(x) = -4 + 2x \qquad L_2^2(x) = 2$$

$$L_3(x) = 6 - 18x + 9x^2 - x^3 \qquad L_3^1(x) = -18 + 18x - 3x^2 \text{ etc.}$$

Finally, the solutions $H(\rho)$ of equation (13.9) are given as

$$H(\rho) = -L_{n+l}^{2l+1}(\rho) \tag{13.17}$$

Caution: Some authors use a different notation for the associated Laguerre polynomials.† The complete (unnormalized) wave function is given as

$$R_{nl}(\rho) = -e^{-\rho/2}\rho^l L_{n+l}^{2l+1}(\rho) \tag{13.18}$$

†These authors prefer a definition of associated Laguerre polynomials that is consistent with the definition of the confluent hypergeometric function. In this case, one writes

$$L_q^p(x) = (-1)^p \frac{d^p}{dx^p} L_{q+p}(x)$$

and the radial function is then proportional to

$$(-1)^{2l+1} L_{n-l-1}^{2l+1}(x).$$

Normalization

Now we will normalize the wave function of equation (13.18). The normalization integral is complicated but can be evaluated from tables of integrals of the associated Laguerre polynomials:

$$\int_0^\infty \rho^2\,d\rho\, e^{-\rho}\rho^{2l}[L_{n+l}^{2l+1}(\rho)]^2 = \frac{2n[(n+l)!]^3}{(n-l-1)!} \tag{13.19}$$

Consequently, the normalized hydrogenic eigenfunctions are given as

$$\psi_{nlm}(r,\theta,\phi) = R_{nl}(r)Y_{lm}(\theta,\phi) \tag{13.20}$$

with $R_{nl}(r)$ given by (upon substitution back to the radial coordinate r)

$$R_{nl}(r) = -\left[\left(\frac{2Z}{na_0}\right)^3 \frac{(n-l-1)!}{2n[(n+l)!]^3}\right]^{1/2}\left(\frac{2Zr}{na_0}\right)^l e^{-Zr/na_0} L_{n+l}^{2l+1}\left(\frac{2Zr}{na_0}\right) \tag{13.21}$$

where we have used

$$\rho = 2Zr/na_0 \tag{13.22}$$

with $a_0 = \hbar^2/\mu e^2$. The first few hydrogenic radial functions are:

$$\begin{aligned}
R_{10}(r) &= 2\left(\frac{Z}{a_0}\right)^{3/2} e^{-Zr/a_0} \\
R_{20}(r) &= \left(\frac{Z}{2a_0}\right)^{3/2} 2\left(1 - \frac{Zr}{2a_0}\right)e^{-Zr/2a_0} \\
R_{21}(r) &= \left(\frac{Z}{2a_0}\right)^{3/2} 3^{-1/2}\frac{Zr}{a_0} e^{-Zr/2a_0} \\
R_{30}(r) &= \left(\frac{Z}{3a_0}\right)^{3/2} 2\left[1 - \frac{2Zr}{3a_0} + \frac{2(Zr)^2}{27a_0^2}\right]e^{-Zr/3a_0} \\
R_{31}(r) &= \left(\frac{Z}{3a_0}\right)^{3/2} \frac{4\sqrt{2}}{3}\frac{Zr}{a_0}\left(1 - \frac{Zr}{6a_0}\right)e^{-Zr/3a_0} \\
R_{32}(r) &= \left(\frac{Z}{3a_0}\right)^{3/2} \frac{2\sqrt{2}}{27\sqrt{5}}\left(\frac{Zr}{a_0}\right)^2 e^{-Zr/3a_0}
\end{aligned} \tag{13.23}$$

Comments on the Wave Functions

As we will see in the chapters to come, the hydrogenic wave functions are very useful in the study of atoms. It is therefore advantageous to take a close look at them.

1. The hydrogenic wave functions of equation (13.20) are an orthonormal set. Their orthogonality for $n = n'$, $l = l'$, but $m \neq m'$, follows from the properties of $\exp(im\phi)$. When the principal and the azimuthal quantum numbers n and m are equal, but $l \neq l'$, the orthogonality of the wave functions follows from the properties of the Legendre polynomials. Finally, when the l and m quantum numbers are the same, but $n \neq n'$, the orthogonality is a consequence of the properties of the associated Laguerre polynomials.
2. However, the eigenfunctions we have determined above do not form a complete set. Why? Because they only pertain to the bound states of the Hamiltonian, which also has continuum states of positive energy. The bound and continuum states together form a complete set.
3. All the $R_{nl}(r)$ contain the factor r^l; therefore, they all vanish at the origin except the S-states ($l = 0$). The credit goes to the centrifugal potential, which repels the electrons from coming too close to the nucleus.
4. We can see that the number of nodes of each radial function listed above is indeed given by $n_r = n - l - 1$. In the general case, the recursion relation of equation (13.10) shows $H(r)$ is a polynomial of degree n_r and thus has that many radial nodes.
5. Similarly, we can speak of bumps of the radial probability distribution function, places where

$$P(r) = r^2 R_{nl}^2 \tag{13.24}$$

is a maximum. (Note that the factor r^2 adjusts for the greater volume of spherical shell available for larger r.) Clearly, for each state nl, $P(r)$ has $n - l$ bumps. Figure 13.2 plots the radial probability distribution for the first few hydrogenic states.

6. The angular probability distribution, the probability for finding the electron at the angles (θ, ϕ) in the element $d\Omega$ is given as

$$|Y_{lm}(\theta, \phi)|^2 \, d\Omega \sim |P_l^m(\cos\theta)|^2 \, d\Omega$$

These probabilities are independent of the azimuth ϕ. Figure 13.3 shows the angular probability distribution for the first few lm's. You can see that the result of increasing m is a shift of the probability distribution from the z-axis toward the equitorial plane. At $|m| = l$, the shift is complete; the probability is entirely peaked at $\theta = \pi/2$.

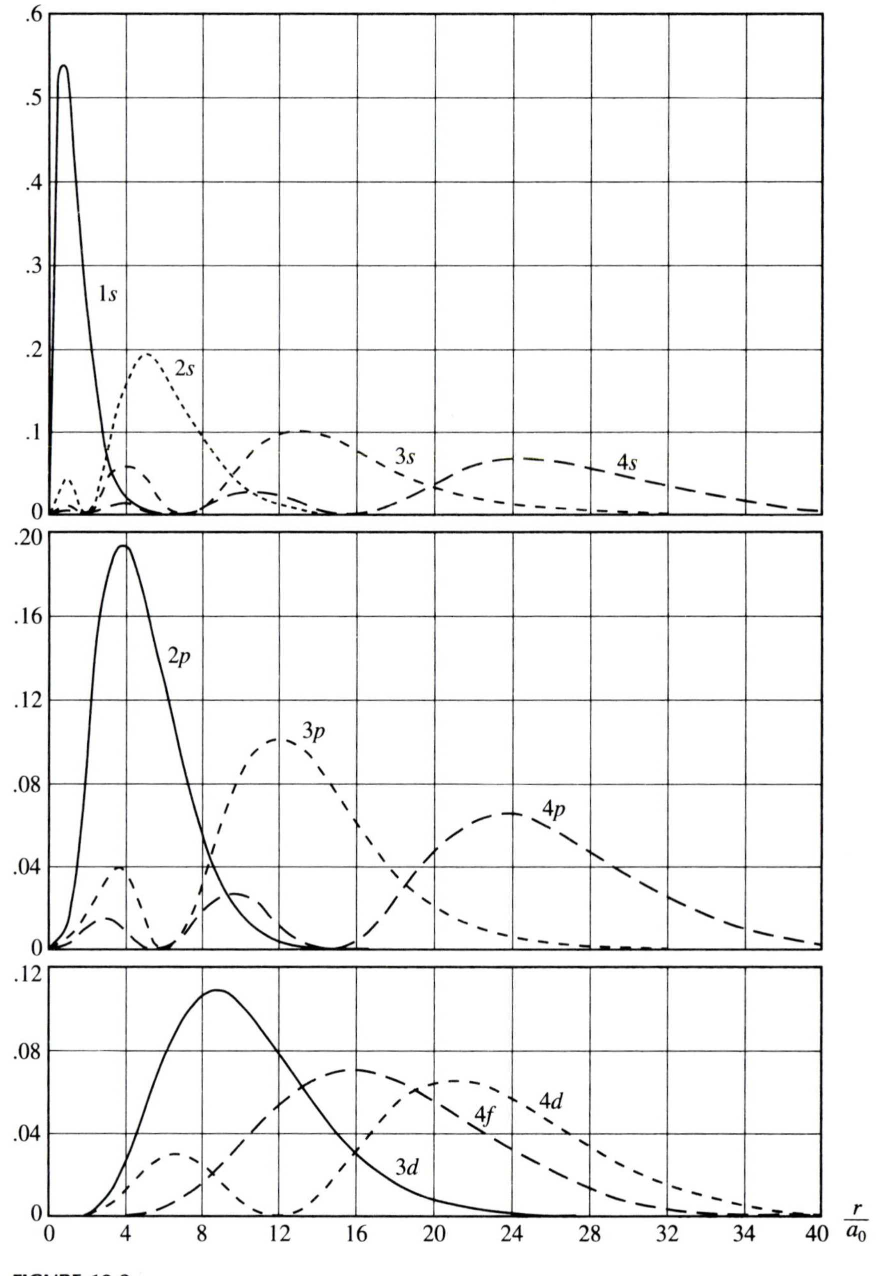

FIGURE 13.2
The radial probability distribution function $|rR_{nl}|^2$ for several values of nl. (Adapted with permission from E. U. Condon and G. H. Shortley, The Theory of Atomic Spectra. *Cambridge, U.K.: Cambridge University Press, 1953.)*

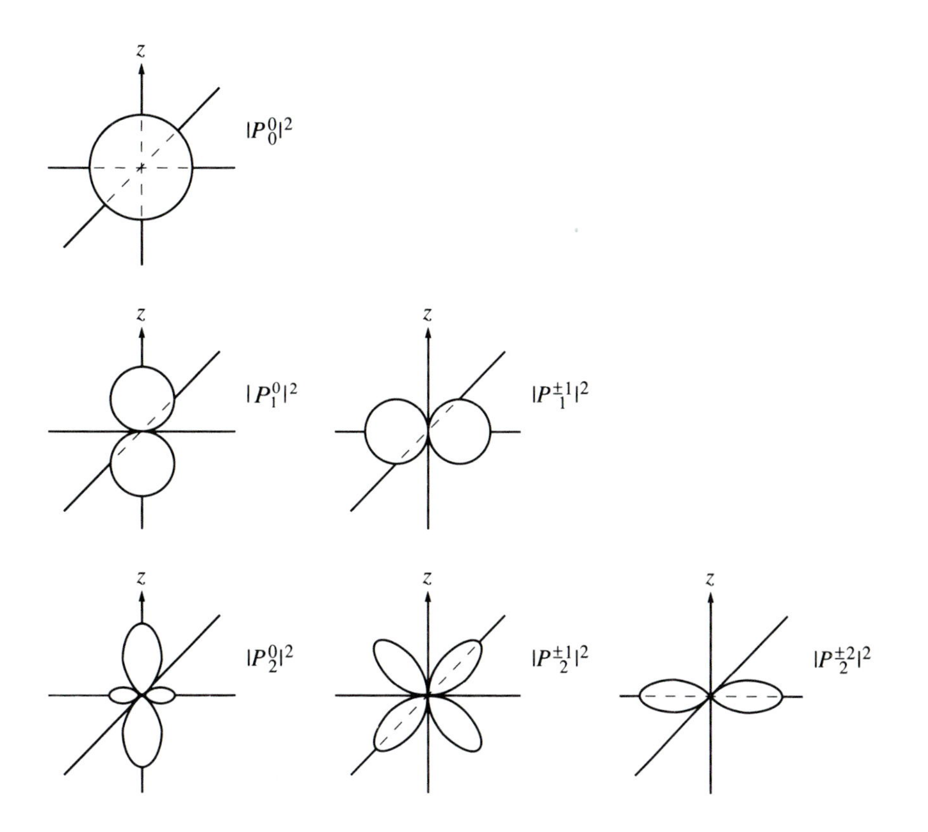

FIGURE 13.3
The angular probability distribution $|P_l^m(\cos\theta)|^2$ plotted for several values of lm.

7. The width of the angular probability distributions for $|m| = l$ falls off $\sim 1/\sqrt{l}$ (a property of the associated Legendre functions which are $(\sin\theta)^l$ in this case). It follows that for very large l, the orbits will be planar as in classical physics.
8. Since all central potentials are invariant under the parity operation, it is also possible to label the eigenstates of the hydrogenic atom by the eigenvalues of parity; that is, the states are states of definite parity. Since

$$Y_{lm}(\theta,\phi) \rightarrow (-1)^l Y_{lm}(\theta,\phi)$$

under the parity operation (see chapter 11), the parity of a hydrogenic state is completely determined by its l-value, it is $(-1)^l$.

To facilitate working out some of the assigned problems and for future reference, the following is a useful list of radial expectation values:

$$\langle r \rangle = \frac{a_0}{2Z} [3n^2 - l(l+1)]$$

$$\langle r^2 \rangle = \frac{a_0^2 n^2}{2Z^2} [5n^2 + 1 - 3l(l+1)]$$

$$\left\langle \frac{1}{r} \right\rangle = \frac{Z}{a_0 n^2}$$

$$\left\langle \frac{1}{r^2} \right\rangle = \frac{Z^2}{a_0^2 n^3 (l + 1/2)} \tag{13.25}$$

Finally, the following recursion formula, called Kramer's rule, is also useful:

$$\frac{Z^2}{a_0^2}\left[\frac{k+1}{n^2}\right]\langle r^k \rangle - \frac{Z}{a_0}(2k+1)\langle r^{k-1}\rangle + k\left(l + \frac{k+1}{2}\right)\left(l - \frac{k-1}{2}\right)\langle r^{k-2}\rangle = 0 \tag{13.26}$$

13.3 CIRCULATING CURRENT AND MAGNETIC MOMENT

Now let's calculate the probability current density using the hydrogenic wave functions. The three-dimensional generalization of the definition (1.26) for the probability current is given as

$$\mathbf{j} = (\hbar/2\mu i)(\psi^* \nabla \psi - \psi \nabla \psi^*) \tag{13.27}$$

In spherical coordinates, the gradient operator is given by

$$\nabla = \mathbf{e}_r \frac{\partial}{\partial r} + \mathbf{e}_\theta \frac{1}{r}\frac{\partial}{\partial \theta} + \mathbf{e}_\phi \frac{1}{r \sin\theta}\frac{\partial}{\partial \phi} \tag{11.17}$$

The time-dependent wave function corresponding to $\psi_{nlm}(\mathbf{r})$ of equation (13.20) is given as

$$\psi_{nlm}(\mathbf{r}, t) = R_{nl}(r) Y_{lm}(\theta, \phi) \exp(-iE_n t/\hbar) \tag{13.28}$$

But Y_{lm} (eq. [11.40]) is the product of a real function in θ and $\exp(im\phi)$. Let's write it as $\Theta(\theta)\exp(im\phi)$. Then

$$\psi_{nlm}(\mathbf{r},t) = R_{nl}(r)\Theta_{lm}(\theta)\exp(im\phi)\exp(-iE_n t/\hbar) \tag{13.29}$$

The probability current for this wave function is easily calculated. First we note that neither the radial component nor the θ-component of the wave function has any imaginary part; therefore, they do not contribute to the current. Consequently, we need only calculate the ϕ-term:

$$\begin{aligned}\mathbf{j} &= \mathbf{e}_\phi \frac{\hbar}{2\mu i} R_{nl}^2\Theta_{lm}^2\left(e^{-im\phi}\frac{1}{r\sin\theta}\frac{\partial}{\partial\phi}e^{im\phi} - e^{im\phi}\frac{1}{r\sin\theta}\frac{\partial}{\partial\phi}e^{-im\phi}\right)\\ &= \mathbf{e}_\phi \frac{|\psi_{nlm}|^2 m\hbar}{\mu r\sin\theta}\end{aligned} \tag{13.30}$$

The current is a latitudinal current, and it is stationary, independent of time. The interpretation is obvious: The current elements circulate around the z-axis while maintaining the symmetry about it.

But how can we interpret the current itself? We can convert the probability current into a charge current by multiplying with the electronic charge $-e$. When Schrödinger first solved the quantum mechanical problem of the hydrogen atom, he thought there were real circulating charges and currents generating electromagnetic fields that radiate light in some way. You see, Born's probability interpretation was not out there yet! With Born's interpretation, we correctly realize that there is no smeared out charged density or real physical currents circulating around the nucleus; whenever we observe, we find the electron at one place. The charge density is a probability density! But what about the current density? The ring of current circulating around the z-axis is to be interpreted as a magnetic moment for the electron. (As you know, we can always replace a current loop or circulating current by a magnetic dipole.) So the current is not a smeared circulating charge or anything like that; when we find the electron, we find not only its charge localized at one place but also its magnetic moment.

The element of magnetic moment arising from the ring of current density is given as

$$d\mathbf{M} = (-e/c)\tfrac{1}{2}(\mathbf{r}\times\mathbf{j})\,d\tau$$

where $d\tau$ is a volume element. Take the z-component

$$d\mathbf{M}_z = -(e/2c)(\mathbf{r}\times\mathbf{j})_z\,d\tau = -(e/2c)r\sin\theta j_\phi d\tau$$

since $\mathbf{j}_r$ and $\mathbf{j}_\theta$ are zero. The total magnetic moment is

$$M_z = -\frac{e}{2c}\int r\sin\theta\,\frac{m\hbar|\psi|^2}{\mu r\sin\theta}\,d\tau$$
$$= -\frac{e\hbar}{2\mu c}\,m\int|\psi|^2\,d\tau$$
$$= -\frac{e\hbar}{2\mu c}\,m \qquad (13.31)$$

where we have used equation (13.30) and the normalization of ψ. In this way, we see that the magnetic moment of the electron (or rather its z-component) is related to the quantum number m representing the z-component of $\mathbf{L}$, the orbital angular momentum; now we can understand why m is called a magnetic quantum number. Furthermore, since the nucleus is so much heavier than the electron, $\mu \approx m_e$, the electronic mass; and then the factor $e\hbar/2m_e c$ is the familiar unit of Bohr magneton.

What we have calculated here is the expectation value of an operator that represents the magnetic moment in the state represented by ψ. Clearly, the magnetic moment operator itself must be proportional to the angular momentum operator $\mathbf{L}$,

$$\mathbf{M} = -(e/2\mu c)\mathbf{L} \qquad (13.32)$$

In chapter 14, we will further investigate this connection in the context of a more complete investigation of the motion of an electron in an electromagnetic field.

A final comment on the quantum mechanics of the hydrogen atom that we have worked out here. This simple calculation gives us the basis for the calculation for more complicated atoms containing many electrons, and eventually a theory for the whole periodic table of elements—a theory that works—that has been validated by myriads of experiments.

PROBLEMS

1. The positronium is an "atom" made up of a positron and an electron bound together by their mutual Coulomb potential. Sketch the energy spectrum of the positronium.
2. Discuss the motion of the moon about the earth using quantum mechanics, treating both objects as mass points. Hence estimate the quantum numbers of the moon's present orbit.
3. Discuss the energy eigenvalues and eigenfunctions of the mesic atom formed by a muon (of mass 105 MeV/c^2 compared to the electron's mass

of 0.51 MeV/c^2) and the hydrogen nucleus. Calculate the wavelength of the mesic X-ray for the $2P \to 1S$ transition. Also find the probability that in the ground state, the muon is inside the nucleus. (Take the radius of the nucleus to be 1.3 F; 1 F $= 10^{-13}$ cm.)

4. An electron is in the ground state of the tritium atom for which the nucleus consists of one proton and two neutrons. Suppose the nucleus instantaneously changes into He^3 consisting of two protons and a neutron. What is the probability that the electron is found in the ground state of the He^3 ion?

5. For hydrogenlike atoms, such as the alkali atoms, the screening effect of the "closed-shell" electrons can be accounted for by considering the electron to move in the potential

$$V(r) = -(e^2/r)(1 + \alpha/r)$$

where α is a constant. Find the energy eigenvalues for this potential.

6. Calculate $\langle r \rangle$ for the ψ_{100} and ψ_{210} states of the hydrogen atom. For the same states find the values of r for which the radial probability distribution $P(r)$ is a maximum. Compare your results with that of the Bohr theory. Prove that, in general, when $l = n - 1$, $P(r)$ peaks at the Bohr-atom value for circular orbits.

7. Calculate the value of

$$\langle T \rangle_{nl} = \langle p^2/2m \rangle_{nl}$$

for a hydrogenic state. (*Hint*: You can use the expression for $\langle 1/r \rangle_{nl}$ for the purpose.) Hence, show that the virial theorem

$$\langle T \rangle = -\tfrac{1}{2}\langle V \rangle$$

holds for the Coulomb potential.

8. An electron in the Coulomb field of a proton is in the following state of coherent superposition of hydrogenic states:

$$\tfrac{1}{3}[\psi_{100} + \sqrt{3}\psi_{210} - \sqrt{5}\psi_{320}]$$

Calculate the expectation values of energy, $\mathbf{L}^2$, and L_z for this state.

9. Calculate the eigenvalues and eigenfunctions of the three-dimensional harmonic oscillator using spherical coordinates:

$$H = \frac{\mathbf{p}^2}{2m} + \frac{1}{2} m\omega^2 r^2$$

(*Hint*: If you find the workings too difficult, feel free to consult chapter 21 where this problem is worked out in detail.)

10. Find the momentum-space wave function for an electron in the hydrogenic ground state.

11. Find the eigenvalues and eigenfunctions of the hydrogen atom for motion in two dimensions. (*Hint*: Use cylindrical coordinates.)

ADDITIONAL PROBLEMS

A1. At time $t = 0$, a hydrogen atom is in the following state of coherent superposition:

$$\psi(\mathbf{r}, 0) = A[\psi_{100} + \sqrt{5}\psi_{210}]$$

(a) Normalize the wave function.

(b) What is the probability that a measurement of L^2 gives the value $6\hbar^2$?

(c) What is the wave function at time t?

(d) What is $\psi(\mathbf{r}, t)$ if the measurement of L^2 at $t = 0$ gives $6\hbar^2$?

A2. Show that for the hydrogen atom for any principal quantum number n, in the state corresponding to maximum l $(= n - 1)$, we have

$$\langle n, n-1|r|n, n-1\rangle = a_0 n(n + \tfrac{1}{2})$$

A3. For the state $\psi_{nlm}(\mathbf{x})$ of the hydrogen atom, determine the matrix element

$$\langle \psi_{nlm}|\delta(\mathbf{x})|\psi_{nlm}\rangle$$

A4. Express the Schrödinger equation for the hydrogen atom in terms of the parabolic coordinates ξ, ζ, ϕ, defined as:

$$\xi^2 = r - z = r(1 - \cos\theta)$$

$$\zeta^2 = r + z = r(1 + \cos\theta)$$

$$\phi = \phi$$

Separate the variables and solve for the eigenvalues and eigenfunctions. Discuss the parities, if any, of the eigenfunctions.

REFERENCES

P. Fong. *Elementary Quantum Mechanics.*

S. Gasiorowicz. *Quantum Physics.*

L. Pauling and E. Wilson. *Introduction to Quantum Mechanics.*

R. Shankar. *Principles of Quantum Mechanics.*

14

Electrons in the Electromagnetic Field

Now we will quantum mechanically treat the motion of electrons (and other charged particles) in an external electromagnetic field, a very important problem when we consider the versatile use of electromagnetic fields in the study of atoms and other quantum systems. However, our treatment will fall short of being completely kosher; we will continue to treat the electromagnetic field classically, but the electron's motion in the field will be given quantum handling.

Let's briefly review the classical theory of the electromagnetic field, in particular Maxwell's equations, the formalism of the scalar and vector potentials, and the gauge considerations. The vacuum Maxwell's equations are given by†

$$\nabla \cdot \mathbf{B}(\mathbf{r}, t) = 0$$

$$\nabla \times \mathcal{E}(\mathbf{r}, t) + (1/c)\partial \mathbf{B}/\partial t = 0$$

$$\nabla \cdot \mathcal{E} = 4\pi\rho(\mathbf{r}, t)$$

$$\nabla \times \mathbf{B} - (1/c)\partial \mathcal{E}/\partial t = 4\pi \mathbf{j}(\mathbf{r}, t)/c \tag{14.1}$$

where $\mathbf{B}$ and $\mathcal{E}$ are the magnetic and electric fields, respectively, and ρ and $\mathbf{j}$ are the charge and current densities; they are the sources of the $\mathcal{E}$ and $\mathbf{B}$ fields. (Alas! We have to use $\mathcal{E}$ for the electric field rather than its natural notation

†We will use Gaussian units, as is customary for many textbooks and especially for the research literature in quantum mechanics.

E in order to avoid confusion with the notation for energy.) The continuity equation

$$\partial\rho/\partial t + \nabla\cdot\mathbf{j} = 0$$

is obeyed by ρ and $\mathbf{j}$.

The scalar and vector potentials, $\phi(\mathbf{r}, t)$ and $\mathbf{A}(\mathbf{r}, t)$, respectively, are defined via the equations

$$\mathbf{B} = \nabla \times \mathbf{A}$$

$$\mathcal{E} = -\nabla\phi - (1/c)\partial\mathbf{A}/\partial t \tag{14.2}$$

However, these are not unique definitions. We can always define new potentials via the *gauge transformation*

$$\mathbf{A}' = \mathbf{A} + \nabla f(\mathbf{r}, t)$$

$$\phi' = \phi - (1/c)\partial f/\partial t \tag{14.3}$$

that yield the same $\mathcal{E}$ and **B** fields where the physics is. A particular choice of the function f defines a particular *gauge*. For static charges, it is most convenient to choose f so that

$$\nabla\cdot\mathbf{A} = 0 \tag{14.4}$$

This is the *Coulomb gauge*. For nonstatic charge distributions, Maxwell's equations, when expressed in terms of ϕ and **A**, simplify the most with the choice of the *Lorentz gauge*:

$$\nabla\cdot\mathbf{A} + (1/c)\partial\phi/\partial t = 0 \tag{14.5}$$

As stated above, we want to treat the motion of the electron in the electromagnetic field quantum-mechanically; the way to do it is to start with the corresponding classical Hamiltonian, which is obtained from the free-field Hamiltonian

$$H_0 = \mathbf{p}^2/2\mu$$

(notice that we continue to use the more general notation μ for the mass of the electron) by making the change (see, for example, J. D. Jackson, *Classical Electrodynamics*)

$$\mathbf{p} \to \mathbf{p} + e\mathbf{A}/c \tag{14.6}$$

and by adding $-e\phi$ (for dealing with static potentials). This gives for the Hamiltonian of an electron in an electromagnetic field

$$H = (1/2\mu)\,[\mathbf{p} + e\mathbf{A}/c]^2 - e\phi \tag{14.7}$$

For a positive charge, change $e \to -e$. In quantum mechanics, we take over the same Hamiltonian with the usual prescription of "operatorizing" it.

First, we will consider the historically important problem of an electron under the influence of a uniform static magnetic field. Afterward, we will discuss the gauge invariance of the Schrödinger equation of the Hamiltonian, equation (14.7).

14.1 MOTION OF AN ELECTRON IN A UNIFORM STATIC MAGNETIC FIELD

Suppose, besides the static Coulomb field, there is a static, uniform magnetic field in which the electron moves. The straightforward gauge condition is

$$\nabla \cdot \mathbf{A} = 0 \tag{14.8}$$

Now, for any function $f(x)$, we have in quantum mechanics the relationship

$$[f(x), p] = i\hbar \frac{df(x)}{dx}$$

It follows that

$$\mathbf{A}\cdot\mathbf{p} - \mathbf{p}\cdot\mathbf{A} = i\hbar\nabla\cdot\mathbf{A} = 0$$

by virtue of equation (14.8). Therefore, the momentum and the vector potential commute. The Hamiltonian H then simplifies further:

$$H = \frac{p^2}{2\mu} + \frac{e}{\mu c}\,\mathbf{A}\cdot\mathbf{p} + \frac{e^2}{2\mu c^2}\,A^2 - e\phi \tag{14.9}$$

Additionally, the vector potential for a uniform constant magnetic field can be conveniently chosen as

$$\mathbf{A} = -\tfrac{1}{2}\mathbf{r} \times \mathbf{B} \tag{14.10}$$

Consequently,

$$\mathbf{A}\cdot\mathbf{p} = -\tfrac{1}{2}[\mathbf{r} \times \mathbf{B}]\cdot\mathbf{p} = \tfrac{1}{2}\mathbf{B}\cdot[\mathbf{r} \times \mathbf{p}] = \tfrac{1}{2}\mathbf{B}\cdot\mathbf{L} \tag{14.11}$$

where we have used the fact that **B** commutes with **r**. Furthermore,

$$A^2 = \tfrac{1}{4}(\mathbf{r} \times \mathbf{B})^2 = \tfrac{1}{4}[r^2B^2 - (\mathbf{r}\cdot\mathbf{B})^2] \tag{14.12}$$

Substituting into equation (14.9), we obtain

$$H = \frac{p^2}{2\mu} + \frac{e}{2\mu c}\,\mathbf{L}\cdot\mathbf{B} + \frac{e^2}{8\mu c^2}\,[r^2B^2 - (\mathbf{r}\cdot\mathbf{B})^2] - e\phi \tag{14.13a}$$

Notice that $e\mathbf{L}/2\mu c$ is also the magnetic moment **M** of the electron, introduced in chapter 13. As a result, the second term in the Hamiltonian, equation (14.13a), can also be written as $\mathbf{M}\cdot\mathbf{B}$. For small fields, the B^2 term is negligible (see below), and thus the energy of a system in a magnetic field is $\sim B$, and the coefficient of proportionality is the magnetic moment. This is the quantum mechanical definition of the magnetic moment.

Finally, let's make the further simplifying assumption that **B** lies in the z-direction. This gives

$$H = \frac{p^2}{2\mu} + \frac{e}{2\mu c}\,BL_z + \frac{e^2B^2}{8\mu c^2}\,(x^2 + y^2) - e\phi \tag{14.13b}$$

The Normal Zeeman Effect

What about the relative magnitude of the two terms in the Hamiltonian, respectively, $\sim B$ and $\sim B^2$? For atomic systems, we can compare the magnitude of the linear and quadrupole terms by taking $\langle L_z\rangle \sim \hbar$ and $\langle x^2 + y^2\rangle \sim a_0^2$, a_0 being the Bohr radius. We have for the ratio of B^2 and the B terms

$$\frac{(e^2/8\mu c^2)a_0^2B^2}{(e\hbar/2\mu c)B} \approx \frac{B}{9 \times 10^9 \text{ gauss}}$$

The magnetic fields available for atomic experiments seldom exceed 10^4 gauss. Naturally, then, we will first consider the effect of the linear term $\sim B$ in the Hamiltonian of equation (14.13b); that is, we will ignore the term $\sim B^2$. Call the rest of the Hamiltonian H_0:

$$H_0 = (p^2/2\mu) - e\phi$$

But this is the hydrogenic Hamiltonian that commutes with L^2 and L_z. The energy eigenstates of H_0 are simultaneously eigenstates of L^2 and L_z. What is the effect of the interaction

$$H_{\text{int}} = (e/2\mu c)BL_z \tag{14.14}$$

on such eigenstates? H_{int} acting on such eigenstates yields just a number proportional to the eigenvalue of L_z. Define the *Larmor frequency*

$$\omega_L = eB/2\mu c \tag{14.15}$$

Then

$$H_{\text{int}}\psi_{nlm}(\mathbf{r}) = \hbar\omega_L m\psi_{nlm}(\mathbf{r}) \tag{14.16}$$

where $m\hbar$ is the eigenvalue of L_z, $-l \le m \le l$. Thus the interaction breaks up the $(2l+1)$-fold m-degeneracy of the central-potential energy levels into $2l+1$ equally spaced components, each with energy

$$E = -\tfrac{1}{2}\mu Z^2e^4/n^2\hbar^2 + \hbar\omega_L m \tag{14.17}$$

Again you can see why m is called the magnetic quantum number; it takes a magnetic field to split up the m-degeneracy.

The magnitude of the splitting is

$$eB\hbar/2\mu c = (B/2.4 \times 10^9 \text{ gauss}) \times 13.6 \text{ eV}$$

For a moderate B-field, the splitting is very small compared to the energy spacing of the hydrogenic levels of H_0.

For radiative transitions between the perturbed energy levels, there is a *selection rule* that operates in such a way that only those transitions for which the m-value changes by 0 or ± 1 are allowed. In effect, this means that a single spectral line in the absence of a **B**-field will split into three lines when the **B**-field is turned on (see below). This is called the *normal Zeeman effect*, which indeed is sometimes observed. A much more common phenomenon, however, is the so-called *anomalous Zeeman effect* for which the pattern is much more complicated. This suggests that we have been ignoring heretofore a fundamental aspect of the interaction of the electron with the magnetic field; this turns out to be the phenomenon of spin angular momentum of the electron, which we will examine further in chapter 15. The anomalous Zeeman effect will be treated in chapter 19.

■ **PROBLEM** Calculate the normal Zeeman effect for transitions between the $l = 2$ and $l = 1$ states of hydrogen and show that there are only three Zeeman lines.

SOLUTION An $l = 2$ state splits into five states and an $l = 1$ state into three (fig. 14.1). The selection rules allow for nine transitions; however, all the tran-

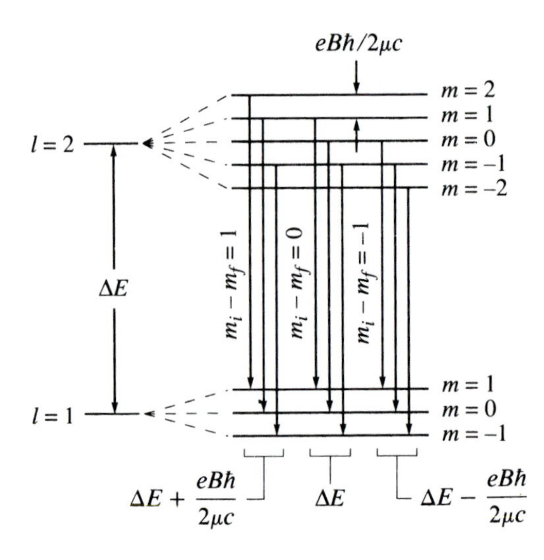

FIGURE 14.1
The normal Zeeman effect for transition from an $l = 2$ state to an $l = 1$ state in hydrogen illustrated. For the nine possible transitions permitted by the dipole selection rule, there are only three distinct energy values resulting in a threefold spectral decomposition of a given line in the presence of a magnetic field.

sitions in each of the groups, $\Delta m = -1$, $\Delta m = 0$, and $\Delta m = +1$, have the same energy

$$E_0 - eB\hbar/2\mu c, \qquad E_0, \qquad E_0 + eB\hbar/2\mu c$$

respectively, where E_0 is the energy of the original transition. Therefore, only three lines are seen. ■

The Interaction $\sim B^2$

We have seen that for ordinary fields used in the laboratory, the quadratic term is utterly negligible for atomic systems. However, there are physical systems, pulsars, for example, in which the magnetic field is reliably determined to be 10^{11} to 10^{12} gauss, for which the B^2 term is important. The B^2 term is also important for the large-scale motion of electrons in external B-fields. We will, therefore, treat the quadratic term.

Let's put $\phi = 0$. The Hamiltonian of equation (14.13b) with $\phi = 0$ consists of three parts, each of which can be thought of separately: (1) the motion in the z-direction; (2) the term involving L_z and $\sim B$; and (3) the motion in the xy-plane, which consists of the rest of the Hamiltonian. Let's call this last part H_3:

$$H_3 = \frac{1}{2\mu}\left(p_x^2 + p_y^2\right) + \frac{1}{2}\left(\frac{e^2B^2}{4\mu c^2}\right)(x^2 + y^2) \tag{14.18}$$

Note that it is an old friend—the two-dimensional oscillator Hamiltonian with oscillator frequency equal to the Larmor frequency $eB/2\mu c$. For the whole Hamiltonian, we notice that the operators L_z and p_z commute with one another and also with H_3. The total wave function then is a simultaneous eigenfunction of H_3, L_z, and p_z and can be labeled as ψ_{nmk}. We have

$$H_3\psi_{nmk} = (n+1)\hbar\omega\psi_{nmk}$$

$$L_z\psi_{nmk} = \hbar m\psi_{nmk}$$

$$p_z\psi_{nmk} = \hbar k\psi_{nmk} \tag{14.19}$$

and the energy eigenvalues are given by

$$E_{nmk} = (n+m+1)\hbar\omega + \hbar^2k^2/2\mu \tag{14.20}$$

The total energy of the electron consists of the kinetic energy of its free z-motion and the energy associated with its xy-motion under the magnetic field. Using symmetry, you can also show that the quantum numbers n and m are either both even or both odd (the proof is left as an exercise). Note also that since the total energy cannot be negative (H is the square of a hermitian operator, as given by eq. [14.7] with $\phi = 0$), we must have

$$n + m \geq 0$$

and thus the lowest energy of the xy-motion of the electron in a magnetic field is not zero. Surprise! There is a finite zero-point energy even though the electron by no means is confined to a small volume of space by the magnetic field.

14.2 THE GAUGE INVARIANCE OF THE SCHRÖDINGER EQUATION AND RELATED PROBLEMS

For the complete Hamiltonian of equation (14.7) of the electromagnetic field, the time-independent Schrödinger equation is given by

$$\frac{1}{2\mu}\left[-i\hbar\nabla + \frac{e}{c}\mathbf{A}\right]^2\psi(\mathbf{r}) = [E + e\phi(\mathbf{r})]\psi(\mathbf{r}) \tag{14.21}$$

where we have transferred the $e\phi$-term to the right-hand side. If we make a special gauge transformation according to equation (14.3), so that the vector potential changes from $\mathbf{A} \to \mathbf{A} + \nabla f(\mathbf{r})$, the Hamiltonian changes accordingly, thus changing the left-hand side of equation (14.21) to

$$\frac{1}{2\mu}\left[-i\hbar\nabla + \frac{e}{c}\mathbf{A} + \frac{e}{c}\nabla f\right]^2\psi$$

If we require gauge invariance of the Schrödinger equation, which is an appropriate requirement since the physics is the same for two different gauges, we must let ψ change too under the gauge transformation:

$$\psi(\mathbf{r}) \rightarrow \psi'(\mathbf{r}) = \exp[i\Lambda(\mathbf{r})]\psi \tag{14.22}$$

To find Λ, let's substitute back into the Schrödinger equation. We have

$$\frac{1}{2\mu}\left[-i\hbar\nabla + \frac{e}{c}\mathbf{A} + \frac{e}{c}\nabla f\right]\cdot\left[-i\hbar\nabla + \frac{e}{c}\mathbf{A} + \frac{e}{c}\nabla f\right]e^{i\Lambda}\psi = (E + e\phi)e^{i\Lambda}\psi$$

The left-hand side is seen to be

$$\frac{1}{2\mu}\left[-i\hbar\nabla + \frac{e}{c}\mathbf{A} + \frac{e}{c}\nabla f\right]\cdot\left(e^{i\Lambda}\left[-i\hbar\nabla\psi + \frac{e}{c}\mathbf{A}\psi + \frac{e}{c}\nabla f\psi + \hbar\nabla\Lambda\psi\right]\right)$$
$$= \frac{1}{2\mu}e^{i\Lambda}\left[-i\hbar\nabla + \frac{e}{c}\mathbf{A} + \frac{e}{c}\nabla f + \hbar\nabla\Lambda\right]^2\psi$$

Clearly the original equation (14.21) is restored if we choose

$$\Lambda = -(e/\hbar c)f \tag{14.23}$$

In this way, it follows that under the gauge transformation, the wave function must transform according to

$$\psi \rightarrow \psi' = e^{-(ie/\hbar c)f(\mathbf{r})}\psi(\mathbf{r}) \tag{14.24}$$

The Aharonov-Bohm Effect

Suppose the **B**-field is 0. Then since $\mathbf{B} = \nabla \times \mathbf{A}$, **A** is given as

$$\mathbf{A} = \nabla f \tag{14.25}$$

This is the gauge-transformed value of an **A** that is originally zero. Thus it makes sense that we can write the Schrödinger equation for a zero **B**-field in two different ways:

1. Forget the field completely and consider

$$\frac{1}{2\mu}(-i\hbar\nabla)^2\psi + V(\mathbf{r})\psi = E\psi$$

2. Write the Schrödinger equation in the transformed gauge, with **A** being given by equation (14.25) and the wave function given by equation (14.24):

$$\frac{1}{2\mu}\left[-i\hbar\nabla + \frac{e}{c}\mathbf{A}\right]^2 \psi' + V(\mathbf{r})\psi' = E\psi' \tag{14.26}$$

with

$$\psi' = \exp(-ief/\hbar c)\psi \tag{14.27}$$

and where f may be written in terms of $\mathbf{A}$ as the line integral

$$f(\mathbf{r}, t) = \int_s \mathbf{A}\cdot d\mathbf{l} \tag{14.28}$$

where s denotes the path of integration (the path that the electron follows). Thus by following a path through a region where the field is zero, but not the vector potential, the electron wave function acquires an additional phase given by

$$\delta = -\frac{e}{\hbar c}\int_s \mathbf{A}\cdot d\mathbf{l} \tag{14.29}$$

Can such a phase be observed? In 1959, Bohm and Aharonov suggested the following experiment. Suppose a beam of electrons is diffracted from a crystal (fig. 14.2, in which the crystal is shown only schematically as two slits). Suppose further that a **B**-field perpendicular to the diffraction plane is established in a magnetized filament behind the plane; the field is confined to the filament and no magnetic force reaches out to the electrons. However, a vector potential **A** is directed along concentric circles around the filament such that the elec-

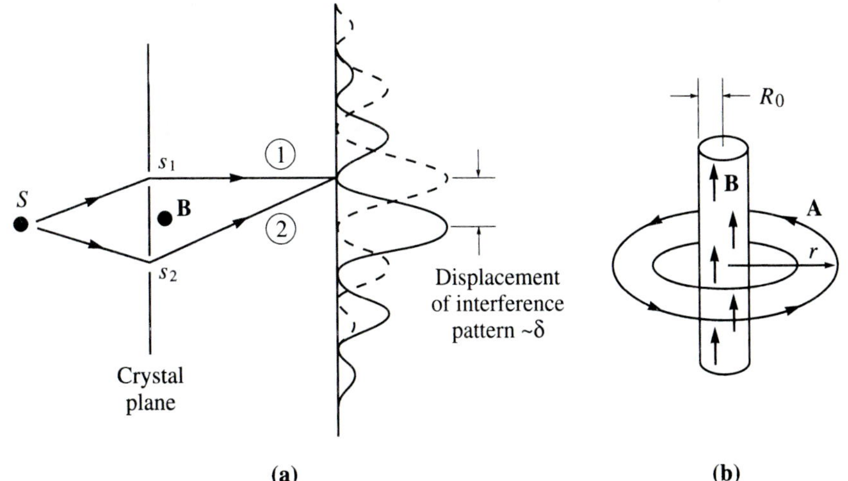

FIGURE 14.2
The Aharonov-Bohm effect. (a) The path of the electron through a crystal (schematically represented by a double-slit) does not intersect the confined axial magnetic-field lines perpendicular to the plane of the paper shown separately in (b) where the vector potential of the field is also shown.

trons do travel through a region of space where **A** is not zero although **B** is. Since $\mathbf{B} = \nabla \times \mathbf{A}$, for a constant cylindrical (axial) **B**-field in the region $r < R_0$ (R_0 = radius of cylinder),

$$\mathbf{A} = \tfrac{1}{2}(\mathbf{B} \times \mathbf{r})R_0^2/r^2$$

for $r > R_0$.

Consider now the electronic paths 1 and 2 that go around the filament as shown in figure 14.2. The phase difference accruing from traveling the different paths is given as

$$\delta_2 - \delta_1 = \frac{e}{\hbar c}\left[\int_1 \mathbf{A}\cdot d\mathbf{l} - \int_2 \mathbf{A}\cdot d\mathbf{l}\right]$$

$$= \frac{e}{\hbar c}\oint \mathbf{A}\cdot d\mathbf{l} \tag{14.30}$$

since the combination of path 1 and back by path 2 makes a closed loop. We now can use Stokes' theorem

$$\oint \mathbf{A}\cdot d\mathbf{l} = \iint (\nabla \times \mathbf{A})\cdot d\mathbf{S}$$

$$= \iint \mathbf{B}\cdot d\mathbf{S} = \Phi \tag{14.31}$$

where Φ is the flux ($= B\pi R_0^2$) and we have assumed that **B** is directed out of the paper in figure 14.2. Clearly there is an extra shift in phase (of magnitude $e\Phi/\hbar c$) due to the supposedly "inaccessible" flux Φ of the magnetic field that the electrons never traverse. But this phase shift is bound to displace the interference pattern, and this is the case found experimentally. This effect is called the *Aharonov-Bohm effect*. Such an effect could not exist in classical physics, since classically an electron either traverses one path or the other, and therefore, it cannot know about the phase difference; but in quantum mechanics, an electron travels both paths 1 and 2 (albeit in potentia), and therefore, acquires the phase shift.

The Aharonov-Bohm effect clearly demonstrates the limitations of the concept of local fields in quantum mechanics; as such, it presents us another case (like EPR) of quantum nonlocality, and this aspect has provoked (like EPR) quite a bit of controversy. How an electron in the Aharonov-Bohm experiment can be affected by a magnetic field with which it has no local contact via the field lines will seem mysterious until we recognize the nonlocal nature of certain quantum-phase relationships. More recently, Berry has shown that both EPR and Aharonov-Bohm belong to a more general class of such nonlocal phase correlations (see M. V. Berry, *Proc. Roy. Soc. Lond., A392*, 45–57, 1984).

Flux Quantization

It should also be clear from the above discussion that if an electron travels in a field-free region that surrounds a "hole" in which there is trapped magnetic flux, then upon completing a closed loop the electron wave function will acquire the additional phase factor $\exp(ie\Phi/\hbar c)$. But this conflicts with our requirement that all electronic wave functions be single valued. The conflict is resolved with the idea of *quantized flux*: If the flux in the enclosed hole is quantized

$$e\Phi/\hbar c = 2\pi n, \quad \text{with } n = 0, \pm 1, \pm 2, \ldots \tag{14.32}$$

then the extra acquired phase factor becomes unity, and there is no problem.

This quantization of enclosed magnetic flux has been observed by the ingenious use of superconductors. As you know, superconductors, when operating below a certain critical temperature, allow a current to flow without resistance. Superconductivity is theorized as due to a special correlation between pairs of electrons that extends over the whole body of the superconductor. It takes energy for the electrons to break away from this correlated state, thus the state is relatively immune to the random thermal motion present in an ordinary conductor. And it is this special correlation that is responsible for many astounding properties of the superconductor. One of these astounding properties is the *Meissner effect*—superconductors exclude all magnetic flux from inside their volumes.

To demonstrate the quantization of magnetic flux, first, a superconducting ring is placed in an external magnetic field while above the critical temperature for the onset of superconductivity. When the ring is cooled below the critical temperature, it becomes superconducting and excludes all flux from within its body by virtue of the Meissner effect, thus trapping some flux in its hole in the middle (fig. 14.3). According to our discussion before, this trapped

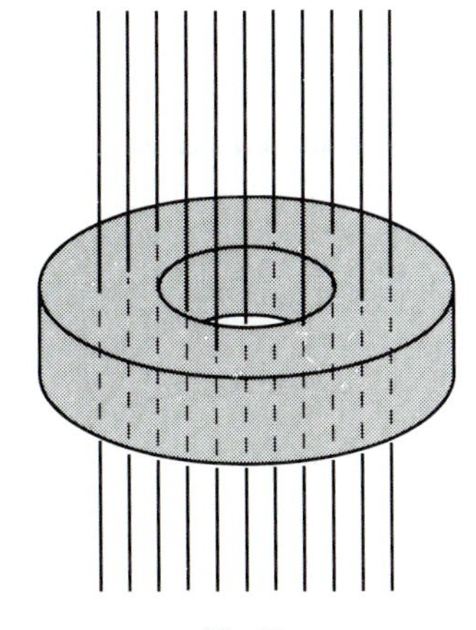

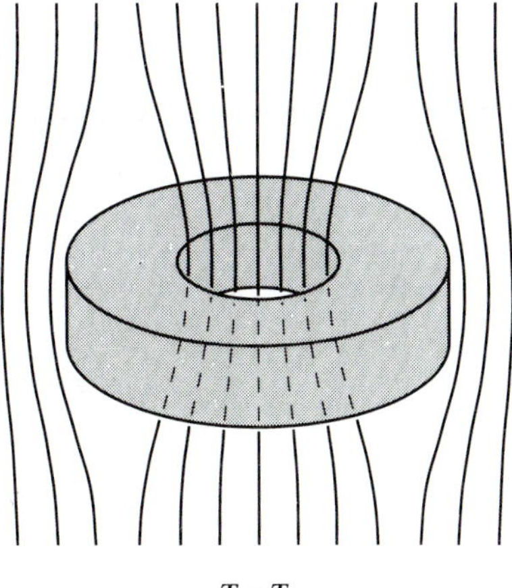

FIGURE 14.3
The Meissner effect and quantization of magnetic flux. When a superconductor with a hole in the middle placed in a magnetic field is cooled below the critical temperature, it traps flux in the hole, which is found to be quantized.

flux must be quantized. Indeed, careful measurements reveal that the flux *is* quantized, however, with an interesting twist. The flux Φ is given by

$$\Phi = (2\pi\hbar c/2e)n$$

instead of equation (14.32). This confirms that the charge carriers in superconductors are indeed correlated electron pairs of charge $2e$, as assumed in current theories of superconductivity.

The Philosophy of Macrorealism and SQUID

In chapter 10, we discussed the troubles that the philosophy of physical realism faces when confronted with the quantum measurement paradox. One of the new philosophies created in defense of physical realism is called macrorealism (see chapter 10). Macrorealism assumes an explicitly stated classical/quantum dichotomy, a division of the world into two categories of objects, some obeying quantum mechanics, some classical (as you know, Bohr's Copenhagen interpretation equivocates on this issue). The idea is ingenious and moreover testable, and it has inspired an interesting experimental program called SQUID (an acronym for superconducting quantum interference device) that employs quantized magnetic flux.

The SQUID is a piece of superconductor with two holes in it that very nearly touch at a point called the weak link (fig. 14.4). Suppose we build a current in the loop around one of the holes. A current sets up a magnetic field just like any old electromagnet, and the field lines representing the magnetic field pass through the hole—that too is usual. What is unusual for a superconductor is that the magnetic flux, the number of field lines per unit area, is quantized as we noted above; the magnetic flux passing through the hole is discrete. This is the key idea.

Suppose we establish such a small current that there is only one quantum of flux. Then we have set up a double-slit type interference question. If there is only one hole, then obviously the flux quantum can be anywhere in it. If the link between the two holes is too thick, then the flux will just be localized in one hole. But with just the right weak link, might we set up quantum interference such that the flux quantum is in both holes at the same time—nonlocalized? If so, quantum coherent superpositions clearly persist even at the scale of macrobodies. But if no such nonlocalization is seen, then we may be able to conclude

FIGURE 14.4
The SQUID—when it is successfully built—will it exhibit quantum coherent superposition even at the macroscopic level?

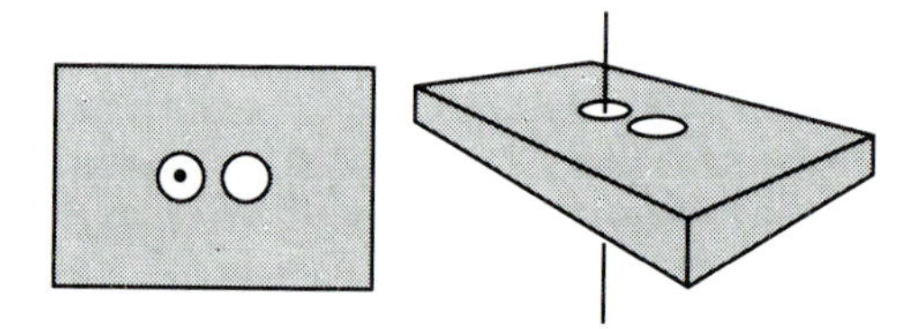

that macrobodies really are classical, they do not permit coherent superpositions as their allowed states.

So far, there is no evidence of any breakdown of quantum mechanics with SQUID, but further research is needed to clearly rule out a classical/quantum dichotomy in the fashion of macrorealism. Read the paper by Leggett referred to at the end of the chapter for further information.

14.3 ANGULAR MOMENTUM MEASUREMENT: THE STERN-GERLACH DEVICE

Since angular momentum is an observable, it must be measurable. Angular momentum, as we have seen, is related to the magnetic moment. Therefore, it can be measured by measuring magnetic moment. The Stern-Gerlach device is a device to measure magnetic moment and thus the angular momentum.

The apparatus (fig. 14.5) is made up of two magnetic pole pieces shaped in such a way as to produce an extremely inhomogeneous field in the (say) z-direction, a field that sharply depends on z, increasing with increasing z. For a magnetic dipole traveling perpendicular to the magnetic field (say, in the x-direction), the field experienced by one end of the dipole is vastly different

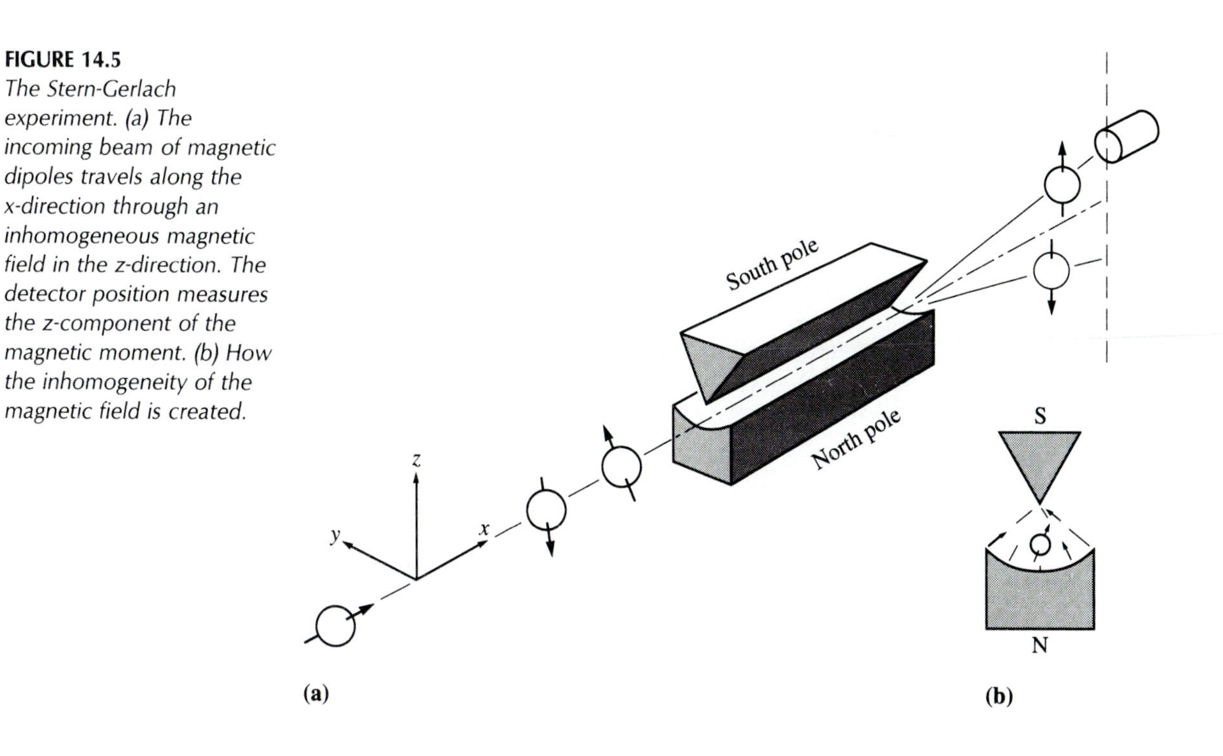

FIGURE 14.5
The Stern-Gerlach experiment. (a) The incoming beam of magnetic dipoles travels along the x-direction through an inhomogeneous magnetic field in the z-direction. The detector position measures the z-component of the magnetic moment. (b) How the inhomogeneity of the magnetic field is created.

from that experienced by the other. This works out to provide a net deflecting force on the dipole. Classically, this force is given by

$$\mathbf{F} = \nabla(\mathbf{M} \cdot \mathbf{B})$$

where **M** is the magnetic dipole moment and **B** is the magnetic field. With our choice of axis, the force of deflection acts along the z-direction and we have

$$\mathbf{F}_z = M_z \frac{\partial B}{\partial z} \tag{14.33}$$

Accordingly, particles should be deflected up or down in direct proportion to the z-component of the dipole moment; by looking at the trace of the deflected particle on a screen, therefore, we can measure the component of its magnetic moment in the direction of the magnetic field.

In this way, the Stern-Gerlach apparatus does to a particle beam of magnetic dipoles what a prism does to white light: It refracts the magnetic dipoles and displays the spectrum of magnetic moment that the particles of the beam possess.

What would we expect classically if a beam of atomic dipoles is shot through the Stern-Gerlach prism? Classically, the spectrum of the z-component of magnetic moment—its allowed values—is continuous, ranging from $-M$ to $+M$. What did Stern and Gerlach find when they first did their experiment? A discrete line spectrum, of course! The spectrum of magnetic moment is quantized; it should be obvious! M_z is proportional to L_z, and L_z is quantized.

To probe M_z is to probe L_z. The number of discrete traces gives the number of allowed values of L_z. If there is only one line, just the undeflected beam, then $L_z = 0$. If there are three lines, the undeflected beam and two more, one deflected up, the other down, then L_z is 1. The multiplicity of the line traces is the multiplicity of L_z, which is $2l + 1$, where l is the quantum number corresponding to angular momentum.

Since l is an integer, naturally this reasoning leads us to the prediction that the number of lines in a magnetic moment spectrum is always odd. We could never, never have a two-line or a six-line spectrum, if this mechanism were all there is.

But unexpectedly (to Stern and Gerlach, who did their experiment in 1922, and to most physicists of the time), when a beam of silver atoms with one valence electron in an S-state was sent through the Stern-Gerlach apparatus, it broke up into two beams, giving two traces (fig. 14.6). And the same thing happens when we send a beam of hydrogen atoms through a Stern-Gerlach device; again we obtain two lines for the magnetic moment spectrum. (It is experimentally hazardous to use free electrons because of the Lorentz force on them and quantum fluctuations; the heavy protons help to keep the trajectory approxi-

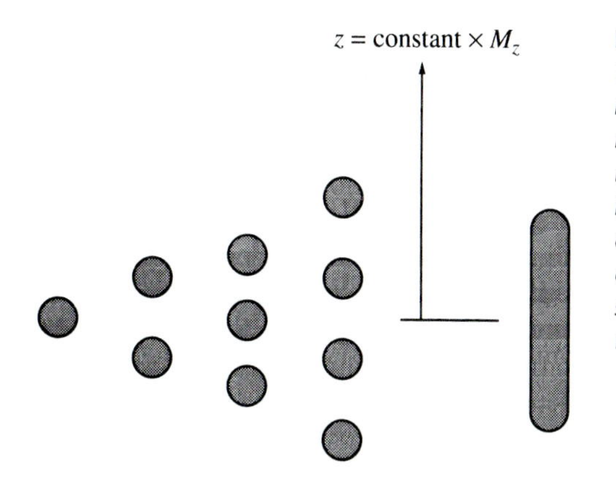

FIGURE 14.6
The Stern-Gerlach spectra—the number of lines denotes the multiplicity of the angular momentum. If the number of lines is even, the angular momentum can only be half-integer. The classical expectation of a continuous spectrum is shown on the extreme right.

mately classical as the electrons undergo Stern-Gerlach deflection. Also, the proton's Stern-Gerlach effect is negligible compared to the electron's.)

The interpretation is unambiguous (although scientists of the time had an extremely hard job in breaking through their classical prejudices before Ulenbeck and Goudsmit wrote their classic paper proclaiming the existence of spin): Beside the angular momentum **L**, for which there is a classical analog (we will refer to **L** as the orbital angular momentum), the electron must have an intrinsic angular momentum, a two-valuedness without any classical parallel. A free electron does not have any orbital angular momentum; thus the two-valuedness must be due to an intrinsic angular momentum, a purely quantum mechanical attribute of many elementary particles. It is called the spin angular momentum or simply *spin*. Since the multiplicity is 2, we can identify the angular momentum associated with spin as 1/2. Thus electrons have spin 1/2.

Don't run pell-mell into the trap of visualizing the spinning electron as some sort of spinning top. Such a classical depiction is simply wrong. In truth, we cannot describe the spin of the electron as any kind of motion in space-time. Rather it is a two-valuedness that is an inborn tendency that behaves like angular momentum. Spin is one more strangeness of the quantum world that you have to accommodate.

Through moderately accurate measurements, Stern and Gerlach also determined quantitative values for the magnetic moments. For the magnetic moment coming from the orbital angular momentum, the proportionality constant between the two (the so-called *gyromagnetic ratio g*) is 1 Bohr magneton, $e\hbar/2\mu c$, as we have seen. However, for spin, the proportionality constant is 2 Bohr magnetons, g for spin is 2; the spin has double the magnetism of orbital motion!

Since spin is a clearly quantum mechanical phenomenon, it is interesting to rediscover quantum mechanics, the matrix mechanics version this time, in the context of spin. And this is what we will proceed to do in chapters 15 and 16, following the elegant treatment given by Richard Feynman in his *Feynman Lectures on Physics*.

PROBLEMS

1. Calculate the wavelength of the normal Zeeman lines for the $4F \rightarrow 3D$ transition in hydrogen for a magnetic field of 10^4 gauss.
2. Calculate the expression for the quantum mechanical current $\mathbf{j}$ of particles whose Hamiltonian is given as

$$H = \frac{[\mathbf{p} + (e/c)\mathbf{A}(\mathbf{r}, t)]^2}{2\mu}$$

3. Prove from symmetry considerations that the quantum numbers m and n in equation (14.19) must be either both even or both odd.
4. Discuss the gauge invariance of the time-dependent Schrödinger equation.
5. Consider once again the problem of the motion of an electron in a uniform external magnetic field $\mathbf{B} = B\mathbf{e}_z$ with the choice of a gauge so that

$$\mathbf{A}_x = -By, \qquad \mathbf{A}_y = \mathbf{A}_z = 0$$

What are the constants of the motion? Solve the eigenvalue problem and show that the eigenvalues are the same as those obtained in the text using a different gauge.

6. (a) In the original Stern-Gerlach experiment using silver atoms, each atom was found to have a magnetic moment whose component in the direction of the **B**-field was 1 Bohr magneton. The kinetic energy of the atoms was 1/5 eV, and they traveled 3 cm through an inhomogeneous magnetic field of gradient 200 kilogauss/cm. Calculate the deflection of the beam at 20 cm from the magnets.
(b) Charged particles are also deflected by magnetic fields because of their charge. Does this have any effect on their Stern-Gerlach measurements? (For a good discussion on this, see G. Baym, *Lectures on Quantum Mechanics*.)

ADDITIONAL PROBLEM

A1. A charged particle moving in the x-y plane in the potential of a two-dimensional harmonic oscillator is placed in a magnetic field $\mathbf{B} = (0, 0, B)$. Find the exact spectrum. Draw a diagram of the result as a function of B and discuss the results in the limits of small and large B.

REFERENCES

Y. Aharonov and D. Bohm, *Phys. Rev.*, *115*, 485, 1959.

S. Gasiorowicz. *Quantum Physics.*

A. M. Leggett. In *The Lessons of Quantum Theory*, J. De Boer, E. Dal, and O. Ulfbeck (eds.).

15

Spin and Matrices

In this chapter we are going to introduce matrices and matrix representation of operators using the spinning electron as our object of study. Heisenberg was the first to hit upon the idea of using matrices in quantum mechanics, although he didn't know that's what he was doing. This is why Heisenberg's discovery is much like a miracle; science historians have very little clue as to how he got the idea of matrix mechanics the way he did it! But that's the way of creative geniuses; their ideas often defy rationality.

Using hindsight, we are going to re-enact a pedagogically improved version of Heisenberg's discovery. Heisenberg was dealing with infinite matrices (see the final section of the chapter); with spin, we will need to handle only finite matrices. So we will begin our discussion with beams of electrons and their mysterious two-valuedness, or spin, and Stern-Gerlach apparatuses to separate the two components; then we will analyze the results further; and so on. All the laws of quantum mechanics can be discovered from this little beginning. And that too is a mini miracle!

So in this chapter we are going to mostly study spin, and spin alone, ignoring the motion in x, y, and z altogether. This separation is only temporary, and later we will put all the motion together.

15.1 BASIC STATES, OPERATORS, AND MATRICES

If a beam of electrons travels through a Stern-Gerlach apparatus, it splits up into two distinct beams. (In practice, we may have to send the electrons in the company of protons to get good classical trajectories, but no matter. Think of what follows as a gedanken experiment.) Clearly, there must be two distinct states associated with the two distinct beams, and our first task is to name them. It is customary to label the two beams as "up" and "down" along the z-axis (the direction of the magnetic field that defines the axis of the Stern-Gerlach apparatus). That is, we picture the two-valuedness of the spin as an arrow that can point either up or down. Electrons in the upper beam are assumed to have their spin arrows pointing up; we will specify their state by the notation

$$|\text{up}\rangle \quad \text{or} \quad |S_z = +\rangle$$

S_z denoting the two-valued spin variable. It's the opposite for those electrons that are in the lower beam: They are caught with their arrows down. We will use the notation

$$|\text{down}\rangle \quad \text{or} \quad |S_z = -\rangle$$

to denote the state of these electrons. If we are interested only in the spin orientations, these states of the electron are enough to describe any arbitrary spin state of the electron, which must be a linear combination of the two. So these states are called basic states or base kets. (Recognize the Dirac notation that we are introducing in this manner?)

It is also clear that the electrons are starting in some initial state, which we will denote by $|i\rangle$, and after passing through the Stern-Gerlach apparatus they are ending in some final state, which we will denote by $|f\rangle$. Of course, $|f\rangle$ can be either $|S_z = +\rangle$ or $|S_z = -\rangle$ depending on whether we find the electron in the upper or the lower beam. We will denote the amplitude for starting as $|i\rangle$ and ending as $|S_z = +\rangle$ with the notation

$$\langle S_z = + | i\rangle$$

where the notation $\langle S_z = +|$ you recognize as the bra corresponding to the ket $|S_z = +\rangle$. The absolute square of this amplitude gives the probability of starting with $|i\rangle$ and ending in the "up" beam after passing through the Stern-Gerlach apparatus. We can do the same thing when the final state is $|S_z = -\rangle$. Again $|\langle S_z = - | i\rangle|^2$ gives the probability for the electron to end in the "down" beam.

Now suppose we put two Stern-Gerlach apparatuses oriented with the same axis in a series, as shown in figure 15.1. Now we do one more trick—we put, say, a plate or something to mask the lower ($S_z = -$) beam. We will call a Stern-Gerlach apparatus with a mask like this a Stern-Gerlach filter. Compared to an open Stern-Gerlach apparatus, a Stern-Gerlach filter *prepares* the beam in one definite state of S_z. Now, if both filters block the down beam, since only the up beam comes out of the first filter and goes undaunted through the second apparatus, we will find that all the electrons are now deflected up and none are

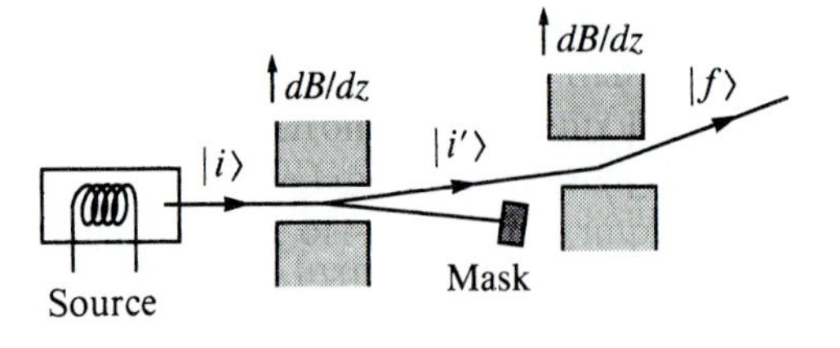

FIGURE 15.1
Stern-Gerlach apparatuses in series; dB/dz is oriented along the same axis in both apparatuses. Both apparatuses mask the "down" beam in this picture.

deflected down. Similarly, if we block the up beam in both filters, the probability is 100% that the beam passing through the second apparatus is deflected down. But if the first filter prepares an up beam, while the second filter is set up to block the up beam, we get no particle at all. These results can be summarized as

$$\langle S_z = + | S_z = + \rangle = 1 = \langle S_z = - | S_z = - \rangle$$

$$\langle S_z = - | S_z = + \rangle = 0 = \langle S_z = + | S_z = - \rangle \qquad (15.1)$$

You will recognize these equations as the normalization and orthogonality conditions for our basic states.

The states $|S_z = \pm\rangle$ are also called polarized states because they have taken on a specific direction for the spin. In comparison, an unpolarized state, such as the initial state $|i\rangle$ referred to before, is described as a linear combination

$$|i\rangle = a|S_z = +\rangle + b|S_z = -\rangle \qquad (15.2)$$

If we start with this state, the amplitude for ending with $|S_z = +\rangle$ is given as

$$\langle S_z = + | i \rangle = a\langle S_z = + | S_z = + \rangle + b\langle S_z = + | S_z = - \rangle = a \qquad (15.3)$$

Similarly,

$$\langle S_z = - | i \rangle = a\langle S_z = - | S_z = + \rangle + b\langle S_z = - | S_z = - \rangle = b \qquad (15.4)$$

But $|a|^2$ and $|b|^2$ are, respectively, the probability that an electron starts in the state $|i\rangle$ and ends up or down upon passing through a Stern-Gerlach device. Since an electron cannot just disappear—it must end up either in the up beam or in the down beam—we have

$$|a|^2 + |b|^2 = 1 \qquad (15.5)$$

Using equations (15.3) and (15.4), we can also write equation (15.5) as

$$|\langle S_z = + | i \rangle|^2 + |\langle S_z = - | i \rangle|^2 = 1$$

or equivalently as

$$\langle S_z = + | i \rangle \langle S_z = + | i \rangle^* + \langle S_z = - | i \rangle \langle S_z = - | i \rangle^* = 1 \tag{15.6}$$

Stern-Gerlach Filters in Series with a Tilt Between Them

To obtain some further properties of the base states, we have to invent more complicated experiments with the Stern-Gerlach devices. Now let's consider two Stern-Gerlach filters (device plus mask) in series (fig. 15.2), but now the second filter is at an angle with respect to the first; for example, suppose we rotate the apparatus about the y-axis through 90°, changing the z-axis into the x-axis. Because of the tilt, the second filter defines a different basis altogether, call it $|S_x = \pm\rangle$; the states $|S_x = \pm\rangle$ are not the same as $|S_z = \pm\rangle$ and both amplitudes $\langle S_x = + | S_z = +\rangle$ and $\langle S_x = - | S_z = +\rangle$ are nonzero. Even if the S_z-filter prepares a beam in the up or $|S_z = +\rangle$ state, the S_x-filter can be used to obtain either an up or a down beam, $|S_x = +\rangle$ or $|S_x = -\rangle$, depending on where we place the mask.

Let's examine in some detail what happens if we start with a base ket of one filter, say $|S_z = +\rangle$, and end up with a base ket of another, such as $|S_x = -\rangle$. Now if we have a third Stern-Gerlach device identical to the first S_z-filter (i.e., axis parallel to the z-direction), will it in some way "know" that the beam started as a $|S_z = +\rangle$ beam? No, the history is wiped out by the insertion of the S_x-basis (except for the attenuation of the intensity). For example, the amplitudes for starting with $|S_z = +\rangle$, going through the intermediate $|S_x = -\rangle$, and ending up with $|S_z = \pm\rangle$ upon passing through the third filter are given, respectively, as

$$\langle S_z = + | S_x = -\rangle \langle S_x = - | S_z = +\rangle$$

and

$$\langle S_z = - | S_x = -\rangle \langle S_x = - | S_z = +\rangle.$$

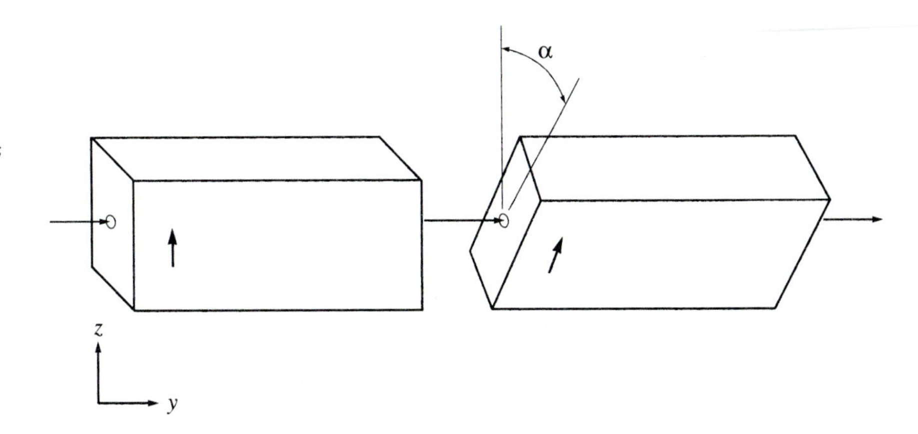

FIGURE 15.2
Two Stern-Gerlach filters in series, but the inhomogeneous magnetic field of the second filter has been tilted by an angle α with respect to that of the first.

If we take the ratio, the common history $\langle S_x = - | S_z = + \rangle$ of the two measurements cancels out.

Now here is something very interesting. Suppose we unmask the S_x-filter, using instead just an open Stern-Gerlach device with axis along the x-direction, and repeat the above measurements; what then? In terms of amplitudes we are now taking the following sums:

$$\langle S_z = + | S_x = + \rangle \langle S_x = + | S_z = + \rangle + \langle S_z = + | S_x = - \rangle \langle S_x = - | S_z = + \rangle \quad (15.7a)$$

and

$$\langle S_z = - | S_x = + \rangle \langle S_x = + | S_z = + \rangle + \langle S_z = - | S_x = - \rangle \langle S_x = - | S_z = + \rangle \quad (15.7b)$$

Experimentally, we *should* get the result 1 (100% probability) for the first sum and 0 for the second. But how should we interpret the result? Certainly none of the individual amplitudes vanishes, yet a combination of them gives zero—this must be interpreted as a result of quantum mechanical interference. We cannot have quantum mechanics without interference! It also follows that an open Stern-Gerlach apparatus has no effect at all—when we sum over both base states of the intermediate device, the result is

$$\sum_{\alpha} | \alpha \rangle \langle \alpha | = 1 \quad (15.8)$$

We have rederived another important quantum mechanical rule for a basic set: Ket-bra sum equals unity.

There is one final rule that we can derive. Consider the unpolarized state $| i \rangle$; let's say it's normalized. We have

$$1 = \langle i | i \rangle = \sum_{\alpha} \langle i | \alpha \rangle \langle \alpha | i \rangle$$

Now compare this equation with equation (15.6), which we can also write as

$$\sum_{\alpha} \langle \alpha | i \rangle \langle \alpha | i \rangle^* = 1$$

The two equations are compatible only if we have

$$\langle \alpha | i \rangle^* = \langle i | \alpha \rangle \quad (15.9)$$

And it is easily seen that this holds for any two general states. This is another one of our previous rules: Backward amplitude = (forward amplitude)*. We are rediscovering the rules followed by quantum mechanical state vectors forming a basic set. The only difference from what we had before is that now the base kets define a finite dimensional Hilbert space.

Operators and Matrices

The next step is to introduce the concept of an operator. Suppose we have an electron beam in some initial state $|i\rangle$ and we pass it through some unspecified "mess" of Stern-Gerlach devices, ending in the final state $|f\rangle$. The amplitude for this process can be written as $\langle f|A|i\rangle$, where A is an operator that represents the Stern-Gerlach mess, and that takes us to the final state upon operating on the initial state. But this way of representing a process is helpful only if we can figure out a concise way to represent A that we can apply with any arbitrary states between which A is sandwiched; otherwise we would have to evaluate the amplitude of A afresh every time we have two new states. Fortunately, and here is the power of defining the base kets, we can take advantage of our basic set defined above and write

$$\langle f|A|i\rangle = \sum_{\alpha\beta} \langle f|\alpha\rangle\langle\alpha|A|\beta\rangle\langle\beta|i\rangle \tag{15.10}$$

where we have used the ket-bra sum-equals-1 rule twice. The four numbers $\langle\alpha|A|\beta\rangle$ are called matrix elements of A in the basis defined by the set $|\alpha\rangle$ (which are $|S_z = \pm\rangle$) for the following reason. We write these elements as an array called a square matrix:

	$\|S_z = +\rangle$	$\|S_z = -\rangle$
$\|S_z = +\rangle$	$\langle S_z = +\|A\|S_z = +\rangle$	$\langle S_z = +\|A\|S_z = -\rangle$
$\|S_z = -\rangle$	$\langle S_z = -\|A\|S_z = +\rangle$	$\langle S_z = -\|A\|S_z = -\rangle$

Note that we are ordering the columns and rows in descending order of the component of spin—this is an important convention. Once this matrix of A is known, we can determine the matrix element of A sandwiched between any two arbitrary states by using equation (15.10). This is a fantastic achievement.

The simplest case, of course, is the operator that defines the base states $|S_z = \pm\rangle$. Let's call this particular operator—the operation of passing through an S_z-filter—the S_z operator. For the S_z-operator, obviously the off-diagonal elements of the matrix above are zero, and taking cognizance of Stern and Gerlach's measurement, we have the matrix

$$\begin{pmatrix} +\frac{1}{2}\hbar & 0 \\ 0 & -\frac{1}{2}\hbar \end{pmatrix} \tag{15.11}$$

The matrix of S_z in its own basis is a *diagonal* matrix.

Sometimes we need to consider an operator that is the product of two operators

$$C = AB$$

and we want to calculate $\langle i|C|j\rangle$. We can do it by using the matrix elements of A and B. For momentary simplicity, suppose $|i\rangle$ and $|j\rangle$ belong to the same basic set $|\alpha\rangle$ that we use to define the matrices of A and B. Then we have

$$\langle i|C|j\rangle = \sum_{\alpha} \langle i|A|\alpha\rangle\langle\alpha|B|j\rangle = \sum_{\alpha} A_{i\alpha}B_{\alpha j}$$

The ijth matrix element of C is the sum of the element-by-element product of matrix elements of the ith row of the matrix of A and the jth column of the matrix of B (fig. 15.3; also see below for a specific example); this is the rule of matrix multiplication. That is to say that the matrix representation of C is the matrix product of the matrices that represent A and B. However, there is an important new element here. Hamilton, who discovered matrices, is supposed to have said in answer to his son's question that he knew how to add matrices, but not how to multiply them. Addition of matrices is simple; you just add the matrices element by element without paying any attention to the order of addition. But matrix multiplication must have confused Hamilton, because the matrix of the product BA is not the same as the matrix of AB. Convince yourself by carrying out the appropriate multiplications, if you are not familiar with this already. So matrices don't commute! Thus they are suitable for representing quantum mechanical operators which, in general, don't commute either.

Coming back to equation (15.10), note that the quantities $\langle\beta|i\rangle$ can also be represented as an array, a column of two numbers:

$$\begin{pmatrix} \langle S_z = +|i\rangle \\ \langle S_z = -|i\rangle \end{pmatrix}$$

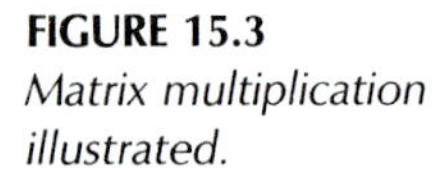

FIGURE 15.3
Matrix multiplication illustrated.

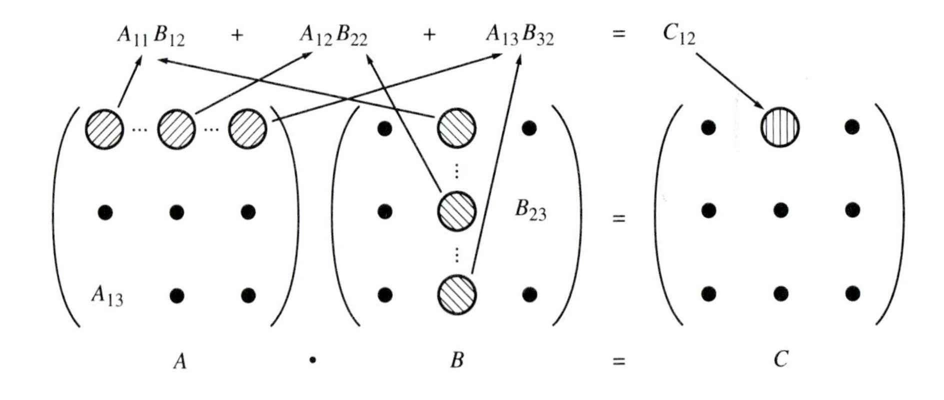

This is called a *column matrix* and it is a representation of the ket $|i\rangle$ in the basis $|S_z = \pm\rangle$. Similarly, the bra $\langle f|$ can be represented by a *row matrix*

$$(\langle f|S_z = +\rangle \quad \langle f|S_z = -\rangle)$$

Therefore, whereas operators are represented as square matrices, kets and bras are represented by column and row matrices respectively. The two-component column matrix that we are considering here in connection with spin is called a *spinor.*

In particular, for the base states themselves, the matrix representations of the kets $|S_z = \pm\rangle$ (since they will be useful later, we will give them special notations) are given as

$$\chi_+ = \begin{pmatrix} 1 \\ 0 \end{pmatrix} \quad \text{and} \quad \chi_- = \begin{pmatrix} 0 \\ 1 \end{pmatrix} \tag{15.12}$$

respectively, and for the bras $\langle S_z = \pm|$ we get

$$\chi_+^\dagger = (1 \quad 0) \quad \text{and} \quad \chi_-^\dagger = (0 \quad 1) \tag{15.13}$$

respectively. Here the † denotes the *hermitian conjugate* of a matrix—a matrix obtained from the original by interchanging rows and columns and taking the complex conjugate; hermitian conjugation changes a column matrix into a row matrix.

Let's examine what happens when we multiply the spinor χ_+ by the matrix of S_z, equation (15.11), from the left:

$$\frac{\hbar}{2}\begin{pmatrix} 1 & 0 \\ 0 & -1 \end{pmatrix}\begin{pmatrix} 1 \\ 0 \end{pmatrix} = \frac{\hbar}{2}\begin{pmatrix} 1 \\ 0 \end{pmatrix}$$

This must be interpreted as an eigenvalue-eigenvector equation. The spinor

$$\begin{pmatrix} 1 \\ 0 \end{pmatrix}$$

is an *eigenspinor* of S_z corresponding to the eigenvalue $\hbar/2$; it is the matrix representation of the eigenstate of S_z denoted by $|S_z = +\rangle$. The measurement of S_z in the state represented by the spinor χ_+ will always result in the value $\frac{1}{2}\hbar$. Similarly, the spinor χ_- is an eigenspinor of S_z with eigenvalue $-\hbar/2$.

With all these definitions, it should now be clear that equation (15.10) is itself a product of matrices—the ket-bra sums are the same sums involved in matrix multiplication; and the sums can be carried out by multiplying a row matrix, a square matrix, and a column matrix together, giving a single number. Check it out.

■ **PROBLEM** The matrix representations of two operators A and B are given by

$$A = \begin{pmatrix} 1 & 2 \\ 2 & 1 \end{pmatrix} \qquad B = \begin{pmatrix} 3 & 2 \\ 2 & 1 \end{pmatrix}$$

Find the matrix representation of the product $C = AB$.

SOLUTION Following the rule above, to find C_{ij}, we take the ith row of A and the jth column of B, multiply them element by element, and add them up. We get

$$C = \begin{pmatrix} 7 & 4 \\ 8 & 5 \end{pmatrix}$$ ■

Spinors

The spinors can be thought of as vectors in a two-dimensional Hilbert space. The particular spinors defined by equation (15.12) act like basis vectors in the spin-space.

A general state in spin-space given by equation (15.2)

$$a|S_z = +\rangle + b|S_z = -\rangle$$

can be represented by a spinor

$$\chi = a\chi_+ + b\chi_-$$
$$= a\begin{pmatrix} 1 \\ 0 \end{pmatrix} + b\begin{pmatrix} 0 \\ 1 \end{pmatrix} = \begin{pmatrix} a \\ b \end{pmatrix}$$

The normalization requirement is

$$\chi^\dagger \chi = 1 = (a^* \quad b^*)\begin{pmatrix} a \\ b \end{pmatrix} = |a|^2 + |b|^2$$

as before.

The scalar product of two spinors χ and χ' is defined by

$$\chi^\dagger \chi' = (a^* \quad b^*)\begin{pmatrix} a' \\ b' \end{pmatrix} = a^*a' + b^*b' \qquad (15.14)$$

Two spinors χ and χ' are orthogonal if their scalar product vanishes.

The spinors that form a basis are an orthonormal set. You can easily verify that the base spinors χ_+ and χ_- defined in equation (15.12) are an orthonormal set.

15.2 ANALOGY OF ELECTRON SPIN AND POLARIZATION OF LIGHT

In spin, we are seeing a two-valuedness similar to the two-valued polarization of light. Let's use this analogy to shed further light on the situation with spin.

Consider x-polarized light propagating in the z-direction; its electric field vector oscillates in the x-direction:

$$\boldsymbol{\mathcal{E}} = \mathcal{E}_0 \mathbf{e}_x \cos(kz - \omega t)$$

For y-polarized light, we have

$$\boldsymbol{\mathcal{E}} = \mathcal{E}_0 \mathbf{e}_y \cos(kz - \omega t)$$

Let's call a polaroid filter that produces respectively x- or y-polarized light an x- or y-filter. A combination of an x- and a y-filter, if 100% efficient, will quench a light beam completely (fig. 15.4). But suppose we insert a polaroid filter between the two with its axis at 45° to the x-(y-)direction (in the xy-plane), call it an x'- (y'-) filter; now there will be light pouring out from the combination of the three (fig. 15.5) because the x'-(y'-) filter wipes out the earlier history of the passage of the light beam through the x-filter (except, of course, for the total intensity). The analogy to the combination of the three Stern-Gerlach filters in series, an S_x-filter sandwiched between two S_z-filters, is easily made if we make the following correspondences:

$$S_z{=}+ \text{ electron} \rightleftharpoons x\text{-polarized light}$$

$$S_z{=}- \text{ electron} \rightleftharpoons y\text{-polarized light}$$

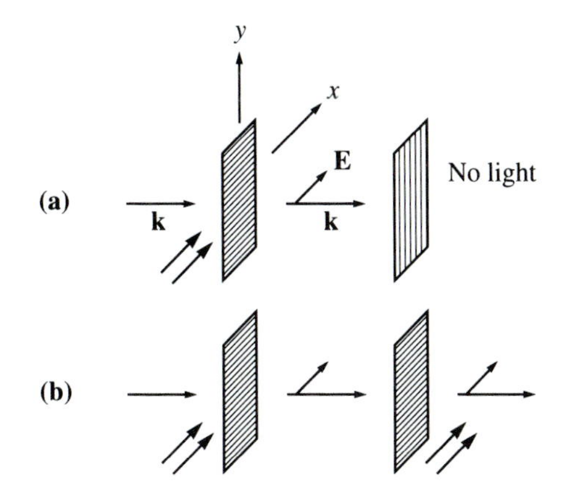

FIGURE 15.4
Polaroids in series. (a) When the polaroids are crossed (i.e., aligned with axes perpendicular to each other) no light gets through the combination (double arrows denote directions of polarization); (b) when the polaroid axes are parallel, all the light that goes through the first polaroid also goes through the second.

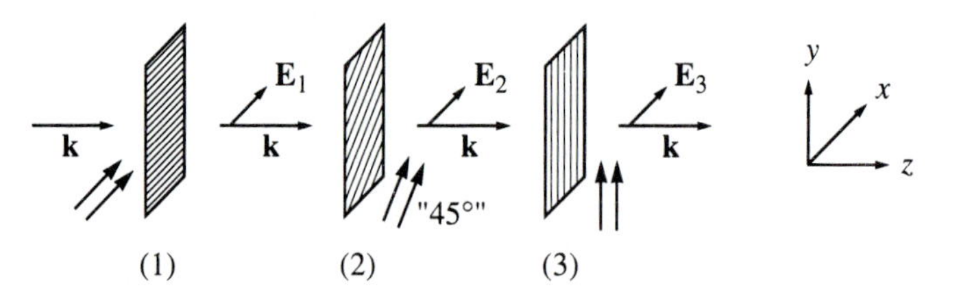

FIGURE 15.5
A 45°-filter placed between the crossed filters of figure 15.4(a) wipes out the history and light finds its way out again.

and

$$S_x = + \text{ electron} \rightleftharpoons x'\text{-polarized light}$$

$$S_x = - \text{ electron} \rightleftharpoons y'\text{-polarized light}$$

How can we put these ideas down quantitatively? In classical electrodynamics, a 45°-polarized beam, x'- or y'-polarized light, is described as superpositions of x- and y-polarized light:

$$\mathcal{E}_0 \mathbf{e}_{x'} \cos(kz - \omega t) = (\mathcal{E}_0/\sqrt{2})\,[\mathbf{e}_x \cos(kz - \omega t) + \mathbf{e}_y \cos(kz - \omega t)]$$

$$\mathcal{E}_0 \mathbf{e}_{y'} \cos(kz - \omega t) = (\mathcal{E}_0/\sqrt{2})\,[-\mathbf{e}_x \cos(kz - \omega t) + \mathbf{e}_y \cos(kz - \omega t)]$$

Analogously, we will write the base ket $|S_x = \pm\rangle$ as similar linear combinations of $|S_z = \pm\rangle$ kets. We get

$$|S_x = +\rangle = (1/\sqrt{2})\,[|S_z = +\rangle + |S_z = -\rangle]$$

$$|S_x = -\rangle = (1/\sqrt{2})\,[-|S_z = +\rangle + |S_z = -\rangle] \tag{15.15}$$

These are the eigenstates of the operator S_x that, let's say, represents the rotated Stern-Gerlach apparatus we are considering. In contrast, the base states $|S_z = \pm\rangle$ are the eigenstates of the operator S_z.

Now we can see why it is that passing an electron through an S_x-filter must wipe out its history. An S_x-filter brings about a measurement of S_x whence the electron becomes an eigenstate of S_x; that is, its state collapses leaving no trace of previous initial conditions. This is, of course, the fundamental measurement postulate of quantum mechanics.

In spinor representation, we now can write down the "eigenspinors" of S_x; for the S_x-spin-up state, it is

$$\begin{pmatrix} \langle S_z = + | S_x = + \rangle \\ \langle S_z = - | S_x = + \rangle \end{pmatrix} = \begin{pmatrix} 1/\sqrt{2} \\ 1/\sqrt{2} \end{pmatrix} \tag{15.16}$$

Similarly, for the down state $|S_x=-\rangle$, we have the eigenspinor

$$\begin{pmatrix} -1/\sqrt{2} \\ 1/\sqrt{2} \end{pmatrix} \tag{15.17}$$

But there is one more problem. Instead of rotating the S_z-Stern-Gerlach apparatus about the y-axis, we can also rotate it 90° about the x-axis, converting z- $\rightarrow$ y-axis. How do we represent $|S_y=\pm\rangle$ polarized states obtained in this fashion? The problem is nontrivial, because these states must be distinguishable from the $|S_x=\pm\rangle$-states obtained above.

Fortunately, we can invoke still one more analogy with polarized light. There is the phenomenon of circular polarization—the result of passing linearly polarized light through a quarter-wave plate. If we pass such circularly polarized light through an x- or y-filter, we get an x- or y-polarized beam; and yet circularly polarized light is a very different beast from 45° polarized light! Mathematically, circularly polarized light is also a linear combination of x- and y-polarized light, except that the electric field of the y-polarized light is 90° out of phase with the electric field of the x-polarized light. For example, for right-circularly polarized light, we have

$$\boldsymbol{\mathcal{E}} = (\mathcal{E}_0/\sqrt{2})\,[\mathbf{e}_x \cos(kz - \omega t) + \mathbf{e}_y \cos(kz - \omega t + \pi/2)]$$

We can look upon this as the real part of a complex electric field

$$\boldsymbol{\mathcal{E}}' \equiv (\mathcal{E}_0/\sqrt{2})\,[\mathbf{e}_x \exp(ikz - i\omega t) + i\mathbf{e}_y \exp(ikz - i\omega t)]$$

where $\mathrm{Re}(\boldsymbol{\mathcal{E}}') = \boldsymbol{\mathcal{E}}$, and where we have used $i = \exp(i\pi/2)$. For left-circularly polarized light, we get

$$\boldsymbol{\mathcal{E}}' \equiv (\mathcal{E}_0/\sqrt{2})\,[\mathbf{e}_x \exp(ikz - i\omega t) - i\mathbf{e}_y \exp(ikz - i\omega t)]$$

Now make the analogy

$$S_y = + \text{ electron} \rightleftharpoons \text{right-circularly polarized light}$$

$$S_y = - \text{ electron} \rightleftharpoons \text{left-circularly polarized light}$$

Now the choice for the appropriate linear combination for $|S_y=\pm\rangle$ should be obvious; it is

$$|S_y=\pm\rangle = (1/\sqrt{2})\,[|S_z=+\rangle \pm i\,|S_z=-\rangle] \tag{15.18}$$

which are distinct from $|S_x=\pm\rangle$. These then constitute the eigenstates of the operator S_y that represents the operation of passing an electron through an

S_y-Stern-Gerlach filter. In the spinor notation, the eigenspinors of S_y are then given as

$$\begin{pmatrix} 1/\sqrt{2} \\ i/\sqrt{2} \end{pmatrix} \quad \text{and} \quad \begin{pmatrix} 1/\sqrt{2} \\ -i/\sqrt{2} \end{pmatrix} \tag{15.19}$$

respectively, for the S_y-spin-up and S_y-spin-down states.

It is also interesting, because now it is clear that the spin states span a complex vector space; but this is no surprise because we are dealing with vectors in Hilbert space here, albeit a finite-dimensional one.

15.3 THE MATRICES OF S_x AND S_y AND SOLUTION OF EIGENVALUE EQUATIONS

We will now construct the matrices for the spin operators S_x and S_y that correspond to the S_x- and S_y-Stern-Gerlach filters. Since spin is a kind of angular momentum, intrinsic and half-integer as it is, we will postulate that the spin matrices obey the angular momentum commutation relations:

$$[S_x, S_y] = i\hbar S_z, \qquad xyz \text{ cyclic} \tag{15.20}$$

We can then calculate the matrices for the spin operators S_x and S_y using equation (11.62) derived from the angular-momentum commutation relations for the special case of $J_\pm = S_\pm$ and with $j = s = 1/2$. We have

$$\langle +\tfrac{1}{2}|S_+|-\tfrac{1}{2}\rangle = \hbar$$

$$\langle -\tfrac{1}{2}|S_-|+\tfrac{1}{2}\rangle = \hbar$$

All other matrix elements are zero. Now using

$$S_x = \tfrac{1}{2}(S_+ + S_-) \quad \text{and} \quad S_y = (1/2i)(S_+ - S_-)$$

we get the following matrices for S_x and S_y, respectively:

$$S_x = \frac{\hbar}{2}\begin{pmatrix} 0 & 1 \\ 1 & 0 \end{pmatrix} \tag{15.21}$$

$$S_y = \frac{\hbar}{2}\begin{pmatrix} 0 & -i \\ i & 0 \end{pmatrix} \tag{15.22}$$

An important class of matrices is called *hermitian* whose matrix elements satisfy the property

$$A_{\alpha\beta} = A^*_{\beta\alpha} \tag{15.23a}$$

As defined before, the hermitian conjugate $A^\dagger$ of a matrix is obtained by first taking its *transpose*, which is obtained by interchanging rows and columns, and then taking the complex conjugate of the elements. Thus, for hermitian matrices, the hermitian conjugate is the same as the original matrix

$$A^\dagger = (A^T)^* = A \tag{15.23b}$$

where the superscript T denotes transpose. You can check that the matrices corresponding to S_x, S_y, and S_z are all hermitian. In general, physical observables in quantum mechanics are represented by hermitian operators whose matrix representations are hermitian matrices.

The Eigenvalue Equation

We will now obtain the eigenstates of S_x, say—or rather the spinor representation of the eigenstates—by solving its eigenvalue equation. By definition, the eigenvalue equation for S_x in the matrix form is obtained when the matrix of S_x acting on its eigenspinor gives back the eigenspinor multiplied by its eigenvalue:

$$\frac{\hbar}{2}\begin{pmatrix} 0 & 1 \\ 1 & 0 \end{pmatrix}\begin{pmatrix} a \\ b \end{pmatrix} = \lambda\begin{pmatrix} a \\ b \end{pmatrix} \tag{15.24}$$

where λ denotes the eigenvalue of S_x. Now the matrix equation (15.24) is really a condensed way of writing two algebraic equations: It stands for two equations obtained by equating each element of the product matrix on the left-hand side to the corresponding element of the matrix on the right-hand side. We also (since $2\lambda/\hbar$ can be thought of as $(2\lambda/\hbar)I$, where I is the *unit matrix* in the 2×2 space whose diagonal elements are 1 and off-diagonal elements 0) can write the equation as

$$\begin{pmatrix} -2\lambda/\hbar & 1 \\ 1 & -2\lambda/\hbar \end{pmatrix}\begin{pmatrix} a \\ b \end{pmatrix} = 0$$

These are two homogeneous equations for two unknowns, a and b (actually λ is also unknown, but for the moment let's pretend that it is not). The condition for a solution is that the determinant (called the secular determinant) of the 2×2 matrix on the right-hand side vanishes. This gives the *secular equation*

$$\begin{vmatrix} -2\lambda/\hbar & 1 \\ 1 & -2\lambda/\hbar \end{vmatrix} = 0 \tag{15.25}$$

that determines the eigenvalues λ. Evaluating the determinant, we get

$$(2\lambda/\hbar)^2 - 1 = 0$$

or

$$\lambda = \pm\hbar/2$$

The eigenspinors corresponding to each of the eigenvalues are now easily determined by going back to equation (15.24); we have for $\lambda = \hbar/2$,

$$a = b = 1/\sqrt{2}$$

where we have used the normalization condition

$$a^2 + b^2 = 1$$

For $\lambda = -\hbar/2$, we get $a = -b = -1/\sqrt{2}$. Therefore, the eigenspinors of S_x are

$$\begin{pmatrix} 1/\sqrt{2} \\ 1/\sqrt{2} \end{pmatrix} \quad \text{and} \quad \begin{pmatrix} -1/\sqrt{2} \\ 1/\sqrt{2} \end{pmatrix}$$

for eigenvalues $+\frac{1}{2}\hbar$ and $-\frac{1}{2}\hbar$, respectively, the same as obtained in section 15.2 via the analogy with polarized light. The eigenspinors of S_y, equation (15.19), can also be determined by the above procedure.

We should note in passing that this way of obtaining the eigenvalues and eigenvectors of matrices is easily generalizable for any $N \times N$ matrix.

Pauli Matrices

It is convenient to define three useful matrices in the 2×2 spin space via the relations

$$S_x = \tfrac{1}{2}\hbar\sigma_x$$

$$S_y = \tfrac{1}{2}\hbar\sigma_y$$

$$S_z = \tfrac{1}{2}\hbar\sigma_z \qquad (15.26)$$

The matrices σ_x, σ_y, and σ_z are called *Pauli matrices*. Explicitly, they are given as

$$\sigma_x = \begin{pmatrix} 0 & 1 \\ 1 & 0 \end{pmatrix} \quad \sigma_y = \begin{pmatrix} 0 & -i \\ i & 0 \end{pmatrix} \quad \sigma_z = \begin{pmatrix} 1 & 0 \\ 0 & -1 \end{pmatrix} \qquad (15.27)$$

Together with the unit matrix I

$$I = \begin{pmatrix} 1 & 0 \\ 0 & 1 \end{pmatrix} \tag{15.28}$$

they form a complete set of matrices for the 2×2 space in the following sense: Any 2×2 hermitian matrix can be expressed in terms of these four matrices. The Pauli matrices satisfy the following important properties:

$$\begin{aligned} \sigma_i^2 &= I, \qquad i = x,\, y, \text{ or } z \\ \sigma_i \sigma_j + \sigma_j \sigma_i &= 2\delta_{ij} I, \qquad i,j = x,\, y, \text{ or } z \\ [\sigma_x, \sigma_y] &= 2i\sigma_z, \qquad xyz \text{ cyclic} \end{aligned} \tag{15.29}$$

For convenience, using vector notation, we also write

$$\mathbf{S} = \hbar \boldsymbol{\sigma}/2 \tag{15.30}$$

Note that

$$\sigma^2 = \sigma_x^2 + \sigma_y^2 + \sigma_z^2 = 3$$

Therefore,

$$\begin{aligned} S^2 &= \hbar^2 \sigma^2/4 = 3\hbar^2/4 \\ &= s(s+1)\hbar^2 \quad \text{with } s = \tfrac{1}{2} \end{aligned} \tag{15.31}$$

as befits the square of any angular-momentum operator.

15.4 THE TRANSFORMATION MATRIX

You may already have noticed one intriguing aspect of all the various things we have done with the spin amplitudes. One Stern-Gerlach filter prepares a beam in a state $|S_z = \alpha\rangle$, another rotated Stern-Gerlach apparatus measures the probability $|\langle S_y = i | S_z = \alpha\rangle|^2$ of finding the prepared beam in a specific S_y-state. All possible data of the experiment are given by the elements of what we can call the *transformation matrix*, the $\langle S_y = i | S_z = \alpha\rangle$; if these four matrix elements are known, all the probabilities of finding any particular outcome can be predicted. This then is the job of quantum mechanics: It figures out the spectrum of outcomes (in the present case that there is a $+\frac{1}{2}$ or $-\frac{1}{2}$ spin beam) and the probabilities of obtaining a particular outcome. The first is the determination of the eigenvalue spectrum, the second is the evaluation of the elements of the transformation matrix.

Another way of seeing the importance of the transformation matrix is to recognize that both the $|S_y = i\rangle$ set and the $|S_z = \alpha\rangle$ set are complete basic sets, and we can choose either of these bases for our calculation. How do we compare the calculation based on one basis with that based on another? The transformation matrix from the S_z-representation to the S_y-representation gives us the necessary information.

Note that what we call wave functions can also be thought of as elements of a transformation matrix, albeit infinite dimensional. For example, the coordinate wave function $\psi_E(x) = \langle x|E\rangle$ is the transformation matrix element from the energy E-basis to the position x-basis.

As an example, let's construct the transformation matrix U from the S_z-basis to the S_y-basis (admittedly, this will seem like pedagogy right now, but what you learn here will be very useful later on). We already have the eigenspinor of S_y in the S_z-basis, equation (15.19). In its own home base, the eigenspinors of S_y obviously are

$$\begin{pmatrix} 1 \\ 0 \end{pmatrix} \quad \text{and} \quad \begin{pmatrix} 0 \\ 1 \end{pmatrix}$$

for S_y-spin up and down, respectively. Now by definition, U must be a matrix such that

$$\begin{pmatrix} 1 \\ 0 \end{pmatrix} = U \begin{pmatrix} 1/\sqrt{2} \\ i/\sqrt{2} \end{pmatrix}$$

and

$$\begin{pmatrix} 0 \\ 1 \end{pmatrix} = U \begin{pmatrix} 1/\sqrt{2} \\ -i/\sqrt{2} \end{pmatrix}$$

Remembering the ordering of the matrices starting from the highest to the lowest eigenvalue and the normalization and orthogonality of spinors, we find that the choice of U that satisfies both of the above requirements is

$$U = \begin{pmatrix} 1/\sqrt{2} & -i/\sqrt{2} \\ 1/\sqrt{2} & i/\sqrt{2} \end{pmatrix} \tag{15.32}$$

We see that once the eigenspinors of a certain operator are known in a given basis, the transformation matrix to the operator's home base is determined row by row by the hermitian conjugate of the eigenspinors ordered in decreasing order of the eigenvalues (i.e., the first row is the hermitian conjugate of the eigenspinor belonging to the highest eigenvalue, and so forth).

Now formally, the transformation matrix U from the S_z-basis to the S_y-basis is

$$U = \begin{array}{c|cc} & |S_z=+\rangle & |S_z=-\rangle \\ \hline |S_y=+\rangle & \langle S_y=+|S_z=+\rangle & \langle S_y=+|S_z=-\rangle \\ |S_y=-\rangle & \langle S_y=-|S_z=+\rangle & \langle S_y=-|S_z=-\rangle \end{array}$$

and notice that the elements of this matrix are complex conjugates of the elements that define the S_y-basis states in terms of the S_z-basis states:

$$\begin{aligned} |S_y=i\rangle &= \sum_\alpha |S_z=\alpha\rangle\langle S_z=\alpha|S_y=i\rangle \\ &= \sum_\alpha |S_z=\alpha\rangle\langle S_y=i|S_z=\alpha\rangle^* \\ &= \sum_\alpha |S_z=\alpha\rangle U_{i\alpha}^* \end{aligned} \tag{15.33}$$

Consequently, once the transformation matrix is known we can write down the relation between the two sets of base states. Note, however, that equation (15.33) is not a matrix equation. It will be left up to the reader to show that equation (15.33) leads to equation (15.18), previously obtained using the analogy of circularly polarized light.

Let's define the inverse of a matrix U via the equation

$$U^{-1}U = UU^{-1} = I$$

The elements of U^{-1} are given by the cofactor of the corresponding transposed element of U divided by the determinant of U:

$$(U^{-1})_{\alpha i} = (1/\det U)C_{i\alpha} \tag{15.34}$$

where det U is the determinant of U, and C denotes cofactor. The proof is as follows: We have

$$\sum_\alpha U_{j\alpha}(U^{-1})_{\alpha i} = (1/\det U)\sum_\alpha U_{j\alpha}C_{i\alpha}$$

Now if $j = i$ in the above equation, the sum on the right-hand side equals det U, by definition, and we have the desired result $UU^{-1} = I$; if $j \neq i$, the sum on the right-hand side represents the determinant of a matrix with two rows identical, and therefore is zero.

Using equation (15.34), the matrix U^{-1} is easily calculated:

$$U^{-1} = \begin{pmatrix} 1/\sqrt{2} & 1/\sqrt{2} \\ i/\sqrt{2} & -i/\sqrt{2} \end{pmatrix} \tag{15.35}$$

The fact that the matrix U^{-1} is the same as the hermitian conjugate of the matrix U:

$$U^{-1} = U^{\dagger} \tag{15.36}$$

is very important. A matrix satisfying equation (15.36) is called a *unitary matrix*; the transformation matrix U is unitary. This is true of all transformations between quantum mechanical base states—they are *unitary transformations.*

A more general proof of the unitarity of U can be constructed as follows. We can always write (why?)

$$1 = \sum_i \langle S_z=\alpha | S_y=i \rangle \langle S_y=i | S_z=\alpha \rangle$$

If we interpret the sum as a matrix product and realize that the second term in the sum is the $i\alpha$th matrix element of U, then from the definition that $U^{-1}U = I$, the first term in the sum must be the αith matrix element of U^{-1}. In other words,

$$(U^{-1})_{\alpha i} = \langle S_z=\alpha | S_y=i \rangle$$

But

$$\langle S_z=\alpha | S_y=i \rangle = \langle S_y=i | S_z=\alpha \rangle^*$$

which is the same as saying $U^{-1} = U^{\dagger}$.

There is a profound reason that U has to be unitary. A unitary matrix corresponds to a unitary operator in Hilbert space, and operation by a unitary operator keeps the normalization of any state vector $|\Psi\rangle$ unchanged. Let $|\Psi'\rangle = U|\Psi\rangle$. Then

$$\langle \Psi' | \Psi' \rangle = \langle U\Psi | U\Psi \rangle = \langle \Psi | U^{\dagger} U | \Psi \rangle = \langle \Psi | \Psi \rangle$$

only if $U^{\dagger}U = 1$, if U is unitary. Thus unitary operations keep the normalizations, and therefore probabilities, unchanged.

Likewise, the probability amplitude $\langle f | i \rangle$ from an arbitrary initial state $|i\rangle$ to an arbitrary final state $|f\rangle$ is independent of the basis (S_z or S_y) in which it is calculated. Let χ_i and χ_f denote the spinor representations of $|i\rangle$ and $|f\rangle$, respectively, in the S_z-representation. Then their spinors in the S_y-representation are given by

$$\chi_i' = U\chi_i \qquad \chi_f' = U\chi_f \tag{15.37}$$

In the old representation

$$\langle f|i\rangle = \chi_f^\dagger \chi_i \tag{15.38}$$

In the new representation

$$\langle f|i\rangle = \chi_f'^\dagger \chi_i' \tag{15.39}$$

But

$$\chi_f'^\dagger = (U\chi_f)^\dagger = \chi_f^\dagger U^\dagger$$

For $\langle f|i\rangle$ in equation (15.39) this gives

$$\langle f|i\rangle = \chi_f^\dagger U^\dagger U\chi_i = \chi_f^\dagger \chi_i$$

since $U^\dagger U = I$.

Now notice something interesting: If we construct the matrix product US_yU^{-1}, the result is a diagonal matrix:

$$\frac{\hbar}{2}\begin{pmatrix} 1/\sqrt{2} & -i/\sqrt{2} \\ 1/\sqrt{2} & i/\sqrt{2} \end{pmatrix}\begin{pmatrix} 0 & -i \\ i & 0 \end{pmatrix}\begin{pmatrix} 1/\sqrt{2} & 1/\sqrt{2} \\ i/\sqrt{2} & -i/\sqrt{2} \end{pmatrix} = \frac{\hbar}{2}\begin{pmatrix} 1 & 0 \\ 0 & -1 \end{pmatrix}$$

Moreover, the diagonal matrix is the matrix of the eigenvalues of S_y. The product US_yU^{-1} is called the *similarity transform* of S_y. Therefore S_y is *diagonalized* by the similarity transformation with the aid of the same unitary matrix U (a similarity transformation with a unitary matrix is called a *unitary transformation*) that is the transformation matrix between the S_z-basis and the S_y-basis. The transformation between two basis representations is intimately connected with the diagonalization of the matrix (in the old basis) of the operator that defines the new basis as its home basis—where it is diagonal.

It is important to note that all the mathematics we have worked out here for a simple example in spin-space applies to bigger matrices as well.

We can now state the whole scheme of matrix mechanics. It is to find the eigenstates of the Hamiltonian by diagonalizing the matrix of the Hamiltonian (the *energy matrix*) in some suitable starting basis. The diagonal matrix of the Hamiltonian in the new basis gives the eigenvalue spectrum, and the transformation matrix to the new basis determines all the probabilities for physically measurable quantities. If we knew all the suitable base states of the world, and all the Hamiltonians to go with them, quantum mechanics would presumably give us the method to calculate all that we can measure!

15.5 INFINITE DIMENSIONAL MATRICES

We can see by now that all the matrix elements of operators that we have previously calculated by the operator method can be displayed as matrices. Let's illustrate with an example.

Consider the matrix elements of the angular-momentum operator **L**, which we call the orbital angular momentum. In chapter 11, we constructed the eigenstates of $|lm\rangle$ of L^2 and L_z, whose (θ, ϕ)-representations are the Y_{lm}. The matrix elements of L^2 and L_z in the basis $|lm\rangle$ are given by

$$\langle l'm'|L^2|lm\rangle = l(l+1)\hbar^2\delta_{ll'}\delta_{mm'}$$

$$\langle l'm'|L_z|lm\rangle = m\hbar\delta_{ll'}\delta_{mm'}$$

In display form, the matrix of L^2 looks like this:

	$\|00\rangle$	$\|11\rangle$	$\|10\rangle$	$\|1\text{-}1\rangle$	$\|22\rangle$	...
$\|00\rangle$	0	0	0	0	0	...
$\|11\rangle$	0	$2\hbar^2$	0	0	0	...
$\|10\rangle$	0	0	$2\hbar^2$	0	0	...
$\|1\text{-}1\rangle$	0	0	0	$2\hbar^2$	0	...
$\|22\rangle$	0	0	0	0	$6\hbar^2$	...
⋮	⋮	⋮	⋮	⋮	⋮	⋮

(15.40)

The matrix of L_z is similarly diagonal:

	$\|00\rangle$	$\|11\rangle$	$\|10\rangle$	$\|1\text{-}1\rangle$	$\|22\rangle$	...
$\|00\rangle$	0	0	0	0	0	...
$\|11\rangle$	0	$\hbar$	0	0	0	...
$\|10\rangle$	0	0	0	0	0	...
$\|1\text{-}1\rangle$	0	0	0	$-\hbar$	0	...
$\|22\rangle$	0	0	0	0	$2\hbar$	...
⋮	⋮	⋮	⋮	⋮	⋮	⋮

(15.41)

The matrix elements of L_x and L_y can be obtained using equation (11.62). They are

$$\begin{aligned}\langle l'm'|L_x|lm\rangle &= \langle l'm'|\tfrac{1}{2}(L_+ + L_-)|lm\rangle \\ &= (\hbar/2)\{\delta_{ll'}\delta_{m',m+1}[(l-m)(l+m+1)]^{1/2} + \delta_{ll'}\delta_{m',m-1} \\ &\quad \times [(l+m)(l-m+1)]^{1/2}\}\end{aligned} \tag{15.42}$$

$$\begin{aligned}\langle l'm'|L_y|lm\rangle &= \langle l'm'|(L_+ - L_-)/2i|lm\rangle \\ &= (\hbar/2i)\{\delta_{ll'}\delta_{m',m+1}[(l-m)(l+m+1)]^{1/2} - \delta_{ll'}\delta_{m',m-1} \\ &\quad \times [(l+m)(l-m+1)]^{1/2}\}\end{aligned} \tag{15.43}$$

In display form the matrix of L_x is

(15.44)

	$\|00\rangle$	$\|11\rangle$	$\|10\rangle$	$\|1\text{-}1\rangle$	$\|22\rangle$	...
$\|00\rangle$	0	0	0	0	0	...
$\|11\rangle$	0	0	$\hbar/\sqrt{2}$	0	0	...
$\|10\rangle$	0	$\hbar/\sqrt{2}$	0	$\hbar/\sqrt{2}$	0	...
$\|1\text{-}1\rangle$	0	0	$\hbar/\sqrt{2}$	0	0	...
$\|22\rangle$	0	0	0	0	0	...
⋮	⋮	⋮	⋮	⋮	⋮	⋰

Likewise, let's display the matrix of L_y:

(15.45)

	$\|00\rangle$	$\|11\rangle$	$\|10\rangle$	$\|1\text{-}1\rangle$	$\|22\rangle$	...
$\|00\rangle$	0	0	0	0	0	...
$\|11\rangle$	0	0	$-i\hbar/\sqrt{2}$	0	0	...
$\|10\rangle$	0	$i\hbar/\sqrt{2}$	0	$-i\hbar/\sqrt{2}$	0	...
$\|1\text{-}1\rangle$	0	0	$i\hbar/\sqrt{2}$	0	0	...
$\|22\rangle$	0	0	0	0	0	...
⋮	⋮	⋮	⋮	⋮	⋮	⋰

It is interesting to notice that the matrices of L_x and L_y may not be diagonal, but they are the next best thing: They are *block diagonal*, meaning that there is no matrix element between one value of l and another. These blocks do not mix even when we multiply these matrices, and you can verify that the angular-momentum commutation relations check out for each block. Indeed, although in principle we are dealing with infinite matrices, since the angular-momentum matrices are either diagonal or block diagonal, we are able to stay within a fixed l and consider each block separately for these matrices.

Finally, the same matrix elements obtain for the total angular-momentum operators, J^2, J_z, J_x, and J_y, except that we have to replace l by j and remember that j can have both integer and half-integer values.

PROBLEMS

1. What is the probability of an exiting $S_y = +\frac{1}{2}$ spin if an electron enters the appropriately rotated S_y-Stern-Gerlach filter with S_z-spin down?
2. An electron is in a state described by the spinor given in the S_z-basis as

$$\chi = \begin{pmatrix} i/\sqrt{5} \\ 2/\sqrt{5} \end{pmatrix}$$

What is the probability that the electron has spin up?

3. How would you represent the eigenspinors of S_z in the S_z-representation for a spin 1 particle? How would you represent the matrices of S_z, S_x, and S_y?
4. (a) Show that any hermitian matrix that commutes with all three σ's is a multiple of the unit matrix. (b) Show that a hermitian matrix that anticommutes with all three σ's does not exist.
5. If **A** and **B** are two vectors that commute with all the components of the Pauli matrices $\boldsymbol{\sigma}$ (but not necessarily with one another), show that

$$(\boldsymbol{\sigma}\cdot\mathbf{A})(\boldsymbol{\sigma}\cdot\mathbf{B}) = (\mathbf{A}\cdot\mathbf{B}) + i\boldsymbol{\sigma}\cdot(\mathbf{A}\times\mathbf{B})$$

6. If TrA (trace of A) denotes the sum of all the diagonal elements of a matrix A, and if **A**, **B**, and **C** are vectors that commute with $\boldsymbol{\sigma}$, prove that
(a) $Tr(\boldsymbol{\sigma}\cdot\mathbf{A}) = 0$
(b) Calculate $Tr[(\boldsymbol{\sigma}\cdot\mathbf{A})(\boldsymbol{\sigma}\cdot\mathbf{B})(\boldsymbol{\sigma}\cdot\mathbf{C})]$
7. Show directly by matrix multiplication that the spinors of equation (15.19) are eigenspinors of S_y, given by equation (15.22), belonging to the appropriate eigenvalue.
8. Write down the transformation matrix U from the S_z-basis to the S_x-basis and show that US_xU^{-1} is diagonal giving the eigenvalue spectrum of S_x.
9. A certain state is given in the S_z-basis as the spinor

$$\begin{pmatrix} 1/\sqrt{3} \\ \sqrt{2}/\sqrt{3} \end{pmatrix}$$

Find the spinor representation of the state in the S_x-basis.

10. Two states $|1\rangle$ and $|2\rangle$ are represented by the spinors

$$\chi_1 = \begin{pmatrix} 1/\sqrt{2} \\ i/\sqrt{2} \end{pmatrix} \quad \text{and} \quad \chi_2 = \begin{pmatrix} -i/\sqrt{3} \\ \sqrt{2}/\sqrt{3} \end{pmatrix}$$

in the S_z-basis. Determine the spinors of the same states in the S_y-basis. Determine the amplitude $\langle 1|2\rangle = \chi_1^\dagger \chi_2$ in the S_z-basis and show by direct matrix multiplication that this amplitude remains unchanged when calculated in the S_y-basis.

11. Show that the matrix

$$M = \begin{pmatrix} 1 & 2 \\ 1 & 2 \end{pmatrix}$$

is diagonalized by a similarity transformation with the matrix U given by

$$U = \begin{pmatrix} \frac{1}{\sqrt{2}} & -\frac{1}{\sqrt{2}} \\ \frac{1}{\sqrt{5}} & \frac{2}{\sqrt{5}} \end{pmatrix}$$

What are the eigenvalues? Is M hermitian? Is U unitary?

12. Consider the 3×3 matrix

$$M = \begin{bmatrix} 0 & -i & 0 \\ i & 0 & -i \\ 0 & i & 0 \end{bmatrix}$$

Is the matrix hermitian? Find the eigenvalues and eigenvectors (which are now column matrices with three rows) of M and normalize the eigenvectors. Find the matrix U that diagonalizes M. Is U unitary?

13. Consider an atom in a state of $l = 1$. Find the eigenvectors and eigenvalues for $L_x L_y + L_y L_x$.

14. Using the matrix elements of the raising and lowering operators in the energy representation of the harmonic oscillator given in chapter 7, calculate the top left 4×4 corner (corresponding to $n = 0, 1, 2, 3$) of the matrix of x^2 and p^2. (*Hint*: You have to figure out the matrix elements of x^2 from the matrix of x by using the rules of matrix multiplication. This was a crucial step in Heisenberg's discovery. Heisenberg did not know matrices; but he intuited what the multiplication rule must be so that a theory could be developed that agreed with experiment.)

ADDITIONAL PROBLEMS

A1. Find the transpose, the complex conjugate, and the hermitian conjugate of the following matrix:

$$\begin{pmatrix} 0 & -i & 0 & i \\ i & 0 & i & 0 \\ 0 & -i & 0 & -i \\ i & 0 & -i & 0 \end{pmatrix}$$

A2. Find the bra spinor corresponding to the ket

$$\begin{pmatrix} 0 \\ i \end{pmatrix}$$

A3. Show that any two diagonal matrices commute.

A4. If M is a diagonal matrix $M_{ij} = m_i \delta_{ij}$, what is the matrix representation of $\exp M$?

A5. Show that the trace of a matrix does not change under a unitary transformation.

A6. Prove by direct matrix multiplication that the Pauli matrices anticommute and that they follow the commutation relations given in equation (15.29).

A7. The operator a satisfies the equations:

$$a^2 = 0, \quad aa^\dagger + a^\dagger a = 0. \qquad \text{Define } N = a^\dagger a$$

(a) Show that $N^2 = N$.
(b) Find 2×2 matrix representations of the operators a and N.

A8. Write down the matrix of the Hamiltonian for the particle in a box in the energy representation.

A9. For the $l = 1$ state, find the three eigenvectors (column matrices with three rows) of L_x.

A rigid rotator is found to be in the state

$$\psi = \frac{1}{\sqrt{6}} \begin{pmatrix} 2 \\ 1 \\ 1 \end{pmatrix}$$

What is the probability that the measurement of L_x of the rotator in this state will give $\hbar$?

A10. The scalar product in the space of the spherical polar coordinates θ, ϕ is given as

$$\langle f_i | f_j \rangle = \int_0^\pi \sin\theta \, d\theta \int_0^{2\pi} d\phi \, f_i^*(\theta,\phi) f_j(\theta,\phi)$$

The matrix of the operator O in this space is defined by the elements $O_{ij} = \langle f_i | O | f_j \rangle$. Consider the subspace spanned by the three functions

$$f_1 = \sin\theta \exp(i\phi), \quad f_2 = \cos\theta, \quad f_3 = \sin\theta \exp(-i\phi).$$

(a) Normalize these functions.
(b) Determine the matrices of the following operators

$$-i\partial/\partial\phi, \quad -\partial^2/\partial\phi^2, \quad \exp(\pm i\phi)\left(\pm\frac{\partial}{\partial\theta} + i\cot\theta\,\frac{\partial}{\partial\phi}\right)$$

A11. The general solution of the stationary-state Schrödinger equation for a particle moving in the one-dimensional square-well potential of figure 4.4 is given as:

$$\begin{aligned} \psi_k(x) &= a_l \exp(ikx) + b_l \exp(-ikx) \qquad &\text{for } x < -a \\ &= a_r \exp(ikx) + b_r \exp(-ikx) \qquad &\text{for } x > a \end{aligned}$$

where l (r) stands for left (right), and k is the wave number.

The scattering matrix (S-matrix) S is defined through the relation:

$$\begin{pmatrix} a_r \\ b_l \end{pmatrix} = \begin{pmatrix} S_{11} & S_{12} \\ S_{21} & S_{22} \end{pmatrix} \begin{pmatrix} a_l \\ b_r \end{pmatrix}$$

(a) Express the reflection coefficient $|R_{r(l)}|^2$ and the transmission coefficient $|T_{r(l)}|^2$ for a particle incident from the right (left), respectively (see chapter 4), in terms of the a's and the b's.
(b) Consider the scattering matrix as a function of k. Show that $S(-k) = S^{-1}(k)$
(c) Determine the scattering matrix in terms of R and T of the square-well solution in chapter 4. Show that S is unitary, that is, $S^\dagger S = 1$.

REFERENCES

M. Chester. *Primer of Quantum Mechanics.*
A. Das and A. C. Melissinos. *Quantum Mechanics: A Modern Introduction.*
R. P. Feynman, R. B. Leighton, and M. Sands. *The Feynman Lectures in Physics*, vol. 3.
T. F. Jordan. *Quantum Mechanics in Simple Matrix Form.*
J. J. Sakurai. *Quantum Mechanics.*

16

Matrix Mechanics: Two-State Systems

The way to learn something new is to use it. The way to learn matrix mechanics is to set up some energy matrices and solve them. And that is what we will do in this chapter. But this is not a purely pedagogical exercise. There are important physical systems, such as the ammonia molecule, for which this approach gives us useful insight.

In the new language of matrices, the Schrödinger equation in the coordinate space is expressed in an infinite-dimensional representation. For the type of problem we will solve presently by the matrix method, we first need to find a finite matrix representation of the Schrödinger equation. That is to say, our first job is to rewrite the Schrödinger equation in a matrix form in a basis of finite dimensions. We have

$$i\hbar \frac{d}{dt} |\psi\rangle = H|\psi\rangle \tag{16.1}$$

where we have replaced the usual partial time derivative by the total time derivative, since we are interested only in the matrix representation of this equation where H is a matrix, not a differential operator. Let's denote our base states as $|j\rangle$. Now multiply equation (16.1) from the left with $\langle j|$:

$$i\hbar \frac{d}{dt} \langle j|\psi\rangle = \langle j|H|\psi\rangle = \sum_i \langle j|H|i\rangle\langle i|\psi\rangle$$

Call $\langle j | \psi \rangle = c_j$. We have

$$i\hbar \frac{d}{dt} c_j = \sum_i H_{ji} c_i \tag{16.2}$$

We can recognize this as a matrix equation with c as a column matrix and H a square matrix; therefore, this is the desired finite matrix representation of the Schrödinger equation.

In this chapter, we will consider only the case of two-base states forming a complete set. This two-state Schrödinger equation is all we need to understand some important basic physics regarding the ammonia molecule, even how the ammonia maser works! For the two-state case, we can write equation (16.2) as

$$i\hbar \frac{dc_1}{dt} = H_{11} c_1 + H_{12} c_2$$

$$i\hbar \frac{dc_2}{dt} = H_{21} c_1 + H_{22} c_2 \tag{16.3a}$$

which we can write also in the matrix form

$$i\hbar \frac{d}{dt} \begin{pmatrix} c_1 \\ c_2 \end{pmatrix} = \begin{pmatrix} H_{11} & H_{12} \\ H_{21} & H_{22} \end{pmatrix} \begin{pmatrix} c_1 \\ c_2 \end{pmatrix} \tag{16.3b}$$

Suppose first that the off-diagonal matrix elements $H_{12} = H_{21} = 0$. Then equation (16.3a) becomes decoupled:

$$i\hbar \frac{dc_1}{dt} = H_{11} c_1$$

$$i\hbar \frac{dc_2}{dt} = H_{22} c_2 \tag{16.4}$$

These equations are easily integrated:

$$c_1(t) = A \exp(-iE_1 t/\hbar)$$

$$c_2(t) = B \exp(-iE_2 t/\hbar)$$

where E_1 and E_2 are the energy eigenvalues of H for the base states $|1\rangle$ and $|2\rangle$, respectively, and where A and B are constants. Consequently, the state $|\psi(t)\rangle$ is given as

$$\begin{aligned}|\psi(t)\rangle &= \sum_i |i\rangle\langle i|\psi(t)\rangle \\ &= \sum_i c_i(t)|i\rangle \\ &= A\exp(-iE_1t/\hbar)|1\rangle + B\exp(-iE_2t/\hbar)|2\rangle \qquad (16.5)\end{aligned}$$

Clearly, the amplitudes in equation (16.4) correspond to stationary states, that is, the probabilities are independent of time. Therefore, whenever we find a representation basis for the Hamiltonian matrix that is diagonal, that is, $H_{ij} = 0$ for $i \neq j$, we have found our stationary states with eigenvalues of H giving us the energy of the stationary states. And if the matrix of H is not diagonal in our starting basis, $H_{ij} \neq 0$ even for $i \neq j$, we can always find a basis, using the procedure of chapter 15, in which H is diagonal.

So why do we call the matrix of the Hamiltonian an energy matrix? In order to obtain the energy eigenstates of the system we have to diagonalize the matrix of H, that's why.

In the rest of the chapter, we will consider a couple of two-state systems and diagonalize their energy matrices. One of these systems is the ammonia molecule; the other is a spinning particle in a magnetic field. In this latter connection, we will also consider some further properties of spin states.

16.1 THE AMMONIA MOLECULE

In the ammonia molecule, NH_3, the three hydrogen atoms are symmetrically located in a plane, and the lone nitrogen can lie either above or below the symmetry plane of the H's, as shown in figure 16.1. Let the ket $|1\rangle$ denote the state of the system corresponding to the configuration shown in figure 16.1(a), and let the configuration shown in figure 16.1(b) be represented by the state vector $|2\rangle$. We will assume that these two states are all the base states we need for calculating the ground state. This is a reasonable assumption about the lowest energy states of NH_3; of course, there are more complex possibilities—rotations, vibrations, or electronic motions—but they all require considerably more energy, and we will ignore them to keep things simple. This simplicity will enable us to learn something about the system, and we don't even have to know the Hamiltonian exactly!

We proceed by *parameterizing* the matrix of the Hamiltonian. By symmetry, the energy of the system in the two configurations must be the same, call it E_0. Consequently, we have

$$H_{11} = H_{22} = E_0$$

Now how about the matrix elements H_{12} and H_{21}? What do they represent? They represent the interaction that mixes the two-base states. Physically, the sit-

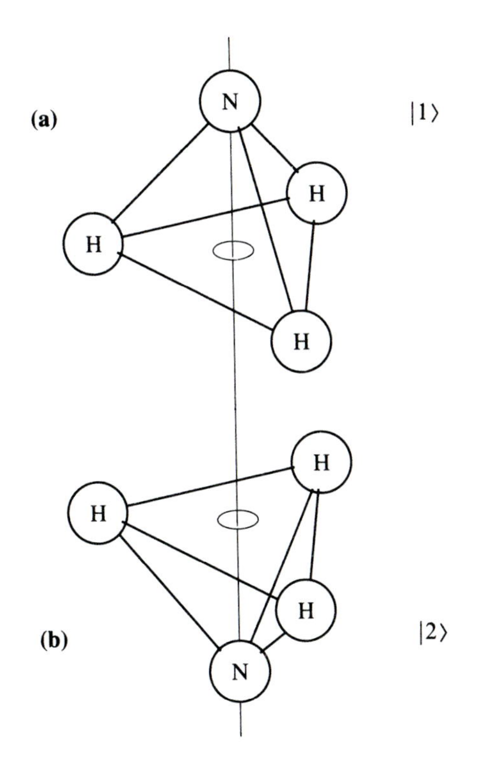

FIGURE 16.1
The two base states of the NH_3 molecule—two equivalent geometries.

uation is that, although the nitrogen is strongly repulsed from the symmetry plane of the H's because there is a high potential barrier, there still is a small probability that the nitrogen tunnels through the plane. And so whichever configuration the molecule is in initially, there is a probability that it ends up in the other configuration, which means that the matrix element of H between the two states cannot be zero. We can, however, use the symmetry argument again to assert that

$$H_{12} = H_{21}$$

For convenience, we choose the interaction matrix element to be a negative number, call it $-A$. A has to be real, of course, since H must be hermitian. Thus the energy matrix is parameterized in terms of two real quantities, E_0 and A:

$$H = \begin{array}{c|cc} & |1\rangle & |2\rangle \\ \hline |1\rangle & E_0 & -A \\ |2\rangle & -A & E_0 \end{array} \tag{16.6}$$

To diagonalize the energy matrix, the first step is to solve the secular equation obtained by putting the secular determinant equal to zero; calling the eigenvalue λ, we have

$$\begin{vmatrix} E_0 - \lambda & -A \\ -A & E_0 - \lambda \end{vmatrix} = 0$$

Evaluating the determinant, we get

$$(E_0 - \lambda)^2 - A^2 = 0$$

The two solutions are

$$\lambda_1 = E_0 + A \qquad \lambda_2 = E_0 - A \tag{16.7}$$

This gives us the eigenvalues of H. Let's label the eigenstates with the subscripts I and II and write

$$E_I = E_0 + A \qquad E_{II} = E_0 - A \tag{16.8}$$

There is, in this particular case, another way to find the eigenvalues. The energy matrix, equation (16.6), can be expressed in terms of Pauli matrices (as can any hermitian matrix in a 2×2 space); we get

$$H = E_0 I - A\sigma_x \tag{16.9}$$

The two terms commute; moreover, the first term is just a constant add-on, and we need to diagonalize only the second term. But the eigenvalues of σ_x, as you know from chapter 15, are ± 1, and consequently, the eigenvalues of H are given by equation (16.8).

But this last way of doing things has another advantage. The matrix that diagonalizes σ_x must then also diagonalize the entire energy matrix (note that the fact that the two terms commute is crucial here). But we can easily construct the matrix U that diagonalizes $-A\sigma_x$ from the eigenspinors of S_x, equations (15.16) and (15.17). In keeping with the convention of writing the first row of U to correspond to the hermitian adjoint of the eigenvector belonging to the highest eigenvalue, it is

$$U = \begin{array}{c|cc} & |1\rangle & |2\rangle \\ \hline |I\rangle & \langle I|1\rangle & \langle I|2\rangle \\ |II\rangle & \langle II|1\rangle & \langle II|2\rangle \end{array} = \begin{pmatrix} -1/\sqrt{2} & 1/\sqrt{2} \\ 1/\sqrt{2} & 1/\sqrt{2} \end{pmatrix} \tag{16.10}$$

You can verify by direct matrix multiplication that with U given by equation (16.10) and H given by equation (16.6), it is indeed the case that

$$UHU^{-1} = UHU^{\dagger} = \begin{pmatrix} E_0 + A & 0 \\ 0 & E_0 - A \end{pmatrix} \tag{16.11}$$

Now we can write the new base kets $|I\rangle$ and $|II\rangle$ in terms of the old ones $|1\rangle$ and $|2\rangle$, taking help from the matrix elements of U. For example,

$$\begin{aligned} |I\rangle &= \sum_i |i\rangle\langle i|I\rangle \\ &= |1\rangle\langle 1|I\rangle + |2\rangle\langle 2|I\rangle \\ &= |1\rangle\langle I|1\rangle^* + |2\rangle\langle I|2\rangle^* \\ &= (1/\sqrt{2})\,[-|1\rangle + |2\rangle] \end{aligned} \tag{16.12}$$

In the same way, we find that

$$|II\rangle = (1/\sqrt{2})\,[|1\rangle + |2\rangle] \tag{16.13}$$

You can also easily verify that the states $|I\rangle$ and $|II\rangle$ are an orthonormal set; they are our new base states.

Let's see how these new base states relate to a concept that you already know—stationary states. The amplitudes of a general state $|\psi\rangle$ to be in our new base states, namely $c_I = \langle I|\psi\rangle$ and $c_{II} = \langle II|\psi\rangle$, must satisfy equations similar to equation (16.4) (whereas the amplitudes of $|\psi\rangle$ in the old base would have to satisfy eq. [16.3]):

$$i\hbar\, dc_I/dt = (E_0 + A)c_I = E_I c_I \tag{16.14a}$$

and

$$i\hbar\, dc_{II}/dt = (E_0 - A)c_{II} = E_{II} c_{II} \tag{16.14b}$$

These equations are integrated easily, and we get

$$\begin{aligned} c_I &= C\exp[-i(E_0 + A)t/\hbar] \\ c_{II} &= D\exp[-i(E_0 - A)t/\hbar] \end{aligned} \tag{16.15}$$

In this way it follows that any general state $|\psi\rangle$ of the system can be expressed as

$$|\psi\rangle = C\exp[-i(E_0 + A)t/\hbar]|I\rangle + D\exp[-i(E_0 - A)t/\hbar]|II\rangle$$
$$= C|\psi_I\rangle + D|\psi_{II}\rangle \tag{16.16}$$

where

$$|\psi_I\rangle = \exp[-i(E_0 + A)t/\hbar]|I\rangle$$

and

$$|\psi_{II}\rangle = \exp[-i(E_0 - A)t/\hbar]|II\rangle \tag{16.17}$$

are the stationary states of the system. (Of course, we are already familiar with this time factor, which is implied to accompany all the time-independent states that we have calculated before, and we are not always careful to distinguish between $|\psi_I\rangle$ and $|I\rangle$.)

There is a parallel to the relation of the new basis $|I\rangle$ and $|II\rangle$ with the old basis $|1\rangle$ and $|2\rangle$ in the classical motion of a coupled pendulum—two identical pendulums connected by a string (fig. 16.2). If we lift one up and set it in motion, it will swing (fig. 16.2a), but gradually the other one takes up the motion, and pretty soon it is the second pendulum that is swinging away, while the first one comes to rest (fig. 16.2b). But that too is only momentary, because the process reverses back and forth. Clearly, these modes (figs. 16.2a and b) are not stationary; they are akin to the states $|1\rangle$ and $|2\rangle$ of the ammonia molecule. But with the coupled pendulum there are also two normal modes. If we set both pendulums simultaneously in motion so that they oscillate in phase (fig. 16.2c), they will dance in this mode forever (if there is no friction, that is); additionally, if we set them in motion in opposite phases, then too the motion continues forever with a definite frequency (fig. 16.2d). These normal modes are the parallels of the eigenstates $|II\rangle$ and $|I\rangle$, respectively, of the NH_3 system.

The splitting of the two originally degenerate states of the ammonia molecule has a practical application, the ammonia maser, which we will take up next.

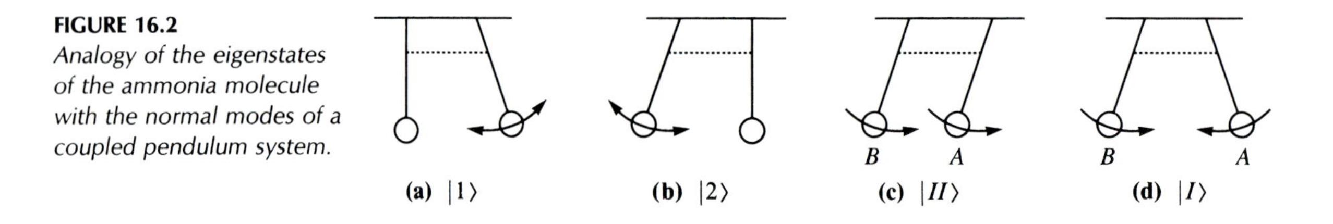

FIGURE 16.2
Analogy of the eigenstates of the ammonia molecule with the normal modes of a coupled pendulum system.

16.2 THE AMMONIA MASER

The maser is a device for the amplification of microwave radiation via stimulated emission. The energy difference between the states $|I\rangle$ and $|II\rangle$ above is such that the photon emitted in a quantum jump from the higher to the lower state is in the microwave range. For the maser to operate, we need a way to get the ammonia molecules predominantly populating the excited state and then for them to deexcite, donating the radiation to a cavity so that microwave energy can build up in it. The first step involves the interaction of an ammonia molecule with a static electric field, and the second, its interaction with a time-dependent electric field. These interactions will be considered below.

The Ammonia Molecule in an External Static Electric Field

In the ammonia molecule, the valence electrons have a slight preference for the N atom over the hydrogen; they stay on the average a little closer to the N atom. The resulting unbalanced charge distribution gives the molecule an *electric dipole moment*, directed from the N atom toward the H plane. When there is an external electric field, there is an additional contribution to the Hamiltonian arising from the interaction of the dipole moment with the electric field. The interaction term can be written as $-\mathbf{D}\cdot\boldsymbol{\mathcal{E}}$, where $\mathbf{D}$ is the electric dipole moment and $\boldsymbol{\mathcal{E}}$ is the static electric field.

Let's go back to the states $|1\rangle$ and $|2\rangle$ of section 16.1. An important point is that when the nitrogen flips over to the other side, as from state $|1\rangle$ to state $|2\rangle$, the electric dipole moment also flips over (fig. 16.3). Consequently, the dipole interaction adds with a different sign in the two states:

$$-\mathbf{D}\cdot\boldsymbol{\mathcal{E}} = D\mathcal{E} \quad \text{for state } |1\rangle$$

$$-\mathbf{D}\cdot\boldsymbol{\mathcal{E}} = -D\mathcal{E} \quad \text{for state } |2\rangle \tag{16.18}$$

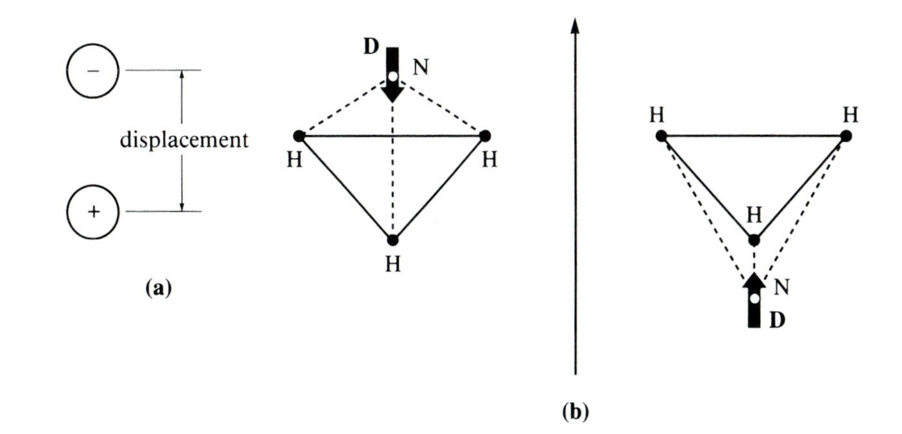

FIGURE 16.3
(a) In the ammonia molecule, the centers of the positive and negative charge distributions are slightly displaced from one another on the average, producing a net dipole moment. (b) The two base states of the ammonia molecule give rise to opposite interaction with an external electric field.

Now, it's too messy to calculate such things as the interaction term A, of the last subsection, and the D that we are encountering here, so we have to gain our guidance from experiment. The data suggest that the dipole interaction term, for large laboratory electric fields of $\sim 10^4$ V/m, is quite comparable to the magnitude of A. Thus it is important to treat both interactions on the same footing.

Since there is only a diagonal contribution to the energy matrix from the dipole interaction term, we have

$$H = \begin{array}{c|cc} & |1\rangle & |2\rangle \\ \hline |1\rangle & E_0 + D\mathcal{E} & -A \\ |2\rangle & -A & E_0 - D\mathcal{E} \end{array} \tag{16.19}$$

The states $|1\rangle$ and $|2\rangle$ are not stationary; there are nonzero off-diagonal matrix elements of H in this basis. The energies of the stationary states are to be found by solving the secular equation:

$$\begin{vmatrix} E_0 + D\mathcal{E} - \lambda & -A \\ -A & E_0 - D\mathcal{E} - \lambda \end{vmatrix} = 0$$

Let's call the stationary states $|\alpha\rangle$ and $|\beta\rangle$; their energies E_α and E_β are given from the solution of the secular equation above as

$$E_\alpha = E_0 + (A^2 + D^2\mathcal{E}^2)^{1/2}$$

$$E_\beta = E_0 - (A^2 + D^2\mathcal{E}^2)^{1/2} \tag{16.20}$$

Figure 16.4 shows the plot of these two energies. The curves are hyperbolic; for large fields the splitting becomes proportional to $\mathcal{E}$ as the curves approach their asymptotes; the splitting does not depend much on A at all because the strong electric field inhibits tunneling. For small fields, on the other hand, the effect of the tunneling interaction A dominates, and the energies vary with the electric field via a term $\sim\mathcal{E}^2$.

We are finally ready to tackle the ammonia maser. Suppose we take a beam of ammonia molecules and pass it through a region where there is a strong transverse electric field; the electrodes are shaped so that $\mathcal{E}$ varies sharply in the direction perpendicular to the beam. What then? A molecule in the state $|\alpha\rangle$ will seek out the region of lower $\mathcal{E}$ in order to minimize its energy (since its energy, see eq. [16.20], increases with $\mathcal{E}$). The opposite happens to a molecule in state $|\beta\rangle$. Therefore, molecules in the different states will be deflected in opposite directions (fig. 16.5) and can be separated. Now we take the beam in the

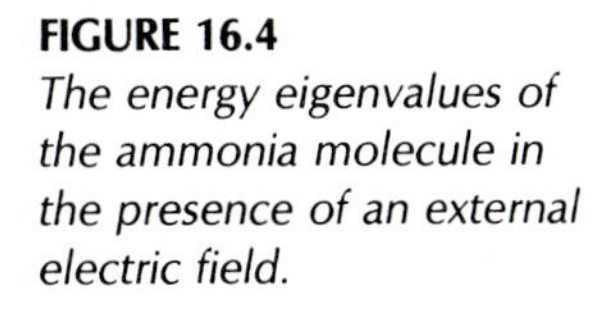

FIGURE 16.4
The energy eigenvalues of the ammonia molecule in the presence of an external electric field.

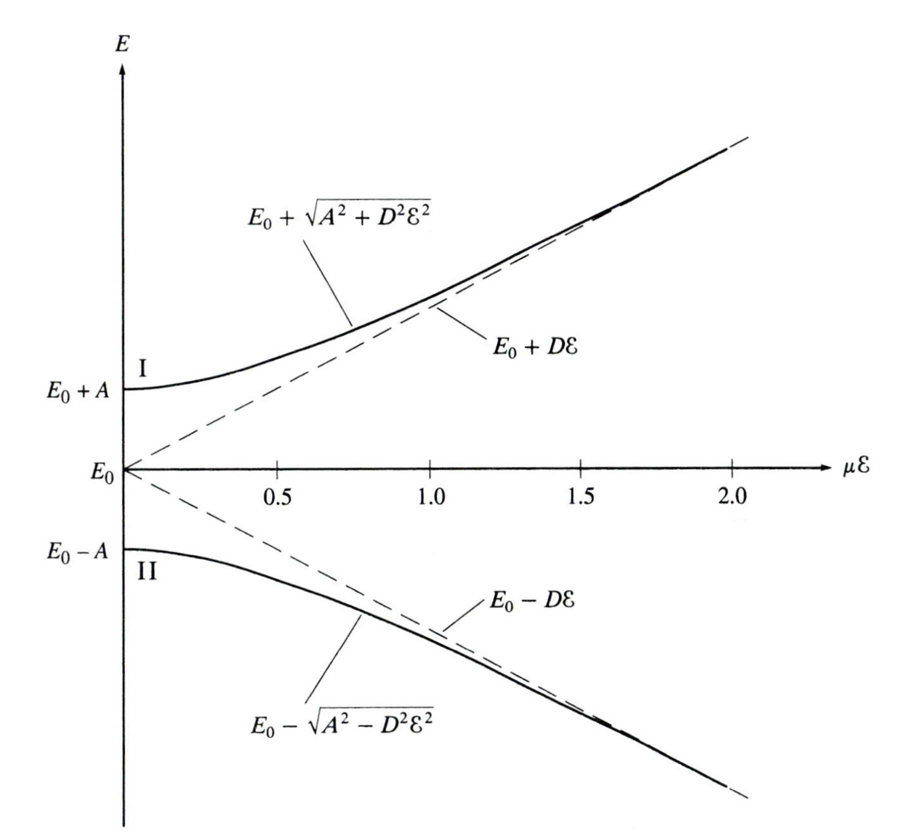

higher energy state $|\alpha\rangle$ through a resonant cavity where there is a time-varying electric field. By interacting with this time-varying field, the molecules surrender their excitation energy $E_\alpha - E_\beta$ to the cavity as they deexcite to the state $|\beta\rangle$. But energy builds up in the electromagnetic field of the maser cavity (fig. 16.6).

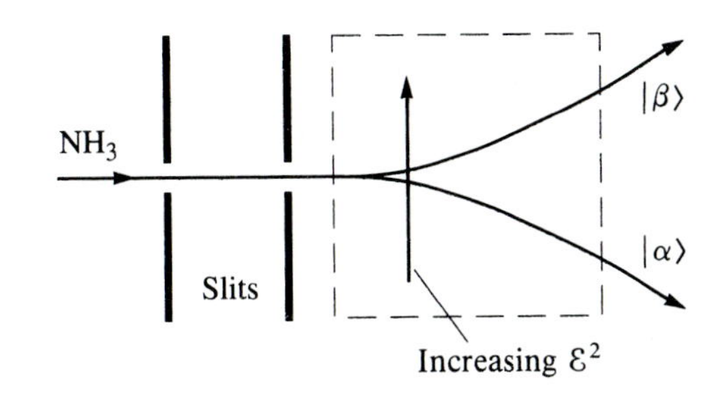

FIGURE 16.5
Separation of the two states of an ammonia beam by an electric field with a gradient in a direction perpendicular to the beam.

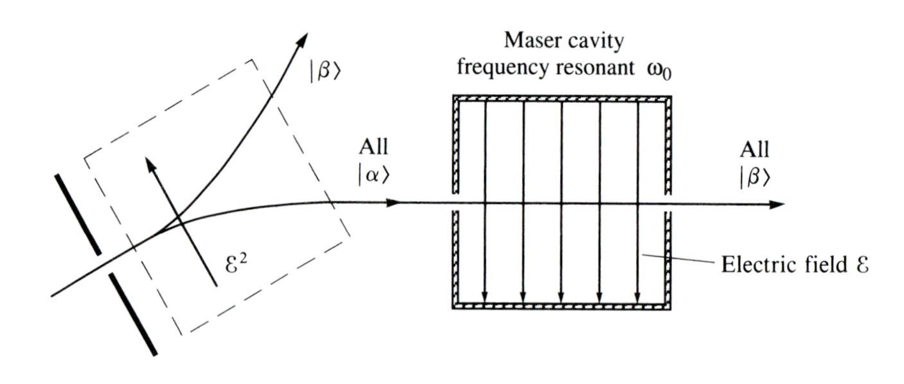

FIGURE 16.6
In the ammonia maser, ammonia molecules in the higher energy state $|\alpha\rangle$ deexcite to the lower state $|\beta\rangle$, building up the electromagnetic energy inside a resonant cavity.

How Ammonia Molecules Interact with a Time-Varying Electric Field

When the electric field oscillates with time, its interaction with the molecular dipole is a time-dependent interaction, and we need to go back to the time-dependent equations to find out how an ammonia molecule behaves in this situation. The equations to be solved in the basis $|1\rangle$ and $|2\rangle$ (using eqs. [16.3] and [16.18]) are

$$i\hbar\, dc_1/dt = (E_0 + D\mathcal{E})c_1 - Ac_2$$

$$i\hbar\, dc_2/dt = (E_0 - D\mathcal{E})c_2 - Ac_1 \qquad (16.21)$$

where $\mathcal{E}$ now is

$$\mathcal{E} = \mathcal{E}_0 \cos \omega t = \tfrac{1}{2}\mathcal{E}_0[\exp(i\omega t) + \exp(-i\omega t)] \qquad (16.22)$$

It still behooves us to use the basic states $|I\rangle$ and $|II\rangle$ defined by equations (16.12) and (16.13). Since

$$c_I = \langle I|\psi\rangle = (1/\sqrt{2})(-\langle 1|\psi\rangle + \langle 2|\psi\rangle)$$
$$= (1/\sqrt{2})(-c_1 + c_2)$$

and

$$c_{II} = \langle II|\psi\rangle = (1/\sqrt{2})(\langle 1|\psi\rangle + \langle 2|\psi\rangle)$$
$$= (1/\sqrt{2})(c_1 + c_2)$$

we can subtract the two equations in (16.21) one from the other, divide by $\sqrt{2}$, and get

$$i\hbar\, dc_I/dt = (E_0 + A)c_I - D\mathcal{E}c_{II} \qquad (16.23a)$$

Similarly, by adding the two equations and dividing by $\sqrt{2}$, we get the equation for c_{II}:

$$i\hbar\, dc_{II}/dt = (E_0 - A)c_{II} - D\mathcal{E}c_I \tag{16.23b}$$

These equations are not easy to solve for a general $\mathcal{E}(t)$. However, if $D\mathcal{E}_0 \ll A$, we expect the solution to be not too different from the solution in the absence of the electric field, namely $c \sim \exp(-iEt/\hbar)$. This prompts us to write the solution in this *perturbation* limit as a product of a slowly varying function of time (slow compared to the exponential) and the exponential:

$$c_I = \gamma_I(t)\exp(-iE_I t/\hbar)$$

$$c_{II} = \gamma_{II}(t)\exp(-iE_{II} t/\hbar) \tag{16.24}$$

Now we will use the idea that the $\gamma(t)$'s vary slowly with time to obtain a solution. To that end, we take the time derivative of equation (16.24):

$$i\hbar\, dc_I/dt = E_I\gamma_I \exp(-iE_I t/\hbar) + i\hbar \exp(-iE_I t/\hbar)\, d\gamma_I/dt$$

Substituting in the differential equation (16.23a) for c_I, we get

$$(E_I\gamma_I + i\hbar\, d\gamma_I/dt)\exp(-iE_I t/\hbar) = E_I\gamma_I \exp(-iE_I t/\hbar) - D\mathcal{E}\gamma_{II}\exp(-iE_{II} t/\hbar)$$

This equation looks unduly complicated, but we can simplify it by canceling out identical terms on both sides and putting $E_I - E_{II} = 2A = \hbar\omega_0$; we get

$$i\hbar\, d\gamma_I/dt = -D\mathcal{E}(t)\exp(i\omega_0 t)\gamma_{II} \tag{16.25a}$$

In the same way, the equation for dc_{II}/dt above gives

$$i\hbar\, d\gamma_{II}/dt = -D\mathcal{E}(t)\exp(-i\omega_0 t)\gamma_I \tag{16.25b}$$

Now invoke the time dependence of the oscillating electric field from equation (16.22). We get

$$i\hbar \frac{d\gamma_I}{dt} = -\tfrac{1}{2}D\mathcal{E}_0[e^{i(\omega+\omega_0)t} + e^{-i(\omega-\omega_0)t}]\gamma_{II}$$

$$i\hbar \frac{d\gamma_{II}}{dt} = -\tfrac{1}{2}D\mathcal{E}_0[e^{i(\omega-\omega_0)t} + e^{-i(\omega+\omega_0)t}]\gamma_I \tag{16.26}$$

These equations are still exact. In our search for a suitable, easily solvable approximation, it helps to formally integrate these equations. For example, we can write

$$\gamma_I(t) = \frac{i}{2\hbar} D\mathcal{E}_0 \left[\int_0^t dt' e^{i(\omega+\omega_0)t'} \gamma_{II}(t') + \int_0^t dt' e^{-i(\omega-\omega_0)t'} \gamma_{II}(t') \right]$$

The importance of our assumption that γ_I and γ_{II} are slowly varying is now revealed. Essentially, this means that the oscillating functions in the integrands will make the above integrals average out to zero unless $\omega \approx \omega_0$. The terms in equation (16.26) containing $\exp[\pm i(\omega + \omega_0)t]$ can therefore be ignored outright, and the remaining terms need be considered only for $\omega \approx \omega_0$, which is the *resonance* condition. In this way, we see that for a weak electric field, the significant contribution comes only from near-resonance frequencies.

We will solve the equations only at resonance $\omega = \omega_0$, because (1) this is the simple thing to do and (2) this adequately illustrates the basic point we are trying to make, how the ammonia donates its energy to the maser cavity. We have

$$\frac{d}{dt}\gamma_I(t) = \frac{i}{2\hbar} D\mathcal{E}_0 e^{-i(\omega-\omega_0)t}\gamma_{II}(t)$$

$$\frac{d}{dt}\gamma_{II}(t) = \frac{i}{2\hbar} D\mathcal{E}_0 e^{i(\omega-\omega_0)t}\gamma_I(t) \qquad (16.27)$$

At resonance, the exponentials become unity, and we can uncouple the differential equations easily by differentiating the equations and substituting one into the other:

$$\frac{d^2}{dt^2}\gamma_I(t) = -(D\mathcal{E}_0/2\hbar)^2\gamma_I(t)$$

$$\frac{d^2}{dt^2}\gamma_{II}(t) = -(D\mathcal{E}_0/2\hbar)^2\gamma_{II}(t) \qquad (16.28)$$

So both γ's now satisfy the equations of simple harmonic motion. The solutions are

$$\gamma_I(t) = a\cos(D\mathcal{E}_0 t/2\hbar) + b\sin(D\mathcal{E}_0 t/2\hbar)$$

$$\gamma_{II}(t) = -ib\cos(D\mathcal{E}_0 t/2\hbar) + ia\sin(D\mathcal{E}_0 t/2\hbar) \qquad (16.29)$$

as you can verify by direct substitution in equation (16.27). The constants a and b are to be determined from the initial conditions.

Suppose at $t = 0$, an ammonia molecule is in the upper energy state $|I\rangle$. From equation (16.16), it follows that $\gamma_I = 1$ and $\gamma_{II} = 0$; and this in its turn requires that $a = 1$ and $b = 0$. But the system does not stay in the state $|I\rangle$ forever. It has the probability

$$P_{II} = |c_{II}|^2 = |\gamma_{II}|^2 = \sin^2(D\mathcal{E}_0 t/2\hbar) \tag{16.30}$$

to quantum jump to state $|II\rangle$. Of course, it has also the probability

$$P_I = |c_I|^2 = |\gamma_I|^2 = \cos^2(D\mathcal{E}_0 t/2\hbar) \tag{16.31}$$

of staying right where it is. Note that the two probabilities add up to 1. Probability is conserved!

Therefore, so long as the electric field is small and in resonance, the probabilities are given by simple oscillatory functions (fig. 16.7). If at time $t = 0$, the ammonia is in state $|I\rangle$, by the time $D\mathcal{E}_0 t/2\hbar = \pi/2$, the ammonia completely flips over to state $|II\rangle$; and when $D\mathcal{E}_0 t/2\hbar = \pi$, it flips again completely, back to its original state $|I\rangle$.

Coming back to the maser cavity, we now can ask, Under what condition does a molecule entering in state $|I\rangle$ make a transition to the state $|II\rangle$? Well, if the molecule spends a time T long enough so that $D\mathcal{E}_0 T/2\hbar = \pi/2$, it certainly will make the transition (other values of $D\mathcal{E}_0 T/2\hbar$ will do too, except for

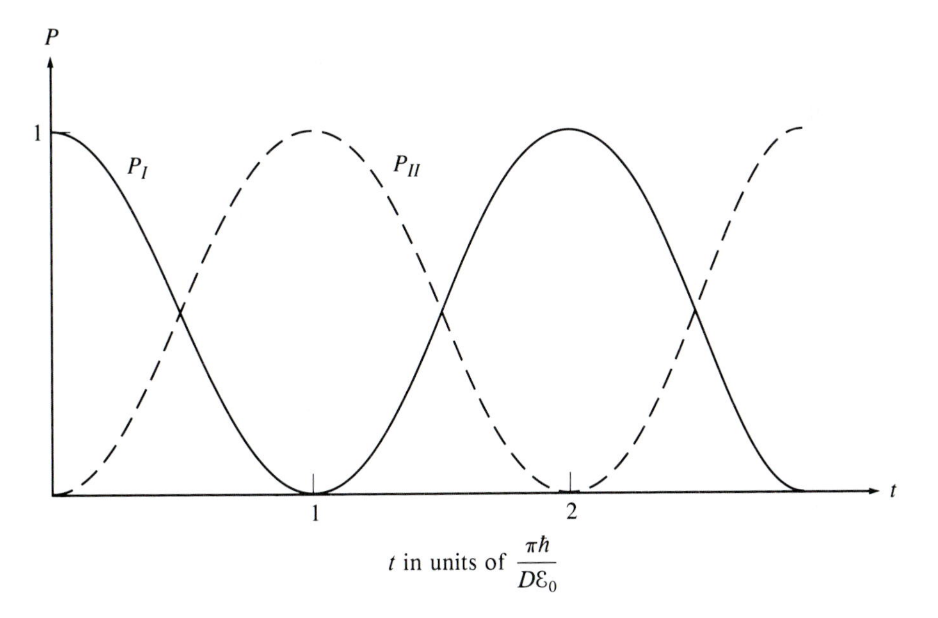

FIGURE 16.7
The probabilities for the two states of the ammonia molecule inside the maser cavity.

π, of course, but they will not be 100% efficient), delivering the energy to the electric field in the cavity because there is no other place for the energy to go (the physics of the details of this energy delivery will be examined in chapter 22). In an operating maser, we make sure that the ammonia molecules deliver enough energy to the resonant cavity not only to offset its losses but also to have a little extra, which then can be taken out from it. Thus the molecular energy is transmuted into the energy of maser radiation.

Note also two more things. Now you can see why it is essential to sift out the molecules, so that only those in the higher energy state enter the cavity. Any molecule in state $|II\rangle$ going through the cavity will absorb energy as it makes the transition to the upper state, defeating the purpose. Second, in case you are getting grandiose ideas, the practical maser cannot achieve 100% efficiency; one reason is that, because of their velocity distribution, the molecules are bound to spend slightly different times in the cavity, thus shortchanging the efficiency.

16.3 SPINNING PARTICLE IN A MAGNETIC FIELD: MAGNETIC RESONANCE

We will now consider the motion of a particle with just a spin degree of freedom in a magnetic field. A case in point is protons placed in a magnetic field as in the phenomenon of nuclear magnetic resonance (NMR). Another case is an electron localized at a crystal lattice site. In both cases the particles interact with the external magnetic field by virtue of their intrinsic magnetic moment arising from spin:

$$\mathbf{M} = (qg/2mc)\mathbf{S} \tag{16.32}$$

where q is the charge ($+e$ or $-e$) and g is the gyromagnetic ratio of the particle. For the electron, g is very nearly 2 so we will always take it to be 2. For protons, $g \approx 5.58$. To be specific, let's consider protons in an NMR situation.

The Hamiltonian of the interaction is given as

$$H = -\mathbf{M}\cdot\mathbf{B} \tag{16.33}$$

If $\mathbf{B}$ is along the z-axis, the Hamiltonian is diagonal in the S_z-basis, producing the familiar split between the $|S_z{=}+\frac{1}{2}\rangle$ and $|S_z{=}-\frac{1}{2}\rangle$ states. In the absence of the magnetic field, the spin-up and spin-down states of the proton are degenerate; the magnetic field removes this degeneracy. If B_0 is the magnitude of the field along the z-axis, the energy matrix is given as

$$H = \begin{array}{c|cc} & |S_z{=}+\rangle & |S_z{=}-\rangle \\ \hline |S_z{=}+\rangle & -\mu_p B_0 & 0 \\ |S_z{=}-\rangle & 0 & +\mu_p B_0 \end{array} \tag{16.34}$$

where $\mu_p = eg\hbar/4m_pc \approx 2.79$ nuclear magnetons (m_p is the proton mass; 1 nuclear magneton $= e\hbar/2m_pc$). The quantity μ_p is therefore the magnetic moment of the proton. Notice that we can write the energy matrix in a concise form using the Pauli matrix σ_z:

$$H = -\mu_p B_0 \sigma_z \tag{16.35}$$

In the phenomenon of nuclear magnetic resonance, the field in the z-direction is static, but an additional time-varying field

$$B_x = B_1 \cos \omega t$$

is applied along the x-direction (fig. 16.8). The z-field is produced by an iron-core electromagnet between whose pole faces is placed the sample of protons (water can be used). The protons are surrounded by a solenoid coil coupled to a tunable radio frequency (RF) circuit; the solenoid coil can be oriented in such a fashion that the RF magnetic field is along the x-axis.

Now we have to add to the Hamiltonian, equation (16.35), the potential energy of interaction of the protons with the radio frequency magnetic field given by

$$-\mu_p B_x \sigma_x = -\mu_p B_1 \tfrac{1}{2}[\exp(i\omega t) + \exp(-i\omega t)]\sigma_x \tag{16.36}$$

Consequently we can write the complete energy matrix as

$$\begin{aligned} H &= -\mu_p B_0 \sigma_z - \mu_p B_x \sigma_x \\ &= \begin{pmatrix} -\mu_p B_0 & -\mu_p B_1 \frac{1}{2}(e^{i\omega t} + e^{-i\omega t}) \\ -\mu_p B_1 \frac{1}{2}(e^{i\omega t} + e^{-i\omega t}) & \mu_p B_0 \end{pmatrix} \end{aligned} \tag{16.37}$$

where we have used the S_z-basis, as before. But this energy matrix is basically the same one that we solved for the ammonia molecule (the absence of the con-

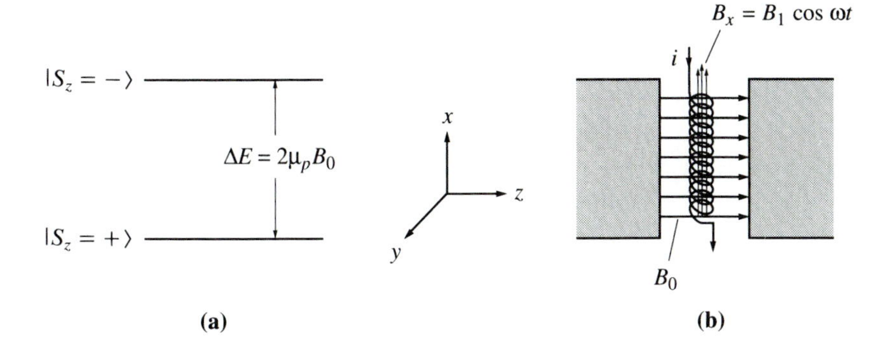

FIGURE 16.8
Nuclear magnetic resonance. (a) The two spin states of a proton placed in a magnetic field have slightly different energy. (b) The experimental arrangement of NMR. There is a time-dependent magnetic field in addition to the constant field and perpendicular to it.

stant add-on does not complicate the solution any) where the perturbation was a time-dependent electric field. Therefore we can use the results obtained in section 16.2: If the radio frequency satisfies the resonance condition

$$\hbar\omega = 2\mu_p B_0 \tag{16.38}$$

there will be transitions induced between the states $|S_z=+\rangle$ and $|S_z=-\rangle$. However, in this case, we have not separated out the spin-up and spin-down components; consequently, at first sight one expects that there will be as many transitions from up to down as there will be from down to up with the probability given by

$$P = \sin^2(\mu_p B_1 t/2\hbar)$$

Suppose there are N_+ protons in the spin-up state and N_- protons in the spin-down state; then the number of protons making transitions per unit time can be written as

$$\Delta N(+\to -) = RN_+$$

$$\Delta N(-\to +) = RN_- \tag{16.39}$$

where $R = P/t$ is the transition probability per unit time. If we have equilibrium conditions, the two ΔN's must be equal, and therefore for a strong enough RF field, the populations of the two states will tend to equalize.

But in actuality, the equalization is never achieved even under equilibrium conditions. Initially, the population of the $|S_z=+\rangle$ state is greater simply by virtue of the Boltzmann factor:

$$\frac{N_-}{N_+} = \frac{N_0\exp(-E_-/kT)}{N_0\exp(-E_+/kT)} = \frac{N_0\exp(-\mu_p B_0/kT)}{N_0\exp(\mu_p B_0/kT)} = e^{-2\mu_p B_0/kT}$$

and $N_+ > N_-$. But can this condition continue in view of the argument presented above?

Well, the population difference continues even in the presence of the RF field because of the so-called *relaxation mechanisms* (such as the interaction of the protons with their nuclear neighbors and collisions in general) that transfer protons from the $|S_z=-\rangle$ to the $|S_z=+\rangle$ state *without* giving the energy back to the RF field.

Because of the population difference, we must conclude that the number of transitions

$$\Delta N(+\to -) > \Delta N(-\to +)$$

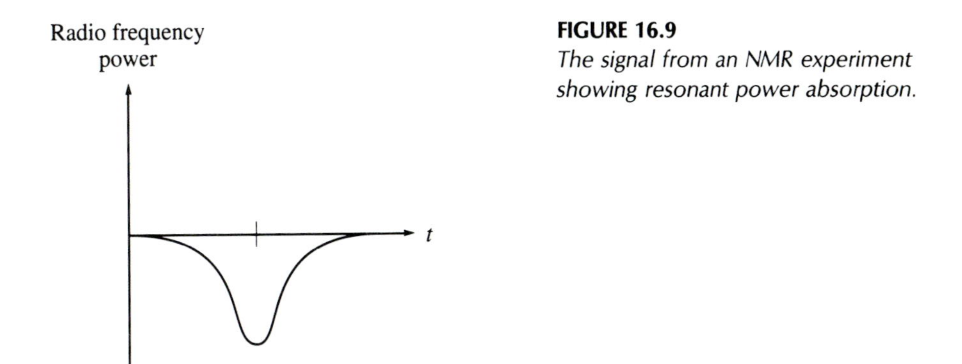

FIGURE 16.9
The signal from an NMR experiment showing resonant power absorption.

Since $|S_z{=}+\rangle$ is at lower energy, this means that the proton sample absorbs more energy than it gives back to the RF field. This absorption of energy is observed as the radio frequency is swept precisely at the resonant frequency of $\omega = 2\mu_p B_0/\hbar$ (fig. 16.9).

16.4 SPINNING PARTICLE IN A MAGNETIC FIELD: SPIN PRECESSION

Now we will look at a different phenomenon. Suppose we have a magnetic field in any spatial direction specified by the unit vector $\mathbf{e}_n$ whose polar angles are θ and ϕ. Suppose also that there is an electron in such a magnetic field whose spin is aligned along $\mathbf{B}$; that is, along $\mathbf{e}_n$. Denote the state of the electron as $|\psi\rangle$. We can expand $|\psi\rangle$ in terms of the S_z-basis states:

$$|\psi\rangle = c_1|S_z{=}+\rangle + c_2|S_z{=}-\rangle \tag{16.40}$$

How do we calculate the amplitudes c_1 and c_2?

The answer can be found using our general formulation for two-state systems. The Hamiltonian is

$$\begin{aligned} H &= -\mathbf{M}_e\cdot\mathbf{B} = (eg\hbar/4mc)\boldsymbol{\sigma}\cdot\mathbf{B} = \mu_B\boldsymbol{\sigma}\cdot\mathbf{B} \\ &= \begin{pmatrix} \mu_B B\cos\theta & \mu_B B\sin\theta(\cos\phi - i\sin\phi) \\ \mu_B B\sin\theta(\cos\phi + i\sin\phi) & -\mu_B B\cos\theta \end{pmatrix} \end{aligned} \tag{16.41}$$

where we have put $g = 2$, and where $\mu_B = e\hbar/2mc$ is the Bohr magneton. The eigenvalues are to be obtained by solving the secular equation. However, in this case, we can guess that the eigenvalues must be $\pm\mu_B B$, because when the electron is in the eigenstate of H, its spin is parallel (or antiparallel) to $\mathbf{B}$. If we denote the eigenspinor corresponding to the eigenvalue $+\mu_B B$ as

$$\begin{pmatrix} a \\ b \end{pmatrix}$$

the eigenvalue equation gives

$$\begin{pmatrix} \mu_B B \cos\theta & \mu_B B \sin\theta(\cos\phi - i\sin\phi) \\ \mu_B B \sin\theta(\cos\phi + i\sin\phi) & -\mu_B B\cos\theta \end{pmatrix}\begin{pmatrix} a \\ b \end{pmatrix} = \mu_B B\begin{pmatrix} a \\ b \end{pmatrix} \tag{16.42}$$

We get

$$\frac{a}{b} = \frac{\sin\theta e^{-i\phi}}{1-\cos\theta} = \frac{\cos(\theta/2)e^{-i\phi}}{\sin(\theta/2)}$$

We must also have $|a|^2 + |b|^2 = 1$. Clearly one possible solution is

$$a = \cos(\theta/2)\exp(-i\phi/2)$$

$$b = \sin(\theta/2)\exp(i\phi/2) \tag{16.43}$$

Therefore, the eigenspinor of H corresponding to the eigenvalue $\mu_B B$ is

$$\begin{pmatrix} \cos(\theta/2)e^{-i\phi/2} \\ \sin(\theta/2)e^{i\phi/2} \end{pmatrix} \tag{16.44}$$

As we saw in section 16.2, the amplitudes c_1 and c_2 of equation (16.40) are just the amplitudes a and b multiplied by $\exp(-i\mu_B Bt/\hbar)$; we have figured them out!

Suppose now that at some initial time $t = 0$, the electron has its spin in some direction $\mathbf{e}_n$. The corresponding spinor is given by equation (16.44). If we call the state $|\psi(0)\rangle$, then

$$|\psi(0)\rangle = a|S_z{=}+\rangle + b|S_z{=}-\rangle \tag{16.45}$$

where a and b are given by equation (16.43). Now suppose we turn on a magnetic field in the z-direction for a time T, and then turn it off. What can we say about the final state? For this problem, the base states $|S_z{=}\pm\rangle$ are also states of definite energy; thus they evolve with time with the usual $\exp(-iEt/\hbar)$ factor. Clearly, we have

$$|\psi(T)\rangle = a\exp(-i\mu_B BT/\hbar)|S_z{=}+\rangle + b\exp(i\mu_B BT/\hbar)|S_z{=}-\rangle \qquad (16.46)$$

This function is the same as $|\psi(0)\rangle$ if the angle ϕ is changed to $\phi' = \phi + 2\mu_B BT/\hbar$ but θ is unchanged. The action of the magnetic field for a time T is therefore to introduce a phase shift of $2\mu_B BT/\hbar$ that is equivalent to a rotation about the z-axis since only a change in ϕ is involved. Since the phase shift is proportional to T, it is equally valid to say that the direction of the spin *precesses* about the z-axis. The angular velocity of precession is equal to

$$2\mu_B B/\hbar = \omega_0$$

This is the quantum analog of the classical Larmor precession.

It is also interesting to calculate the expectation values of S_x, S_y, and S_z in the state $|\psi(T)\rangle$ above (the steps will be left as an exercise). We get

$$\langle\psi(T)|S_x|\psi(T)\rangle = (\hbar/2)\sin\theta\cos(\phi + \omega_0 T)$$

$$\langle\psi(T)|S_y|\psi(T)\rangle = (\hbar/2)\sin\theta\sin(\phi + \omega_0 T)$$

$$\langle\psi(T)|S_z|\psi(T)\rangle = (\hbar/2)\cos\theta \qquad (16.47)$$

In this way we can see that the average values of S_x and S_y oscillate with angular frequency ω_0 just like the components of a classical angular momentum undergoing Larmor precession.

These equations of Larmor precession check out well in experiments with negatively charged muons whose magnetic properties are the same as the electrons', although they are 105 times heavier. In the experiment, as an external magnetic field causes the muon spin to precess, the spin direction is analyzed by looking at the decay electrons, taking advantage of the fact that the decay electrons are emitted preferentially in a direction opposite to the muon spin. The measured angular frequency of precession agrees well with the angular frequency calculated above in terms of the muon magnetic moment, which is determined from an independent experiment.

Does the quantum particle precess about the direction of the magnetic field then with the angular velocity ω_0 just as a classical particle does? We should not get too carried away with analogy. Remember that there is a big difference between the precession of an angular momentum vector in classical physics and in quantum physics. In classical physics, all three components of the vector have known values, always. In quantum mechanics, however, only one of the spin components can be measured at any given time.

■ **PROBLEM** An electron is initially in a state represented by the spinor

$$\begin{pmatrix} \alpha(0) \\ \beta(0) \end{pmatrix} \tag{16.48}$$

A magnetic field in the y-direction $\mathbf{B} = B\mathbf{e}_y$ is turned on at $t = 0$ for a time T. Calculate the spinor at $t = T$.

SOLUTION Suppose the spinor at time t is given as

$$\begin{pmatrix} \alpha(t) \\ \beta(t) \end{pmatrix}$$

The matrix Schrödinger equation for the time evolution of the spinor is given as

$$i\hbar \frac{d}{dt}\begin{pmatrix} \alpha(t) \\ \beta(t) \end{pmatrix} = \mu_B B \begin{pmatrix} 0 & -i \\ i & 0 \end{pmatrix}\begin{pmatrix} \alpha(t) \\ \beta(t) \end{pmatrix}$$

Simplifying, we get

$$d\alpha/dt = -\omega\beta$$

$$d\beta/dt = \omega\alpha \tag{16.49}$$

where $\omega = \mu_B B/\hbar$. Eliminating $\beta(t)$, we get the equation for $\alpha(t)$ as

$$\frac{d^2\alpha}{dt^2} + \omega^2\alpha = 0$$

for which the solution is

$$\alpha(t) = a\cos\omega t + b\sin\omega t$$

where a and b are constants to be determined from the initial conditions. Substituting $\alpha(t)$ back into equation (16.49), we get

$$\beta(t) = a\sin\omega t - b\cos\omega t$$

According to this solution, the spinor at $t = 0$ is

$$\begin{pmatrix} a \\ -b \end{pmatrix} = \begin{pmatrix} \alpha(0) \\ \beta(0) \end{pmatrix}$$

This gives $a = \alpha(0)$ and $b = -\beta(0)$. Therefore, the spinor at $t = T$ is

$$\begin{pmatrix} \alpha(0)\cos\omega T - \beta(0)\sin\omega T \\ \alpha(0)\sin\omega T + \beta(0)\cos\omega T \end{pmatrix}$$

■

Spinor in an Arbitrary Direction

One interesting thing about the eigenspinor

$$\begin{pmatrix} \cos(\theta/2) & e^{-i\phi/2} \\ \sin(\theta/2) & e^{i\phi/2} \end{pmatrix} \tag{16.44}$$

is that the magnetic field does not make an appearance in it; this means that the result is the same even in the limit $B \to 0$. What we have determined is the eigenspinor of $\boldsymbol{\sigma}\cdot\mathbf{e}_n$; we have found out how to represent in the spinor form the state of an electron with spin in an arbitrary direction. You can easily check that our previous results for the spinors when the spin lies along the x- or y-axis are simply special cases of this general spinor.

A fascinating aspect of the spinor given by equation (16.44) is that if we change ϕ by 2π, the spinor changes sign; it does not return to its original value. This is a common characteristic of half-integer spin-wave functions. However, the minus sign is just a phase factor and does not contradict any of the principles of quantum mechanics. But it draws our attention once again to the fact that we cannot simulate spin as angular momentum of any kind of classical rotation.

PROBLEMS

1. The neutral K-meson K^0 and its antiparticle $\bar{K}^0$ form a two-state system whose energy matrix is not diagonal, but is given as

$$\begin{pmatrix} mc^2 & c^2\Delta m \\ c^2\Delta m & mc^2 \end{pmatrix}$$

Define a basis $|K_1\rangle$ and $|K_2\rangle$ in which the energy matrix is diagonal. What is the relation between the two bases?

2. Write the two-state Hamiltonian matrix in a certain basis $|1\rangle$, $|2\rangle$ in a general form as

$$\begin{pmatrix} H_{11} & H_{12} \\ H_{21} & H_{22} \end{pmatrix}$$

Impose hermiticity of H. Find the eigenvalues and the unitary transformation that diagonalizes the Hamiltonian. Express the eigenstates in terms of the old base states.

3. Find the eigenvalues and eigenvectors of $a\sigma_z + b\sigma_x + cI$, where a, b, and c are constants. Hence find the eigenstates $|\alpha\rangle$ and $|\beta\rangle$ of the ammonia molecule in a static electric field.
4. Consider the following energy matrix:

	$|1\rangle$	$|2\rangle$	$|3\rangle$	$|4\rangle$
$|1\rangle$	a	0	0	$-b$
$|2\rangle$	0	c	id	0
$|3\rangle$	0	$-id$	$-c$	0
$|4\rangle$	$-b$	0	0	$-a$

Find the eigenvalues and eigenvectors of this Hamiltonian and the matrix that diagonalizes it. (*Hint:* If you rearrange the ordering of the basis, you can find an ordering in which the matrix becomes block diagonal.)

5. Derive equation (16.47) for $\langle S_x\rangle$, $\langle S_y\rangle$, and $\langle S_z\rangle$ for an electron precessing in a magnetic field.
6. Consider a spin $\frac{1}{2}$ particle. At time $t = 0$, the particle is in the state $|S_z = +\rangle$.
   (a) If S_x is measured at $t = 0$, what is the probability of getting a value $\hbar/2$?
   (b) Suppose instead of performing the above measurement, the system is allowed to evolve in a magnetic field $\mathbf{B} = B_0\mathbf{e}_y$. Using the S_z-basis, calculate the state of the system after a time t.
   (c) At time t, suppose we measure S_x; what is the probability that a value $\frac{1}{2}\hbar$ will be found?
7. Consider the spin half particle of problem 6 once again.
   (a) At time $t = 0$, we measure S_x and find a value $\hbar/2$. What is the state vector immediately after the measurement?
   (b) At the same instant of the measurement, we apply a magnetic field $\mathbf{B} = B_0\mathbf{e}_z$ on the particle and allow the particle to precess for a time T. What is the state of the system at $t = T$?
   (c) At $t = T$, the magnetic field is very rapidly rotated so that it is now $\mathbf{B} = B_0\mathbf{e}_y$. After another time interval T, a measurement of S_x is carried out once more. What is the probability that a value $\hbar/2$ is found?
8. Consider a spin half particle placed in a magnetic field whose components are $B_x = 0$ and $B_y = B_0/\sqrt{2} = B_z$.
   (a) Calculate the energy matrix in the S_z-basis.
   (b) Calculate the eigenvalues and eigenvectors of the Hamiltonian.
   (c) The particle is in the state $|S_z = -\rangle$ at time $t = 0$; what values will result if the energy is measured and with what probabilities?

(d) Calculate the state vector at time t. At that instant, S_y is measured; what is the average value of the results that can be obtained?

9. The matrices by which state vectors in quantum mechanics transform under rotation are given by the general expression

$$D(\mathbf{e}_n, \theta) = \exp(-i\mathbf{J}\cdot\mathbf{e}_n\theta/\hbar)$$

where $\mathbf{J}$ is total angular momentum, $\mathbf{e}_n$ is the axis of rotation, and θ is the finite rotation angle. It follows that for a spin $\frac{1}{2}$ particle, the rotation matrix is given by

$$D(\mathbf{e}_n, \theta) = \exp(-i\boldsymbol{\sigma}\cdot\mathbf{e}_n\theta/2)$$

(a) Expand the exponential and show that

$$D(\mathbf{e}_n, \theta) = I\cos\theta/2 - i\boldsymbol{\sigma}\cdot\mathbf{e}_n\sin\theta/2$$

(b) Show that a 2π rotation is equivalent to multiplying the state vector by -1.

(c) Find the rotation matrix for a rotation about the y-axis by an angle θ followed by a rotation of an angle ϕ about the z-axis.

(d) Show that if we operate on the spinor

$$\begin{pmatrix} 1 \\ 0 \end{pmatrix}$$

by the rotation matrix you found in part (c), we get

$$\begin{pmatrix} \cos(\theta/2)\exp(-i\phi/2) \\ \sin(\theta/2)\exp(i\phi/2) \end{pmatrix}$$

Interpret the result. (*Hint:* You may find the expression in problem 5 of chapter 15 useful.)

ADDITIONAL PROBLEMS

A1. Determine $\langle i|j\rangle$ and $|i\rangle\langle j|$ for each of the following pairs of states:

(a) $\begin{pmatrix} i \\ 0 \end{pmatrix}$ $\begin{pmatrix} 0 \\ i \end{pmatrix}$

(b) $\begin{pmatrix} i \\ 0 \end{pmatrix}$ $\begin{pmatrix} i \\ 0 \end{pmatrix}$

(c) $\begin{pmatrix} 1 \\ 0 \end{pmatrix}$ $\begin{pmatrix} 0 \\ 1 \end{pmatrix}$

Also compute $\Sigma_i |i\rangle\langle i|$ for each set of states.

A2. In an NMR experiment, a sample of water displays resonant absorption when the frequency of the transverse magnetic field goes through the value 42.3 MHz. If the parallel magnetic field (B_0) is 10,000 gauss, what value of g for the proton obtains?

REFERENCES

C. Cohen-Tannoudji, B. Div, and F. Laloe. *Quantum Mechanics.*
A. Das and A. C. Melissinos. *Quantum Mechanics: A Modern Introduction.*
R. P. Feynman, R. B. Leighton, and M. Sands. *The Feynman Lectures in Physics*, vol. 3.
S. Gasiorowicz. *Quantum Physics.*
J. J. Sakurai. *Quantum Mechanics.*

17

The Addition of Angular Momenta

The electron is a four-dimensional particle! In addition to the three measurements we need to tell where the electron is, we need another measurement to tell which way its spin is pointing, the value of the z-component of its spin, S_z. Four measurements, four dimensions.

We need to know if its spin is up or down before we can completely specify the state of the electron. The two sets of information, space and spin, are independent; locating the spatial coordinates does not resolve the additional ambiguity about whether the spin is up or down, and vice versa. The position basis $|xyz\rangle$ must now be extended to include spin. We can write the extended base states as

$$|xyz, S_z{=}\pm\rangle \equiv |xyz\rangle|S_z{=}\pm\rangle$$

Consider a free electron whose spatial wave function is $\exp(\pm i\mathbf{k}\cdot\mathbf{r})$ with a twofold degeneracy. The spin degree of freedom adds another twofold degeneracy. Therefore, when we include spin, the eigenfunctions of the Hamiltonian

$$H = p^2/2m$$

are

$$\psi_{\mathbf{k}}\chi_+ = (2\pi)^{-3/2}\exp(i\mathbf{k}\cdot\mathbf{r})\begin{pmatrix}1\\0\end{pmatrix}$$

$$\psi_{\mathbf{k}}\chi_- = (2\pi)^{-3/2}\exp(i\mathbf{k}\cdot\mathbf{r})\begin{pmatrix}0\\1\end{pmatrix}$$

$$\psi_{-\mathbf{k}}\chi_+ = (2\pi)^{-3/2}\exp(-i\mathbf{k}\cdot\mathbf{r})\begin{pmatrix}1\\0\end{pmatrix}$$

$$\psi_{-\mathbf{k}}\chi_- = (2\pi)^{-3/2}\exp(-i\mathbf{k}\cdot\mathbf{r})\begin{pmatrix}0\\1\end{pmatrix} \tag{17.1}$$

Next consider the bound states $|nlm\rangle$. We can include spin by defining kets that include the information about spin

$$|nlm_l, m_s\rangle \equiv |nlm_l\rangle|m_s\rangle$$

where we use m_s (which can take on the value $\frac{1}{2}$ or $-\frac{1}{2}$) to denote the quantum numbers of S_z to bring it the benefit of the same kind of notation that we use for the orbital angular momentum; this is only fair, since spin is angular momentum too. Notice, however, that there is no need for our base states to carry an additional quantum number s analogous to the orbital quantum number l; l can vary, but s will be assumed to be $\frac{1}{2}$ unless otherwise specified. That way we avoid extra baggage as long as we can!

However, this is not the only way that the base states can be defined. To join space and spin, we can just juxtapose them as above; or we can be more imaginative. For example, we can think of adding the two angular-momentum vectors, the orbital angular momentum $\mathbf{L}$ and the spin angular momentum $\mathbf{S}$, to define a total angular momentum $\mathbf{J}$ and use the eigenvalues of J^2 and J_z (denote them as j and m, respectively) to specify our base states:

$$|nljm\rangle$$

The quantum mechanical tricks of such *addition of angular momenta* are applicable to other situations as well. For example, suppose we have two spinning electrons; how do we specify the base states of the combined system? Again we can just juxtapose the two individual quantum numbers

$$|m_s, m_s'\rangle \equiv |m_s\rangle|m_s'\rangle$$

or we can add the two spin vectors to define the total spin, $\mathbf{S}_1 + \mathbf{S}_2 = \mathbf{S}$, and use the total spin and its projection to define the base states as

$$|S, m\rangle$$

But now that we are considering the states of two fermions, identical fermions, the important considerations of chapter 9 regarding the antisymmetry of the wave function will enter, too, making the treatment even more novel.

Angular-momentum coupling is an important recurring problem in the application of quantum mechanics, important enough to warrant special treatment (introduction of special angular-momentum addition coefficients called Clebsch-Gordan coefficients) even at the level of this book. Learning this will avoid some otherwise unnecessary-to-learn alternative derivations, and besides, use of this technique in the derivation of selection rules (see chapter 22) will prove to be insightfully rich.

In addition to all this, in this chapter we will study some recent developments regarding the EPR paradox, which is facilitated by the consideration of spin.

One more thing—admittedly, the joining of space and spin that we accomplish here is a little ad hoc; it is more like living together than a marriage. A proper wedding of these degrees of freedom, as Dirac showed long ago, requires a relativistic treatment, which is beyond the scope of this book.

17.1 ADDING ORBITAL AND SPIN ANGULAR MOMENTUM

Since the orbital angular momentum depends only on space coordinates and the spin $\mathbf{S}$ has nothing to do with space, $\mathbf{L}$ and $\mathbf{S}$ commute. It follows that the components of the total angular momentum defined by

$$\mathbf{J} = \mathbf{L} + \mathbf{S}$$

satisfy the angular-momentum commutation relations (hence all the matrix elements derived for J^2 and the components of $\mathbf{J}$ in chapter 11 hold true for $\mathbf{J}$ as defined here). We want to construct eigenstates of J^2 and J_z ($= L_z + S_z$) by taking suitable linear combinations of the states $|lm_l, m_s\rangle$ (where we now suppress the additional quantum n). Note that J_z commutes with L^2, L_z, S^2, and S_z, but J^2 fails to commute with L_z and S_z (why?).

To this end, consider the linear combination

$$|jm\rangle = \alpha\,|lm-\tfrac{1}{2},\tfrac{1}{2}\rangle + \beta\,|lm+\tfrac{1}{2},-\tfrac{1}{2}\rangle \tag{17.2}$$

constructed in such a way that it is already an eigenstate of J_z belonging to the eigenvalue m. Now we will show that by suitably fixing α and β we can make this into an eigenstate of J^2 as well, belonging to the eigenvalue $j(j+1)\hbar^2$; in the process we will also determine the possible values of j.

Now $J^2 = (\mathbf{L} + \mathbf{S})^2$ can be written as

$$\begin{aligned} J^2 &= L^2 + S^2 + 2\mathbf{L}\cdot\mathbf{S} \\ &= L^2 + S^2 + 2L_zS_z + L_+S_- + L_-S_+ \end{aligned} \tag{17.3}$$

We will use the following results for the operation of the angular-momentum operators $L_\pm$ and $S_\pm$ on angular-momentum states:

$$L_+|lm\rangle = [(l+m+1)(l-m)]^{1/2}\hbar|lm{+}1\rangle$$

$$L_-|lm\rangle = [(l-m+1)(l+m)]^{1/2}\hbar|lm{-}1\rangle$$

$$S_+|+\tfrac{1}{2}\rangle = 0 = S_-|-\tfrac{1}{2}\rangle \qquad S_+|-\tfrac{1}{2}\rangle = \hbar|+\tfrac{1}{2}\rangle \qquad S_-|+\tfrac{1}{2}\rangle = \hbar|-\tfrac{1}{2}\rangle \quad (17.4)$$

Now we can see the effect that J^2 has when it operates on the linear combination of equation (17.2):

$$\begin{aligned} J^2|jm\rangle = \alpha\hbar^2\{&[l(l+1) + (\tfrac{3}{4}) + 2(m-\tfrac{1}{2})(\tfrac{1}{2})]|lm{-}\tfrac{1}{2},\tfrac{1}{2}\rangle \\ &+ [(l+m+\tfrac{1}{2})(l-m+\tfrac{1}{2})]^{1/2}|lm{+}\tfrac{1}{2},-\tfrac{1}{2}\rangle\} \\ + \beta\hbar^2\{&[l(l+1) + (\tfrac{3}{4}) + 2(m+\tfrac{1}{2})(-\tfrac{1}{2})]|lm{+}\tfrac{1}{2},-\tfrac{1}{2}\rangle \\ &+ [(l-m+\tfrac{1}{2})(l+m+\tfrac{1}{2})]^{1/2}|lm{-}\tfrac{1}{2},\tfrac{1}{2}\rangle\} \end{aligned}$$

But if $|jm\rangle$ is an eigenstate of J^2, then

$$J^2|jm\rangle = j(j+1)\hbar^2|jm\rangle = j(j+1)\hbar^2(\alpha|lm{-}\tfrac{1}{2},\tfrac{1}{2}\rangle + \beta|lm{+}\tfrac{1}{2},-\tfrac{1}{2}\rangle)$$

Both equations can be satisfied if

$$\alpha[l(l+1) + (\tfrac{3}{4}) + m - \tfrac{1}{2}] + \beta[(l-m+\tfrac{1}{2})(l+m+\tfrac{1}{2})]^{1/2} = j(j+1)\alpha$$

$$\beta[l(l+1) + (\tfrac{3}{4}) - m - \tfrac{1}{2}] + \alpha[(l+m+\tfrac{1}{2})(l-m+\tfrac{1}{2})]^{1/2} = j(j+1)\beta \tag{17.5}$$

Therefore, what is required is

$$\begin{aligned}(l+m+\tfrac{1}{2})(l-m+\tfrac{1}{2}) = {} & [j(j+1) - l(l+1) - (\tfrac{3}{4}) - m + \tfrac{1}{2}] \\ & \times [j(j+1) - l(l+1) - (\tfrac{3}{4}) + m + \tfrac{1}{2}]\end{aligned}$$

This equation has two solutions; the first one is

$$j(j+1) - l(l+1) - (\tfrac{1}{4}) = l + \tfrac{1}{2} \tag{17.6}$$

which gives $j = l + \tfrac{1}{2}$. The second solution is

$$j(j+1) - l(l+1) - (\tfrac{1}{4}) = -l - \tfrac{1}{2}$$

which gives $j = l - \tfrac{1}{2}$.

For $j = l + \frac{1}{2}$, substituting equation (17.6) into equation (17.5), we get

$$\frac{\beta}{\alpha} = \left(\frac{l + \frac{1}{2} - m}{l + \frac{1}{2} + m}\right)^{1/2} \tag{17.7}$$

Combining with the normalization condition,

$$\alpha^2 + \beta^2 = 1$$

we find

$$\alpha = (2l + 1)^{-1/2}(l + \tfrac{1}{2} + m)^{1/2}$$

$$\beta = (2l + 1)^{-1/2}(l + \tfrac{1}{2} - m)^{1/2} \tag{17.8}$$

In this way we see that the coupled state $|j{=}l{+}\frac{1}{2}, m\rangle$ is

$$\begin{aligned} |j{=}l{+}\tfrac{1}{2}, m\rangle = (2l + 1)^{-1/2}[&(l + \tfrac{1}{2} + m)^{1/2}|lm{-}\tfrac{1}{2}, \tfrac{1}{2}\rangle \\ &+ (l + \tfrac{1}{2} - m)^{1/2}|lm{+}\tfrac{1}{2}, -\tfrac{1}{2}\rangle] \end{aligned} \tag{17.9}$$

We can guess that the state $|j{=}l{-}\frac{1}{2}, m\rangle$ must have the form

$$\begin{aligned} |j{=}l{-}\tfrac{1}{2}, m\rangle = (2l + 1)^{-1/2}[&-(l + \tfrac{1}{2} - m)^{1/2}|lm{-}\tfrac{1}{2}, \tfrac{1}{2}\rangle \\ &+ (l + \tfrac{1}{2} + m)^{1/2}|lm{+}\tfrac{1}{2}, -\tfrac{1}{2}\rangle] \end{aligned} \tag{17.10}$$

You may think that we went through a lot of algebra to define a coupled representation for the state vectors, and wonder if it is worth it. Yes, the coupled basis has physical use. True, when the Hamiltonian is just coulombic, the uncoupled representation is enough. But as we will see later in chapter 19, the total $|jm\rangle$ basis comes in very handy when, as is the case with real atoms, there is a spin-orbit interaction $\sim \mathbf{L}\cdot\mathbf{S} = \frac{1}{2}(J^2 - L^2 - S^2)$, which is diagonal in this representation.

17.2 ADDING TWO SPINS

Let's consider two electrons with spin angular momentum denoted by operators $\mathbf{S}_1$ and $\mathbf{S}_2$, respectively. The components of each satisfy the usual angular-momentum commutation relations, but the members of each set commute with the members of the other set, since the degrees of freedom of the two particles are independent:

$$[\mathbf{S}_1, \mathbf{S}_2] = 0 \tag{17.11}$$

Now define the total spin **S** as

$$\mathbf{S} = \mathbf{S}_1 + \mathbf{S}_2 \tag{17.12}$$

By virtue of equation (17.11), the components of **S** are easily seen to obey the angular-momentum commutation rules. We are interested in the eigenvalues and eigenvectors of S^2 and S_z, which define the coupled representation.

The uncoupled representation in the present case consists of four base states all condensed into the notation

$$|m_s, m_s'\rangle$$

Explicitly, we have $|+\frac{1}{2},+\frac{1}{2}\rangle$, $|+\frac{1}{2},-\frac{1}{2}\rangle$, $|-\frac{1}{2},+\frac{1}{2}\rangle$, and $|-\frac{1}{2},-\frac{1}{2}\rangle$. Since S_z, the z-component of the total spin, commutes with S_1^2, S_2^2, S_{1z}, and S_{2z}, we expect S_z to be diagonal already in the uncoupled basis. We can easily check this out:

$$S_z|\tfrac{1}{2},\tfrac{1}{2}\rangle = (S_{1z} + S_{2z})|\tfrac{1}{2},\tfrac{1}{2}\rangle = (\tfrac{1}{2}\hbar + \tfrac{1}{2}\hbar)|\tfrac{1}{2},\tfrac{1}{2}\rangle = \hbar|\tfrac{1}{2},\tfrac{1}{2}\rangle$$

$$S_z|\tfrac{1}{2},-\tfrac{1}{2}\rangle = S_z|-\tfrac{1}{2},\tfrac{1}{2}\rangle = 0$$

$$S_z|-\tfrac{1}{2},-\tfrac{1}{2}\rangle = -\hbar|-\tfrac{1}{2},-\tfrac{1}{2}\rangle \tag{17.13}$$

It is instructive to display the matrix representation of S_z in the uncoupled basis:

	$\lvert\frac{1}{2},\frac{1}{2}\rangle$	$\lvert\frac{1}{2},-\frac{1}{2}\rangle$	$\lvert-\frac{1}{2},\frac{1}{2}\rangle$	$\lvert-\frac{1}{2},-\frac{1}{2}\rangle$
$\lvert\frac{1}{2},\frac{1}{2}\rangle$	$\hbar$	0	0	0
$\lvert\frac{1}{2},-\frac{1}{2}\rangle$	0	0	0	0
$\lvert-\frac{1}{2},\frac{1}{2}\rangle$	0	0	0	0
$\lvert-\frac{1}{2},-\frac{1}{2}\rangle$	0	0	0	$-\hbar$

There is a degeneracy that can be removed by forming orthogonal linear combinations of the states $|\frac{1}{2},-\frac{1}{2}\rangle$ and $|-\frac{1}{2},\frac{1}{2}\rangle$; note, however, that any such linear combination is not an eigenstate of either S_{1z} or S_{2z}. Could this be an indication that such linear combinations are eigenstates of an operator, such as S^2, that does not commute with S_{1z} and S_{2z}?

Let's examine the operator S^2 and write it as

$$\begin{aligned} \mathbf{S}^2 &= (\mathbf{S}_1 + \mathbf{S}_2)^2 = S_1^2 + S_2^2 + 2\mathbf{S}_1 \cdot \mathbf{S}_2 \\ &= S_1^2 + S_2^2 + 2S_{1z}S_{2z} + S_{1+}S_{2-} + S_{1-}S_{2+} \end{aligned} \tag{17.14}$$

Using equation (17.14), the matrix of S^2 in the uncoupled basis (the explicit calculation will be left up to the reader) is seen to be

$$\hbar^2 \begin{pmatrix} 2 & 0 & 0 & 0 \\ 0 & 1 & 1 & 0 \\ 0 & 1 & 1 & 0 \\ 0 & 0 & 0 & 2 \end{pmatrix}$$

We can see that although the base states $|\frac{1}{2},\frac{1}{2}\rangle$ and $|-\frac{1}{2},-\frac{1}{2}\rangle$ are already eigenstates of S^2 with eigenvalue $S(S+1)\hbar^2 = 2\hbar^2$, the other two base states are not. The matrix, however, is block diagonal; therefore, the middle block can be diagonalized separately to find out the appropriate linear combinations of $|\frac{1}{2},-\frac{1}{2}\rangle$ and $|-\frac{1}{2},\frac{1}{2}\rangle$ that are eigenstates of S^2.

But of course the middle block, the matrix

$$\begin{pmatrix} 1 & 1 \\ 1 & 1 \end{pmatrix}$$

is all too familiar; it is $I + \sigma_x$, for which we know the eigenstates. The appropriate linear combinations are then given as

$$(1/\sqrt{2})\,[|\tfrac{1}{2},-\tfrac{1}{2}\rangle + |-\tfrac{1}{2},\tfrac{1}{2}\rangle]$$

which corresponds to the eigenvalue $S = 1$, and

$$(1/\sqrt{2})\,[|\tfrac{1}{2},-\tfrac{1}{2}\rangle - |-\tfrac{1}{2},\tfrac{1}{2}\rangle]$$

(where we have chosen the phase slightly differently in accordance with convention) which corresponds to the eigenvalue $S = 0$.

The solution to the eigenvalue problem we sought is now complete. To summarize, the eigenvalues of S^2 are given by the two allowed values of the *total spin quantum number* S, 1 and 0 (the actual value of S^2 is, of course, $S(S+1)\hbar^2$). The $S = 1$ eigenstate

$$|S{=}1, M\rangle$$

is the *triplet* state (corresponding to the threefold value that M can take) given in the uncoupled basis as

$$\begin{aligned} |S{=}1, M{=}1\rangle &= |\tfrac{1}{2},\tfrac{1}{2}\rangle \\ |S{=}1, M{=}0\rangle &= (1/\sqrt{2})\,[|\tfrac{1}{2},-\tfrac{1}{2}\rangle + |-\tfrac{1}{2},\tfrac{1}{2}\rangle] \\ |S{=}1, M{=}-1\rangle &= |-\tfrac{1}{2},-\tfrac{1}{2}\rangle \end{aligned} \tag{17.15}$$

The $S=0$ eigenstate $|S{=}0, M{=}0\rangle$ is the *singlet* state given in the uncoupled basis as

$$|S{=}0, M{=}0\rangle = (1/\sqrt{2})\,[|\tfrac{1}{2},-\tfrac{1}{2}\rangle - |-\tfrac{1}{2},\tfrac{1}{2}\rangle] \tag{17.16}$$

As in the case of spin-orbit coupling, the spin-coupled base states are also useful when the Hamiltonian of the system warrants it. For example, there are examples of spin-spin interaction in nature; the two-nucleon potential has an $\mathbf{S}_1 \cdot \mathbf{S}_2$ term and so does the interaction between quarks. The spin-coupled basis comes in handy in these situations.

Spin and the Pauli Principle

The states of total spin have another interesting property; they come with a definite symmetry under the exchange of the particle labels $1 \to 2$. The triplet states are symmetric under particle exchange (as can be verified from eq. [17.15]); the singlet state is antisymmetric, changing sign under particle exchange as can be verified from equation (17.16).

In chapter 9, we defined the two-particle exchange operator P_{12}. Since the particles have both spatial and spin labels, P_{12} must involve interchanging both sets of labels. Let P_{12}^{σ} denote the exchange operator that interchanges the spin labels alone; it thus acts upon the spin part of the states. We can write the results above as

$$P_{12}^{\sigma}|S{=}1, M\rangle = |S{=}1, M\rangle$$

$$P_{12}^{\sigma}|S{=}0, M{=}0\rangle = -|S{=}0, M{=}0\rangle \tag{17.17}$$

Now the states of spin $\frac{1}{2}$ particles must be antisymmetric under particle exchange; this is how the Pauli principle is incorporated in quantum mechanics (see chapter 9). It follows that P_{12} acting on the total state vector of two particles $|\psi(1,2)\rangle$ must give a minus sign:

$$P_{12}|\psi(1,2)\rangle = -|\psi(1,2)\rangle \tag{17.18}$$

Let's write $P_{12} = P_{12}^{r} P_{12}^{\sigma}$, where P_{12}^{r} is the *space-exchange operator*—it exchanges the spatial coordinates of the two particles. Moreover, we have seen that by going to the center-of-mass system, we can express the spatial state of two particles also in the form $|nlm\rangle$ where

$$\langle r\theta\phi|nlm\rangle = R_{nl}(r)Y_{lm}(\theta\phi)$$

An interchange P_{12}^{r} of the two particles' spatial coordinates is equivalent to $\mathbf{r} \to -\mathbf{r}$, or

$$r \to r$$

$$\theta \to \pi - \theta$$

$$\phi \to \pi + \phi$$

This is the same as the parity operation. Therefore, under P_{12}^r the radial function remains unchanged, but Y_{lm} transforms as

$$Y_{lm}(\theta\phi) \to (-1)^l Y_{lm}(\theta\phi)$$

Even l's are symmetric under space exchange, odd l's are antisymmetric. Consequently, triplet states that are spin-symmetric must be accompanied by odd orbital angular-momentum (odd l) states so that the total state can be antisymmetric under two-particle exchange. On the other hand, singlet states go with even l's for the spatial functions.

An immediate application of the above result is seen in proton-proton scattering at low energy, which is dominated by the S-wave of the two protons (in the center-of-mass system) because of the absence of a centrifugal barrier. According to the above, the S-wave must come with the singlet state, and this is indeed found to be the case.

The general antisymmetry of the two-particle wave function has an important application to the ground state of He. If we use a hydrogenic model (with $Z = 2$), the lowest spatial state of both electrons is $|nlm\rangle = |100\rangle$, where n is now the principal quantum number; thus the spatial part of the two-electron wave function, the product of two S-state functions, is symmetric under exchange. The spin part, accordingly, must be antisymmetric, must be the singlet state.

17.3 A GENERAL METHOD FOR ANGULAR-MOMENTUM COUPLING: CLEBSCH-GORDAN COEFFICIENTS

From the two examples considered so far, it is clear that whenever we consider two angular momenta, there are two sets of mutually commuting operators, and we can use the eigenstates of either set to define a basis. The two sets then must be connected by a unitary transformation. The coefficients of this unitary transformation are called Clebsch-Gordan coefficients (also Wigner 3-j symbols or vector addition coefficients).

Let $\mathbf{J}_1$ and $\mathbf{J}_2$ be the two angular momenta, and let

$$[\mathbf{J}_1, \mathbf{J}_2] = 0 \tag{17.19}$$

Define the total angular momentum $\mathbf{J}$ as

$$\mathbf{J} = \mathbf{J}_1 + \mathbf{J}_2 \tag{17.20}$$

The two commuting sets of operators are (1) J_1^2, J_{1z}, J_2^2, J_{2z} and (2) J_1^2, J_2^2, J^2, J_z. Let's label the corresponding eigenkets (1) $|j_1 m_1 j_2 m_2\rangle$ and (2) $|j_1 j_2 JM\rangle$. The Clebsch-Gordan coefficients are the unitary transformation matrix elements that connect the two sets:

$$|j_1 j_2 JM\rangle = \sum_{m_1 m_2} \langle j_1 m_1 j_2 m_2 | j_1 j_2 JM\rangle | j_1 m_1 j_2 m_2\rangle \tag{17.21}$$

As you can see, this is just an application of the ket-bra sum-equals-one rule once again; the coefficients

$$\langle j_1 m_1 j_2 m_2 | j_1 j_2 JM\rangle \equiv \langle j_1 m_1 j_2 m_2 | JM\rangle \tag{17.22}$$

are called the Clebsch-Gordan coefficients where we usually write them in the form on the right-hand side of equation (17.22). The inverse transformation can be written as

$$|j_1 m_1 j_2 m_2\rangle = \sum_{JM} \langle j_1 j_2 JM | j_1 m_1 j_2 m_2\rangle | j_1 j_2 JM\rangle \tag{17.23}$$

But since the transformation matrix is real (by choice) and unitary (real unitary matrices are called orthogonal), we have

$$\begin{aligned}\langle j_1 j_2 JM | j_1 m_1 j_2 m_2\rangle &= \langle j_1 m_1 j_2 m_2 | j_1 j_2 JM\rangle^* = \langle j_1 m_1 j_2 m_2 | j_1 j_2 JM\rangle \\ &\equiv \langle j_1 m_1 j_2 m_2 | JM\rangle \end{aligned} \tag{17.24}$$

Accordingly, we can write equation (17.23) as

$$|j_1 m_1 j_2 m_2\rangle = \sum_{JM} \langle j_1 m_1 j_2 m_2 | JM\rangle | j_1 j_2 JM\rangle \tag{17.25}$$

The following properties of the coefficients immediately follow:

$$\sum_{JM} \langle j_1 m_1 j_2 m_2 | JM\rangle \langle j_1 m_1' j_2 m_2' | JM\rangle = \delta_{m_1 m_1'} \delta_{m_2 m_2'} \tag{17.26}$$

$$\sum_{m_1 m_2} \langle j_1 m_1 j_2 m_2 | JM\rangle \langle j_1 m_1 j_2 m_2 | J'M'\rangle = \delta_{JJ'} \delta_{MM'} \tag{17.27}$$

These are referred to as the orthogonality relations for the Clebsch-Gordan coefficients.

It is obvious that a Clebsch-Gordan coefficient must vanish unless $m_1 + m_2 = M$ (fig. 17.1). Therefore, the summations in equation (17.21) or (17.27) do not really run over both m_1 and m_2; in truth there is only one independent summation index.

Another most important property of the Clebsch-Gordan coefficients is that

$$\langle j_1 m_1 j_2 m_2 | JM \rangle \neq 0, \quad \text{only if } |j_1 - j_2| \leq J \leq j_1 + j_2 \tag{17.28}$$

This is called the *triangle inequality* since geometrically the relation means that we must be able to form a triangle with sides j_1, j_2, and J. This is the same as the old prescription of the vector model of the atom.

The triangle inequality tells us the possible range of values of total J when we add two angular momenta j_1 and j_2 whose values are fixed. This is the range over which we carry out the summation in equation (17.26).

For a proof of the triangle rule, let's consider the possible values of M. Since $M = m_1 + m_2$, its maximum value is $j_1 + j_2$. This value is realized for the single state $|j_1 j_1 j_2 j_2\rangle$. The total J of this state is $J = j_1 + j_2$. The next highest value of M, $j_1 + j_2 - 1$, is realized for the linear combinations of two states in the uncoupled representations

$$|j_1 j_1{-}1 j_2 j_2\rangle \quad \text{and} \quad |j_1 j_1 j_2 j_2{-}1\rangle$$

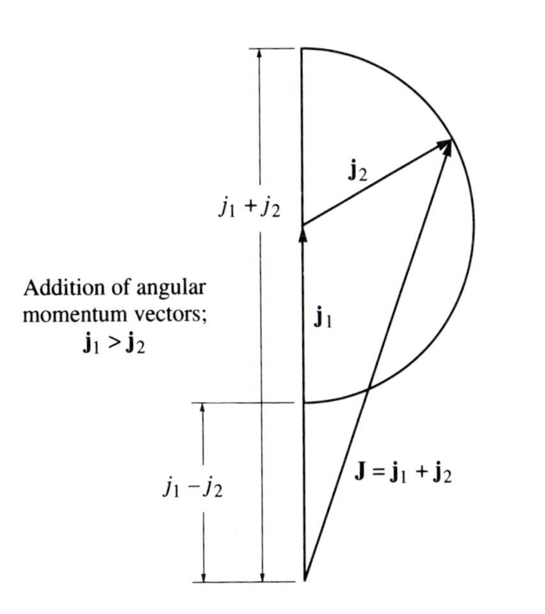

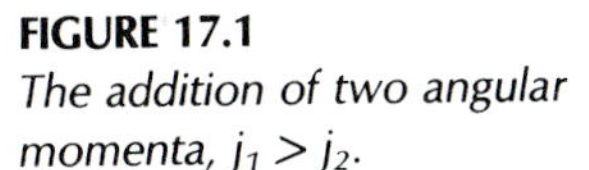
FIGURE 17.1
The addition of two angular momenta, $j_1 > j_2$.

One of the linear combinations belongs to $J = j_1 + j_2$, and the second one to $J = j_1 + j_2 - 1$. In the same vein, the value of $M = j_1 + j_2 - 2$ will be realized for the three linear combination of the three uncoupled states

$$|j_1 j_1{-}2 j_2 j_2\rangle, \quad |j_1 j_1{-}1 j_2 j_2{-}1\rangle, \quad \text{and} \quad |j_1 j_1 j_2 j_2{-}2\rangle$$

corresponding to three values of the total J

$$j_1 + j_2, \quad j_1 + j_2 - 1, \quad \text{and} \quad j_1 + j_2 - 2$$

If we continue this process, we can see that each time we lower M by 1, a new value of J appears; the process terminates when we arrive at a value for which either $m_1 = -j_1$ or $m_2 = -j_2$. It follows that the minimum value of J is $|j_1 - j_2|$.

Let's do a further plausibility check. The total number of states in the $|j_1 m_1 j_2 m_2\rangle$ basis is obviously $(2j_1 + 1)(2j_2 + 1)$. Is this the case with the total number of states in the JM-representation? The total number of states for each J is $2J + 1$ (the number of possible M). Assume for convenience that $j_1 > j_2$. Now

$$\begin{aligned}\sum_{J=j_1-j_2}^{j_1+j_2} (2J + 1) &= [2(j_1 + j_2) + 1] + [2(j_1 + j_2 - 1) + 1] \\ &\quad + \ldots + [2(j_1 - j_2) + 1] \\ &= \sum_{n=0}^{2j_2} [2(j_1 - j_2 + n) + 1] \\ &= \tfrac{1}{2}(4j_1 + 2)(2j_2 + 1) \\ &= (2j_1 + 1)(2j_2 + 1) \end{aligned} \tag{17.29}$$

It checks out.

We choose the phase convention of Condon and Shortley (see reference at the end of chapter) to fix an overall sign that occurs in the evaluation of the Clebsch-Gordan coefficients. The following set of symmetry relations also holds (for derivation, see the Condon and Shortley reference at the end of the chapter):

$$\begin{aligned}\langle j_1 m_1 j_2 m_2 | JM\rangle &= (-1)^{j_1+j_2-J}\langle j_2 m_2 j_1 m_1 | JM\rangle \\ &= (-1)^{j_1+j_2-J}\langle j_1{-}m_1 j_2{-}m_2 | J{-}M\rangle \\ &= (-1)^{j_1-m_1}\left(\frac{2J + 1}{2j_2 + 1}\right)^{1/2} \langle j_1 m_1 J{-}M | j_2{-}m_2\rangle \end{aligned} \tag{17.30}$$

TABLE 17.1
A few Clebsch-Gordan coefficients

$\langle \frac{1}{2} m_s l m_l \mid jm \rangle$		
$j \backslash m_s$	$+\frac{1}{2}$	$-\frac{1}{2}$
$l+\frac{1}{2}$	$\sqrt{\left(\frac{l+\frac{1}{2}+m}{2l+1}\right)}$	$\sqrt{\left(\frac{l+\frac{1}{2}-m}{2l+1}\right)}$
$l-\frac{1}{2}$	$\sqrt{\left(\frac{l+\frac{1}{2}-m}{2l+1}\right)}$	$-\sqrt{\left(\frac{l+\frac{1}{2}+m}{2l+1}\right)}$

$\langle 1 m_s l m_l \mid jm \rangle$			
$j \backslash m_s$	$+1$	0	-1
$l+1$	$\sqrt{\left(\frac{(l+m)(l+m+1)}{(2l+1)(2l+2)}\right)}$	$\sqrt{\left(\frac{(l-m+1)(l+m+1)}{(2l+1)(l+1)}\right)}$	$\sqrt{\left(\frac{(l-m)(l-m+1)}{(2l+1)(2l+2)}\right)}$
l	$\sqrt{\left(\frac{(l+m)(l-m+1)}{2l(l+1)}\right)}$	$\frac{-m}{\sqrt{(l(l+1))}}$	$-\sqrt{\left(\frac{(l-m)(l+m+1)}{2l(l+1)}\right)}$
$l-1$	$\sqrt{\left(\frac{(l-m)(l-m+1)}{2l(2l+1)}\right)}$	$-\sqrt{\left(\frac{(l-m)(l+m)}{l(2l+1)}\right)}$	$\sqrt{\left(\frac{(l+m+1)(l+m)}{2l(2l+1)}\right)}$

The advantage of the Clebsch-Gordan formalism is that these coefficients are easily evaluated using numerical codes that employ recursion relations; most importantly for the user, they are tabulated in the literature. Table 17.1 gives some Clebsch-Gordan coefficients that will be useful in this book.

Example: Coupling of Two-Spin $\frac{1}{2}$ Particles

For the case of the coupling of two-spin $\frac{1}{2}$ particles, $j_1 = \frac{1}{2}$, $j_2 = \frac{1}{2}$. We have for the coupled state, now using Clebsch-Gordanology,

$$|SM\rangle = \sum_{m_s m_{s'}} \langle \tfrac{1}{2} m_s \tfrac{1}{2} m_s' | SM\rangle | m_s m_s'\rangle \tag{17.31}$$

Suppose we want to calculate the state for $S = 1$, $M = 1$. There is only one term contributing to the sum since we cannot get $M = 1$ unless $m_s = m_{s'} = \frac{1}{2}$. We have

$$|S{=}1 M{=}1\rangle = \langle \tfrac{1}{2}\tfrac{1}{2}\tfrac{1}{2}\tfrac{1}{2}|11\rangle | m_s{=}\tfrac{1}{2} m_s'{=}\tfrac{1}{2}\rangle$$

Now a short lesson in using table 17.1. The first problem we face is that the Clebsch-Gordan coefficient above cannot be obtained directly from the table—the table requires the second angular momentum to be integer or zero (the allowed values of l), but no matter. We use the third relation of equation (17.30) to rewrite our Clebsch-Gordan as follows:

$$\langle \tfrac{1}{2}\tfrac{1}{2}\tfrac{1}{2}\tfrac{1}{2}|11\rangle = (-1)^{1/2-1/2}(\tfrac{3}{2})^{1/2}\langle \tfrac{1}{2}\tfrac{1}{2}1-1|\tfrac{1}{2}-\tfrac{1}{2}\rangle$$

The Clebsch-Gordan on the right-hand side above can be found in the table under $j = l - \frac{1}{2}$ and $m_s = \frac{1}{2}$ (with $l = 1$ and $m = -\frac{1}{2}$) and equals $(2/3)^{1/2}$. In this way it follows that

$$\langle \tfrac{1}{2}\tfrac{1}{2}\tfrac{1}{2}\tfrac{1}{2}|11\rangle = 1$$

Therefore,

$$|11\rangle = |m_s{=}\tfrac{1}{2}m_s'{=}\tfrac{1}{2}\rangle$$

as before. If we evaluate the rest of the Clebsch-Gordan coefficients the same way (do it! it's good practice), equation (17.31) can be put in the following matrix form:

$$\begin{matrix} |SM\rangle & & & |m_s m_{s'}\rangle \\ \begin{pmatrix} |11\rangle \\ |10\rangle \\ |1-1\rangle \\ |00\rangle \end{pmatrix} & = & \begin{pmatrix} 1 & 0 & 0 & 0 \\ 0 & 1/\sqrt{2} & 1/\sqrt{2} & 0 \\ 0 & 0 & 0 & 1 \\ 0 & 1/\sqrt{2} & -1/\sqrt{2} & 0 \end{pmatrix} & \begin{pmatrix} |\frac{1}{2}\frac{1}{2}\rangle \\ |\frac{1}{2}-\frac{1}{2}\rangle \\ |-\frac{1}{2}\frac{1}{2}\rangle \\ |-\frac{1}{2}-\frac{1}{2}\rangle \end{pmatrix} \end{matrix} \tag{17.32}$$

Note that in equation (17.32), the columns represent the basis vectors themselves, not their components.

Since Clebsch-Gordan coefficients work both ways, we can also write

$$|m_s m_s'\rangle = \sum_{SM} \langle \tfrac{1}{2}m_s\tfrac{1}{2}m_s'|SM\rangle |SM\rangle \tag{17.33}$$

For example, the state $|\frac{1}{2}-\frac{1}{2}\rangle$ is given in terms of the coupled representation as

$$|\tfrac{1}{2}-\tfrac{1}{2}\rangle = (1/\sqrt{2})(|10\rangle + |00\rangle) \tag{17.34}$$

■ **PROBLEM** Two identical particles each have orbital angular momentum l. For what values of total L is the two-particle state symmetric under exchange?

SOLUTION The trick is to use the appropriate symmetry relation of the relevant Clebsch-Gordan coefficient. We write the coupled state as

$$|l_1 l_2 LM\rangle = \sum_{m_1} \langle l_1 m_1 l_2 M-m_1 | LM\rangle |l_1 m_1\rangle |l_2 M-m_1\rangle$$

If we apply the particle exchange operator from the left, we get

$$\begin{aligned} P_{12}|l_1 l_2 LM\rangle = |l_2 l_1 LM\rangle &= \sum_{m_1} \langle l_2 M-m_1 l_1 m_1 | LM\rangle |l_1 m_1\rangle |l_2 M-m_1\rangle \\ &= \sum_{m_1} (-1)^{l_1+l_2-L} \langle l_1 m_1 l_2 M-m_1 | LM\rangle |l_1 m_1\rangle |l_2 M-m_1\rangle \\ &= (-1)^{l_1+l_2-L} |l_1 l_2 LM\rangle \end{aligned}$$

It follows that the condition for symmetry under exchange is

$$(-1)^{l_1+l_2-L} = 1 \tag{17.35}$$

But $l_1 = l_2 = l$. Therefore, symmetry under exchange will prevail only when L = even (0, 2, 4, and so forth). ■

Spectroscopic Notation

In the good old days, when we did not worry about spin, we introduced the notations S-state, P-state, and the like to denote the orbital angular momentum of the object. Now we have to include spin in our notation, and if we are smart we will also include total J in it. Therefore, we will introduce the following conventions for the spectroscopic notation of a state:

1. Continue to use the capital letters S, P, D, F, G, and all that to denote the value of the orbital angular momentum (call a typical letter L).
2. Append a superscript $2S + 1$ to the left of L to denote the spin multiplicity, the possible values of the spin projection.
3. Append a subscript J to the right of L to denote the value of the total angular momentum.

An example will clarify the notation completely:

$${}^{2S+1}L_J = {}^1S_0$$

It means that the spin is 0, the orbital angular momentum is 0, and the total angular momentum is also 0 (as it has to be). Of course, we are talking about a multielectron state. For a single electron, the $2S + 1$ superscript is redundant since it is always 2.

17.4 THE SINGLET STATE AND THE EPR PARADOX REVISITED

Recall the EPR paradox? EPR pointed out that if two particles interact and then stop interacting, they become correlated in such a fashion that measurement of one's position will collapse the other's wave function in a definite position state as well (and this messes up the probabilities of subsequent measurement results on the second particle). This is a paradox if we treasure locality, because the two particles are not interacting in any local fashion at the time of the measurement. And yet quantum mechanics says there is an effect; quantum mechanics seems to insist on nonlocality.

However, this is still theory. Can we do an experiment in the fashion of the EPR gedanken experiment that will resolve the issue of locality versus nonlocality? With the help of the concept of the singlet state, we can.

The total spin of the singlet state is zero, and this necessitates that the component particles always have opposite S_z-spins; if one is spin up, the other must be spin down, and vice versa. Does this correlation persist even when the two partners joined in the singlet state fly apart?

To be concrete, we can think of the following experiment. If we scatter low-energy protons from a proton target (hydrogen gas), only protons in S-wave ($l = 0$) will scatter; the low energy will not be enough to penetrate the centrifugal barrier that higher orbital angular momentum will present. The antisymmetry of the wave function then guarantees that the two particles are in a singlet state since their total orbital angular momentum is zero and symmetric under space exchange. Now we allow the interacting particles to move apart until they are vastly separated, and then we measure the S_z-spin of each. Of course, because of their correlation implied by the singlet state, their S_z-spins will always obtain as $\pm\frac{1}{2}\hbar$. So measuring both is really unnecessary; measuring one's spin guarantees that the other's spin must be opposite, no matter what the orientation of the z-axis is. Indeed, quantum mechanics, in the way we conventionally interpret it, says that after the measurement of S_{1z} of the first particle, the state of the second particle also collapses, becoming an eigenstate of S_{2z}. Hence the probabilities of all subsequent measurements on particle 2 become affected, and yet we have not done anything to it directly. All the physical measurement is on particle 1 at a distance. It is thus that quantum mechanics is insisting upon nonlocality, that the properties of a particle are affected by measurements carried out on another, even though they are widely separated.

In chapter 10, we briefly discussed hidden variables—locality can be saved by postulating hidden variables, hidden parameters that allow the particles to always be in a definite state, yet probabilistic results are retained for an ensemble. You see, the consternation that quantum mechanics creates comes from the coherent superposition; the singlet state is a coherent superposition, we cannot tell which particle has spin up and which down. Only measurement can settle that by collapsing a particular state of the coherent superposition. And this seems strange, because collapsing one particle in a definite spin state seems to have fixed the other's spin state as well. But hidden-variables theory has a simpler explanation (or so it seems) of all this—suppose the particles have definite spin states to begin with.

Can we ever rule out this hidden-variables possibility? Indeed, one of the greatest achievements of recent years in our understanding of quantum nonlocality has come from our ability to show that no local hidden-variables theory can ever reproduce all the quantum mechanical predictions in the situation described above. But before we give the proof, we must delve further into the quantitative predictions of quantum mechanics in the experimental situation presented above.

Suppose we first make a measurement of S_{1z} of the first particle; the measurement fixes the z-axis. Now we make a second measurement on the correlated partner for its spin component S_n in a direction $\mathbf{e}_n$ that lies at an angle θ to the z-axis. Suppose the result is $|S_{1z}=+\rangle$, spin up in the first measurement (for which the probability is $\frac{1}{2}$). What is the probability that it is also up (along $\mathbf{e}_n$) in the second measurement?

The singlet state correlation tells us this: The first measurement also collapses the state of the second particle to be the spin-down eigenstate of S_{2z}; if we made the second measurement along the same (z-) axis, we would have certainly found it to be represented by the spinor

$$\chi_- = \begin{pmatrix} 0 \\ 1 \end{pmatrix} \tag{17.36}$$

So what we are asking for is to express this spinor in terms of the eigenvectors of $\sigma \cdot \mathbf{e}_n$ that we calculated in chapter 16 (eq. [16.44] with $\phi = 0$). These eigenspinors are

$$\begin{pmatrix} \cos(\theta/2) \\ \sin(\theta/2) \end{pmatrix}$$

for the eigenvalue $\frac{1}{2}\hbar$ and

$$\begin{pmatrix} -\sin(\theta/2) \\ \cos(\theta/2) \end{pmatrix}$$

for the eigenvalue $-\frac{1}{2}\hbar$ (we did not calculate this one in chapter 16, but surely you can see that this is the right eigenspinor from orthogonality alone). Since these are a complete set of eigenspinors, we can express the eigenspinor of equation (17.36) in terms of them as follows:

$$\begin{pmatrix} 0 \\ 1 \end{pmatrix} = \sin(\theta/2) \begin{pmatrix} \cos(\theta/2) \\ \sin(\theta/2) \end{pmatrix} + \cos(\theta/2) \begin{pmatrix} -\sin(\theta/2) \\ \cos(\theta/2) \end{pmatrix} \tag{17.37}$$

The probability we want—that the second measurement gives spin up along $\mathbf{e}_n$ after the first measurement has given spin up—is given by the square of the expansion coefficient multiplying the first spinor on the right-hand side above, $\sin^2(\theta/2)$. The total probability P_{++}, of observing S_{1z} as up and S_{2n} as up, is given as the product of the $\frac{1}{2}$ and $\sin^2(\theta/2)$,

$$P_{++} = \tfrac{1}{2}\sin^2(\theta/2)$$

We can calculate in the same manner the probabilities P_{+-} (spin up along z for particle 1, spin down along $\mathbf{e}_n$ for particle 2), and so forth, and we get

$$\begin{aligned} P_{++} &= \tfrac{1}{2}\sin^2(\theta/2) \\ P_{+-} &= \tfrac{1}{2}\cos^2(\theta/2) \\ P_{-+} &= \tfrac{1}{2}\cos^2(\theta/2) \\ P_{--} &= \tfrac{1}{2}\sin^2(\theta/2) \end{aligned} \tag{17.38}$$

We are now ready to show that a local hidden-variables theory will not always agree with the quantum mechanical prediction, and consequently the ideas of local hidden variables and quantum mechanics are fundamentally incompatible ideas. The proof was first given by Bell (hence the incompatibility is referred to as *Bell's theorem*). Below we will consider in a simple way how the locality assumption gives an inequality (called a Bell inequality) that is not compatible with quantum mechanics.

A Bell Inequality

Let's return to the set-up we considered earlier, but with some added twists. We have two beams of spin-correlated protons that move apart after scattering from one another; call the members of a correlated pair of protons Joe and Moe (J and M), and we have two experimenters who are set to observe them with Stern-Gerlach apparatuses—call them J-detector and M-detector. As before, whenever the J-detector and the M-detector are set up parallel to each other at whatever angle to the vertical (fig. 17.2), if the Joe-observer sees one of the correlated protons with spin up (down), the Moe observer sees her proton with the opposite value of S_z, down (up).

Thus, a typical synchronized sequence of detection by two distant observers with parallel settings of their detectors will show a perfect "hit" pattern like this (U stands for spin up, D for spin down):

Joe: *U D U U D D U D U D U U U D U D D D*

Moe: *D U D D U U D U D U D D D U D U U U*

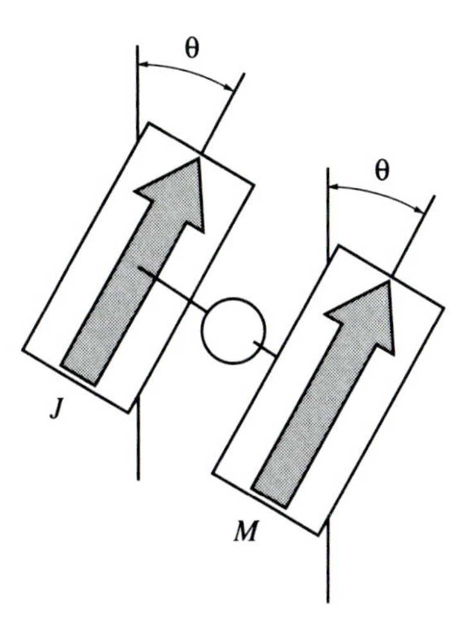

FIGURE 17.2
Parallel arrangement of the J- and M-detectors.

since $[P_{+-}(\theta = 0) + P_{-+}(\theta = 0)] = 1$. (Behold: Since the individual beams are unpolarized, in a long sequence each observer sees a 50-50 admixture of U and D protons.)

But of course, with the assumption of hidden variables, this is not surprising at all, because the hidden variables predetermine the value of the spin vector of each observed particle to be equal and opposite and along the z-axis. This presupposes more information than quantum mechanics permits us (all the components of spin, whereas quantum mechanics only allows us definite information about one), but that's the whole idea of hidden variables.

Next let's consider the case when the two observers set their Stern-Gerlach apparatuses at an angle θ. For example, for an angle $\theta = 60°$, the value of $P_{+-} + P_{-+}$ is 3/4. This means that with this set-up of the detectors (fig. 17.3), for every four proton pairs, the number of hits (on the average) is 3 and the number of misses is 1, as in this detection sequence:

Joe: *U D D D D U D D U D U U D U U U*

Moe: *D D U U U U U U D D D D U D U D*

If you think of the spin polarizations as binary-code messages, the messages are no longer the same for the two observers: There is an error (a miss) in Moe's message (compared to Joe's) once in every four observations.

Now comes Bell's inequality. Start with both detectors parallel. The sequences observed are now completely correlated. But change Joe's setting by the angle $\theta = 60°$ and the sequences are no longer the same; now they contain "errors"—one miss in every four observations on the average. And likewise, come back to the parallel setting, and this time, change Moe's setting by the same angle; again there will be a miss for every four observations on the aver-

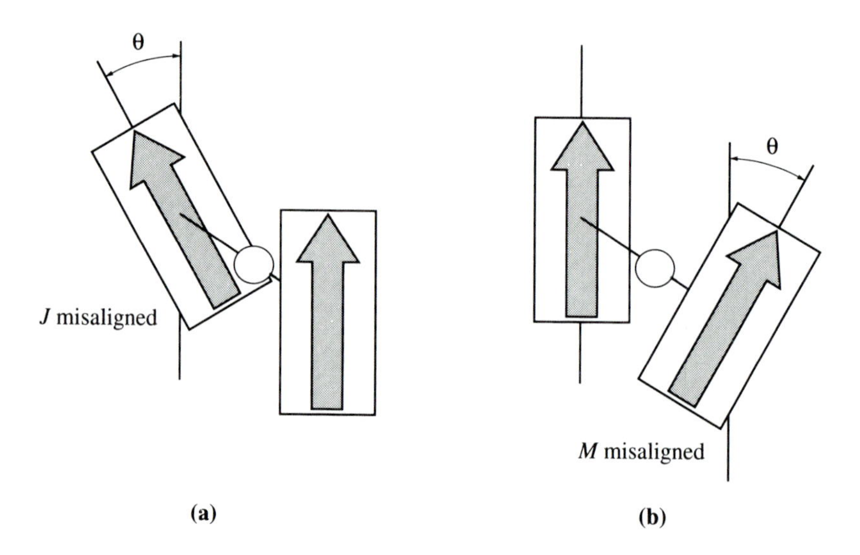

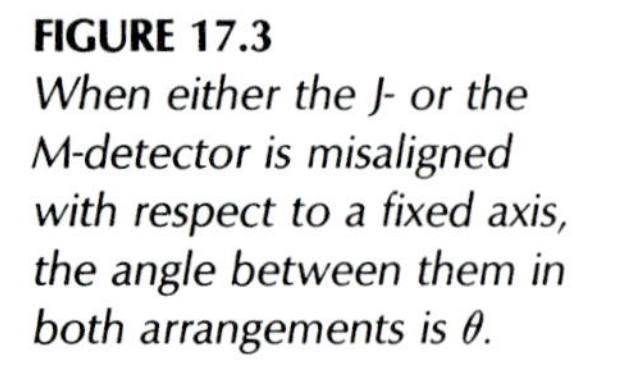

FIGURE 17.3
When either the J- or the M-detector is misaligned with respect to a fixed axis, the angle between them in both arrangements is θ.

age. And this is irrespective of how far apart the two detectors and their observers are. One could be in New York, the other in Los Angeles, with the scattering taking place somewhere in between.

If locality is valid, if the posited hidden variables that manipulate the protons to take on the particular spin orientation that is demanded by the situation are local, we can say this much with certainty: What we do with Joe's detector cannot mess up Moe's message, at least not instantly, and vice versa. Thus, after starting with parallel settings, if the Joe-observer turns the Joe-detector by the angle $\theta = 60°$, and at the same time, the Moe-observer turns the Moe-detector in the opposite direction by the same angle (so that the two detectors are now at an angle of $2\theta = 120°$, fig. 17.4), what will the error rate be? If locality of the hidden variables is valid, each maneuver causes an error rate of one out of every four observations, but cannot affect what happens at the other location at a distance; so the total error rate will be two out of four. However, it may happen that every now and then Joe's error cancels out Moe's (you know, two wrongs sometimes make a right). Thus, the error rate will be less than or equal to $\frac{2}{4}$. This gives you some flavor of a Bell inequality.

But according to quantum mechanics, the error rate for $\theta = 120°$ is $(1 - \cos^2 60°) = \frac{3}{4}$, three misses out of four. It differs considerably from the value predicted by local hidden-variable theories. Hence Bell's theorem: Local hidden variables are not compatible with quantum mechanics.

More sophisticated inequalities have been proven (with very general, minimum assumptions on the hidden variables) that also have the added merit of allowing experimental test. All the tests to date (the most famous of which is the Aspect experiment) support quantum mechanics; that is, the Bell inequalities,

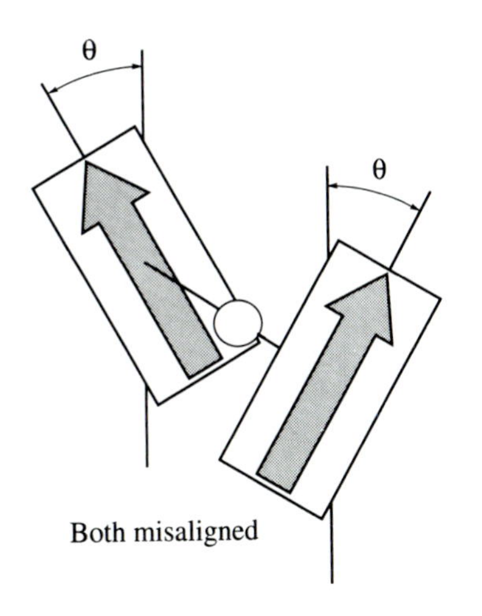

FIGURE 17.4
When both detectors are tilted by θ with respect to a fixed axis, but in opposite directions, the angle between the two detectors is 2θ.

and therefore the locality assumption, are not upheld by any of the experiments. Quantum systems have experimentally proven nonlocal phase correlations exactly as revealed by quantum mechanics.

We need to look at a couple of things about the Aspect experiment. First, the experiment was conducted with polarization-correlated photons in a state analogous to the spin-singlet of the protons obtained from the decay of laser-irradiated calcium atoms. Second, the touchy aspect of this kind of experiment is the problem of switching the axis of the detectors while the photons are in flight; otherwise one can always argue that there is some sort of local information leak. That is, one really has to prove that the phase-information is traveling faster than light! Aspect solved this problem by switching the photon beams instead of the detectors; the photon beams were switched every 10^{-10} s with the help of acousto-optical switches between pre-set detectors. The 10^{-10} s switching time was shorter than the photon's travel time across the laboratory set-up. (For further details, see A. Aspect, J. Dalibard, and G. Roger, *Physical Review Letters*, *49*, 1804, 1982.)

Incidentally, Bell's theorem has not exactly stopped the propounders of hidden variables who now advocate nonlocal hidden variables. We will return to the subject in chapter 24.

Notice the far-reaching ingenuity of the EPR-Bell work. First, EPR revealed the nonlocality of quantum correlations and of quantum collapse. Then, can we avoid nonlocality by invoking hidden variables? No, Bell showed that the hidden variables suffer from the same problem, nonlocality, and thus they cannot save the EPR-favored philosophy of physical realism based on independent separate objects; physical realism, like a captain with his ship, goes down with locality.

PROBLEMS

1. Show that for the state represented by the spinor

$$\begin{pmatrix} \exp(i\mathbf{k}\cdot\mathbf{r}) \\ \exp(-i\mathbf{k}\cdot\mathbf{r}) \end{pmatrix}$$

the measurement of momentum will give $\hbar\mathbf{k}$ and $-\hbar\mathbf{k}$ with equal probabilities. Further show that a simultaneous measurement of momentum and the z-component of spin will never give $+\hbar\mathbf{k}$ *and* $-\frac{1}{2}\hbar$. Also show that a simultaneous measurement of momentum and y-component of spin in the state represented by

$$\begin{pmatrix} \exp(i\mathbf{k}\cdot\mathbf{r}) \\ -i\exp(i\mathbf{k}\cdot\mathbf{r}) \end{pmatrix}$$

will positively yield $\hbar\mathbf{k}$ and $-\frac{1}{2}\hbar$, respectively.

2. Work out the coupling of spin $\mathbf{S}$ with orbital angular momentum $\mathbf{L}$ for a spin $\frac{1}{2}$ particle using Clebsch-Gordan coefficients (use table 17.1).
3. Work out the addition of orbital angular momentum $\mathbf{L}$ with spin $\mathbf{S}$ for a spin-1 object using the elaborate method of section 17.1. Next work out the same coupling using Clebsch-Gordan coefficients (use table 17.1).
4. Consider the coupling of two deuterons, which are particles of spin 1. Suppose the orbital angular momentum of their relative motion is L. What are the possible states of total spin and total angular momentum (orbital plus total spin) after accounting for the symmetrization rules?
5. The nucleus of ^{3}He consists of two protons and a neutron. Ignore the spatial degrees of freedom. Obtain the eigenstates of S^2 and S_z for the combined system. How do you account for the appropriate antisymmetrization rules? (*Hint:* Couple the two protons first.)
6. A particle of spin $\frac{1}{2}$ is in a D-state of orbital angular momentum. What are its possible states of total angular momentum? Suppose the single particle Hamiltonian is

$$H = A + B\mathbf{L}\cdot\mathbf{S} + C\mathbf{L}\cdot\mathbf{L}$$

What are the values of the energy for each of the different states of total angular momentum in terms of the constants A, B, and C?

7. Consider the two-particle Hamiltonian for a spin system (i.e., ignore spatial coordinates) given as

$$H = A + B\mathbf{S}_1\cdot\mathbf{S}_2$$

Calculate the eigenstates and eigenvalues of H for two identical spin $\frac{1}{2}$ particles (a) using the uncoupled basis, matrices, and spinors and (b) using the coupled basis.

8. Show that the values of $\boldsymbol{\sigma}_1 \cdot \boldsymbol{\sigma}_2$ (where $\boldsymbol{\sigma}$ refers to Pauli matrices) in the triplet and singlet states are given by $+1$ and -3, respectively. Hence show that the spin exchange operator P_{12}^{σ} can be written as

$$P_{12}^{\sigma} = (1 + \boldsymbol{\sigma}_1 \cdot \boldsymbol{\sigma}_2)/2$$

9. Show that the expectation value $C(\theta)$ of $(\boldsymbol{\sigma}_1 \cdot \mathbf{e})(\boldsymbol{\sigma}_2 \cdot \mathbf{e}')$ in the singlet state of two particles labeled by 1 and 2, where $\mathbf{e}$ and $\mathbf{e}'$ are two arbitrary unit vectors, is given by

$$C(\theta) = -\cos\theta$$

$C(\theta)$ is called the *correlation coefficient*. What is the relation of $C(\theta)$ with the probabilities in equation (17.37)?

10. Discuss whether or not the nonlocal resolution of EPR paradox violates the causality principle—that cause always precedes effect. (*Hint:* Ask yourself, Can nonlocality of the EPR-Bell arrangement be used to transfer messages?)

ADDITIONAL PROBLEMS

A1. Write a ket for the state of the nucleus helium-3 (see problem 5 above) in the uncoupled spin representation. How many quantum numbers are needed to completely specify the state?

A2. Pauli discovered the neutrino in the nineteen-thirties by noting that the decay

$$n \rightarrow p + e$$

is no-go because of angular momentum conservation. Explain. Demonstrate how if another spin $\frac{1}{2}$ object is emitted in neutron decay, the problem is solved.

A3. Show that the total angular momentum quantum number J of two fermions is always an integer.

A4. Write the antisymmetric wave function of two neutrons in a one-dimensional box, including spin degrees of freedom in the form of a determinant.

A5. Show that the following superposition of states in the $|l\ m_l,\ m_s\rangle$ basis:

$$-(2/3)^{1/2}|1\ -1,\ 1/2\rangle + (1/3)^{1/2}|1\ 0,\ -1/2\rangle$$

is an eigenstate of J^2 and $\mathbf{L}\cdot\mathbf{S}$ and determine the corresponding eigenvalues.

A6. Consider two electrons of spin $\frac{1}{2}$ and $l = 1$.

(a) What are the possible values of the quantum number for the total orbital angular momentum $\mathbf{L} = \mathbf{L}_1 + \mathbf{L}_2$?

(b) What are the possible values of the quantum number for the total spin $\mathbf{S} = \mathbf{S}_1 + \mathbf{S}_2$?

(c) Now find the possible quantum numbers j for the total angular momentum $\mathbf{J} = \mathbf{L} + \mathbf{S}$.

(d) What are the possible values of the total angular momentum of each particle j_1 and j_2?

(e) Find the possible values of j from the combination of j_1 and j_2 and compare with the result of part (c).

A7. Using Table 17.1 for Clebsch-Gordan coefficients, write down the explicit θ_1, ϕ_1, θ_2, ϕ_2-representation of the coupled state $|1\ 1\ 1\ 1\rangle$ of two p electrons.

REFERENCES

D. Bohm. *Quantum Theory.*

E. U. Condon and G. H. Shortley. *The Theory of Atomic Spectra.*

S. Gasiorowicz. *Quantum Physics.*

N. Herbert. *Quantum Reality.*

A. I. M. Ray. *Quantum Mechanics.*

R. Shankar. *Principles of Quantum Mechanics.*

18

Approximation Methods for Stationary States

We have come a long way. We have gotten the basics of quantum mechanics down; we have even calculated a few things that can be compared with experimental data. But we also have pretty much exhausted what can be calculated exactly in quantum mechanics. The next step in dealing with real-world quantum systems is to develop approximation techniques. And it is to these techniques that this and a couple of subsequent chapters of the book will be devoted.

The approximation methods we will develop all use our repertoire of exact solutions—the infinite-potential well, the harmonic oscillator, and the hydrogen atom—but now the Hamiltonian will include interaction terms that cannot be treated exactly, interaction terms that are called a perturbation on the starting exact solution. These perturbations can be imposed externally via an electromagnetic field or they may represent interactions within the system itself. The whole idea works so long as we can treat the perturbation as small, that is, so long as the perturbation is really a perturbation (suspend the question, what is small? for the moment; it will be clear a little later).

There are two kinds of perturbation, time independent and time dependent. In this and chapters 19, 20, and 21, we will deal with time-independent perturbations of Hamiltonians whose stationary eigenstates form a discrete set. Time-dependent perturbations and continuum states will be treated later.

One question is pertinent here. With the advent of computers, we certainly can use numerical methods to calculate some of the properties of real physical systems, and we do. But in spite of that, the perturbation theory developed here retains its usefulness because of a certain physical clarity it gives us. Thus whenever it is applicable, we still use perturbation theory to gain physical insight, if not necessarily very accurate numbers.

Besides perturbation theory, we will develop a second approximation method called the variational method (also called the Rayleigh-Ritz method), which, although often not as reliable as the perturbation theory, has the virtue of being more versatile.

18.1 TIME-INDEPENDENT PERTURBATION THEORY FOR NONDEGENERATE STATES

Suppose we know the eigenvalue spectrum as well as the eigenstates $|\phi_n\rangle$ of what seems from physical considerations the major part of the Hamiltonian of a system, call it H_0:

$$H_0|\phi_n\rangle = E_n^{(0)}|\phi_n\rangle \tag{18.1}$$

where we will assume in this section that the $|\phi_n\rangle$'s are nondegenerate. Call the remaining part of the Hamiltonian, the time-independent perturbation, H_1, and write the total Hamiltonian of the system as

$$H = H_0 + \lambda H_1 \tag{18.2}$$

where we add the parameter λ for convenience. The perturbation parameter λ will be assumed to be between 0 and 1. When λ is 0, the Hamiltonian becomes the unperturbed Hamiltonian where we begin; and at the end of the calculation, we let $\lambda \to 1$, when the Hamiltonian goes over to the Hamiltonian proper for the system under consideration. (H_1 itself, of course, may have a small parameter, which we don't tamper with.)

The perturbation theory we will develop is a way to approximate the eigenvalues and eigenfunctions of H in terms of the eigenvalues and eigenfunctions of H_0 using the matrix elements of the perturbation H_1 in the basis defined by $|\phi_n\rangle$. The desired quantities will find expression as power series expansions in powers of λ. Unfortunately, we cannot prove that such a power series is convergent; the higher-order terms of the expansion get much too complicated to facilitate thorough examination. Consequently, we have to be satisfied with calculating the first few terms of the expansion; our experience is that these first few terms do properly describe physical systems to which we apply our theory. In other words, the theory seems to work; therefore, we will not unduly worry about the convergence of the perturbation series.

Let's denote the eigenstates of the complete Hamiltonian H as $|\psi_n\rangle$ and the eigenvalues as E_n; we have

$$(H_0 + \lambda H_1)|\psi_n\rangle = E_n|\psi_n\rangle \tag{18.3}$$

Our goal is to develop power series expansions of $|\psi_n\rangle$ and E_n. We will assume that there is a one-to-one correspondence between the set of states $|\psi_n\rangle$ and $|\phi_n\rangle$; in other words, as $\lambda \to 0$, $|\psi_n\rangle \to |\phi_n\rangle$ and $E_n \to E_n^{(0)}$.

Since the eigenstates $|\phi_n\rangle$ form a complete set, we can expand $|\psi_n\rangle$ as (while keeping in the spirit of the $\lambda \to 0$ limit above)

$$|\psi_n\rangle = |\phi_n\rangle + \sum_{k \neq n} c_{nk}(\lambda)|\phi_k\rangle \tag{18.4}$$

where the coefficient c_{nk} represents the amount of admixture of the kth unperturbed state in the state $|\psi_n\rangle$. From the $\lambda \to 0$ limit, it follows that $c_{nk}(0)$ must be 0. Let's also choose the phase of $|\psi_n\rangle$ so that

$$c_{nk} = \langle \phi_k | \psi_n \rangle$$

is real and positive. We now can write the power series expansions as

$$c_{nk}(\lambda) = \lambda c_{nk}^{(1)} + \lambda^2 c_{nk}^{(2)} + \cdots \tag{18.5}$$

and

$$E_n = E_n^{(0)} + \lambda E_n^{(1)} + \lambda^2 E_n^{(2)} + \cdots \tag{18.6}$$

Then we substitute these expansions in the Schrödinger equation (18.3). This gives

$$\begin{aligned}(H_0 + \lambda H_1)&\left[|\phi_n\rangle + \sum_{k \neq n} \lambda c_{nk}^{(1)} |\phi_k\rangle + \sum_{k \neq n} \lambda^2 c_{nk}^{(2)} |\phi_k\rangle + \cdots \right] \\ &= (E_n^0 + \lambda E_n^{(1)} + \lambda^2 E_n^{(2)} + \cdots) \\ &\quad \times \left[|\phi_n\rangle + \sum_{k \neq n} \lambda c_{nk}^{(1)} |\phi_k\rangle + \sum_{k \neq n} \lambda^2 c_{nk}^{(2)} |\phi_k\rangle + \cdots \right]\end{aligned} \tag{18.7}$$

In order that equation (18.7) can be satisfied for arbitrary λ, the coefficient of each power of λ on both sides must be equal. This gives us a series of equations representing various orders of perturbation corrections. The very first one of the series is obtained by equating terms proportional to λ:

$$H_0 \sum_{k \neq n} c_{nk}^{(1)} |\phi_k\rangle + H_1 |\phi_n\rangle = E_n^{(0)} \sum_{k \neq n} c_{nk}^{(1)} |\phi_k\rangle + E_n^{(1)} |\phi_n\rangle \tag{18.8}$$

But $H_0|\phi_k\rangle = E_k^{(0)}|\phi_k\rangle$. Substituting into equation (18.8), we get an equation that is a little closer to our desired goal—the first order perturbation correction $E_n^{(1)}$ to the energy:

$$E_n^{(1)}|\phi_n\rangle = H_1|\phi_n\rangle + \sum_{k\neq n}(E_k^{(0)} - E_n^{(0)})c_{nk}^{(1)}|\phi_k\rangle \tag{18.9}$$

All that remains is to multiply both sides with the bra $\langle\phi_n|$ from the left and to use the orthogonality relation of the unperturbed states, $\langle\phi_n|\phi_k\rangle = \delta_{nk}$. We get

$$E_n^{(1)} = \langle\phi_n|H_1|\phi_n\rangle \tag{18.10}$$

The first-order energy correction to any given unperturbed state is just the expectation value of the perturbing interaction potential in the state under consideration; in the language of matrices, it is the diagonal matrix element.

We can squeeze out one more important relationship from equation (18.9) above. Multiply the equation with the bra $\langle\phi_m|$ from the left ($m \neq n$) and use orthogonality of the unperturbed set once again. We obtain

$$\langle\phi_m|H_1|\phi_n\rangle + (E_m^{(0)} - E_n^{(0)})c_{nm}^{(1)} = 0$$

This gives the mixing coefficient $c_{nk}^{(1)}$ in first order:

$$c_{nk}^{(1)} = \frac{\langle\phi_k|H_1|\phi_n\rangle}{E_n^{(0)} - E_k^{(0)}} \tag{18.11}$$

This expression tells us a lot about the nature of the mixing produced by the perturbation. Whenever there is a nonvanishing off-diagonal matrix element that $|\phi_n\rangle$ shares with another basis state $|\phi_k\rangle$, it acquires an admixture of that state via the perturbation. However, the amount of the mixing depends not only on the value of the relevant matrix element, but also crucially on the *energy denominator*. If all the off-diagonal matrix elements of H_1 are more or less equal, the nearby states mix more than faraway ones.

Furthermore, you now can see what "small" means for a perturbation. The smallness of a perturbation is not only about the magnitudes of the matrix elements, but also about how large their magnitudes are relative to the level separations; it is the ratio of the two that must be small.

How about normalization of the state? From equation (18.4), we get

$$\langle\psi_n|\psi_n\rangle = 1 + \lambda^2 \sum_{k\neq n} |c_{nk}^{(1)}|^2 + \cdots$$

We can see that there is no term proportional to λ in this normalization product; to first order, the state is already normalized. Therefore the first-order perturbed eigenstate is given as

$$|\psi_n\rangle = |\phi_n\rangle + \sum_{k\neq n}\frac{\langle\phi_k|H_1|\phi_n\rangle}{E_n^{(0)} - E_k^{(0)}}|\phi_k\rangle \tag{18.12}$$

Let's now consider second-order corrections. Go back to equation (18.7) and equate terms proportional to λ^2. This gives

$$H_0 \sum_{k \neq n} c_{nk}^{(2)} |\phi_k\rangle + H_1 \sum_{k \neq n} c_{nk}^{(1)} |\phi_k\rangle = E_n^{(0)} \sum_{k \neq n} c_{nk}^{(2)} |\phi_k\rangle + E_n^{(1)} \sum_{k \neq n} c_{nk}^{(1)} |\phi_k\rangle + E_n^{(2)} |\phi_n\rangle \tag{18.13}$$

Now take the scalar product with $\langle \phi_n |$ from the left; this gives us the expression for the second-order correction to energy:

$$E_n^{(2)} = \sum_{k \neq n} \langle \phi_n | H_1 | \phi_k \rangle c_{nk}^{(1)}$$

Substituting for $c_{nk}^{(1)}$ from equation (18.11), we obtain

$$E_n^{(2)} = \sum_{k \neq n} \frac{|\langle \phi_k | H_1 | \phi_n \rangle|^2}{E_n^{(0)} - E_k^{(0)}} \tag{18.14}$$

where we have used $\langle \phi_n | H_1 | \phi_k \rangle = \langle \phi_k | H_1 | \phi_n \rangle^*$. The second-order correction to energy due to a perturbation is obtained as a sum of terms whose strengths are given by the square of the off-diagonal matrix elements of the perturbing potential connecting the given state with all the other states of the unperturbed system, weighted down by the reciprocal of the energy difference between the states.

The expression for the second-order perturbation correction to the energy derives special importance because the first-order correction that we calculated above often vanishes. In first order, the perturbation has to be able to connect the state with itself in order to contribute; but this may be precluded because of a symmetry principle operating for the state under consideration. In second order, on the other hand, the perturbation has to connect the given state with some other state and back (fig. 18.1), and therefore, its contribution is no longer inhibited by symmetry. The following additional comments are pertinent:

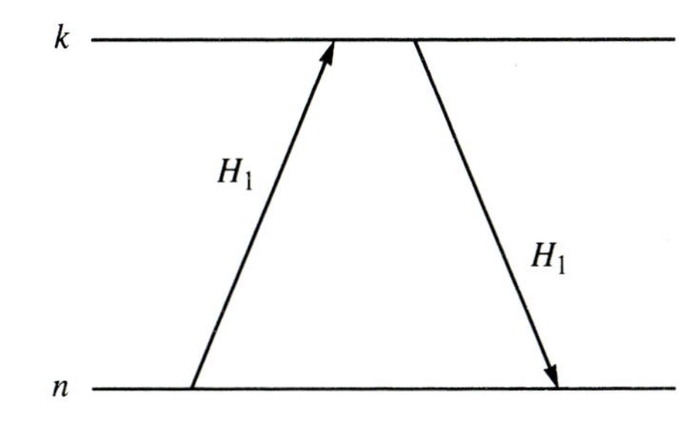

FIGURE 18.1
How the second-order perturbation theory works. There is no inhibition from symmetry since two different states are involved.

1. For the ground state (i.e., $|\phi_n\rangle = |\phi_0\rangle$), the second-order correction is always negative, it always lowers the energy of the ground state.
2. The energy denominator counts; if the off-diagonal matrix elements are roughly equal, the nearby levels will make a bigger contribution to the second-order energy shift than levels far away.
3. There is a tendency of levels to repel one another. If a state $|\phi_k\rangle$ shares a large matrix element of H_1 with the given state $|\phi_n\rangle$ and is energetically close but lies above $E_n^{(0)}$, the second-order energy shift for $|\phi_n\rangle$ is downward; if the level $|\phi_k\rangle$ lies below, the state $|\phi_n\rangle$ is pushed up.

It is not difficult to obtain the second-order mixing coefficients $c_{nk}^{(2)}$ from equation (18.13) by taking a suitable scalar product, but we will not have any occasion to use the second-order wave function in this book; therefore, the second-order correction to the wave functions will be left as an exercise.

Example: The Anharmonic Oscillator

In chapter 7, we commented on how any general potential can be expanded about an equilibrium point and how the first nonvanishing term of the expansion is equivalent to the harmonic oscillator potential. Let's now consider the next higher-order term in such an expanded potential, which is $\sim x^3$. The resulting Hamiltonian of the *anharmonic oscillator* can be written as

$$H = H_0 + H_1 \tag{18.15}$$

where H_0 is the unperturbed oscillator Hamiltonian

$$H_0 = (p^2/2m) + \tfrac{1}{2}m\omega^2 x^2 \tag{7.1}$$

and H_1 is the perturbation

$$H_1 = Cx^3 \tag{18.16}$$

assuming C is small. Let's calculate the perturbation correction to the energy of the ground state.

First we calculate the first-order correction to the ground state energy given by equation (18.10). We have

$$E_0^{(1)} = \langle\phi_0|Cx^3|\phi_0\rangle = \int_{-\infty}^{\infty} dx\, u_0^*(x)Cx^3u_0(x) = 0$$

since the oscillator wave functions u_n are functions of definite parity and the perturbing potential is an odd parity function of x. The integral vanishes because the particle has as much probability to be where the interaction potential is positive ($x > 0$) as where it is negative ($x < 0$), and the total interaction energy just cancels out.

So it is necessary to calculate the second-order energy shift using equation (18.14) where the main task is to calculate the off-diagonal matrix elements $\langle \phi_k | Cx^3 | \phi_0 \rangle$. Fortunately, this can be done easily by using the matrix elements of x, equation (7.44), that we calculated with the help of the creation and annihilation operators a and $a^\dagger$ for the oscillator and by noting that

$$\langle \phi_{n'} | Cx^3 | \phi_n \rangle = C \sum_{m,k} \langle \phi_{n'} | x | \phi_m \rangle \langle \phi_m | x | \phi_k \rangle \langle \phi_k | x | \phi_n \rangle$$

Now substitute the matrix elements of x from equation (7.44). This gives

$$\begin{aligned} \langle \phi_{n'} | Cx^3 | \phi_n \rangle = C(\hbar/2m\omega)^{3/2}[\{(n+1)(n+2)(n+3)\}^{1/2}\delta_{n',n+3} \\ + \{n(n-1)(n-2)\}^{1/2}\delta_{n',n-3} \\ + 3(n+1)(n+1)^{1/2}\delta_{n',n+1} \\ + 3nn^{1/2}\delta_{n',n-1}] \end{aligned} \tag{18.17}$$

For the ground state $|\phi_0\rangle$, we can see that only two states, $|\phi_1\rangle$ and $|\phi_3\rangle$, share matrix elements of the perturbation Cx^3 with $|\phi_0\rangle$; these matrix elements are

$$\langle \phi_1 | H_1 | \phi_0 \rangle = 3C(\hbar/2m\omega)^{3/2}$$

and

$$\langle \phi_3 | H_1 | \phi_0 \rangle = \sqrt{6}C(\hbar/2m\omega)^{3/2}$$

Finally, from equation (18.14), using the now familiar oscillator eigenvalue expression $E_n = (n + \frac{1}{2})\hbar\omega$ in the energy denominator, we get

$$E_0^{(2)} = -11C^2\hbar^2/8m^3\omega^4 \tag{18.18}$$

And the ground-state energy of the anharmonic oscillator up to second order in perturbation theory is given as

$$E_0 = \tfrac{1}{2}\hbar\omega - 11C^2\hbar^2/8m^3\omega^4 \tag{18.19}$$

Stark Effect in Hydrogen

It was Stark who first studied the effect of an external electric field on atomic states, hence the name Stark effect. The unperturbed Hamiltonian for the hydrogen atom is given by

$$H_0 = \frac{p^2}{2\mu} - \frac{e^2}{r}$$

with eigenfunctions $\phi_{nlm}(\mathbf{r}) = \langle \mathbf{r} | \phi_{nlm} \rangle$ given by equation (13.20). The perturbing potential for the Stark effect is

$$H_1 = e\boldsymbol{\mathcal{E}} \cdot \mathbf{r} = e\mathcal{E}z \tag{18.20}$$

choosing the z-axis along the electric field. Note that $e\mathcal{E}$ plays the role of a small perturbation parameter here.

The nondegenerate perturbation theory that we have developed can be applied only to the ground state, since all the excited states are degenerate. The first-order perturbation correction is found to be

$$E_{100}^{(1)} = e\mathcal{E}\langle \phi_{100} | z | \phi_{100} \rangle = e\mathcal{E} \int |\phi_{100}|^2 z \, d^3r$$

and this vanishes because the state $|\phi_{100}\rangle$ is a state of definite parity. Therefore, for the ground state, there is no energy shift proportional to the electric field. Classically, if there is a permanent electric dipole moment $\mathbf{D}$, then an electric field always leads to an energy shift of $-\mathbf{D} \cdot \boldsymbol{\mathcal{E}}$. The absence of such an energy shift means that if the ground state of an atom is nondegenerate, the atom can have no permanent electric dipole moment. Of course, the situation is different when there is a degeneracy and the ground state is a linear combination of even and odd parity states. However, actual systems never quite achieve such a degeneracy, and the situation that prevails is that of the ammonia molecule. For that case, we already saw, in chapter 16, that for strong fields, the energy shift indeed is close to being proportional to the electric field, and so we may as well agree that the molecule has a permanent electric dipole moment (see later).

So we have to go to second order to find any Stark effect on the energy. The second-order energy shift is

$$E_{100}^{(2)} = e^2\mathcal{E}^2 \sum_{nlm \neq 100} \frac{|\langle \phi_{nlm} | z | \phi_{100} \rangle|^2}{E_{100}^{(0)} - E_{nlm}^{(0)}} \tag{18.21}$$

The matrix element in equation (18.21) is given as

$$\begin{aligned}
\langle \phi_{nlm} | z | \phi_{100} \rangle &= \int d^3r \, R_{nl}(r) Y_{lm}^*(\theta\phi) r \cos\theta R_{10}(r) Y_{00}(\theta\phi) \\
&= \int r^2 \, dr \, R_{nl}(r) r R_{10}(r) \\
&\quad \times \int d\Omega \, Y_{lm}^*(\theta\phi)(4\pi/3)^{1/2} Y_{10}(\theta\phi)(1/4\pi)^{1/2} \\
&= (1/\sqrt{3})\delta_{l1}\delta_{m0} \int r^2 \, dr \, R_{nl}(r) r R_{10}(r)
\end{aligned} \tag{18.22}$$

where we have used the relations $\cos\theta = (4\pi/3)^{1/2} Y_{10}$ and $Y_{00} = (1/4\pi)^{1/2}$, and also have utilized the orthogonality of the spherical harmonics to perform the angular integral.

Behold! In order for a state to contribute to the energy shift of another state via the electric dipole interaction, the magnetic quantum number m of the two states must be the same; that is, the perturbation operates with a *selection rule*

$$\Delta m = 0 \tag{18.23}$$

The selection rule is a consequence of the fact that the perturbing interaction commutes with L_z,

$$[H_1, L_z] = 0 \tag{18.24}$$

In this way we see that the calculation of the second-order Stark energy shift boils down to the evaluation of a radial integral, which can be done numerically for each n. The result can be substituted into equation (18.21), giving us the second-order energy shift of the ground state. In principle, even the continuum states can make a contribution (remember, the hydrogenic bound states alone do not a complete set make!). Fortunately, an order of magnitude estimate can be found just by calculating the first term.

Electric Polarizability

To first order in the perturbing electric field, the hydrogen ground state is given as

$$|\psi_0\rangle = |\phi_{100}\rangle + e\mathcal{E} \sum_{nlm \neq 100} \frac{\langle \phi_{nlm}|z|\phi_{100}\rangle}{E_{100} - E_{nlm}} |\phi_{nlm}\rangle \tag{18.25}$$

The expectation value of the z-component of the electric dipole moment operator, $-ez$, in this state is given as

$$D = -e \int d^3r\, \phi^*_{100} z \phi_{100} + 2e^2\mathcal{E} \sum_{nlm \neq 100} \frac{|\langle \phi_{nlm}|z|\phi_{100}\rangle|^2}{E_{nlm} - E_{100}} \tag{18.26}$$

The first term on the right-hand side is zero. Therefore, the electric polarizability, defined as

$$\alpha = D/\mathcal{E} \tag{18.27}$$

is given by

$$\alpha = 2e^2 \sum_{nlm \neq 100} \frac{|\langle \phi_{nlm}|z|\phi_{100}\rangle|^2}{E_{nlm} - E_{100}} \tag{18.28}$$

The same expression obtains from the energy calculation if we equate the second-order energy shift calculated above to $-\frac{1}{2}\alpha\mathcal{E}^2$.

The sum in equation (18.28) goes over all states including the continuum states, and consequently a large amount of computational effort is necessary for its evaluation. However, an upper bound can be calculated using the so-called *Unsold closure principle.* We simply replace all the energy denominators by their lowest possible value, namely, the energy difference ΔE between the first excited state and the ground state; this enables us to remove the energy denominators out of the sum, which now can be evaluated using completeness of the set, also called closure. We have

$$\alpha = \frac{2e^2}{\Delta E} \sum_{nlm \neq 100} |\langle \phi_{nlm} | z | \phi_{100} \rangle|^2$$

$$= \frac{2e^2}{\Delta E} \left[\sum_{nlm} |\langle \phi_{nlm} | z | \phi_{100} \rangle|^2 - |\langle \phi_{100} | z | \phi_{100} \rangle|^2 \right]$$

The second term in the bracket is zero; in the first term we can use completeness—ket-bra sum equals one; this gives

$$\sum \langle \phi_{100} | z | \phi_{nlm} \rangle \langle \phi_{nlm} | z | \phi_{100} \rangle$$

$$= \langle \phi_{100} | z^2 | \phi_{100} \rangle = \int d^3r \, \phi_{100}^* z^2 \phi_{100}$$

$$= \frac{1}{\pi a_0^3} \int_0^\infty \int_0^\pi \int_0^{2\pi} \exp(-2r/a_0) r^4 \cos^2\theta \sin\theta \, dr \, d\theta \, d\phi = a_0^2$$

which also follows from the spherical symmetry of the ground state (how so?). The energy difference between the ground and first excited states for hydrogen is given as

$$\Delta E = 3e^2/8a_0$$

Therefore the upper limit for the electric polarizability α for hydrogen is given as

$$\alpha \leq 2e^2 a_0^2/\Delta E = 16a_0^3/3 \tag{18.29}$$

18.2 DEGENERATE PERTURBATION THEORY

If there is degeneracy (or even near degeneracy), the energy denominators of the perturbation expansions developed above blow up, signifying that the basis we have chosen for our zero-order Hamiltonian is not the right basis. The question is, can we find a right basis?

You can see the solution when you think about it. Notice that the energy denominators always arise in combination with the matrix element: What we have is

$$\frac{\langle \phi_k | H_1 | \phi_n \rangle}{E_n^{(0)} - E_k^{(0)}}$$

and the denominator blows up when there are degenerate states for which there is a nonzero off-diagonal matrix element. If there is no off-diagonal matrix element, there is no problem with degeneracy or nearby states.

So the trick is to redefine the basis such that there are no off-diagonal matrix elements of the perturbed Hamiltonian between degenerate states. In other words, diagonalize the submatrix of the full-energy matrix that belongs to the space of degenerate (or nearly degenerate) states. If $|\phi_k\rangle$ and $|\phi_n\rangle$ are degenerate, this means that we diagonalize the submatrix

$$\begin{pmatrix} \langle \phi_n | H_0 + H_1 | \phi_n \rangle & \langle \phi_n | H_1 | \phi_k \rangle \\ \langle \phi_k | H_1 | \phi_n \rangle & \langle \phi_k | H_0 + H_1 | \phi_k \rangle \end{pmatrix} \tag{18.30}$$

following the method outlined in chapter 15 (solve the secular equation, etc.) and use the basis in which the submatrix is diagonal as our new perturbation basis.

Notice the saving grace, that we need to diagonalize only the submatrix that belongs to the degenerate space, not the energy matrix in the entire Hilbert space. So we have to work a little harder than for the nondegenerate case, but it is nothing intractable, such as an infinite dimensional energy matrix!

Needless to mention, after we have redefined the basis and removed the degeneracy, the matrix element of any additional perturbation will have to be expressed in terms of the new basis before using the nondegenerate perturbation theory above. An example will clarify all these procedural matters.

The Ammonia Molecule in an Electric Field by Perturbation Theory

As our example, we will treat the ammonia molecule in the two-state model by means of perturbation theory. As we saw in chapter 16, this problem can be solved exactly; therefore, the advantage of taking this as our pedagogical example is that we will be able to evaluate the validity of the perturbation theory against the exact solution.

Remember, there are two degenerate states of energy E_0 (say), which we denote as $|1\rangle$ and $|2\rangle$, that differ by the relative position of the nitrogen atom with respect to the plane of the three H's (fig. 16.1). Then there are the tunneling interaction A and the interaction of the ammonia's electric dipole (fig. 16.3) with the external electric field ($+D\mathcal{E}$ for state $|1\rangle$ and $-D\mathcal{E}$ for state $|2\rangle$). To start, we have to redefine the basis in some fashion to remove the degeneracy. But it all depends on what the relative magnitude of the dipole interaction term is compared to A.

The Weak-Field Case

When the external electric field is weak so that $D\mathcal{E} \ll A$, we remove the degeneracy by diagonalizing A in the degenerate bases $|1\rangle$ and $|2\rangle$ and treat $D\mathcal{E}$ as perturbation in the redefined basis. Therefore, the matrix to be diagonalized is

$$\begin{array}{c|cc} & |1\rangle & |2\rangle \\ \hline |1\rangle & E_0 & -A \\ |2\rangle & -A & E_0 \end{array} \tag{18.31}$$

and the perturbation H_1 is (in the same basis)

$$\begin{pmatrix} D\mathcal{E} & 0 \\ 0 & -D\mathcal{E} \end{pmatrix} \tag{18.32}$$

We have already diagonalized the matrix (18.31) (see chapter 16), which gives the redefined unperturbed basis as $|I\rangle$ and $|II\rangle$:

$$|I\rangle = (1/\sqrt{2})\,[-|1\rangle + |2\rangle]$$

$$|II\rangle = (1/\sqrt{2})\,[|1\rangle + |2\rangle] \tag{18.33}$$

The unperturbed Hamiltonian H_0' in the new basis is diagonal and is

$$\begin{array}{c|cc} & |I\rangle & |II\rangle \\ \hline |I\rangle & E_0 + A & 0 \\ |II\rangle & 0 & E_0 - A \end{array}$$

Our next task is to express the matrix of the perturbation, equation (18.32), in the new basis. From equation (16.10), the transformation matrix between the two bases is

$$U = \begin{pmatrix} -1/\sqrt{2} & 1/\sqrt{2} \\ 1/\sqrt{2} & 1/\sqrt{2} \end{pmatrix}$$

Consequently, the perturbation H' in the new basis is

$$\begin{aligned} H' &= UH_1U^{-1} = UH_1U^{\dagger} \\ &= \begin{pmatrix} -1/\sqrt{2} & 1/\sqrt{2} \\ 1/\sqrt{2} & 1/\sqrt{2} \end{pmatrix} \begin{pmatrix} D\mathcal{E} & 0 \\ 0 & -D\mathcal{E} \end{pmatrix} \begin{pmatrix} -1/\sqrt{2} & 1/\sqrt{2} \\ 1/\sqrt{2} & 1/\sqrt{2} \end{pmatrix} \\ &= \begin{pmatrix} 0 & -D\mathcal{E} \\ -D\mathcal{E} & 0 \end{pmatrix} \end{aligned} \tag{18.34}$$

Now we can use nondegenerate perturbation theory with H' as the perturbation. Clearly, then, the first-order perturbation energy shift in the new basis, which is given by the expectation value or the diagonal matrix element above, vanishes. The second-order correction is

$$E_I^{(2)} = \sum_{m \neq I} \frac{\langle m|H'|I\rangle\langle I|H'|m\rangle}{E_I^{(0)} - E_m^{(0)}}$$

$$= \frac{\langle II|H'|I\rangle\langle I|H'|II\rangle}{(E_0 + A) - (E_0 - A)} = \frac{D^2\mathcal{E}^2}{2A} \tag{18.35}$$

You can calculate $E_{II}^{(2)}$ in the same manner. It is found to be

$$E_{II}^{(2)} = -D^2\mathcal{E}^2/2A \tag{18.36}$$

Thus up to the second order, the energies of the stationary states of the perturbed system are

$$E_I = E_0 + A + D^2\mathcal{E}^2/2A$$

$$E_{II} = E_0 - A - D^2\mathcal{E}^2/2A \tag{18.37}$$

which is in complete agreement with the exact solution, equation (16.20), if we expand the latter appropriately for small $\mathcal{E}$.

The Strong-Field Case

In the strong-field case, $D\mathcal{E} \gg A$; consequently, we must treat tunneling as the perturbation and include the dipole interaction in the diagonal unperturbed energy matrix H_0 (the dipole interaction removes the degeneracy):

	$\|1\rangle$	$\|2\rangle$
$\|1\rangle$	$E_0 + D\mathcal{E}$	0
$\|2\rangle$	0	$E_0 - D\mathcal{E}$

Therefore, this is now a case of nondegenerate perturbation theory. The matrix of the perturbation H_1 is

	$\|1\rangle$	$\|2\rangle$
$\|1\rangle$	0	$-A$
$\|2\rangle$	$-A$	0

The perturbation has no diagonal matrix element, therefore, there is no first-order energy shift due to the perturbation. The second-order energy shift for the state $|1\rangle$ is

$$E_1^{(2)} = \sum_{m \neq 1} \frac{\langle m|H_1|1\rangle\langle 1|H_1|m\rangle}{E_1^{(0)} - E_m^{(0)}} = \frac{A^2}{2D\mathcal{E}} \tag{18.38}$$

and similarly, for the state $|2\rangle$, we get

$$E_2^{(2)} = -A^2/2D\mathcal{E} \tag{18.39}$$

Therefore, the energy eigenvalues for the strong-field case up to second order are

$$E_1 = E_0 + D\mathcal{E} + A^2/2D\mathcal{E}$$

$$E_2 = E_0 - D\mathcal{E} - A^2/2D\mathcal{E} \tag{18.40}$$

which are also in agreement with the exact solution in the appropriate limit. Notice also that as $D\mathcal{E}$ increases, becoming $\gg A$, the energy shift is indeed proportional to the applied field, and in this sense we can say that the molecule has a permanent dipole moment D.

18.3 THE VARIATIONAL METHOD

What if we have a situation where a physical system has a Hamiltonian that is amenable neither to exact solution nor to a perturbation treatment? Under such circumstances, we have a technique called the variational method (also called the Rayleigh-Ritz method) that comes in handy.

The variational method is based on the simple fact that the average energy of a system, that is, the expectation value of its Hamiltonian H, in a state represented by an arbitrary wave function, has got to be greater than or equal to the ground-state energy of the system. Since the ground-state energy acts as a lower bound on an expectation value calculated with any arbitrary trial wave function, we can choose a trial function with a number of variational parameters and then minimize the expectation value of H as a function of these parameters. Now you can see why this method is called the variational method.

Let's prove the contention above. Any arbitrary trial function ψ can, of course, be expanded in terms of the complete set of eigenfunctions u_n of H belonging to eigenvalues E_n:

$$\psi = \sum_n a_n u_n$$

Or, using Dirac notation,

$$|\psi\rangle = \sum_n a_n |u_n\rangle \tag{18.41}$$

The expectation value of H in the state is given as (noting that $|\psi\rangle$ may not be normalized):

$$\langle H\rangle = \frac{\langle\psi|H|\psi\rangle}{\langle\psi|\psi\rangle} = \frac{\sum_n |a_n|^2 E_n}{\sum_n |a_n|^2} \tag{18.42}$$

Now replace all the energies E_n by the lowest eigenvalue of H, the ground state energy E_0; obviously, we get an inequality

$$\langle H\rangle \geq \frac{E_0 \sum_n |a_n|^2}{\sum_n |a_n|^2} = E_0 \tag{18.43}$$

Therefore, the name of the game in the variational method is to (1) choose judiciously a trial wave function that contains a whole bunch of free parameters λ_i:

$$\psi = \psi(\lambda_1, \lambda_2, \ldots) \tag{18.44}$$

and (2) vary the parameters λ_i until the expectation value of H is minimized:

$$\partial\langle H\rangle/\partial\lambda_i = 0 \tag{18.45}$$

Before we even illustrate the method, let's ask what kind of success we expect of such a simple technique. The key thing to note is that the value of $\langle H\rangle$ is roughly proportional to the square of the expansion coefficients, $|a_n|^2$; this means that if our guess for the ground state is contaminated, say by 10%, with an excited-state wave function, the correction to the energy is only $(0.1)^2$ or 1%. Therefore, we expect to do very well on the estimate of energy by this method, although perhaps not so well in determining the wave functions.

Example: The Anharmonic Oscillator by the Variational Method

Let's apply the variational method to the anharmonic oscillator Hamiltonian

$$H = -\frac{\hbar^2}{2m}\frac{d^2}{dx^2} + Cx^4 \tag{18.46}$$

for which no exact analytical solution exists and for which there is no obvious split of the Hamiltonian into an unperturbed part and a perturbation that may work. On the other hand, we can readily intuit what would be a good trial function for the variational method—the harmonic oscillator eigenfunction corresponding to the ground state suitably parameterized. Thus, we choose for the trial function

$$\psi = \lambda^{1/2}\pi^{-1/4}\exp(-\lambda^2 x^2/2) \tag{18.47}$$

which is already normalized (i.e., $\langle\psi|\psi\rangle = 1$). Accordingly, we need only to evaluate

$$\begin{aligned}\langle H\rangle = \langle\psi|H|\psi\rangle &= \lambda\pi^{-1/2}\int_{-\infty}^{\infty} dx\, e^{-\lambda^2x^2/2}\left(-\frac{\hbar^2}{2m}\frac{d^2}{dx^2} + Cx^4\right)e^{-\lambda^2x^2/2} \\ &= \lambda\pi^{-1/2}\int_{-\infty}^{\infty} dx\left[-\frac{\hbar^2}{2m}(\lambda^4x^2 - \lambda^2) + Cx^4\right]e^{-\lambda^2x^2}\end{aligned}$$

which we will have to minimize with respect to λ. The integral can be evaluated with the help of the formulas

$$\begin{aligned}\int_{-\infty}^{\infty} dx\, x^{2n}e^{-\lambda^2x^2} &= \frac{\pi^{1/2}}{\lambda}, \quad \text{for } n = 0 \\ &= \frac{1\cdot 3\cdot 5\cdots(2n-1)\pi^{1/2}}{2^n\lambda^{2n+1}}, \quad \text{for } n = 1, 2, \ldots\end{aligned} \tag{18.48}$$

We get

$$\langle H\rangle = \frac{\hbar^2\lambda^2}{4m} + \frac{3C}{4\lambda^4} \tag{18.49}$$

Putting $\partial\langle H\rangle/\partial\lambda = 0$ gives the value of λ that makes $\langle H\rangle$ a minimum as

$$\lambda = (6mC/\hbar^2)^{1/6}$$

Substituting this value into equation (18.49), we get the desired value of the ground-state energy:

$$\langle H\rangle = 1.082\left(\frac{\hbar^2}{2m}\right)^{2/3} C^{1/3} \tag{18.50}$$

This compares very well with the value 1.060 obtained for the coefficient via numerical integration, showing the appropriateness of the variational technique for energy calculations.

How do we generate better estimates of the ground-state energy with the variational method? You already know—by including more variational parameters. In the case above, an obvious idea to improve the calculation is to try the linear combination of the $n = 0$ and $n = 2$ oscillator functions ($n = 1$ doesn't mix because of parity) and to use the coefficients of the linear combination as well as λ as variational parameters. This will be left as an exercise.

Can we use the variational technique for calculating excited states? For example, how do we apply the variational method for calculating the first and second excited states of the Cx^4 potential above? Since H is invariant under parity, the states are alternately of even and odd parity. For the odd-parity first excited state, we take as our trial function a linear combination of odd-parity oscillator states.

What do we do for the second excited state of the Cx^4 Hamiltonian? Here parity alone does not help. We must somehow use a trial function whose expansion does not contain the first two states of the Hamiltonian. So after we have calculated the first two states, we can use for the second excited state a trial function that is orthogonal to both of these, and so on for higher excited states.

For three-dimensional problems, so long as there is spherical symmetry, for the ground state at least, the variational calculation is quite similar to the above one-dimensional case. Since $l = 0$, we just choose a suitable spherically symmetric radial function as the trial function. For nonzero l, the trial function must have both a radial part and an angular part, which is Y_{lm}. In addition, we pay attention to the number of radial nodes, and so forth.

Most of the examples of the application of approximation techniques so far have been to pedagogical problems. Starting with chapter 19, however, we will consider a series of practical applications to realistic quantum systems—atoms, molecules, and the like. Finally, we really are ready for the real-world application of quantum mechanics.

PROBLEMS

1. Calculate the first-order correction to the one-dimensional harmonic oscillator ground state due to the anharmonic perturbation Cx^3.
2. Calculate the second-order energy shift of the first two excited states of the one-dimensional harmonic oscillator due to the anharmonic perturbation Cx^3. Also calculate the first-order correction to the wave functions of these states.
3. Use second-order perturbation theory to calculate the Stark effect on the energy of a charged linear harmonic oscillator for which the perturbation is $q\mathcal{E}x$, where $\mathcal{E}$ is the electric field and q is the charge. Also find the exact solution and compare your perturbation result with the exact solution.
4. Find the first- and second-order corrections to the energy of the first three states of the linear harmonic oscillator due to the anharmonic perturbation Cx^4.
5. A particle of mass m is in an infinite potential well perturbed as shown in figure 18.2. (a) Calculate the first-order energy shift of the nth eigenvalue

FIGURE 18.2

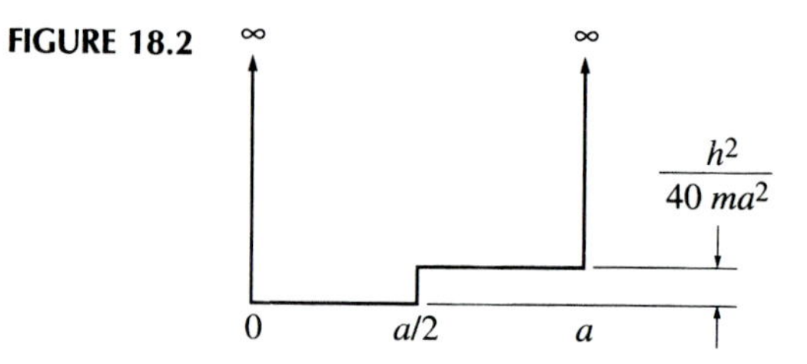

due to the perturbation. (b) Write out the first three nonvanishing terms for the perturbation expansion of the ground state in terms of the unperturbed eigenfunctions of the infinite well. (c) Calculate the second-order energy shift for the ground state.

6. Derive the equation for the second-order perturbation correction to the state vector in nondegenerate perturbation theory. Discuss the normalization of the perturbed state in this order.

7. Given the matrix for H_0

$$\begin{pmatrix} 2 & 0 & 0 \\ 0 & 2 & 0 \\ 0 & 0 & 4 \end{pmatrix}$$

and for the perturbation H_1 as

$$\begin{pmatrix} 0 & 1 & 0 \\ 1 & 0 & 1 \\ 0 & 1 & 0 \end{pmatrix}$$

in the orthonormal basis $|\phi_1\rangle$, $|\phi_2\rangle$, and $|\phi_3\rangle$, determine the energy eigenvalues correct to second order in the perturbation.

8. Consider the so-called *spin Hamiltonian*:

$$H = AS_z^2 + B(S_x^2 - S_y^2), \qquad |B| \ll |A|$$

for a system of spin 1. (Such a Hamiltonian obtains for a spin-1 ion located in a crystal with rhombic symmetry.) Show that the Hamiltonian in the S_z-basis is

$$\hbar^2 \begin{pmatrix} A & 0 & B \\ 0 & 0 & 0 \\ B & 0 & A \end{pmatrix}$$

Find the eigenvalues of this Hamiltonian using degenerate perturbation theory.

9. Calculate the linear Stark effect (that is, first-order energy shift due to the perturbation $e\mathcal{E}z$, where $\mathcal{E}$ is the electric field) for the splitting of the $n = 2$ states of hydrogen. Use the unperturbed eigenfunctions given in chapter 13. Why is angular momentum no longer conserved?

10. Consider the perturbed two-dimensional oscillator Hamiltonian

$$H = \frac{p_x^2 + p_y^2}{2m} + \tfrac{1}{2}k(x^2 + y^2) + \lambda xy$$

Use perturbation theory to calculate the energy shift of the degenerate first excited state to first order due to the perturbation λxy. What are the first-order wave functions?

11. Consider the anharmonic oscillator Hamiltonian

$$H = \frac{p^2}{2m} + \tfrac{1}{2}kx^2 + Cx^4$$

Use the variational method to calculate the ground state using the trial function $\psi = au_0 + bu_2$, where u_0 and u_2 are the linear harmonic oscillator functions corresponding to the ground and second excited states, $n = 0$ and $n = 2$, respectively. Choose b/a and the oscillator frequency ω as your variational parameters. Do not forget normalization.

12. Using the variational method, calculate an approximate expression for the binding energy of the deuteron ground state assuming that the neutron-proton interaction is given by the so-called Yukawa potential

$$V(r) = V_0 \frac{e^{-r/a}}{r/a}$$

where V_0 is the strength and a (= 1.4 F) is the range of the potential. Assume the ground state to be an S-state. Use the variational wave function $R(r) = N\exp(-\alpha r)$, where N is a normalization constant and α the variational parameter. Hence determine the numerical value of V_0 from the observed value of the deuteron binding energy of 2.23 MeV. (Assume $m_n c^2 = m_p c^2 = 938$ MeV; also do not forget reduced mass!)

13. Use the variational method to estimate the ground-state energy of the three-dimensional harmonic oscillator. Take $l = 0$ and use the trial function

$$R(r) = N\exp(-\alpha r)$$

as in the problem above.

ADDITIONAL PROBLEMS

A1. Calculate the first-order energy shift of the first three states of the particle-in-a-one-dimensional box problem (walls at $x = 0$ and $x = a$) due to the perturbing potential (a) $V_1 x/a$ and (b) $V_1 (x/a)^2$, where V_1 is a constant. What is the condition under which the perturbation calculation is valid?

A2. Consider a particle in a two-dimensional box (walls at $x = 0$, a and $y = 0$, b) under the perturbing influence of the potential $V_1 \sin(\pi x/a)$. Calculate the first-order energy shift for an arbitrary eigenstate.

A3. Show how the doubly degenerate eigenenergies of a particle in a two-dimensional box of equal dimensions (faces at $x = 0$, a; $y = 0$, a) separate under the perturbation $V_1 \sin(\pi x/a)$.

A4. Consider the $n = 3$, $l = 2$ state of a hydrogen atom perturbed by a crystalline field

$$V' = C(L_x^2 + L_y^2 - 2L_z^2)$$

where C is a constant. Determine the splitting of the level due to V'. Neglect spin and neglect the interaction with all other states.

A5. Use the variational method for solving the Schrödinger equation for the truncated harmonic oscillator potential

$$\begin{aligned} V(x) &= \tfrac{1}{2} kx^2 \quad \text{for } x > 0 \\ &= \infty \qquad\ \ \text{for } x < 0 \end{aligned}$$

Use the trial function $\psi = x \exp(-bx)$ (b is a variational parameter) to calculate an approximate value for the ground state energy and compare with the exact result (cf. problem 7.5).

REFERENCES

There are no special references to this chapter. Most books on quantum mechanics treat this subject well.

19

Quantum Systems: Atoms with One and Two Electrons

What is the difference between the real hydrogen atom and the hydrogen atom that we solved in chapter 13? Well, that's not too hard to answer—we left out the spin in our earlier treatment. In the real hydrogen atom, the spinning electron does a twist, like Chubby Checker, a tangle with its own orbital motion, that is called the spin-orbit coupling; this leads to a structure in the energy spectrum, known as the fine structure, that breaks up the l-degeneracy that quantum mechanics predicts for the "unreal" hydrogen atom. The spin-orbit coupling is not the only effect that contributes to the *fine structure*; an additional relativistic effect coming from the relativistic increase of mass also plays a role. A much smaller effect due to the coupling of the electron's spin with the nuclear spin produces an additional *hyperfine structure* in the spectrum, but we will not get into that much detail; also beyond the scope of the book is what is called a self-energy effect of the interaction of the electron with its own electromagnetic field.

When we treat the hydrogen atom with both spatial and spin degrees of freedom included in our bookkeeping of the states, the effect of an external magnetic field on the atom manifests itself with a raging complexity that is quite different from the normal Zeeman effect we encountered earlier. This is the anomalous Zeeman effect, and now we can give an explanation for it.

In addition to a realistic accounting for the degrees of freedom of the electron in the hydrogen atom, in this chapter we will treat the helium atom, which is a three-body system; furthermore, two of the bodies, the electrons, are identical particles, and therefore, the Pauli principle plays an interesting new role here. This is the tease about quantum systems: There is always something new, they keep on being intriguing.

19.1 THE FINE STRUCTURE OF HYDROGEN

Let's first examine the origin of spin-orbit coupling. Think classically for a moment. When we say there is only the Coulomb force between the electron and the proton, we are assuming that the proton is static, as viewed from the electron's rest system. But in actuality, the electron moves around the proton; therefore, if you were to sit and watch the proton from the electron's rest system, the proton would be seen as a whirling charge circling the electron, a current loop. A current loop produces a magnetic field at the site of the electron to which the electron responds because of its intrinsic magnetic moment. This is the general idea behind the spin-orbit interaction.

To derive a formula, assume that the electron's velocity is $\mathbf{v}$; then the proton's velocity from the electron's rest-frame point of view is $-\mathbf{v}$. Also, if the radius vector is $\mathbf{r}$ from the proton to the electron, from the electron to the proton it is $-\mathbf{r}$. Now use the Biot-Savart law to write down the magnetic field of the proton at the site of the electron:

$$\mathbf{B} = \frac{e}{c}\frac{\mathbf{r}\times\mathbf{v}}{r^3} \equiv \frac{e}{mc}\frac{\mathbf{L}}{r^3} \tag{19.1}$$

where $\mathbf{L}$ is the orbital angular momentum of the electron. The interaction of this magnetic field with the electron's intrinsic magnetic moment, $\mathbf{M} = -(e/mc)\mathbf{S}$, where $\mathbf{S}$ is the electron's spin vector, leads to the spin-orbit coupling interaction $H_{\text{s.o.}}$:

$$\begin{aligned} H_{\text{s.o.}} &= -\mathbf{M}\cdot\mathbf{B} = -(-e/mc)(e/mc)\mathbf{S}\cdot\mathbf{L}/r^3 \\ &= (e^2/m^2c^2r^3)\mathbf{S}\cdot\mathbf{L} \end{aligned} \tag{19.2}$$

However, the kinematics used above is nonrelativistic. Relativistically, the electron also precesses about the nucleus (this is called the *Thomas precession*) with a certain frequency. The net upshot of this precession is that the magnetic field "seen" by the electron is only half as large as the one assumed in the derivation of equation (19.2), and therefore the spin-orbit coupling term is

$$H_{\text{s.o.}} = (e^2/2m^2c^2r^3)\mathbf{S}\cdot\mathbf{L} \tag{19.3}$$

(If you are a little uneasy about all these classical arguments to settle what the form is for a purely quantum mechanical interaction term in the Hamiltonian, relax. A more rigorous derivation of eq. [19.3] can be given from Dirac's relativistic equation for the electron.)

Before we calculate anything, let's consider the relativistic mass correction to the Hamiltonian. In chapter 13, we took the kinetic energy as $p^2/2m$ for both the proton and the electron. Instead, we should use the relativistic expression for the energy of the electron,

$$E = (p^2c^2 + m^2c^4)^{1/2}$$

For the kinetic energy (ignoring any reduced-mass consideration), this gives

$$(p^2c^2 + m^2c^4)^{1/2} - mc^2 = \frac{p^2}{2m} - \frac{1}{8}\frac{p^4}{m^3c^2}$$

We get an extra term in the Hamiltonian, call it H_{kin}:

$$H_{\text{kin}} = -p^4/8m^3c^2 \tag{19.4}$$

Calculation of the Spin-Orbit Interaction

When we try to calculate the energy shift of hydrogenic states due to the perturbation given by the spin-orbit interaction, equation (19.2), we find that it is an example of degenerate perturbation theory if we keep to the $|nlm_lm_s\rangle$ basis. And since for each nl, there are $2(2l+1)$ degenerate states, the energy matrices we have to diagonalize are fairly large for large n. But all that work can be avoided by realizing that the $\mathbf{L}\cdot\mathbf{S}$ interaction can be written as

$$H_{\text{s.o.}} = (e^2/4m^2c^2r^3)(J^2 - L^2 - S^2) \tag{19.5}$$

and this is already diagonal in the coupled $|nljm\rangle$ representation that we obtained in chapter 17. In that basis, we have

$$\begin{aligned}\langle nl'j'm'|H_{\text{s.o.}}|nljm\rangle &= \delta_{ll'}\delta_{mm'}\delta_{jj'}(e^2/4m^2c^2)\langle 1/r^3\rangle_{nl} \\ &\times \hbar^2[j(j+1) - l(l+1) - \tfrac{1}{2}(\tfrac{1}{2}+1)]\end{aligned} \tag{19.6}$$

which is the first-order perturbation energy shift; and thus for each n, the l-degeneracy breaks up, but in a particular way. For each l, j can be either $l+\frac{1}{2}$ or $l-\frac{1}{2}$. Also, the value of the radial integral $\langle 1/r^3\rangle_{nl}$ can be calculated using Kramer's rule, equation (13.26) with help from equation (13.25), as

$$\langle 1/r^3\rangle_{nl} = \int_0^\infty dr\, r^2[R_{nl}(r)]^2\frac{1}{r^3} = \frac{1}{a_0^3n^3l(l+\frac{1}{2})(l+1)} \tag{19.7}$$

Combining equations (19.6) and (19.7) and substituting $j = l \pm \frac{1}{2}$, we finally obtain the first-order energy shift due to the spin-orbit interaction:

$$E^{(1)}_{\text{s.o.}} = \tfrac{1}{4}mc^2\alpha^4\frac{1}{n^3l(l+\frac{1}{2})(l+1)}\begin{bmatrix} l \\ -l-1\end{bmatrix} \tag{19.8}$$

where the upper and lower values in the bracket correspond to $j = l + \frac{1}{2}$ and $j = l - \frac{1}{2}$, respectively, and where α is the fine structure constant $e^2/\hbar c$. The spin-orbit interaction energy $\sim\alpha^4$, whereas the separation between the unperturbed hydrogen states goes as $\sim\alpha^2$. Thus the spin-orbit splitting of levels is *very* small compared to, say, the energy difference between $1S$ and $2P$ states.

The result above holds true even for the limit $l = 0$, even though for this case $\langle 1/r^3 \rangle \to \infty$ and $\langle \mathbf{L}\cdot\mathbf{S} \rangle = 0$, and we might be getting an indeterminate result. But formula (19.8) checks out when we use Dirac's relativistic theory. Therefore, we may use equation (19.8) for the first-order energy shift of all hydrogenic states due to the spin-orbit interaction, including $l = 0$ states.

Calculation of Energy Shift Due to H_{kin}

For calculating the energy shift due to H_{kin}, we note that for the hydrogen atom, since H_0 is given as

$$H_0 = (p^2/2m) - e^2/r \tag{19.9}$$

we can write $p^2/2m$ as

$$p^2/2m = H_0 + e^2/r \tag{19.10}$$

Substituting into equation (19.4) for H_{kin} we get

$$H_{\text{kin}} = -(1/2mc^2)(p^2/2m)^2 = -(1/2mc^2)(H_0 + e^2/r)^2 \tag{19.11}$$

The first-order energy shift of the state $|nlm\rangle$ due to the relativistic mass correction is then given as

$$\begin{aligned} E_{\text{kin}}^{(1)} &= -\frac{1}{2mc^2}\langle nlm|\left(H_0 + \frac{e^2}{r}\right)\left(H_0 + \frac{e^2}{r}\right)|nlm\rangle \\ &= -\frac{1}{2mc^2}\left[E_n^2 + 2E_n e^2\left\langle\frac{1}{r}\right\rangle_{nl} + e^4\left\langle\frac{1}{r^2}\right\rangle_{nl}\right] \end{aligned}$$

Here E_n is given by the Bohr formula; the radial integrals can be substituted from equation (13.25), and we obtain

$$\begin{aligned} E_{\text{kin}}^{(1)} &= -\frac{1}{2mc^2}\left[\left(\frac{mc^2\alpha^2}{2n^2}\right)^2 - 2e^2\,\frac{mc^2\alpha^2}{2n^2}\left(\frac{1}{a_0 n^2}\right) + e^4\,\frac{1}{a_0^2 n^3(l+\frac{1}{2})}\right] \\ &= -\frac{mc^2\alpha^4}{2}\left[\frac{1}{n^3(l+\frac{1}{2})} - \frac{3}{4n^4}\right] \end{aligned} \tag{19.12}$$

The Fine Structure

Notice that both corrections $E_{\text{s.o.}}^{(1)}$ and $E_{\text{kin}}^{(1)}$ are α^4 effects, and they must be combined in order to obtain the total fine structure splitting of the levels of hydrogen. We find, after some algebra and eliminating l in favor of j,

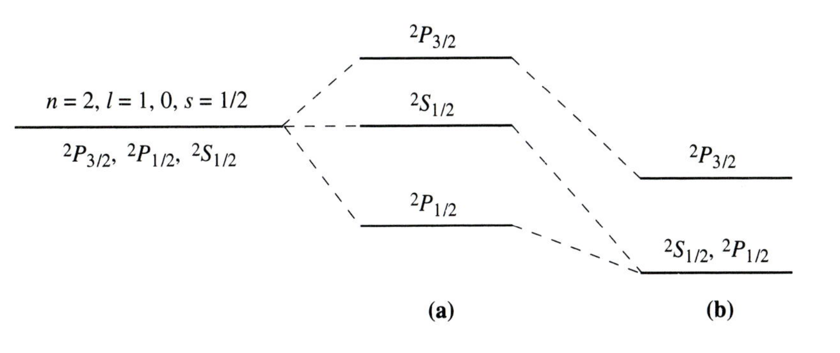

FIGURE 19.1
Splitting of the $n = 2$ hydrogen levels. (a) The effect of spin-orbit coupling alone; (b) The energy levels when both spin orbit and relativistic effects are included. The remaining $^2S_{1/2}$-$^2P_{1/2}$-degeneracy is removed by Lamb shift (not shown).

$$E^{(1)} = E^{(1)}_{\text{s.o.}} + E^{(1)}_{\text{kin}} = -\frac{mc^2\alpha^4}{2n^3}\left[\frac{1}{j + \frac{1}{2}} - \frac{3}{4n}\right] \tag{19.13}$$

for both values of $l = j \pm \frac{1}{2}$. The energy shift depends on the j-value, and each level with $l > 0$ is split into two levels, the level of higher j having the higher energy (fig. 19.1). Note, however, that the splitting of a given l-level is entirely produced by the spin-orbit interaction; the kinetic energy term contributes a constant amount for a given l.

So at this level of realistic calculation, we are still left with some degeneracy. The electron's self-energy effect, which can only be calculated using quantum field theory, breaks up this remaining degeneracy. The resultant splitting of the $^2S_{1/2}$ and the $^2P_{1/2}$ (note the spectroscopic notation, which stands for $^{2S+1}L_J$) levels of hydrogen was first observed by Lamb and Retherford (the name Lamb shift is given to the effect).

19.2 THE ANOMALOUS ZEEMAN EFFECT

We considered the effect of a magnetic field on the electron in an atom in chapter 14, but those were our simple, prespin days. The result was the so-called normal Zeeman effect—normal meaning an effect that could be explained semiclassically. When we immerse the spinning-electron atom (the real atom) in a magnetic field, we get the anomalous or quantum Zeeman effect. And now we are in a position to calculate it.

First we will do our calculation for the weak-field case. For the unperturbed Hamiltonian H_0 we take the previous kinetic + Coulomb terms with the add-ons that we calculated above:

$$H_0 = \frac{p^2}{2m} - \frac{e^2}{r} - H_{\text{kin}} + H_{\text{s.o.}} \tag{19.14}$$

The perturbation is given as

$$H_1 = (e/2mc)(\mathbf{L} + 2\mathbf{S})\cdot\mathbf{B} \tag{19.15}$$

The first term represents the interaction of the orbital magnetic moment of the circulating electron with the **B**-field; the second is the interaction of the intrinsic magnetic moment of a spinning quantum electron of $g = 2$.

It should be clear from our choice of the zero-order Hamiltonian that we have to calculate our perturbation in the coupled $|ljm_j\rangle$-basis (the added subscript is to avoid confusion with mass m). Choose the z-axis in the direction of **B**; then the first-order energy shift due to the perturbation H_1 is given as

$$\begin{aligned}\langle ljm_j|(eB/2mc)(L_z + 2S_z)|ljm_j\rangle &= \langle ljm_j|(eB/2mc)(J_z + S_z)|ljm_j\rangle \\ &= (eB/2mc)(\hbar m_j + \langle ljm_j|S_z|ljm_j\rangle)\end{aligned} \tag{19.16}$$

So it boils down to the calculation of the matrix element of S_z in the coupled jm-representation. But there is no way to carry out the operation of S_z without going back to the uncoupled basis, which we do by using the Clebsch-Gordan coefficients of chapter 17:

$$\begin{aligned}\langle ljm_j|S_z|ljm_j\rangle = \sum_{m_s, m_{s'}} &\langle lm_j{-}m_s \tfrac{1}{2} m_s|jm_j\rangle\langle lm_j{-}m_s' \tfrac{1}{2} m_s'|jm_j\rangle \\ &\times \langle lm_j{-}m_s|lm_j{-}m_s'\rangle\langle m_s|S_z|m_s'\rangle\end{aligned}$$

The S_z-matrix element gives $\hbar m_s$ times a Kronecker delta that eliminates the sum over $m_{s'}$. The sum over m_s gives two terms. For $j = l + \frac{1}{2}$, we get (using the appropriate Clebsch-Gordans from table 17.1),

$$\langle ll{+}\tfrac{1}{2} m_j|S_z|ll{+}\tfrac{1}{2} m_j\rangle = \frac{\hbar}{2}\left(\frac{l + \frac{1}{2} + m_j}{2l + 1} - \frac{l + \frac{1}{2} - m_j}{2l + 1}\right) = \frac{\hbar m_j}{2l + 1} \tag{19.17}$$

And for $j = l - \frac{1}{2}$, we obtain,

$$\langle ll{-}\tfrac{1}{2} m_j|S_z|ll{-}\tfrac{1}{2} m_j\rangle = \frac{\hbar}{2}\left(\frac{l + \frac{1}{2} - m_j}{2l + 1} - \frac{l + \frac{1}{2} + m_j}{2l + 1}\right) = -\frac{\hbar m_j}{2l + 1} \tag{19.18}$$

Substituting the S_z-matrix element in equation (19.16), we get the first-order Zeeman energy shift

$$\Delta E_{ljm_j} = \frac{e\hbar B}{2mc}\, m_j\left(1 \pm \frac{1}{2l + 1}\right) \qquad j = l \pm \tfrac{1}{2} \tag{19.19}$$

for a specific energy level under the influence of a weak magnetic field. The energy shift is proportional to m_j. In the absence of a magnetic field, the energy levels corresponding to all the possible orientations of the total angular momen-

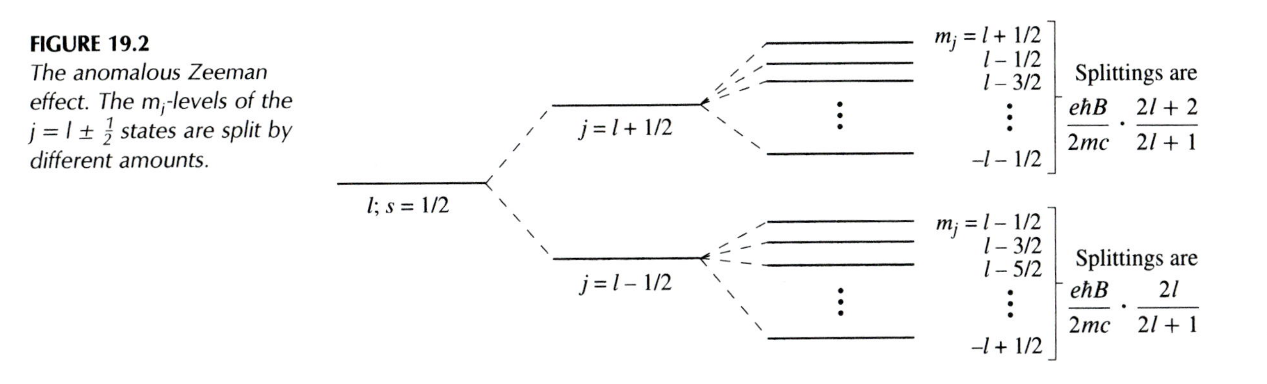

FIGURE 19.2
The anomalous Zeeman effect. The m_j-levels of the $j = l \pm \frac{1}{2}$ states are split by different amounts.

tum are of the same energy; but when the magnetic field is turned on, each level receives a shift in proportion to the component of the total angular momentum in the direction of the applied field.

The splitting of the levels of a single l, first by the spin-orbit interaction and then by the magnetic interaction, is shown in figure 19.2. The transitions obey the selection rules

$$\Delta m_j = \pm 1, 0$$

as in normal Zeeman effect, but now there can be more than three lines because the splittings are not uniformly the same for every multiplet; in general, they are different from one j-multiplet to another, although the same within a multiplet. All the different lines arising from $\Delta l = 1$, $\Delta m_j = 0, \pm 1$ transitions between the $n = 2$ and $n = 1$ states are shown in figure 19.3.

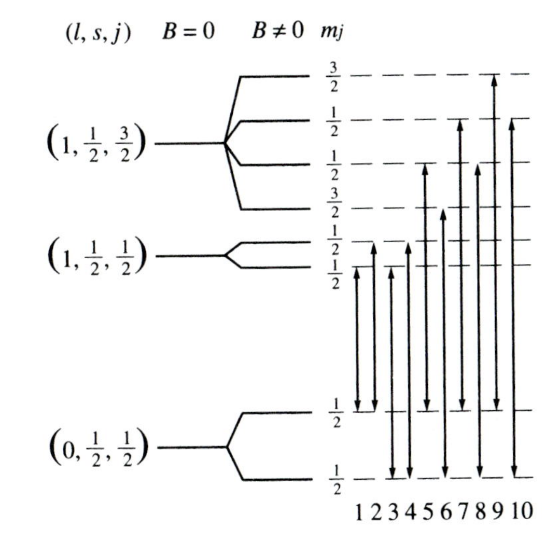

FIGURE 19.3
The anomalous Zeeman splittings of all the $n = 2$, $l = 1$ and $n = 1$, $l = 0$ levels and the allowed dipole transitions among them.

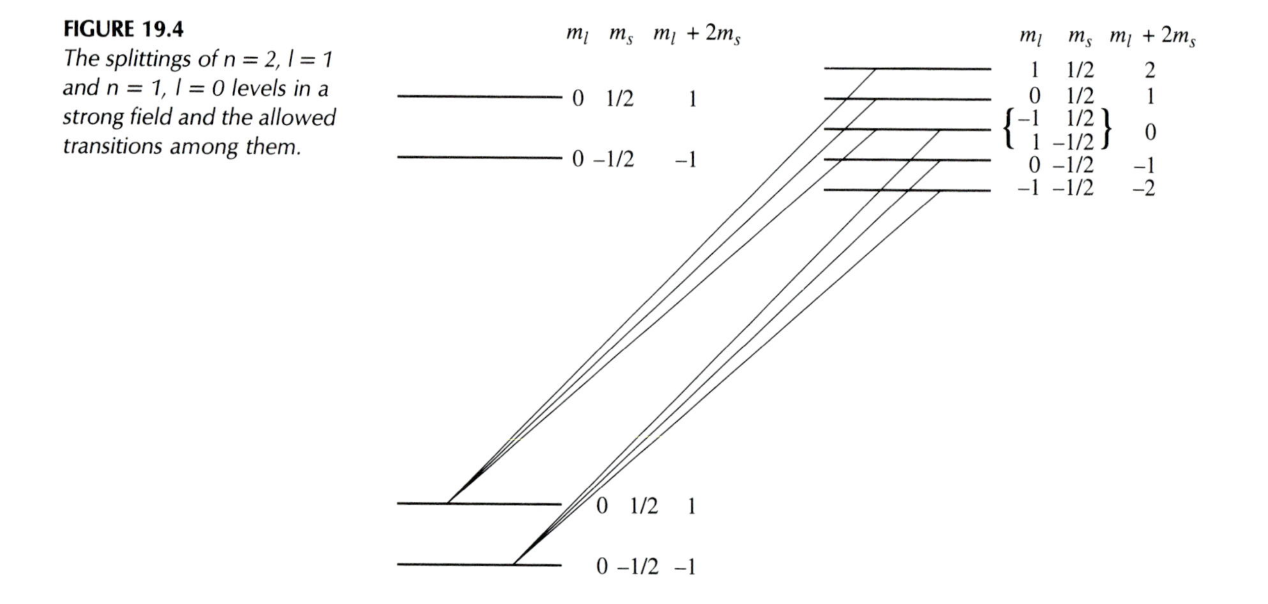

FIGURE 19.4
The splittings of n = 2, l = 1 and n = 1, l = 0 levels in a strong field and the allowed transitions among them.

In the case of a strong magnetic field, the **L** and **S** vectors decouple and precess independently (speaking in a classical analogy) about the **B**-field. We now can ignore the H_{kin} and $H_{\text{s.o.}}$ terms; the interaction H_1 with the magnetic field, equation (19.15), is already diagonal in the appropriate $|nlm_lm_s\rangle$-basis, its expectation value being

$$\langle nlm_lm_s|H_1|nlm_lm_s\rangle = (eB\hbar/2mc)(m_l + 2m_s) \tag{19.20}$$

This gives quite a different situation from the weak-field case. For example, now the $n = 2$, $l = 1$ level of hydrogen splits into five levels. The possible transitions to the two $n = 1$ levels are shown in figure 19.4.

19.3 THE HELIUM ATOM

The helium atom is a three-body problem and, just as in classical physics, there is no way to find an exact solution to the quantum three-body problem. In what follows, we will make a series of approximations to find better and better pictures of the ground state of helium. After we are done with the ground state, we will sketch the treatment for the lowest excited states.

In the crudest possible model, we ignore all interactions between the two electrons and assume that each electron moves independently in the unmodified

Coulomb field of its supposedly point nucleus (in actuality, of course, each screens the nuclear charge somewhat from the other). The Hamiltonian for this assumed system is

$$H = -\frac{\hbar^2}{2\mu}\nabla_1^2 - \frac{2e^2}{r_1} - \frac{\hbar^2}{2\mu}\nabla_2^2 - \frac{2e^2}{r_2} \tag{19.21}$$

where the subscripts 1 and 2 refer to the two electrons. But this Hamiltonian is simply the sum of two hydrogenic one-electron Hamiltonians with $Z = 2$, and the corresponding ground-state energy is clearly given as twice the hydrogenic ground energy E_1 calculated with $Z = 2$; in other words, it is $2Z^2$ times the energy of the hydrogen ground state

$$E_0 = -2Z^2 \cdot 13.6 \text{ eV} = -108.8 \text{ eV}$$

The experimental value of the ground state energy is −78.975 eV. Unfortunately, this indicates that the neglected interaction contributes to the same order as the unperturbed energy above, so its treatment by perturbation theory is not expected to be kosher. Nevertheless, as a first try, we will go ahead and try a perturbation treatment of the Coulomb repulsion between the two electrons:

$$H_1 = e^2/r_{12} \tag{19.22}$$

where $r_{12} = |\mathbf{r}_1 - \mathbf{r}_2|$ is the distance between the two electrons (fig. 19.5).

However, before we calculate the first-order energy shift due to this perturbation, let's inject some real-world physics into the problem, and this means acknowledging that the two electrons are identical particles and that electrons have spin. We have to antisymmetrize the ground state properly against exchange of the two electrons. For the spin part of the state, it is convenient to

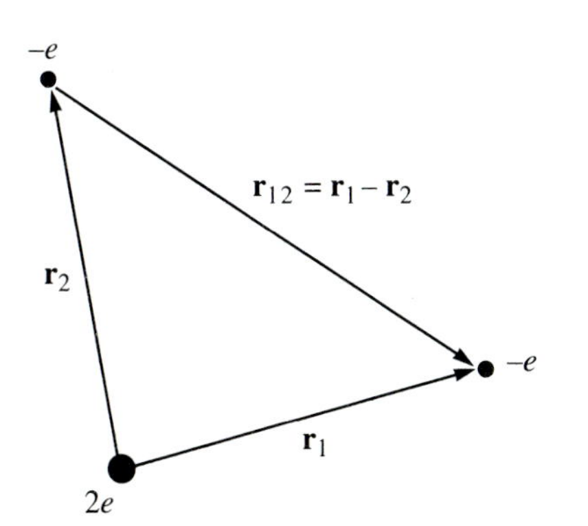

FIGURE 19.5
The geometry of a two-electron atom.

use the coupled representation $|S, m_s\rangle$ with the triplet $S = 1$ state being the symmetric and the singlet $S = 0$ state the antisymmetric state under exchange of the spin coordinates.

How about the spatial part of the state? Since we are concerned with the ground state, both electrons must occupy the lowest hydrogenic state $|\phi_{nlm}\rangle = |\phi_{100}\rangle$. Therefore, the spatial state has to be symmetric (the antisymmetric state identically vanishes, an example of the Pauli principle operating).

So the correct unperturbed ground state is given as

$$|\psi_0\rangle = |\phi_{100}(1)\rangle|\phi_{100}(2)\rangle|S{=}0 m_s{=}0\rangle \tag{19.23}$$

and we want to calculate $\langle\psi_0|H_1|\psi_0\rangle$.

Since H_1 has no spin dependence, the scalar product of the spin states in $\langle H_1\rangle$ gives 1 by virtue of orthonormality. Therefore, the expectation value above is given by the spatial matrix element alone:

$$\langle\psi_0|H_1|\psi_0\rangle = \int d^3r_1\, d^3r_2 |\phi_{100}(\mathbf{r}_1)|^2 \frac{e^2}{|\mathbf{r}_1 - \mathbf{r}_2|} |\phi_{100}(\mathbf{r}_2)|^2 \tag{19.24}$$

We can interpret $e|\phi_{100}(\mathbf{r}_1)|^2$ as the charge density of electron 1 and $e|\phi_{100}(\mathbf{r}_2)|^2$ as the charge density of electron 2; the integral above is then easily seen as the electrostatic interaction energy of two spherically symmetric charge distributions.

Because of the spherical symmetry of the charge distributions, the necessary integration can be carried out directly. It is, however, instructive to do the integration using a more sophisticated method, which applies even when the charge distribution is nonspherical. In other words, let's learn a new trick!

You see, $1/|\mathbf{r}_1 - \mathbf{r}_2|$ can be expanded in terms of Legendre polynomials (this will be left as an exercise):

$$\begin{aligned}\frac{1}{|\mathbf{r}_1 - \mathbf{r}_2|} &= \frac{1}{r_1}\sum_{L=0}^{\infty}\left(\frac{r_2}{r_1}\right)^L P_L(\cos\theta) \qquad r_1 > r_2 \\ &= \frac{1}{r_2}\sum_{L=0}^{\infty}\left(\frac{r_1}{r_2}\right)^L P_L(\cos\theta) \qquad r_2 > r_1\end{aligned} \tag{19.25}$$

Here θ is the angle between the two vectors $\mathbf{r}_1$ and $\mathbf{r}_2$. When we substitute these expansions into equation (19.24), we get an angular integral over the angles of $\mathbf{r}_1$ and $\mathbf{r}_2$, $\theta_1\phi_1$ and $\theta_2\phi_2$, respectively. But we can rewrite $P_L(\cos\theta)$ in terms of the integration variables by using the *spherical harmonic addition theorem* (the proof of which is beyond the scope of this book, but no matter, it's just mathematics):

$$P_L(\cos\theta) = \frac{4\pi}{2L+1}\sum_{M=-L}^{L} Y_{Lm}(\theta_1\phi_1)Y^*_{Lm}(\theta_2\phi_2)$$

$$= \sum_{M=-L}^{L} \frac{(L-M)!}{(L+M)!} P_L^{|M|}(\cos\theta_1)P_L^{|M|}(\cos\theta_2)e^{iM(\phi_1-\phi_2)} \qquad (19.26)$$

The angular integral in equation (19.24) now is doable:

$$\int_0^{\pi} P_L^{|M|}(\cos\theta_1)\sin\theta_1\,d\theta_1 \int_0^{\pi} P_L^{|M|}(\cos\theta_2)\sin\theta_2\,d\theta_2 \int_0^{2\pi} e^{iM\phi_1}\,d\phi_1$$
$$\times \int_0^{2\pi} e^{-iM\phi_2}d\phi_2 = 16\pi^2\delta_{L0}\delta_{M0} \qquad (19.27)$$

where we have used

$$\int_0^{2\pi} d\phi\, e^{iM\phi} = 2\pi\delta_{M0}$$

$$\int_0^{\pi} \sin\theta\, d\theta\, P_L(\cos\theta) = 2\delta_{L0}$$

and where the last integral is a special case of the orthonormality integral of Legendre polynomials.

In this way, upon further substituting into equation (19.24),

$$\phi_{100}(\mathbf{r}) = (1/\sqrt{\pi})(Z/a_0)^{3/2}\exp(-Zr/a_0)$$

for the wave functions and for the spherical harmonic expansions and angular integrals calculated above, we get for the first-order energy shift the following expression:

$$\langle\psi_0|H_1|\psi_0\rangle = 16e^2\left(\frac{Z}{a_0}\right)^6 \int_0^{\infty} r_1\,dr_1\,e^{-2Zr_1/a_0}\left[\int_0^{r_1} r_2^2\,dr_2\,e^{-2Zr_2/a_0} + r_1\int_{r_1}^{\infty} r_2\,dr_2\,e^{-2Zr_2/a_0}\right]$$

These integrals are straightforward. Performing the integration over r_2, we obtain

$$\langle\psi_0|H_1|\psi_0\rangle = 4\,\frac{Z^3}{a_0^3}\,e^2\int_0^{\infty} r_1 e^{-2Zr_1/a_0}[1 - e^{-2Zr_1/a_0}(Zr_1/a_0 + 1)]\,dr_1 \qquad (19.28)$$

Finally, performing the integration over r_1, we obtain

$$\langle \phi_{100} | H_1 | \phi_{100} \rangle = 5Ze^2/8a_0 = (5Z/4)(\tfrac{1}{2}mc^2\alpha^2) \tag{19.29}$$

For $Z = 2$, the first-order energy shift is then 34 eV (of course, the contribution is positive since the interaction between the electrons is repulsive). Adding to the unperturbed energy of $E_0^{(0)} = -108.8$ eV, we get

$$E_0 = -74.8 \text{ eV}$$

which compares unexpectedly well with the experimental value of -78.975 eV.

Screening and Effective Charge in Perturbation Theory

The full nuclear charge Z is not seen by either of the electrons due to the shielding or screening effect of the other. But what is the effective charge that the electrons do see? The perturbation theory developed above also can be used to get an estimate of the effective charge, Z_{eff}.

Let's write the effective Schrödinger equation for electron 1 as

$$\left[-\frac{\hbar^2}{2m} \nabla_1^2 - \frac{Z_{\text{eff}} e^2}{r_1} \right] \phi_{100}(\mathbf{r}_1) = E\phi_{100}(\mathbf{r}_1) \tag{19.30}$$

in which, by definition,

$$-Z_{\text{eff}} e^2/r_1 = (-Ze^2/r_1) + V(r_1) \tag{19.31}$$

where $V(r_1)$ is the shielding potential due to the Coulomb repulsion of electron 2 that blocks the view of the charge of the nucleus from electron 1. In perturbation theory, $V(r_1)$ can be easily calculated from the interaction matrix element $\langle \psi_0 | H_1 | \psi_0 \rangle$ calculated above. We write

$$\langle \psi_0 | H_1 | \psi_0 \rangle = \int V(r_1) \phi_{100}^2(\mathbf{r}_1) r_1^2 \, dr_1 \sin\theta_1 \, d\theta_1 \, d\phi_1$$

Then, comparing with equation (19.28), we readily find

$$V(r_1) = (e^2/r_1)[1 - (1 + Zr_1/a_0)\exp(-2Zr_1/a_0)] \tag{19.32}$$

From the defining equation (19.31), we have

$$\begin{aligned} Z_{\text{eff}} &= Z - V(r_1) r_1/e^2 \\ &= (Z - 1) + (1 + Zr_1/a_0)\exp(-2Zr_1/a_0) \\ &= 1 + (1 + 2r_1/a_0)\exp(-4r_1/a_0) \end{aligned} \tag{19.33}$$

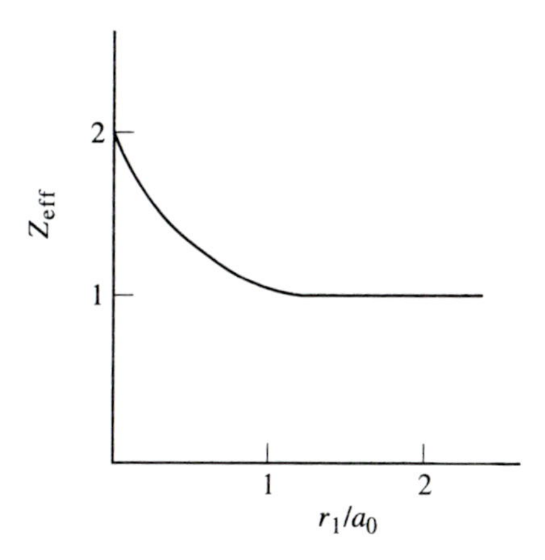

FIGURE 19.6
The effective charge of an electron in the helium atom as a function of its distance (in units of the Bohr radius) from the nucleus.

Z_{eff} is plotted in figure 19.6 as a function of r_1/a_0. Clearly, for small values of r, an electron in helium sees the full nuclear charge with little screening due to the presence of the other electron. But as r increases, the effect of screening is felt; finally, when $r > a_0$, the screening is essentially complete, and the electron sees only one unit of nuclear charge.

Accounting for Screening by a Variational Calculation

Shall we be so bold as to try to explain the remaining discrepancy between theory and experiment by accounting for screening of the nuclear charge distribution that each electron sees due to the presence of the other?

We can approach the question by a variational calculation. Suppose we perform a variational calculation for the ground state, choosing the spatial wave function as the product $\phi_{100}(\mathbf{r}_1)\phi_{100}(\mathbf{r}_2)$ as before, except that the nuclear charge is now a variable in the hydrogenic Hamiltonian for which ϕ is the ground-state solution:

$$\left(-\frac{\hbar^2}{2m}\nabla^2 - \frac{Z_{\text{eff}}e^2}{r}\right)\phi_{100}(\mathbf{r}) = \epsilon_{100}\phi_{100}(\mathbf{r}) \tag{19.34}$$

with $\epsilon_{100} = -\frac{1}{2}mc^2(Z_{\text{eff}}\alpha)^2$. We need to calculate

$$\iint d^3r_1\, d^3r_2\, \phi^*_{100}(\mathbf{r}_1)\phi^*_{100}(\mathbf{r}_2)\left[\frac{p_1^2}{2m} + \frac{p_2^2}{2m} - \frac{Ze^2}{r_1} - \frac{Ze^2}{r_2} + \frac{e^2}{|\mathbf{r}_1 - \mathbf{r}_2|}\right] \times \phi_{100}(\mathbf{r}_1)\phi_{100}(\mathbf{r}_2) \tag{19.35}$$

There are two single-particle Hamiltonian matrix elements to calculate and one two-body element. The single-body matrix elements can be calculated as follows:

$$\int d^3r_1\, \phi^*_{100}(\mathbf{r}_1)\left[\frac{p_1^2}{2m} - \frac{Ze^2}{r_1}\right]\phi_{100}(\mathbf{r}_1)\int d^3r_2\, |\phi_{100}(\mathbf{r}_2)|^2$$
$$= \int d^3r_1\, \phi^*_{100}(\mathbf{r}_1)\left[\frac{p_1^2}{2m} - \frac{Z_{\text{eff}}e^2}{r_1} + \frac{(Z_{\text{eff}}-Z)e^2}{r_1}\right]\phi_{100}(\mathbf{r}_1)$$
$$= \epsilon_{100} + (Z_{\text{eff}} - Z)e^2\int d^3r_1\, \phi^*_{100}(\mathbf{r}_1)\,\frac{1}{r_1}\,\phi_{100}(\mathbf{r}_1)$$
$$= \epsilon_{100} + (Z_{\text{eff}} - Z)e^2(Z_{\text{eff}}/a_0) = \epsilon_{100} + (Z_{\text{eff}} - Z)Z_{\text{eff}}mc^2\alpha^2 \quad (19.36)$$

where we have used equations (19.34) and (13.25). The single-body Hamiltonian for electron 2 makes an identical contribution. Also, the two-body electron-electron interaction term is the same as in equation (19.29), except that we must replace Z by Z_{eff}. Summing up, we get for the expectation of the Hamiltonian in the variational ground state the following expression in terms of the variational parameter Z_{eff}:

$$\langle H\rangle = -\tfrac{1}{2}mc^2\alpha^2[2Z_{\text{eff}}^2 + 4Z_{\text{eff}}(Z - Z_{\text{eff}}) - 5Z_{\text{eff}}/4]$$
$$= -\tfrac{1}{2}mc^2\alpha^2[-2Z_{\text{eff}}^2 + 4ZZ_{\text{eff}} - 5Z_{\text{eff}}/4] \quad (19.37)$$

Now minimize $\langle H\rangle$ with respect to Z_{eff}:

$$\frac{d\langle H\rangle}{dZ_{\text{eff}}} = 0$$

This gives the value of Z_{eff} that minimizes $\langle H\rangle$ as

$$Z_{\text{eff}} = Z - \tfrac{5}{16} \approx 1.7 \quad (19.38)$$

(It's a minimum; $d^2\langle H\rangle/dZ_{\text{eff}}^2 > 0$.) Each electron sees 15% less of the nuclear charge on the average because of the screening by the other electron. Substituting into equation (19.37), we get for the ground state energy

$$E_0 \leq -\tfrac{1}{2}mc^2\alpha^2[2(Z - \tfrac{5}{16})^2]$$

This gives for helium ($Z = 2$), $E_0 \leq -77.4$ eV. Indeed, accounting for screening gives us a better agreement with experiment.

How could we improve such a calculation even further? We would have to introduce more variational parameters and grind out a better result. And we could use physics to improve upon the trial wave function used above; it does not have to be unimaginative. For example, we have assumed above that the two electrons see the same screened charge of the nucleus. But this ignores the fact that the two electrons repel each other, which means that if one draws closer to the nucleus at some instant, it tends to push the other farther out. Accounting for such differences in screening is indeed found to improve the energy estimate.

The Lowest Excited States of Helium: Exchange Energy

We will consider only those excited states that are obtained by lifting one of the electrons to the $n = 2$, $2S$ or $2P$ state while leaving the other in the $1S$. The resultant unperturbed $1S2S$ and $1S2P$ configurations constitute a total of $4 + 12 = 16$ degenerate states, all having the same energy of -68 eV. The Coulomb repulsion breaks up this degeneracy except for the m-degeneracy, which remains intact because the Coulomb potential commutes with L_z.

The antisymmetrization of the unperturbed state will now be of some consequence. The coupled-spin states of the two electrons are the singlet $S = 0$ (called parahelium) and the triplet $S = 1$ (called orthohelium). The singlet is antisymmetric in the spin variables and therefore must carry with it a state symmetric under space exchange, namely the state

$$(1/\sqrt{2})\,[|\phi_{100}(1)\rangle|\phi_{2lm}(2)\rangle + |\phi_{2lm}(1)\rangle|\phi_{100}(2)\rangle] \tag{19.39}$$

On the other hand, the triplet is symmetric under spin exchange, and thus must accompany a space-antisymmetric state:

$$(1/\sqrt{2})\,[|\phi_{100}(1)\rangle|\phi_{2lm}(2)\rangle - |\phi_{2lm}(1)\rangle|\phi_{100}(2)\rangle] \tag{19.40}$$

Furthermore, in the perturbation energy calculation we can take $m = 0$ for even the $2P$ states without any loss of generality, since the energy shift is independent of m. As before, the spin-independence of the perturbing electronic Coulomb repulsion translates into the spin part of the matrix element giving unity.

Again it is the spatial matrix element, but with a twist. We have

$$\begin{aligned}\langle H_1\rangle_{1S,2l} &= \tfrac{1}{2}e^2\left[\left\langle \phi_{100}(1)\phi_{2l0}(2) \pm \phi_{2l0}(1)\phi_{100}(2) \left| \frac{1}{|\mathbf{r}_1 - \mathbf{r}_2|} \right| \right.\right.\\ &\qquad \left.\left. \times\, \phi_{100}(1)\phi_{2l0}(2) \pm \phi_{2l0}(1)\phi_{100}(2)\right\rangle\right]\\ &= e^2 \iint d^3r_1\, d^3r_2\, |\phi_{100}(\mathbf{r}_1)|^2 |\phi_{2l0}(\mathbf{r}_2)|^2 \frac{1}{|\mathbf{r}_1 - \mathbf{r}_2|}\\ &\quad \pm e^2 \iint d^3r_1\, d^3r_2\, \phi^*_{100}(\mathbf{r}_1)\phi^*_{2l0}(\mathbf{r}_2) \frac{1}{|\mathbf{r}_1 - \mathbf{r}_2|} \phi_{2l0}(\mathbf{r}_1)\phi_{100}(\mathbf{r}_2)\end{aligned} \tag{19.41}$$

since the interaction is symmetric under the exchange of $\mathbf{r}_1$ and $\mathbf{r}_2$. The first term above (we will call this integral the *direct integral* $J_{1S,2l}$) is of the same form that we calculated for the interaction energy of the $(1S)^2$-configuration; this one has a classical analogy, as noted already. The second integral above, called the *exchange integral*, $K_{1S,2l}$, however, has no analog in classical physics, it has no classical interpretation. It arises purely as a result of the antisymmetrization of the quantum state, and it comes with different signs for the singlet and triplet states. The exchange interaction term consequently breaks the degeneracy of the singlet (para-) and triplet (ortho-) helium states. Note also that for the exchange integral the integral over both d^3r_1 and d^3r_2 involves both wave functions; indeed, it depends on the overlap of the two wave functions.

Explicitly, we have

$$\langle H_1 \rangle_{1s,2l} = J_{1S,2l} \pm K_{1S,2l} \tag{19.42}$$

where the plus sign goes with the singlet and the minus with the triplet. The evaluation of the actual integrals (which is similar to the one we carried out in the last subsection) will be left as an exercise, but this much can be said about them:

$$J - K < J + K \tag{19.43}$$

That is, the singlet states are raised up more than the triplet states by the perturbation (fig. 19.7), their splitting being $2K$.

You can see this in the following way without doing any integration. The probability of finding one electron at $\mathbf{r}_1$ and the other at $\mathbf{r}_2$ is

$$\tfrac{1}{2}\,|\phi_{100}(\mathbf{r}_1)\phi_{nlm}(\mathbf{r}_2) \pm \phi_{nlm}(\mathbf{r}_1)\phi_{100}(\mathbf{r}_2)|^2$$

where the + sign goes with the singlet and the minus with the triplet. Behold! When in the triplet state, the two electrons are less likely to approach each other than when in the singlet state; in fact as $\mathbf{r}_1 \to \mathbf{r}_2$, the triplet probability goes to zero. You are rediscovering the Pauli principle here, nothing new, but it does mean that the expectation value of $1/|\mathbf{r}_1 - \mathbf{r}_2|$ is smaller for the triplet state than for the singlet state; this is the same result as equation (19.43).

So there is a spin-dependent interaction arising from exchange symmetry that is much stronger than the spin-orbit coupling; in fact, it is of the same order of magnitude as electrostatic forces. It is this spin-dependent interaction due

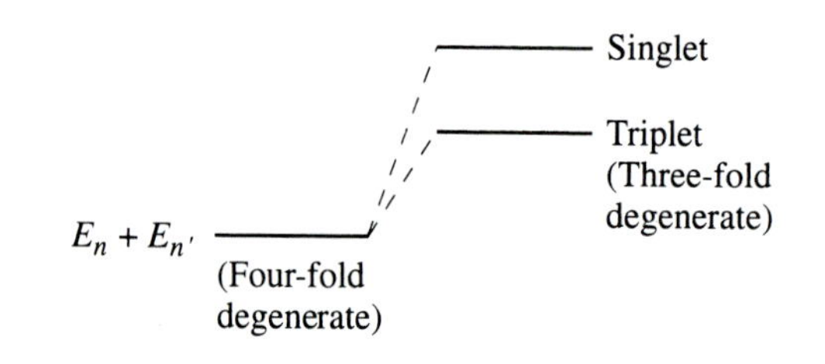

FIGURE 19.7
The exchange part of the Coulomb interaction removes the degeneracy of the singlet and triplet states in helium with the triplet state energetically lower.

to exchange that is responsible for lining up electron spins (akin to the triplet state above) in a ferromagnet, as was first pointed out by Heisenberg.

PROBLEMS

1. Calculate and compare the perturbation energy shift due to the relativistic mass correction, the spin-orbit coupling, and the interaction with an external magnetic field (of 20,000 gauss) for the various $2P$ levels of the hydrogen atom. Calculate all the possible spectral lines from the $2P \to 1S$ transition (don't forget the $\Delta m = \pm 1, 0$ selection rule).
2. If the nucleus is taken to be a uniformly charged sphere of radius R, then the Coulomb potential in the atom is modified to

$$V(r) = \frac{Ze^2}{2R}\left(\frac{r^2}{R^2} - 3\right) \qquad r < R$$
$$= -Ze^2/r \qquad r > R$$

   What is the perturbing potential by which the above differs from the unperturbed Coulomb potential of a point charge? Calculate the energy shift produced by this perturbation for the $1S$ and $2S$ levels of hydrogen. How big are these effects for a muonic hydrogen atom (a negative muon bound to a proton; mass of muon = 207 electron mass)? Take R for the proton to be 0.8×10^{-13} cm.
3. The magnetic moment operator for the deuteron can be written as

$$\mathbf{M}_d = (M_N/\hbar)\mathbf{L}_p + M_p\boldsymbol{\sigma}_p + M_n\boldsymbol{\sigma}_n$$

   where M_N is the nuclear magneton ($e\hbar/2m_p$), $M_p = 2.79M_N$ is the magnetic moment of the proton, and $M_n = -1.91M_N$ is the magnetic moment of the neutron. Explain the rest of the notation.

   Calculate the deuteron magnetic moment when the deuteron is in the $^3S_1$ and $^3D_1$ states, respectively (that is, calculate the interaction of the deuteron with an external magnetic field when in these states). Compare with the experimental value of 0.86 M_N for the deuteron ground state and hence the D-state admixture in the mainly $^3S_1$ ground state of the deuteron.
4. Positronium is an electrically bound system of an electron and a positron. Calculate first the energies of the $n = 1$ and $n = 2$ states of the positronium, and then the relativistic mass correction and the spin-orbit energy shifts.
5. Calculate the ground-state energy of the hydrogen atom using the variational method with the following trial function:

$$\phi_\lambda(r) = (1 - r/\lambda), \qquad r < \lambda$$
$$= 0, \qquad r > \lambda$$

What is the relation of λ_{min} to the Bohr radius?

6. Derive the expansions of $1/|\mathbf{r}_1 - \mathbf{r}_2|$ in terms of the Legendre polynomials, equation (19.25). (*Hint*: Write $1/|\mathbf{r}_1 - \mathbf{r}_2|$ in the form $F(x, \xi) = (x^2 + 1 - 2x\xi)^{-1/2}$ and expand in Taylor series about $x = 0$.)
7. Calculate the integrals J and K for the $1S - 2S$ and $1S - 2P$ configurations of helium.
8. Consider the lowest state of orthohelium, which is a ${}^3S_1$ state. Calculate its magnetic moment.

ADDITIONAL PROBLEMS

A1. Construct the matrix of the perturbation

$$H_1 = (e/2mc)(\mathbf{L} + 2\mathbf{S})\cdot\mathbf{B}$$

in the space consisting of the two states (written in the basis $|l\ m_1,\ m_s\rangle$):

$$-(2/3)^{1/2}|1\ -1,\ 1/2\rangle + (1/3)^{1/2}|1\ 0,\ -1/2\rangle$$

and

$$(2/3)^{1/2}|1\ 1,\ -1/2\rangle - (1/3)^{1/2}|1\ 0,\ 1/2\rangle.$$

You can take the B-field to be in the z-direction.

A2. Consider the doublet P level of the sodium atom in a uniform magnetic field. Taking the Hamiltonian for the problem as

$$H = (2\epsilon/3)\mathbf{S}\cdot\mathbf{L} + \mu(\mathbf{L} + 2\mathbf{S})\cdot\mathbf{B}$$

where ϵ is a constant, calculate the energy levels by using perturbation theory in both the weak and the strong field limits.

A3. Consider a state in a one-electron atom in which the electron's orbital angular momentum quantum number is $l = 4$. For the state with the greatest total angular momentum j and greatest z-component m, calculate

(a) the angle between the total angular momentum vector $\mathbf{J}$ and the z-direction;
(b) the angle between orbital angular momentum vector $\mathbf{L}$ and $\mathbf{J}$ and that between the spin vector $\mathbf{S}$ and $\mathbf{J}$;
(c) the magnitude of the component of the magnetic moment along $\mathbf{J}$.

A4. If the angular momentum of the nucleus is **I** and the total angular momentum of the electrons is **J**, the total angular momentum of the atom is $\mathbf{F} = \mathbf{I} + \mathbf{J}$, and the quantum number f for **F** ranges from $I + J$ to $|I - J|$. Show that the number of possible f values is $2I + 1$ if $I < J$ and $2J + 1$ if $J < I$. Consider the physical significance of this result. (*Hint*: Think of hyperfine splitting.)

REFERENCES

E. E. Anderson. *Modern Physics and Quantum Mechanics.*
S. Gasiorowicz. *Quantum Physics.*
D. Park. *Introduction to the Quantum Theory.*

20

Quantum Systems: Atoms and Molecules

The Schrödinger equation, to paraphrase a similar statement made by Dirac, solves all of chemistry and most of physics. Chemistry at the base level is the physics of atoms, and most chemical properties of elements are determined primarily by the lowest energy states of their atoms. But how do we apply the Schrödinger equation to atoms containing three or more electrons even for the modest purpose of calculating their ground states? Now, accounting for the interaction between the electrons, even via perturbation theory, is a *many-body problem* and, on the face of it, looks like a formidable task. On the other hand, there are such experimental regularities—one of them is the famous periodic table—in the data of complex atoms that it is easy to intuit that the atomic many-body system must be amenable to some simplification. The proof that this is so was first given by Hartree, who showed that the equation an electron obeys even in a complex atom is, to a first approximation, a single-particle equation; the potential of the nucleus and all the many-body potentials add to give an effective central potential—not the Coulomb $1/r$ potential, mind you—in which each electron moves relatively freely like an independent particle. Our first task in this chapter is to get an understanding of Hartree's work.

But can we understand the periodic table on the basis of such an independent particle model? The answer is a qualified yes; qualified, because occasionally we will have to resort to the effect of *residual interactions* between the electrons. But on the whole, the independent particle picture is all we need, along with the Pauli principle, to understand the periodic table, which is our second task.

The final task of the chapter is to give you an introduction to the application of quantum mechanics to molecules.

20.1 THE HARTREE EQUATION FOR ATOMS

Let's start with the Schrödinger equation for an atom with Z electrons:

$$H\psi(\mathbf{r}_1,\mathbf{r}_2,\ldots,\mathbf{r}_Z) = E\psi(\mathbf{r}_1,\mathbf{r}_2,\ldots\mathbf{r}_Z) \tag{20.1}$$

where the Hamiltonian H is a sum of one- and two-body terms:

$$\begin{aligned} H &= \sum_i \left(\frac{\mathbf{p}_i^2}{2m} - \frac{Ze^2}{r_i}\right) + \frac{1}{2}\sum_{i,j\neq i} \frac{e^2}{|\mathbf{r}_i - \mathbf{r}_j|} \\ &= \sum_i H_i + \frac{1}{2}\sum_{i,j\neq i} V_{ij} \end{aligned} \tag{20.2}$$

Equation (20.1) is a partial differential equation in $3Z$ dimensions. To derive a single-particle equation that approximates the Schrödinger equation above is to assume that a solution of equation (20.1) exists in the form of a product of single-particle wave functions:

$$\psi(\mathbf{r}_1,\mathbf{r}_2,\ldots,\mathbf{r}_Z) = \phi_1(\mathbf{r}_1)\phi_2(\mathbf{r}_2)\cdots\phi_Z(\mathbf{r}_Z) \tag{20.3}$$

The trick is to consider this as a trial wave function for a variational calculation and then pop the question: What is the equation that is satisfied by these single-particle functions $\phi_i(\mathbf{r}_i)$ so that the trial function (20.3) minimizes

$$\langle H\rangle = \int \psi^* H\psi\, d\tau \tag{20.4}$$

provided of course ψ is normalized (i.e., $\int d\tau\, \psi^*\psi = 1$)? We can ensure the normalization of ψ by assuming the normalization conditions

$$\int \phi_i^*\phi_i\, d^3\mathbf{r}_i = 1 \tag{20.5}$$

for each of the ϕ's.

Do you see the ingenuity of this method? Ordinarily, we do a variational calculation by picking the wave functions first, then calculating $\langle H\rangle$, and finally minimizing $\langle H\rangle$. Instead, here we are starting with the proposition that the variational procedure itself will tell us how to choose the wave function, so long as it is of the general form, equation (20.3). It is a little like pulling yourself up by your own bootstraps.

Substituting equation (20.3) into equation (20.4) and noting that the single-particle part of H, H_i operates only upon the coordinates of the ith electron and V_{ij} operates only upon the two-body coordinates of both i and j, we get

$$\langle H \rangle = \sum_i \int \phi_i^* H_i \phi_i \, d^3\mathbf{r}_i + \frac{1}{2} \sum_{i,j \neq i} \iint d^3\mathbf{r}_i \, d^3\mathbf{r}_j \, \phi_i^* \phi_j^* V_{ij} \phi_i \phi_j \tag{20.6}$$

All the other coordinates integrate out to one. Now consider the minimization of $\langle H \rangle$ with respect to variation of ϕ_i^*: We have

$$\delta\langle H \rangle = \sum_i \int \delta\phi_i^* \left[H_i + \sum_{j \neq i} \int \phi_j^* V_{ij} \phi_j \, d^3\mathbf{r}_j \right] \phi_i \, d^3\mathbf{r}_i = 0 \tag{20.7}$$

where the variations $\delta\phi_i^*$ satisfy the equations

$$\int \delta\phi_i^* \phi_i \, d^3\mathbf{r}_i = 0 \tag{20.8}$$

by virtue of the normalization condition of equation (20.5).

Equation (20.8) acts as a constraint on the variation of $\langle H \rangle$, and we handle it by the standard method of Lagrangian multipliers—multiply each of the equations (20.8) by a multiplier ϵ_i (the labeling here is highly significant!) and subtract the sum from equation (20.7). This gives

$$\sum_i \int \delta\phi_i^* \left[H_i + \sum_{j \neq i} \int \phi_j^* V_{ij} \phi_j \, d^3\mathbf{r}_j - \epsilon_i \right] \phi_i \, d^3\mathbf{r}_i = 0 \tag{20.9}$$

But the variations $\delta\phi_i^*$ above are independent; therefore, the left-hand side of equation (20.9) cannot possibly vanish unless the coefficient of each $\delta\phi_i^*$ vanishes; that is, we must have

$$\left(H_i + \sum_{j \neq i} \int \phi_j^* V_{ij} \phi_j \, d^3\mathbf{r}_j \right) \phi_i = \epsilon_i \phi_i$$

Or, upon substituting for H_i and V_{ij}, the equation that the single-particle wave functions ϕ_i must satisfy is found to be

$$\left(-\frac{\hbar^2}{2m} \nabla_i^2 - \frac{Ze^2}{r_i} + e^2 \sum_{j \neq i} \int d^3\mathbf{r}_j \, \frac{|\phi_j(\mathbf{r}_j)|^2}{|\mathbf{r}_i - \mathbf{r}_j|} \right) \phi_i(\mathbf{r}_i) = \epsilon_i \phi_i(\mathbf{r}_i) \tag{20.10}$$

This integro-differential equation is called the *Hartree equation*. It is an eigenvalue equation for a single-particle Hamiltonian for the ith electron moving in the Coulomb potential of the nucleus *plus* a potential contributed by all the other electrons. However, the crucial element is that the potential due to the other electrons

$$V_{\text{s.c.}}(i) = e^2 \sum_{j \neq i} \int d^3\mathbf{r}_j \frac{|\phi_j(\mathbf{r}_j)|^2}{|\mathbf{r}_i - \mathbf{r}_j|} \tag{20.11}$$

depends on the charge density $e|\phi_j(\mathbf{r}_j)|^2$, which we can know only after we have solved equation (20.10). Therefore, the procedure calls for a *self-consistent* search for solutions whose insertion back into the equation that generates them iteratively reproduces what we begin with. The potential $V_{\text{s.c.}}$, the effect of all the other particles on each particle, is a *self-consistent* potential.

So by introducing the idea of a self-consistent potential field, the electronic many-body problem has been reduced to a one-body problem. Actually, the self-consistent field does not account for all of the many particle interactions; there still is left a *residual interaction*. More on that later.

For the purpose of practical applications, the self-consistent field is replaced by its average over the angles of the radius vector $\mathbf{r}_i$, thus making it into a central potential. Now the single-particle wave functions acquire the usual *nlm* quantum numbers (to which the spin quantum number m_s is added as before).

The total energy E is obtained when we substitute back into equation (20.6) the wave functions obtained from the solution of the Hartree equation. Beware! This energy is not equal to the sum of the single-particle energies obtained as eigenvalues of the Hartree equation. As you can see from equation (20.10), we have

$$\epsilon_i = \int d^3r_i\, \phi_i^* \left[-\frac{\hbar^2}{2m} \nabla_i^2 - \frac{Ze^2}{r_i} \right] \phi_i + e^2 \sum_{j \neq i} \int d^3r_i\, d^3r_j\, \phi_i^* \frac{|\phi_j|^2}{|\mathbf{r}_i - \mathbf{r}_j|} \phi_i$$

Therefore, the repulsive interelectronic interaction is double counted in the sum over single-particle energies, and the correct expression is

$$E = \sum_i \epsilon_i - \frac{1}{2} \sum_{i, j \neq i} \int d^3r_i\, d^3r_j\, \phi_i^* \frac{|\phi_j|^2}{|\mathbf{r}_i - \mathbf{r}_j|} \phi_i \tag{20.12}$$

As an example, let's consider how a Hartree calculation for the ground state of helium might proceed. In chapter 19, we calculated $V(r_1)$, the potential field that particle 1 sees due to the presence of the other electron. We can solve the Hartree equation with $V(r_1)$ as our $V_{\text{s.c.}}$; the solution of the equation will generate a new single-particle wave function, which we then substitute in the calculation of $V(r_1)$ to get a new $V_{\text{s.c.}}$ to use in the next iteration. Alternatively, we can use a wave function intermediate between the $1S$ hydrogenic wave function and the new wave function obtained above to start the next iteration.

We may have to do this many times until, of course, a repetition brings forth no appreciable change in the wave function. Finally, we obtain the ground-state energy from the eigenvalues and eigenfunctions of the Hartree equation by using equation (20.12).

Antisymmetrization of the Wave Function

The Hartree equation leaves out the effect of the electronic correlations introduced by the antisymmetrization of the total wave function of a many-electron atom. You know how to antisymmetrize a two-particle wave function. How is it done for a many-particle wave function?

The antisymmetrization of a product of N single-particle states

$$\psi = \prod_i \phi_{\alpha_i}(i)$$

can be formally expressed by operating on ψ with the antisymmetrizer operator $\hat{A}$ defined as

$$\hat{A} = (N!)^{-1/2} \sum_i (-1)^p \hat{P}_i \tag{20.13}$$

Here $\hat{P}$ stands for the $N!$ possible permutation operators that interchange a pair of state indices or particle indices (but not both); the subscript i denotes the ith permutation belonging to this set. The parity of the number of permutations (even or odd) is denoted by p. We can easily verify that this definition incorporates the Pauli principle, since if any of the state indices α_i are the same, the antisymmetrization operator will give zero.

It is more convenient, however, to visualize the antisymmetrized wave function as a *Slater determinant* given by

$$\psi_a = \frac{1}{\sqrt{N!}} \begin{vmatrix} \phi_{\alpha_1}(1) & \phi_{\alpha_1}(2) & \cdots & \phi_{\alpha_1}(N) \\ \phi_{\alpha_2}(1) & \phi_{\alpha_2}(2) & \cdots & \phi_{\alpha_2}(N) \\ \cdots & \cdots & & \cdots \\ \phi_{\alpha_N}(1) & \phi_{\alpha_N}(2) & \cdots & \phi_{\alpha_N}(N) \end{vmatrix} \tag{20.14}$$

The change of sign of ψ_a under permutation of any two-particle or state labels immediately follows from the change of sign of a determinant when two columns or two rows are interchanged. Again, if two of the state labels are the same, the determinant vanishes identically; thus the Pauli principle is incorporated.

When the antisymmetrization, and thus the Pauli principle, is taken into account for the trial wave function, we get from the variational procedure above the *Hartree-Fock equation*, which differs from the Hartree equation by the appearance of an additional *exchange-potential term*. The exchange term is important for actual calculations, but for our purpose, which is to understand qualitatively how quantum mechanics is able to explain the main features of atomic structure, we have no need to delve into it.

20.2 THE QUANTUM MECHANICAL EXPLANATION OF THE PERIODIC TABLE

To a good approximation, then, the average potential in which the single electron moves in a complex atom is a central potential; consequently, the electronic states can be characterized by the quantum numbers n, l, m_l, and m_s. The one important difference between this and the Coulomb potential is that the l-degeneracy is removed, and the single-electron eigenvalues are now determined not by n, the principal quantum number, alone, but also by l, the orbital angular momentum. Naturally, we will label the single electronic states by nl (referring as before to l using letters such as S, P, D, F, G, etc.) and call them *orbitals*.

Figure 20.1 shows a typical ordering of electronic energy levels. Although the l-degeneracy is removed, the levels still are found to bunch together, displaying a *shell structure*. Each group of near-degenerate levels corresponds to an

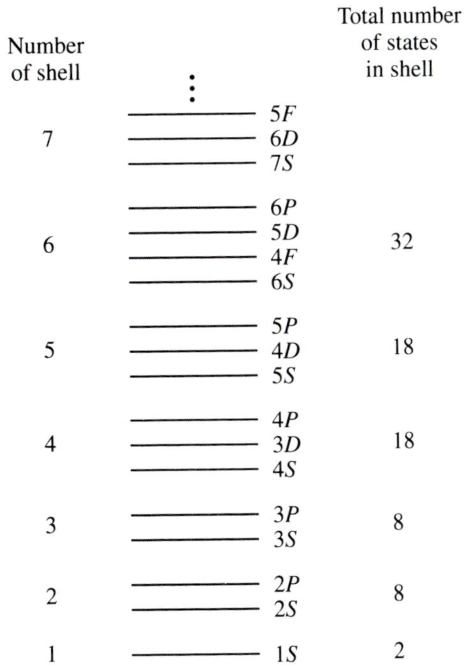

Number of shell	Levels	Total number of states in shell
	⋮	
7	$5F$, $6D$, $7S$	
6	$6P$, $5D$, $4F$, $6S$	32
5	$5P$, $4D$, $5S$	18
4	$4P$, $3D$, $4S$	18
3	$3P$, $3S$	8
2	$2P$, $2S$	8
1	$1S$	2

FIGURE 20.1
Atomic shell structure (drawn schematically).

electronic shell. Whereas there is a large energy gap between levels belonging to two consecutive shells, levels of the same shell differ little in energy.

According to the Pauli principle, we can put only one electron in a state specified by all four quantum numbers n, l, m_l, m_s. By the same token, it takes $2(2l+1)$ electrons to fill up one orbital nl; and when we have filled up all the orbitals in an electronic shell, we have what is called a *closed shell.* In the atomic ground state, the electrons fill the orbitals in order of their energy (this is called the *aufbau principle*); the lowest energy orbital, the $1S$, goes first. It takes two electrons to fill the $1S$, and we get a closed shell—the helium atom. The next shell can take two more electrons in $2S$ and six more in $2P$, and thus, an extra eight electrons. Adding the two from the previous shell closure, we now have ten electrons and the neon atom with two shells closed. The next shell closure comes in Ar ($Z = 18$) with a $(1S)^2(2S)^2(2P)^6(3S)^2(3P)^6$ configuration and three shells closed. The fourth shell closes at Kr ($Z = 36$), the fifth shell at Xe ($Z = 54$), and the sixth with the Rn ($Z = 86$) atom. Because of the shell gap in the electronic energy-level structure, these atoms with closed shells are especially stable, and they are very reluctant to combine chemically with another element. As you know, they are all inert gases.

The shell structure of the electronic energy levels explains the periods of the periodic table, its major feature. But we can understand much more than just the shell structure with the help of the self-consistent potential model. If you compare the energy levels of figure 20.1 with the hydrogenic energy levels of figure 13.1, you will discover that not only is the l-degeneracy gone, but there is an "l effect": Levels of higher l are pushed up one shell (for $l = 2$) or even two shells (for $l = 3$) when compared to the hydrogenic "shell structure." Not only that, $4S$ is often found lower in energy than $3D$; and $6S$ is lower than $4F$. What is going on? Well, there is a simple explanation on the basis of our model. The electrons in the S-orbitals experience more overlap with the small r-region; they can get much closer to the nucleus and thus enjoy the full benefit of the nuclear attraction. In contrast, the D-electrons, for example, are kept away from the nucleus by the extra centrifugal barrier and are deprived of some of its attractive energy. As a result, we find that the first electron in the beginning of a new shell always fills an S-state. Naturally, atoms with one more electron than needed for shell closure have similar chemical properties (which are determined by the last unfilled level); they are the alkali metals: Li, Na, K, and so forth. Likewise, atoms that are one electron short of making a closed-shell configuration have a "hole" in a P-orbital, and their chemical properties seem to be similar; behold the similarity of the chemistry of F, Cl, and Br.

Another interesting effect occurs when we consider the elements in which the $4F$ level is being filled—these are the *rare-earth* elements. Again, it so works out for the solutions of the self-consistent potential that the radius (where the electronic radial-probability distribution peaks) of the $6S$ level is larger than that of the $4F$ level. So once the $6S$ level is filled, the $6S$ electrons make up the outer layer of all the rare-earth elements and shield the inner $4F$ electrons, no matter how many there are. The consequence is the chemical similarity of all the rare earths. The same phenomenon explains the similarity of the properties of atoms that fill the $3D$ level, although to a lesser extent; but at least Mn, Ni, Fe, and Co all have similar chemistry.

Spectroscopic Description or Term Value

The electronic configuration does not tell us about the spectroscopic description of an atomic state, namely, the orbital, spin, and total angular momentum of the state, except for a few cases. For closed shells, clearly, the ground-state wave function is spherically symmetric and the orbital, spin, and total angular momenta are all zero. The same is true for a closed orbital. For a single electron over and above a closed shell, the electron occupies an S-orbital, and thus the total angular momentum is equal to the electron's spin, namely, $\frac{1}{2}$.

For the more general case when there are several electrons, there are several orbital angular momenta to add, and spins, and then the total orbital angular momenta to the total spin to obtain the total J. Or should we add each orbital angular momentum to the spin first to get a total angular momentum j for each electron, and then add these j's together to obtain the final total J's? Which is the right procedure? Decisions, decisions!

It would be a matter of decision if these coupled states were all degenerate, as the simple configuration description suggests. But we have been ignoring some additional interaction among the electrons that now helps us to remove the degeneracy and so makes our choice for us. There are two important interactions, each of which removes the degeneracy in part. These are (1) the residual interaction; for an electron i, this is

$$V_{\text{res}}(i) = \sum_{j \neq i} \frac{e^2}{|\mathbf{r}_i - \mathbf{r}_j|} - V_{\text{s.c.}}(i) \tag{20.15}$$

where $V_{\text{s.c.}}(i)$ is the self-consistent potential given by equation (20.11); and (2) the spin-orbit interaction, which for the electron i is

$$V_{\text{s.o.}}(i) = C(r_i)\mathbf{L}_i \cdot \mathbf{S}_i$$

where $C(r_i)$ depends on the effective central potential in which the electron moves.

The coupling scheme that is appropriate depends on the relative strength of the two interactions above. For example, if the spin-orbit coupling dominates, you already know that the total j-scheme, coupling the orbital and spin angular momenta of individual electrons to a total angular momentum, is the appropriate one. This is called the *j-j coupling scheme*:

$$\mathbf{l}_i + \mathbf{s}_i = \mathbf{j}_i$$

$$\mathbf{J} = \sum_i \mathbf{j}_i$$

In atoms it is usually the case, however, that the residual interactions are the heavies, and they get the nod. Then *L-S coupling*

$$\mathbf{L} = \sum_i \mathbf{l}_i$$

$$\mathbf{S} = \sum_i \mathbf{s}_i$$

$$\mathbf{J} = \mathbf{L} + \mathbf{S}$$

is the appropriate one; then total L, total S, and total J are all good quantum numbers (constants of the motion). You can see now that when we developed the spectroscopic notation, we had the L-S coupling in mind. L-S coupling is also called *Russell-Saunders coupling*.

The set of $(2L + 1)(2S + 1)$ states belonging to an electronic configuration with given values of S and L is called the *spectral term* or simply the *term*. If $L \geq S$, the multiplicity of the term $2S + 1$ determines the number of different total J-values that can occur, which is also the number of levels into which the term will split when the neglected spin-orbit interaction is reintroduced into the picture. For $L < S$, the number of J-values is $2L + 1$, and the number of levels will be less than the (spin) multiplicity.

Consider an example—the case of the carbon atom. The electronic configuration is $(1S)^2(2S)^2(2P)^2$, and it is the coupling of the last two orbitals that matters. The possible total spin $S = 0$, 1; and the possible L values are 0 (S), 1 (P) and 2 (D). The Pauli principle or the antisymmetry of the wave function dictates that $S = 0$ goes with symmetric orbital states, namely L = even, and $S = 1$ goes with states of odd L (cf. eq. [17.35]). Therefore, the terms are

$$^1S_0,\ ^3P_0,\ ^3P_1,\ ^3P_2, \text{ and } ^1D_2$$

Now which one of these will have the lowest energy?

Here is where some rules called *Hund's rules* become useful. The first Hund's rule is: The state with the largest spin is of the lowest energy. We can see the origin of this rule: When the spins are aligned, one gains interaction energy as in the case of helium (remember exchange interaction in chapter 19?). In the current case, the lowest state then has to be one of the ^3P-states. But which one?

Hund has a rule for just this choice: If the incomplete orbital is less than half-full, $J = |L - S|$ is the ground state; if the incomplete orbital is more than half-full, then $J = L + S$ gets the nod for the ground state. So, using this rule, the ground state of the carbon atom must have $J = 0$, since the P-orbital is less than half-full.

What if we have more than one value of L for the same value of S competing to become the ground state? In that case we invoke still another rule of

Hund: Among the levels with a given value of S, the state with the largest value of L has the smallest energy. In the case above, using this rule, we can tell that ${}^1D_2$ has lower energy than ${}^1S_0$, although neither is the ground state. Incidentally, these rules of Hund are not ad hoc; they check out with quantum mechanical calculations.

Table 20.1 summarizes the electronic configurations and the spectroscopic terms for the elements of the periodic table. The spectroscopic term is important in spectroscopy where the L, S, and J quantum numbers are part and parcel of the selection rules that govern electromagnetic transitions between atomic states. Knowing SLJ for the ground states enables us to calculate these quantum numbers for excited states, a very important task.

20.3 INTRODUCTION TO MOLECULAR STRUCTURE

Loosely speaking, we think of molecules as conglomerates of atoms bound by interatomic forces, but this picture is actually valid only for large molecules, if then. As many-body systems, molecules consist of both nuclear degrees of freedom and electronic ones. Moreover, there is no longer any spherical symmetry. Therefore, on the face of it, deciphering molecular structure seems to be a difficult task.

The saving grace is that the nuclei are much heavier than the electrons, and consequently, their motion is much slower than the electrons'. To the zeroth approximation, therefore, we can regard the nuclei as fixed centers of potential and treat the nuclear motion as a perturbation. This way of separating out some of the complexities of molecular dynamics is called the *adiabatic approximation*, also known as the *Born-Oppenheimer approximation*.

To elucidate the basic ideas of the Born-Oppenheimer approximation, let r denote the collection of all electronic coordinates and R all the nuclear coordinates. We can write the Hamiltonian as

$$H = H_0 + T_R \tag{20.16}$$

where T_R is the sum of the kinetic energies of the nuclei and is to be regarded as a perturbation on H_0 given by

$$H_0 = T_r + V(r, R) \tag{20.17}$$

where T_r is the sum of the electronic kinetic energies, and $V(r, R)$ denotes all the interactions—electron-electron, electron-nuclear, and nuclear-nuclear.

Consider, first, the Schrödinger equation of H_0, where now R are fixed parameters:

$$[T_r + V(r, R)]\phi_n(r, R) = \epsilon_n(R)\phi_n(r, R) \tag{20.18}$$

TABLE 20.1
Atomic electronic configurations and spectroscopic term-values

Z	*Element*	*Configuration*	*Spectroscopic term*
1	H	$(1S)$	$^2S_{1/2}$
2	He	$(1S)^2$	$^1S_0$
3	Li	$(\mathrm{He})(2S)$	$^2S_{1/2}$
4	Be	$(\mathrm{He})(2S)^2$	$^1S_0$
5	B	$(\mathrm{He})(2S)^2(2P)$	$^2P_{1/2}$
6	C	$(\mathrm{He})(2S)^2(2P)^2$	$^3P_0$
7	N	$(\mathrm{He})(2S)^2(2P)^3$	$^4S_{3/2}$
8	O	$(\mathrm{He})(2S)^2(2P)^4$	$^3P_2$
9	F	$(\mathrm{He})(2S)^2(2P)^5$	$^2P_{3/2}$
10	Ne	$(\mathrm{He})(2S)^2(2P)^6$	$^1S_0$
11	Na	$(\mathrm{Ne})(3S)$	$^2S_{1/2}$
12	Mg	$(\mathrm{Ne})(3S)^2$	$^1S_0$
13	Al	$(\mathrm{Ne})(3S)^2(3P)$	$^2P_{1/2}$
14	Si	$(\mathrm{Ne})(3S)^2(3P)^2$	$^3P_0$
15	P	$(\mathrm{Ne})(3S)^2(3P)^3$	$^4S_{3/2}$
16	S	$(\mathrm{Ne})(3S)^2(3P)^4$	$^3P_2$
17	Cl	$(\mathrm{Ne})(3S)^2(3P)^5$	$^2P_{3/2}$
18	Ar	$(\mathrm{Ne})(3S)^2(3P)^6$	$^1S_0$
19	K	$(\mathrm{Ar})(4S)$	$^2S_{1/2}$
20	Ca	$(\mathrm{Ar})(4S)^2$	$^1S_0$
21	Sc	$(\mathrm{Ar})(4S)^2(3D)$	$^2D_{3/2}$
22	Ti	$(\mathrm{Ar})(4S)^2(3D)^2$	$^3F_2$
23	V	$(\mathrm{Ar})(4S)^2(3D)^3$	$^4F_{3/2}$
24	Cr	$(\mathrm{Ar})(4S)(3D)^5$	$^7S_3$
25	Mn	$(\mathrm{Ar})(4S)^2(3D)^5$	$^6S_{3/2}$
26	Fe	$(\mathrm{Ar})(4S)^2(3D)^6$	$^5D_4$
27	Co	$(\mathrm{Ar})(4S)^2(3D)^7$	$^4F_{9/2}$
28	Ni	$(\mathrm{Ar})(4S)^2(3D)^8$	$^3F_4$
29	Cu	$(\mathrm{Ar})(4S)(3D)^{10}$	$^2S_{1/2}$
30	Zn	$(\mathrm{Ar})(4S)^2(3D)^{10}$	$^1S_0$
31	Ga	$(\mathrm{Ar})(4S)^2(3D)^{10}(4P)$	$^2P_{1/2}$
32	Ge	$(\mathrm{Ar})(4S)^2(3D)^{10}(4P)^2$	$^3P_0$
33	As	$(\mathrm{Ar})(4S)^2(3D)^{10}(4P)^3$	$^4S_{3/2}$
34	Se	$(\mathrm{Ar})(4S)^2(3D)^{10}(4P)^4$	$^3P_2$
35	Br	$(\mathrm{Ar})(4S)^2(3D)^{10}(4P)^5$	$^2P_{3/2}$
36	Kr	$(\mathrm{Ar})(4S)^2(3D)^{10}(4P)^6$	$^1S_0$
37	Rb	$(\mathrm{Kr})(5S)$	$^2S_{1/2}$
38	Sr	$(\mathrm{Kr})(5S)^2$	$^1S_0$
39	Y	$(\mathrm{Kr})(5S)^2(4D)$	$^2D_{3/2}$
40	Zr	$(\mathrm{Kr})(5S)^2(4D)^2$	$^3F_2$
41	Nb	$(\mathrm{Kr})(5S)(4D)^4$	$^6D_{1/2}$
42	Mo	$(\mathrm{Kr})(5S)(4D)^5$	$^7S_3$
43	Tc	$(\mathrm{Kr})(5S)^2(4D)^5$	$^6S_{5/2}$
44	Ru	$(\mathrm{Kr})(5S)(4D)^7$	$^5F_5$
45	Rh	$(\mathrm{Kr})(5S)(4D)^8$	$^4F_{9/2}$
46	Pd	$(\mathrm{Kr})(4D)^{10}$	$^1S_0$
47	Ag	$(\mathrm{Kr})(5S)(4D)^{10}$	$^2S_{1/2}$
48	Cd	$(\mathrm{Kr})(5S)^2(4D)^{10}$	$^1S_0$

(continued)

TABLE 20.1
(Continued)

Z	*Element*	*Configuration*	*Spectroscopic term*
49	In	$(\mathrm{Kr})(5S)^2(4D)^{10}(5P)$	$^2P_{1/2}$
50	Sn	$(\mathrm{Kr})(5S)^2(4D)^{10}(5P)^2$	$^3P_0$
51	Sb	$(\mathrm{Kr})(5S)^2(4D)^{10}(5P)^3$	$^4S_{3/2}$
52	Te	$(\mathrm{Kr})(5S)^2(4D)^{10}(5P)^4$	$^3P_2$
53	I	$(\mathrm{Kr})(5S)^2(4D)^{10}(5P)^5$	$^2P_{3/2}$
54	Xe	$(\mathrm{Kr})(5S)^2(4D)^{10}(5P)^6$	$^1S_0$
55	Cs	$(\mathrm{Xe})(6S)$	$^2S_{1/2}$
56	Ba	$(\mathrm{Xe})(6S)^2$	$^1S_0$
57	La	$(\mathrm{Xe})(6S)^2(5D)$	$^2D_{3/2}$
58	Ce	$(\mathrm{Xe})(6S)^2(4F)(5D)$	$^3H_5$
59	Pr	$(\mathrm{Xe})(6S)^2(4F)^3$	$^4I_{9/2}$
60	Nd	$(\mathrm{Xe})(6S)^2(4F)^4$	$^5I_4$
61	Pm	$(\mathrm{Xe})(6S)^2(4F)^5$	$^6H_{5/2}$
62	Sm	$(\mathrm{Xe})(6S)^2(4F)^6$	$^7F_0$
63	Eu	$(\mathrm{Xe})(6S)^2(4F)^7$	$^8S_{7/2}$
64	Gd	$(\mathrm{Xe})(6S)^2(4F)^7(5D)$	$^9D_2$
65	Tb	$(\mathrm{Xe})(6S)^2(4F)^9$	$^6H_{15/2}$
66	Dy	$(\mathrm{Xe})(6S)^2(4F)^{10}$	$^5I_8$
67	Ho	$(\mathrm{Xe})(6S)^2(4F)^{11}$	$^4I_{15/2}$
68	Er	$(\mathrm{Xe})(6S)^2(4F)^{12}$	$^3H_6$
69	Tm	$(\mathrm{Xe})(6S)^2(4F)^{13}$	$^2F_{7/2}$
70	Yb	$(\mathrm{Xe})(6S)^2(4F)^{14}$	$^1S_0$
71	Lu	$(\mathrm{Xe})(6S)^2(4F)^{14}(5D)$	$^2D_{3/2}$
72	Hf	$(\mathrm{Xe})(6S)^2(4F)^{14}(5D)^2$	$^3F_2$
73	Ta	$(\mathrm{Xe})(6S)^2(4F)^{14}(5D)^3$	$^4F_{3/2}$
74	W	$(\mathrm{Xe})(6S)^2(4F)^{14}(5D)^4$	$^5D_0$
75	Re	$(\mathrm{Xe})(6S)^2(4F)^{14}(5D)^5$	$^6S_{5/2}$
76	Os	$(\mathrm{Xe})(6S)^2(4F)^{14}(5D)^6$	$^5D_4$
77	Ir	$(\mathrm{Xe})(6S)^2(4F)^{14}(5D)^7$	$^4F_{9/2}$
78	Pt	$(\mathrm{Xe})(6S)(4F)^{14}(5D)^9$	$^3D_3$
79	Au	$(\mathrm{Xe})(6S)(4F)^{14}(5D)^{10}$	$^2S_{1/2}$
80	Hg	$(\mathrm{Xe})(6S)^2(4F)^{14}(5D)^{10}$	$^1S_0$
81	Tl	$(\mathrm{Xe})(6S)^2(4F)^{14}(5D)^{10}(6P)$	$^2P_{1/2}$
82	Pb	$(\mathrm{Xe})(6S)^2(4F)^{14}(5D)^{10}(6P)^2$	$^3P_0$
83	Bi	$(\mathrm{Xe})(6S)^2(4F)^{14}(5D)^{10}(6P)^3$	$^4S_{3/2}$
84	Po	$(\mathrm{Xe})(6S)^2(4F)^{14}(5D)^{10}(6P)^4$	$^3P_2$
85	At	$(\mathrm{Xe})(6S)^2(4F)^{14}(5D)^{10}(6P)^5$	$^2P_{3/2}$
86	Rn	$(\mathrm{Xe})(6S)^2(4F)^{14}(5D)^{10}(6P)^6$	$^1S_0$
87	Fr	$(\mathrm{Rn})(7S)$	$^2S_{1/2}$
88	Ra	$(\mathrm{Rn})(7S)^2$	$^1S_0$
89	Ac	$(\mathrm{Rn})(7S)^2(6D)$	$^2D_{3/2}$
90	Th	$(\mathrm{Rn})(7S)^2(6D)^2$	$^3F_2$
91	Pa	$(\mathrm{Rn})(7S)^2(5F)^2(6D)$	$^4K_{11/2}$
92	U	$(\mathrm{Rn})(7S)^2(5F)^3(6D)$	$^5L_6$
93	Np	$(\mathrm{Rn})(7S)^2(5F)^4(6D)$	$^6L_{11/2}$
94	Pu	$(\mathrm{Rn})(7S)^2(5F)^6$	$^7F_0$
95	Am	$(\mathrm{Rn})(7S)^2(5F)^7$	$^8S_{7/2}$
96	Cm	$(\mathrm{Rn})(7S)^2(5F)^7(6D)$	$^9D_2$

(continued)

TABLE 20.1
(Continued)

Z	*Element*	*Configuration*	*Spectroscopic term*
97	Bk	$(\mathrm{Rn})(7S)^2(5F)^9$	$^6H_{15/2}$
98	Cf	$(\mathrm{Rn})(7S)^2(5F)^{10}$	$^5I_8$
99	Es	$(\mathrm{Rn})(7S)^2(5F)^{11}$	$^4I_{15/2}$
100	Fm	$(\mathrm{Rn})(7S)^2(5F)^{12}$	$^3H_6$
101	Md	$(\mathrm{Rn})(7S)^2(5F)^{13}$	$^2F_{7/2}$
102	No	$(\mathrm{Rn})(7S)^2(5F)^{14}$	$^1S_0$

where the eigenvalues $\epsilon_n(R)$ and the eigenfunctions ϕ_n must be evaluated as functions of the fixed parameter R.

Suppose that somehow we have figured out the solutions of equation (20.18). Since these solutions form a complete set, now we can expand the solution of the full Schrödinger equation

$$H\psi(r,R) = E\psi(r,R) \tag{20.19}$$

in terms of the complete set of the ϕ_n's:

$$\psi(r,R) = \sum_n \Phi_n(R)\phi_n(r,R) \tag{20.20}$$

The expansion coefficients depend only on R, since we are expanding only the r-dependence of the wave function ψ.

Substitute equation (20.20) into equation (20.19), use equation (20.18), multiply by $\phi_m^*(r,R)$, and integrate over all electronic positions r. This gives

$$\sum_n \int dr\, \phi_m^*(r,R)T_R\Phi_n(R)\phi_n(r,R) + \epsilon_m(R)\Phi_m(R) = E\Phi_m(R) \tag{20.21}$$

where we have used the orthogonality of the set $\phi_n(r,R)$. Now the operator T_R, which is a sum of ∇_R^2, operates on both $\Phi_n(R)$ and $\phi_n(r,R)$. But

$$\nabla^2(\Phi\phi) = \phi\nabla^2\Phi + \text{terms containing derivatives of } \phi$$

If we retain only the $\phi\nabla^2\Phi$ term, the ϕ's integrate out of equation (20.21), and we get

$$[T_R + \epsilon_m(R)]\Phi_m(R) = E\Phi_m(R) \tag{20.22}$$

This is just the Schrödinger equation for $\Phi_m(R)$ (which are the probability amplitudes for the nuclei to be at R while the electrons are in the state ϕ_n) in the potential $\epsilon_m(R)$, which is the eigenenergy of the electrons for fixed positions of the nuclei.

It follows that in the Born-Oppenheimer approximation, the coherent superposition ψ reduces to a simple product wave function:

$$\psi_{m\nu}(r,R) = \Phi_{m\nu}(R)\phi_m(r,R) \tag{20.23}$$

where the label ν distinguishes the different solutions of equation (20.22). For each state of the light particles determined by the quantum number m, there corresponds a state of the heavy particles determined by a different quantum number ν. This is a tremendous simplification. Notice that the terms we neglected in order to arrive at equation (20.22) can be included as higher-order approximations, but ordinarily, the lowest order above is enough.

Electronic Motion, Vibration, and Rotation

According to the Born-Oppenheimer approximation, then, we calculate molecular states by (1) solving the electronic motion for fixed nuclear centers and (2) using the electronic energy (as a function of nuclear positions) as the potential function to describe the nuclear motion.

The potential $\epsilon_m(R)$ in the Schrödinger equation (20.22) in the nuclear coordinates will have a set of minima; these are the equilibrium positions of the nuclei. For a small deviation from the equilibrium position R_0, we can expand the potential $\epsilon_m(R)$ as follows:

$$\epsilon_m(R) \approx \epsilon_m(R_0) + \tfrac{1}{2}(R - R_0)^2[d^2\epsilon_m/dR^2]_0 \tag{20.24}$$

The second term is the oscillator potential and gives rise to vibrational motion of the nuclei, manifested as equidistant levels. Moreover, the operator T_R in equation (20.22) involves angular coordinates as well; thus the nuclei will also take on rotational motion, manifested as $E_J \sim J(J + 1)$.

Let's estimate the energies involved in the vibrational and rotational motions compared to the electronic motion. For electronic motion, the order of magnitude of the electronic energies follows from the uncertainty principle: The position uncertainty is of the order of the molecular size a. Therefore $\Delta p \sim \hbar/a$, and this gives

$$\epsilon \approx \Delta p^2/2m = \hbar^2/2ma^2 \tag{20.25}$$

From equation (20.24), the frequency of the vibrational motion is given by

$$M\omega^2 \approx d^2\epsilon/dR^2$$

where M is the nuclear mass. If $\epsilon \sim 1/a^2$, then from a dimensional argument, $d^2\epsilon/dR^2 \sim 1/a^4$. Therefore,

$$d^2\epsilon/dR^2 \approx \hbar^2/ma^4$$

and

$$\hbar\omega \approx (m/M)^{1/2}\hbar^2/ma^2 \approx (m/M)^{1/2}\epsilon \tag{20.26}$$

The vibrational energies are smaller than the electronic ones by the factor $(m/M)^{1/2}$, the square root of the ratio of the electronic mass and the nuclear mass. Similarly, for rotation, we have

$$E_J = J(J+1)\hbar^2/2I \approx \hbar^2/2Ma^2 = (m/M)\epsilon \tag{20.27}$$

since the moment of inertia I is $\approx Ma^2$. The rotational-level energies are smaller even than the vibrational by the factor $(m/M)^{1/2}$.

It follows that the vibrations and rotations of the molecule have a negligible effect on the electronic level structure. Nevertheless, each electronic level will have superimposed on it both rotational and vibrational levels.

Molecular Orbitals

One important question is, When or in what configuration does a molecule bind? To answer this question we have to calculate the eigenvalues $\epsilon_m(R)$ and see if this potential has a minimum as a function of R, a minimum deep enough to hold the nuclei together. Let's calculate the case of the simplest molecule, the H_2^+ ion, to give you a flavor of how the idea works.

The Hamiltonian H_0 is given as

$$H_0 = -\frac{\hbar^2\nabla^2}{2m} - \frac{e^2}{|\mathbf{r}-\mathbf{R}_a|} - \frac{e^2}{|\mathbf{r}-\mathbf{R}_b|} + \frac{e^2}{|\mathbf{R}_a-\mathbf{R}_b|} \tag{20.28}$$

where $\mathbf{r}$ denotes the electron's position and $\mathbf{R}_a$ and $\mathbf{R}_b$ the positions of the two nuclei. The Schrödinger equation of the Hamiltonian can be solved using elliptical coordinates; however, for our purpose, sketching how a variational calculation works in this case will be adequate.

A variational calculation, as you know, begins with the choice of a trial wave function. Physically, the electron can be thought of as being bound to either this proton or that one; therefore, an appropriate trial wave function is the coherent superposition of the electron's hydrogenic $(1S)$ state with each of the protons:

$$\phi(\mathbf{r},\mathbf{R}_a,\mathbf{R}_b) = c_1\phi_a(\mathbf{r},\mathbf{R}_a) + c_2\phi_b(\mathbf{r},\mathbf{R}_b) \tag{20.29}$$

where

$$\phi_a(\mathbf{r},\mathbf{R}_a) = (\pi a_0^3)^{-1/2}\exp(-|\mathbf{r}-\mathbf{R}_a|/a_0)$$

$$\phi_b(\mathbf{r},\mathbf{R}_b) = (\pi a_0^3)^{-1/2}\exp(-|\mathbf{r}-\mathbf{R}_b|/a_0) \tag{20.30}$$

However, since the potential is symmetric under reflection about a plane through the midpoint $(\mathbf{R}_a + \mathbf{R}_b)/2$ of the molecule normal to the axis, more appropriate trial functions are those that reflect this symmetry, namely, even- and odd-parity combinations of the functions ϕ_a and ϕ_b:

$$\phi_{\pm}(\mathbf{r},\mathbf{R}_a,\mathbf{R}_b) = c_{\pm}[\phi_a(\mathbf{r},\mathbf{R}_a) \pm \phi_b(\mathbf{r},\mathbf{R}_b)] \tag{20.31}$$

This defines a *molecular orbital*; in this particular form it is a linear combination of atomic orbitals (LCAO). The normalization factors $c_{\pm}$ are given as

$$\begin{aligned} \frac{1}{c_{\pm}^2} &= \langle \phi_{\pm} | \phi_{\pm} \rangle = 2 \pm 2 \int d^3\mathbf{r}\, \phi_a(\mathbf{r},\mathbf{R}_a)\phi_b(\mathbf{r},\mathbf{R}_b) \\ &= 2 \pm 2S(R) \end{aligned} \tag{20.32}$$

where the integral in equation (20.32) is called the *overlap integral* (for obvious reasons) and is denoted as $S(R)$.

Integrals such as S are most easily evaluated by using elliptical coordinates ρ, σ, and ϕ, defined by

$$\begin{aligned} \rho &= (1/R)\,[|\mathbf{r} - \mathbf{R}_a| + |\mathbf{r} - \mathbf{R}_b|] \\ \sigma &= (1/R)\,[|\mathbf{r} - \mathbf{R}_a| - |\mathbf{r} - \mathbf{R}_b|] \end{aligned} \tag{20.33}$$

where $R = |\mathbf{R}_a - \mathbf{R}_b|$. ϕ is the usual azimuthal angle. Moreover, we have (see a book on advanced calculus)

$$\int d^3\mathbf{r} = \int_1^{\infty} d\rho \int_{-1}^{1} d\sigma \int_0^{2\pi} d\phi\, \frac{R^3}{8}\,(\rho^2 - \sigma^2)$$

Therefore,

$$\begin{aligned} S &= \frac{1}{8\pi}\frac{R^3}{a_0^3} \int_1^{\infty} e^{-(R/a_0)\rho}\, d\rho \int_{-1}^{1} (\rho^2 - \sigma^2)\, d\sigma \int_0^{2\pi} d\phi \\ &= \left[1 + \frac{R}{a_0} + \frac{R^2}{3a_0^2}\right] e^{-R/a_0} \end{aligned} \tag{20.34}$$

The expectation value of H_0 in the two states $|\phi_{\pm}\rangle$ is given as

$$\begin{aligned} \langle H_0 \rangle_{\pm} = \epsilon_{\pm}(R) &= \frac{1}{2 \pm 2S}\,[\langle \phi_a | H_0 | \phi_a \rangle + \langle \phi_b | H_0 | \phi_b \rangle \pm 2\langle \phi_a | H_0 | \phi_b \rangle] \\ &= \frac{1}{1 \pm S}\,[\langle \phi_a | H_0 | \phi_a \rangle \pm \langle \phi_a | H_0 | \phi_b \rangle] \end{aligned} \tag{20.35}$$

where clearly, from symmetry, $\langle\phi_a|H_0|\phi_a\rangle = \langle\phi_b|H_0|\phi_b\rangle$. We have

$$\langle\phi_a|H_0|\phi_a\rangle = \int d^3\mathbf{r}\,\phi_a^*\left[\frac{-\hbar^2\nabla^2}{2m} - \frac{e^2}{|\mathbf{r}-\mathbf{R}_a|} - \frac{e^2}{|\mathbf{r}-\mathbf{R}_b|} + \frac{e^2}{R}\right]\phi_a$$
$$= E_1 + \frac{e^2}{R} - e^2\int d^3\mathbf{r}\,\frac{|\phi_a(\mathbf{r},\mathbf{R}_a)|^2}{|\mathbf{r}-\mathbf{R}_b|} \tag{20.36}$$

where E_1 is the hydrogenic ground-state energy of $-1Ry$, $|\phi_a\rangle$ being the eigenstate of the first two terms in H_0. The remaining integral in equation (20.36) is the potential energy of attraction of the electronic charge distribution about one proton toward the other proton. This integral also can be evaluated using elliptical coordinates:

$$\int d^3\mathbf{r}\,\frac{|\phi_a|^2}{|\mathbf{r}-\mathbf{R}_b|} = \frac{1}{R}\left[1 - e^{-2R/a_0}\left(1+\frac{R}{a_0}\right)\right]$$

Substituting back into equation (20.36), we obtain

$$\langle\phi_a|H_0|\phi_a\rangle = E_1 + (e^2/R)(1 + R/a_0)\exp(-2R/a_0)$$
$$= \langle\phi_b|H_0|\phi_b\rangle \tag{20.37}$$

We also have to calculate $\langle\phi_a|H_0|\phi_b\rangle$, which, following the same reasoning as above, is given as

$$\langle\phi_a|H_0|\phi_b\rangle = (E_1 + e^2/R)S - e^2\int d^3\mathbf{r}\,\frac{\phi_a\phi_b}{|\mathbf{r}-\mathbf{R}_b|} \tag{20.38}$$

The integral here can be evaluated using the same trick as above and will be left as an exercise. The final result is

$$\langle\phi_a|H_0|\phi_b\rangle = [E_1 + (e^2/R)]\,S(R) - (e^2/a_0)\,[1 + (R/a_0)]\exp(-R/a_0) \tag{20.39}$$

It is now an easy matter to combine all the pieces and calculate the eigenvalues $\epsilon_\pm(R)$ as a function of R. Figure 20.2 is the plot of these functions. The result is interesting. The molecule does not bind in the odd-parity state, $\epsilon_-(R)$ does not have a minimum at any R; for this reason the odd-parity state is called the *anti-bonding orbital.* The reason for the failure to bond is related to the symmetry requirement that it has to satisfy, namely, vanish at $(\mathbf{R}_a + \mathbf{R}_b)/2$; consequently it does not reap the benefit of a region of rather large attractive potential energy.

For the even-parity state, on the other hand, the probability reaches a maximum for the electron to be halfway between the nuclei, and the electron gains

FIGURE 20.2
The energy of the hydrogen molecular ion as a function of the nuclear separation for the symmetric (+) and the antisymmetric (−) states. Only the symmetric state binds.

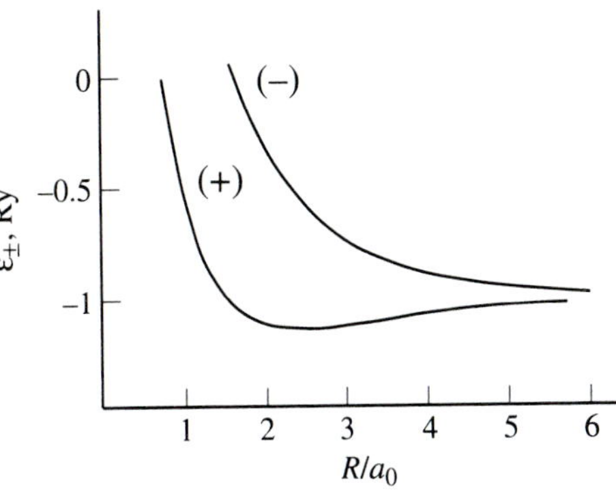

valuable attractive energy from both protons. As a result, the even-parity state does bind, and it is called the *bonding orbital.*

The H_2 Molecule: An Example of the Valence-Bond Method

The H_2 molecule is the simplest molecule that has the complexity of having both more than one center of Coulomb attraction and more than one electron to share those centers (fig. 20.3). If we ignore the spin-orbit coupling, the energy eigenstates can be classified according to their total spin, which can be either singlet or triplet state.

Which one, singlet or triplet, should get the nod for the ground state? Let's attempt an answer by looking at the symmetry. If we use the *molecular-orbital method* in which a trial wave function consisting of the antisymmetrized product of molecular orbitals, such as that of equation (20.31), is used for a variational calculation, the answer is easy. The lowest energy is obtained when we put both electrons in the same bonding orbital ϕ_+; but we can form only a symmetric spatial function then—the antisymmetric combination is identically zero. And this means that the ground state is a singlet state.

Objection! If we display the spatial part of the wave function constructed above, it looks like this:

$$\Psi_{\text{sym}}(1,2) \sim [\phi_a(\mathbf{r}_1) + \phi_b(\mathbf{r}_1)][\phi_a(\mathbf{r}_2) + \phi_b(\mathbf{r}_2)]$$
$$= [\phi_a(\mathbf{r}_1)\phi_a(\mathbf{r}_2) + \phi_b(\mathbf{r}_1)\phi_b(\mathbf{r}_2)] + [\phi_a(\mathbf{r}_1)\phi_b(\mathbf{r}_2) + \phi_b(\mathbf{r}_1)\phi_a(\mathbf{r}_2)]$$

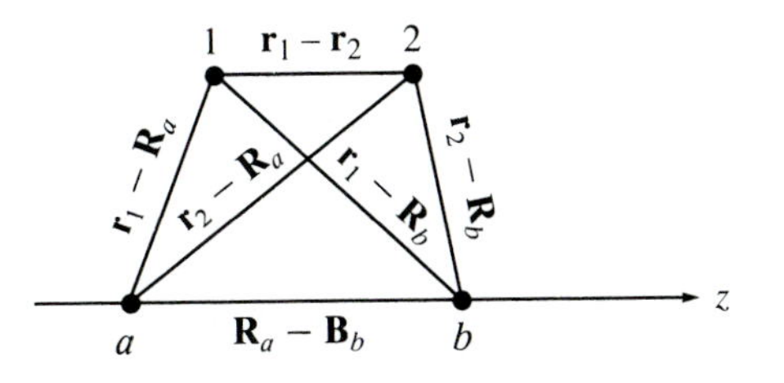

FIGURE 20.3
The geometry of the hydrogen (H_2) molecule.

In the first two terms, both electrons revolve around the same center, a or b; the last two terms have one electron going around each proton. For large separation R between the nuclei, it is obvious that the last term—the depiction of the molecule as two hydrogen atoms—is more appropriate, yet both terms are given equal weight!

So in the *valence-bond method*, we chuck the $\phi_a(1)\phi_a(2) + \phi_b(1)\phi_b(2)$ part and go with a variational wave function that consists of just the second term above. Below we will sketch how this method works but not go into details.

We have to calculate the expectation value of the Hamiltonian

$$H_0 = -\frac{\hbar^2\nabla^2(1)}{2m} - \frac{\hbar^2\nabla^2(2)}{2m} - \frac{e^2}{|\mathbf{r}_1 - \mathbf{R}_a|} - \frac{e^2}{|\mathbf{r}_1 - \mathbf{R}_b|} - \frac{e^2}{|\mathbf{r}_2 - \mathbf{R}_a|} - \frac{e^2}{|\mathbf{r}_2 - \mathbf{R}_b|} + \frac{e^2}{|\mathbf{R}_a - \mathbf{R}_b|} + \frac{e^2}{|\mathbf{r}_1 - \mathbf{r}_2|} \quad (20.40)$$

in the trial wave functions

$$\Psi_\pm(1,2) = [2(1 \pm S^2)]^{-1/2}[\phi_a(\mathbf{r}_1)\phi_b(\mathbf{r}_2) \pm \phi_b(\mathbf{r}_1)\phi_a(\mathbf{r}_2)] \quad (20.41)$$

where the + (−) goes with the singlet (triplet) spin state. Again we use the fact that $\Psi_\pm$ are eigenstates of the single-particle part of H_0. Both eigenvalues are E_1. Therefore, we obtain

$$\epsilon_\pm(R) = \langle H_0\rangle_\pm = 2E_1 + e^2\int d^3r_1\, d^3r_2\, \Psi_\pm^*(1,2)\left[\frac{1}{R} + \frac{1}{|\mathbf{r}_1 - \mathbf{r}_2|} - \frac{1}{|\mathbf{r}_1 - \mathbf{R}_b|} - \frac{1}{|\mathbf{r}_2 - \mathbf{R}_a|}\right]\Psi_\pm(1,2)$$

Now expanding $\Psi_\pm$ and using symmetry, we get

$$\epsilon_\pm(R) = 2E_1 + \frac{e^2}{R} + \frac{V_C(R) \pm V_{ex}(R)}{(1 \pm S^2)} \quad (20.42)$$

where $V_C(R)$, which is called the Coulomb integral, is given as

$$V_C(R) = \int d^3r_1\, d^3r_2\, |\phi_a(\mathbf{r}_1)|^2|\phi_b(\mathbf{r}_2)|^2\left[\frac{e^2}{|\mathbf{r}_1 - \mathbf{r}_2|} - \frac{e^2}{|\mathbf{r}_1 - \mathbf{R}_b|} - \frac{e^2}{|\mathbf{r}_2 - \mathbf{R}_a|}\right] \quad (20.43)$$

and where $V_{ex}(R)$, called the exchange integral, is given as

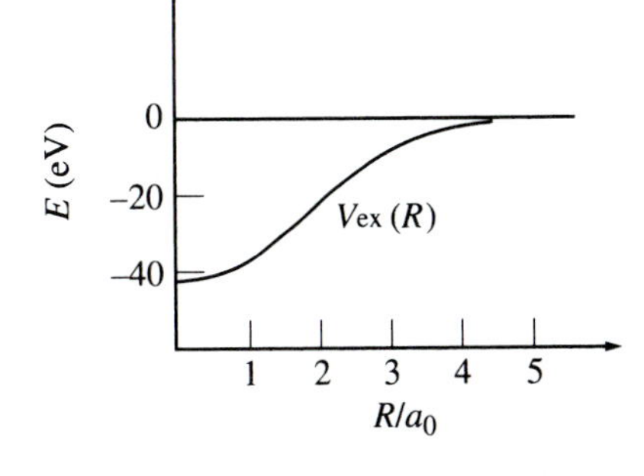

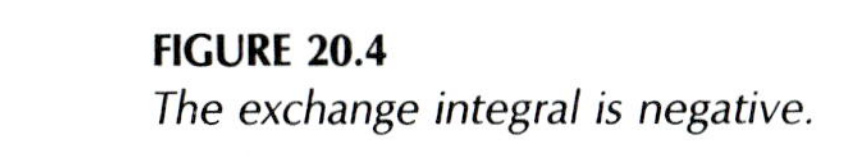

FIGURE 20.4
The exchange integral is negative.

$$V_{\text{ex}}(R) = \int d^3r_1\, d^3r_2\, \phi_a^*(\mathbf{r}_1)\phi_b^*(\mathbf{r}_2)\left[\frac{e^2}{|\mathbf{r}_1 - \mathbf{r}_2|} - \frac{e^2}{|\mathbf{r}_1 - \mathbf{R}_b|} - \frac{e^2}{|\mathbf{r}_2 - \mathbf{R}_a|}\right] \times \phi_b(\mathbf{r}_1)\phi_a(\mathbf{r}_2) \tag{20.44}$$

The exchange integral is, in general, negative (fig. 20.4), and it is this fact that determines that $\epsilon_+(R) < \epsilon_-(R)$; that is, the singlet state is lower in energy than the triplet state (fig. 20.5). Notice, also, how a purely quantum effect, the exchange interaction, plays a crucial role in molecular binding.

So out of the two methods, the molecular orbital and valence bond, which is the superior one, or, at least, which is the more used? The valence bond method has intuitive virtues and has received much support from Linus Pauling, no less. Nevertheless, the popularity of the molecular-orbital calculation has increased over the years. Among other accomplishments, the molecular orbital method has been successfully applied to treat the famous case of circular bonding in the benzene molecule.

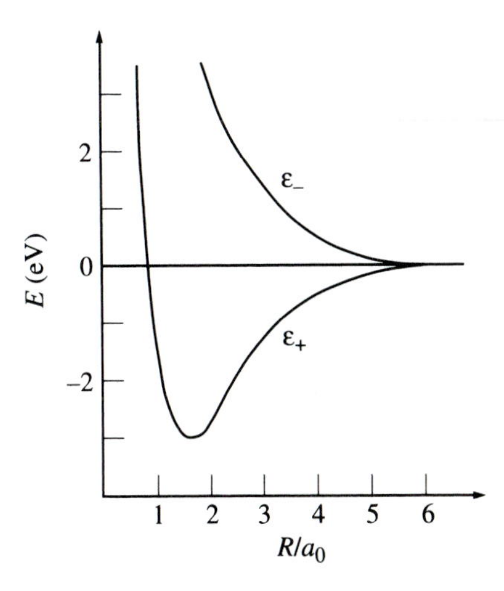

FIGURE 20.5
The energy of symmetric (singlet) and antisymmetric (triplet) states for the H_2 molecule as a function of nuclear separation. Only the symmetric state binds.

A final comment is in order. Neither of these two methods, molecular orbital or valence bond, gives good result for the H_2 molecular binding energy. The reason is that both wave functions describe the behavior for small separation of the protons quite wrongly. The point is that, when the two protons are close, the system looks more like the He atom, with charge $Z = 2$, while both variational wave functions continue to put the electrons in S-states for $Z = 1$.

PROBLEMS

1. Use the antisymmetrization operator (eq. [20.13] in text) to generate the normalized, antisymmetrized wave function for the Li atom assuming the unantisymmetrized product to be $\phi_{1S}\chi_{+}(1)\phi_{1S}\chi_{-}(2)\phi_{2S}\chi_{+}(3)$. Hence, express the function as a Slater determinant.
2. What are the total angular momentum values permitted for the following spectroscopic terms:

$$^3S,\ ^3P,\ ^4D,\ ^3F$$

3. List all the possible spectroscopic terms that can arise from the following electronic configurations:

$$(nS)(n'P),\ (nP)(n'P),\ (3D)^2,\ (nS)(n'P)^2$$

4. Using Hund's rules, determine the ground-state terms for the following atoms (show your reasoning):

O, K, Zr, Co

5. Give an order of magnitude numerical estimate of the energy of electronic, vibrational, and rotational states of molecules.
6. The nuclei of a diatomic molecule are moving in a potential field given as

$$\epsilon(R) = -2D[(a_0/R) - (a_0^2/R^2)] + (\hbar^2/2MR^2)J(J+1)$$

Express this potential near its minimum by a linear oscillator potential and determine the vibrational energies of the molecule.
7. Evaluate the integral in equation (20.38).

ADDITIONAL PROBLEMS

A1. What spectral terms are possible for the atomic state with the configuration $(1S)^2(2S)^2(2P)(3S)$? Show the states in an energy level diagram. The diagram need not be in scale but should reflect the proper ordering of levels in L-S coupling upon consideration of effects due to residual Coulomb interactions and spin-orbit coupling.

A2. For the hydrogen molecule, the nuclear component of the wave function must be antisymmetric with respect to exchange of the two nucleons. The nuclear wave function has a spatial part (rotational component) and a spin component. If the spin is singlet ($S = 0$, parahydrogen) or triplet ($S = 1$, orthohydrogen), what can you say for the symmetry of the rotational states? At ordinary temperatures, what ratio of ortho- to parahydrogen would you expect? Discuss qualitatively how the picture changes as we cool hydrogen molecules toward absolute zero.

REFERENCES

G. Baym. *Lectures on Quantum Mechanics.*
A. S. Davydov. *Quantum Mechanics.*
S. Gasiorowicz. *Quantum Physics.*

21

Quantum Systems: Fermi and Bose Gases

The simplest model of a solid of volume Ω, for example a metal, obtains when we ignore its crystal structure and simply look upon Ω as an infinite-potential box in three dimensions filled with noninteracting electrons. The normalized energy eigenstates can be taken to be three-dimensional plane waves, $\Omega^{-1/2}\exp(i\mathbf{k}\cdot\mathbf{r})$, provided we use periodic boundary conditions (see chapter 3). This means that the possible wave vectors $\mathbf{k}$ for a cubic box of dimension L are given by

$$k_x = 2\pi n_x/L, \qquad k_y = 2\pi n_y/L, \qquad k_z = 2\pi n_z/L \tag{21.1}$$

where n_x, n_y, and n_z are integers in the range $(-\infty,\infty)$.

Suppose we begin to fill the box with N electrons. We can put two electrons (spin up and spin down) in each $\mathbf{k}$-level, and the configuration corresponding to the lowest energy will obtain when all the $\mathbf{k}$-states up to a maximum $\mathbf{k}_F$ are filled; $\mathbf{p}_F = \hbar\mathbf{k}_F$ is called the Fermi momentum. We have

$$N = \sum_{\mathbf{k}\le\mathbf{k}_F} 2 \tag{21.2}$$

You can think of the ground configuration of filled $\mathbf{k}$-states forming a sphere in momentum space—call it the *Fermi sphere* (fig. 21.1); it is also called the Fermi sea, and you can imagine excitations as waves of the sea.

If N is large, and if the size L of the box is of macroscopic order, the levels of discrete $\mathbf{k}$ are spaced very closely. Under this situation, we certainly are entitled to replace the sum above by an integral. The rule for such a conversion

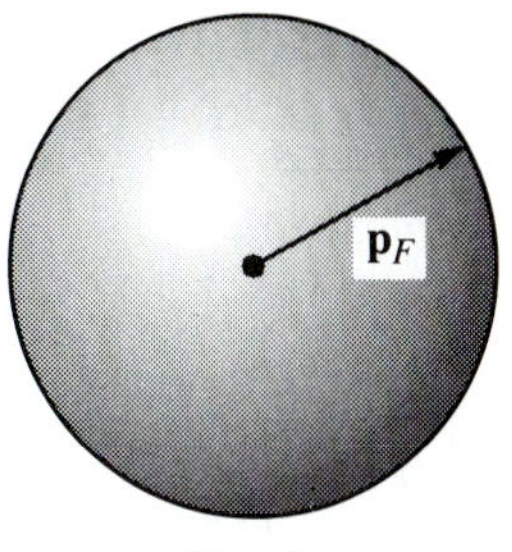

FIGURE 21.1
The Fermi sphere in momentum space called the Fermi sea.

is easily found. From equation (21.1), the interval between two consecutive k_x-values is $2\pi/L$. It follows that

$$\sum_{k_x} \to \frac{L}{2\pi} \int dk_x$$

Extending the same argument to y- and z-dimensions, we get

$$\sum_{\mathbf{k}} \to \frac{\Omega}{(2\pi)^3} \int d^3k \tag{21.3}$$

since $L^3 = \Omega$. Substituting back into equation (21.2), we get

$$N = \sum_k 2 = \frac{2\Omega}{(2\pi)^3} \int d^3k = \frac{2\Omega}{(2\pi)^3} 4\pi \int_0^{k_F} k^2\, dk = \frac{\Omega}{\pi^2} \frac{k_F^3}{3}$$

This gives the value of the Fermi momentum $\hbar k_F$ in terms of the density

$$n = N/\Omega = k_F^3/3\pi^2 \tag{21.4}$$

$$p_F = \hbar k_F = (3\pi^2 n)^{1/3}\hbar \tag{21.5}$$

The energy eigenvalue for the box states is $\hbar^2k^2/2m$. The *Fermi energy* is defined as the energy of a fermion at the top of the Fermi sea; that is,

$$E_F = \hbar^2 k_F^2/2m = (\hbar^2\pi^2/2m)(3n/\pi)^{2/3} \tag{21.6}$$

The total energy of the filled Fermi sphere is given as

$$E = 2\sum_{\mathbf{k}\le\mathbf{k}_F} \frac{\hbar^2k^2}{2m} = \frac{2\Omega}{(2\pi)^3}\frac{\hbar^2}{2m}\int_{\mathbf{k}\le\mathbf{k}_F} d^3k\, k^2 = \frac{8\pi\Omega}{(2\pi)^3}\frac{\hbar^2}{2m}\int_0^{k_F} k^4\, dk = \frac{\hbar^2}{2m}\frac{\Omega k_F^5}{5\pi^2}$$

Substituting from equation (21.4), we get

$$E = N\frac{3}{5}\frac{\hbar^2 k_F^2}{2m} = \frac{\pi^3\hbar^2}{10m}\left(\frac{3n}{\pi}\right)^{5/3}\Omega \tag{21.7}$$

Additionally, we have

$$E/N = (3/5)E_F \tag{21.8}$$

Typically, values of E_F in a metal are 5–10 eV. At normal temperature, kT (don't be confused, this k is Boltzmann's constant) $\approx 1/40$ eV; accordingly, the most easily available states for electronic excitations are all occupied. Only a very few electrons can be excited. Therefore, in a picture of a metal as a crystal lattice of ions with an average of one free electron per atom, only the ions contribute to the specific heat. In this way, the noninteracting Fermi gas model can explain such basic properties about metals.

However, for a more appropriate account, for example, for a theory of the difference between conductors, semiconductors, and nonconductors, we need to bring in further aspects of the real situation, and an interacting Fermi gas model is necessary. In this chapter, we will discuss one such model, the periodic potential model. The particles are assumed to interact with a common potential but still are independent of one another.

The simplest model of nuclei can also be constructed in terms of a noninteracting Fermi gas, and such a model has some limited applications for the bulk properties of nuclei. But for understanding the structure of individual nuclei, one has to resort to a potential model. In this chapter, we will also formulate the shell model of the nucleus based on a three-dimensional harmonic oscillator plus spin-orbit potential model.

The interacting Fermi gas model has also found notable application in the physics of white dwarfs and neutron stars, which are important in astrophysics. We will not treat them here.

The last section of the chapter will be devoted to a Bose gas whose particles construe a total wave function that is symmetric under exchange. We will see how this leads to a very different physical situation from fermions.

21.1 THE PERIODIC POTENTIAL MODEL OF THE SOLID

To achieve an understanding of why some materials are good conductors, while others are insulators, we have to invoke the effect of the crystal lattice structure on the energy levels of the electrons. One of the simplest ways to get the gist of the problem is to consider the motion of an electron in a one-dimensional periodic potential V:

$$V(x + L) = V(x) \tag{21.9}$$

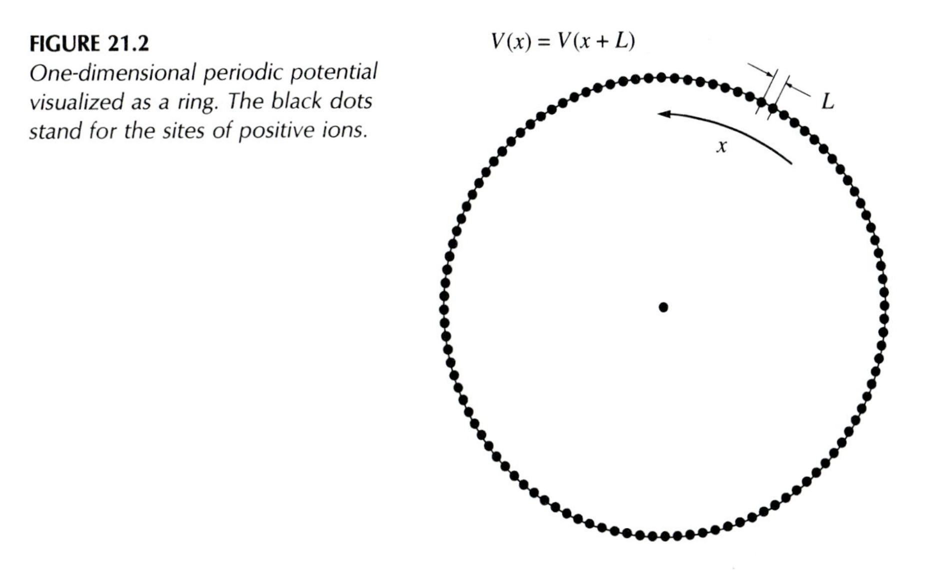

FIGURE 21.2
One-dimensional periodic potential visualized as a ring. The black dots stand for the sites of positive ions.

with periodicity distance L. The periodicity is violated at the end points of a linear lattice, but we can imagine the lattice to be bent in the shape of a snake biting its own tail, becoming a ring of circumference NL where N is the number of cells in the ring (fig. 21.2). We shall be interested only for large N, say 10^{23}.

The Hamiltonian H for the system is given as

$$H = \frac{p^2}{2m} + V(x) \tag{21.10}$$

To find the eigenfunctions of H, we invoke its symmetry under the displacement by L. Literally, for this Hamiltonian system, we cannot tell whether we are at x or at $x + L$; and there must be some advantage to be gained from this symmetry.

Symmetry Under Finite Displacements

Define the displacement operator D such that

$$Df(x) = f(x + L) \tag{21.11}$$

We are interested in finding the eigenfunctions of D

$$D\psi(x) = \psi(x + L) = \tilde{d}\psi(x) \tag{21.12}$$

where $\tilde{d}$ is the eigenvalue of D. Using Floquet's theorem (for a proof, read Mathews and Walker, *Mathematical Methods of Physics*) of differential equations, we can represent the eigenfunctions as

$$\psi(x) = \exp(ikx)v(x) \tag{21.13}$$

where k is arbitrary, and v is periodic:

$$v(x + L) = v(x) \tag{21.14}$$

The eigenvalue $\tilde{d}$ of D, corresponding to the eigenfunction ψ, is given as $\exp(ikL)$. Check it out from equation (21.12).

Since we have a closed ring, we must also impose single-valuedness at $x = NL$. This gives

$$\psi(x + NL) = \psi(x)$$

or

$$\exp[ik(x + NL)] = \exp(ikx)$$

which requires that

$$k = 2\pi n/NL, \quad \text{with } n = 0, \pm 1, \pm 2, \ldots \tag{21.15}$$

k is called the *propagation constant* of the state.

The crucial point is that the system is symmetric under D; in other words, D commutes with H,

$$[D, H] = 0 \tag{21.16}$$

which means that D and H have simultaneous eigenfunctions. It follows that the eigenfunctions of H have the same form as $\psi(x)$ above, equation (21.13): This is called *Bloch's theorem*, and the wave functions of the form of $\psi(x)$ are called *Bloch wavefunctions*.

So what has been gained by all this? The point is this. If $\psi(x)$ is known over any one cell of the periodic lattice, using equation (21.12), we can instantly calculate it for any other cell:

$$\psi(x + L) = \exp(ikL)\psi(x) \tag{21.17}$$

So we need to solve the Hamiltonian of equation (21.10) for only one cell.

Kronig-Penny Model

The Schrödinger equation for the Hamiltonian of equation (21.10)

$$\left[-\frac{\hbar^2}{2m}\frac{d^2}{dx^2} + V(x)\right]\psi(x) = E\psi(x) \tag{21.18}$$

upon substitution of the solution, equation (21.13), in the Bloch form gives the following equation for $v(x)$:

$$\frac{d^2v}{dx^2} + 2ik\frac{dv}{dx} + \frac{2m}{\hbar^2}\left[E - V(x) - \frac{\hbar^2k^2}{2m}\right]v(x) = 0 \tag{21.19}$$

The solution of this equation will be much simplified because of the periodicity conditions $v(x + L) = v(x)$ and $dv(x + L)/dx = dv/dx$.

We will do the explicit calculation only for the Kronig-Penny model consisting of a square potential lattice (fig. 21.3) with wells of depth V_0 and width b and hills of width $c(b + c = L)$. The energy $E < V_0$ is shown by the solid line. Let's introduce the notation

$$k_w = \left(\frac{2mE}{\hbar^2}\right)^{1/2} \qquad k_h = \left[\frac{2m}{\hbar^2}(V_0 - E)\right]^{1/2} \tag{21.20}$$

The solution of equation (21.19) for the square lattice can now be easily written down

$$v_1(x) = A\exp[i(k_w - k)x] + B\exp[-i(k_w + k)x] \tag{21.21}$$

for the region of the well, and

$$v_2(x) = C\exp[(k_h - ik)x] + D\exp[-(k_h + ik)x] \tag{21.22}$$

for the region of the hill.

There are four unknowns in these solutions, but there are also four connecting equations: two for the periodicity conditions and two for the continuity of v and dv/dx. We obtain:

$$A + B = \exp(-ikL)\,[C\exp(k_hL) + D\exp(-k_hL)]$$

$$ik_w(A - B) = k_h\exp(-ikL)\,[C\exp(k_hL) - D\exp(-k_hL)] \tag{21.23}$$

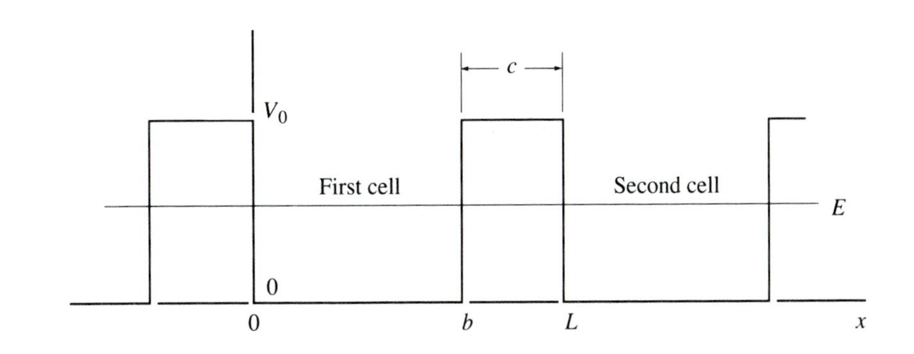

FIGURE 21.3
The Kronig-Penny form of the square potential lattice.

from the periodicity conditions and

$$A \exp(ik_w b) + B \exp(-ik_w b) = C \exp(k_h b) + D \exp(-k_h b)$$

$$ik_w[A \exp(ik_w b) - B \exp(-ik_w b)] = k_h[C \exp(k_h b) - D \exp(-k_h b)] \tag{21.24}$$

from the continuity conditions. The four equations (21.23) and (21.24) have nontrivial solutions only if the determinant of the matrix of the coefficients of A, B, C, and D vanishes. After some algebra this gives the following dispersion relation between the propagation constant k and the wave numbers k_h and k_w:

$$\cos k_w b \cosh k_h c - \frac{k_w^2 - k_h^2}{2k_w k_h} \sin k_w b \sinh k_h c = \cos kL \qquad E < V_0 \tag{21.25}$$

Since k_w and k_h contain the energy E, this dispersion relation is also the eigenvalue equation for E. The equation may look complicated, but it is the same type of equation we obtained for the bound states in the square potential well in chapter 4, equation (4.31). Can eigenstates that are represented by waves that propagate all over the crystal be considered bound? Yes! The domain of these functions extends over the finite interval $0 \le x \le NL$; this makes them normalizable. Thus they may certainly be considered as bound states, bound to the whole crystal, not to any particular localized cell.

When the energy E exceeds the potential V_0, k_h becomes imaginary; let's denote it by $\pm i\kappa_h$. The derivation leading to equation (21.25) still goes through; it has nothing against an imaginary wave number. Consequently, we can get the dispersion relation in this case by simply substituting $k_h \to i\kappa_h$ into equation (21.25):

$$\cos k_w b \cos \kappa_h c - \frac{k_w^2 + \kappa_h^2}{2k_w \kappa_h} \sin k_w b \sin \kappa_h c = \cos kL \qquad E > V_0 \tag{21.26}$$

which is again an eigenvalue equation.

Admittedly, these eigenvalue equations are quite complicated and can be solved only numerically; however, their interesting new feature can easily be deciphered. Behold! The right-hand sides of the eigenvalue equations above are bounded between -1 and $+1$; therefore, not all values of E are allowed, but only those that make the left-hand side of these equations also lie in the same interval. That is, blessed are those values of E for which

$$-1 \le (\text{left-hand sides of eqs. [21.25] and [21.26]}) \le 1 \tag{21.27}$$

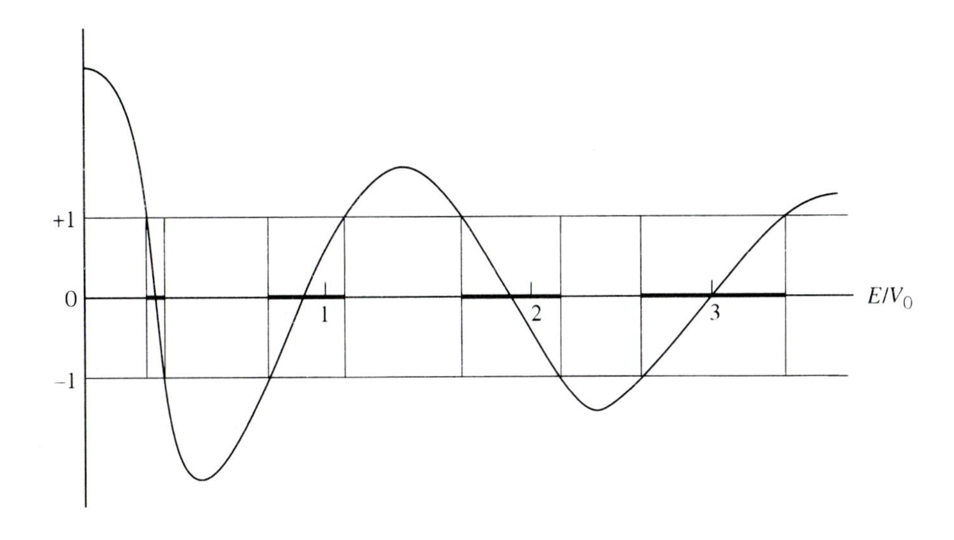

FIGURE 21.4
The left-hand sides of equations (21.25) and (21.26) are plotted as functions of E which join smoothly at $E = V_0$. The heavy lines depict the allowed range of energy values (the conduction bands).

All other values are excluded. This is the origin of *band structure* in solids. The allowed values of E are said to form the *conduction bands*. The values in between make up the *forbidden bands* (fig. 21.4).

If we think in terms of E as a function of the propagation constant k, we gain further insight. The value of k varies from 0 to π/L over the first conduction band, and then from π/L to $2\pi/L$ across the second conduction band, and so forth. At each of the band edges, $kL = \pm n\pi$, there is a gap in the permitted values of the energy E. This is shown in figure 21.5, where the dotted line is the graph of the free-particle dispersion relation.

Physically, the values of k corresponding to the positions of the energy gaps are such that the wave, in partially reflecting back at a well, finds itself to

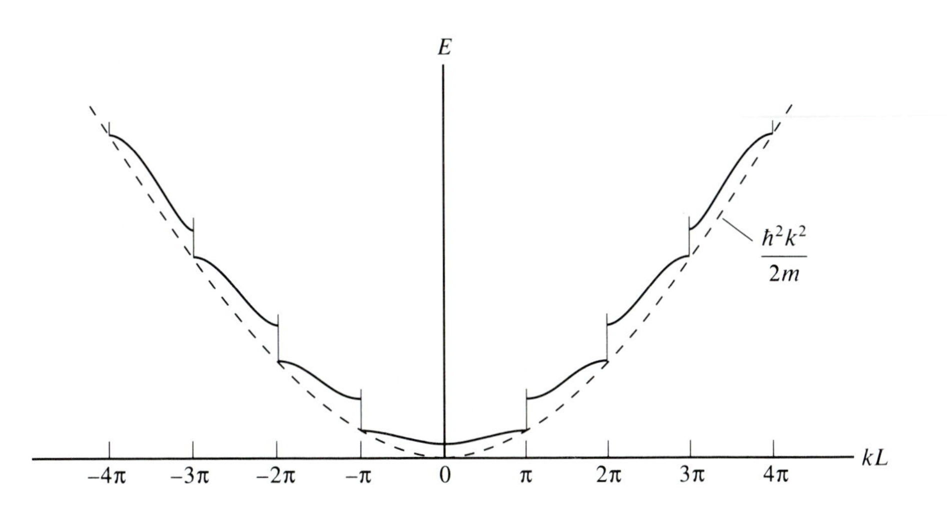

FIGURE 21.5
Energy E as a function of k as modified by the band structure for the square potential lattice; the dotted line represents the free particle E versus k dispersion curve.

be in constructive interference with all the reflected waves from the other wells. This strengthens the reflected wave so much that no traveling wave occurs.

Now we understand the difference between a conductor, a semiconductor, and an insulator. A conductor happens when the electrons partially fill a conduction band; in that situation electrons can accelerate in response to an external electric field because there are neighboring momentum states available to them. In contrast, insulators must consist of filled conduction bands; now electrons are unable to respond to an electric field because they have no adjacent empty momentum state to go to, unless, of course, the field is very strong. For strong fields, the electrons from the filled band jump across the energy gap to an allowed state in the next conduction band—this is the phenomenon of breakdown of an insulator. And what is a semiconductor? A semiconductor is a lucky insulator, as it has a very small forbidden energy abyss for the electrons to cross. So, even a small change of conditions, a rise of temperature, for example, can take electrons across the gap to allowed energy states and the semiconductor behaves like a conductor.

21.2 THE NUCLEAR SHELL MODEL

The most astounding idea that has emerged in our theoretical picture of the nucleus is that nucleons move as basically independent objects inside the nucleus under the auspices of a common central potential. This is surprising, because nucleons are such strongly interacting objects that we would expect them to interact violently all the time. Instead, they seem to settle down in a fairly ho-hum existence, like the electrons in an atom. One consequence is that the energy levels of nuclei also exhibit a shell structure. Nuclei with filled shells are found to be especially stable and are called *magic nuclei*. In this section we will give an explanation of the *magic numbers*—the numbers of nucleons needed to close nuclear shells—based on a harmonic oscillator plus a spin-orbit potential model called the shell model of the nucleus.

The Three-Dimensional Harmonic Oscillator

The three-dimensional harmonic oscillator Hamiltonian is given by

$$H = -\frac{\hbar^2}{2m}\nabla^2 + \frac{1}{2}m\omega^2 r^2 \tag{21.28}$$

and as such it can be solved both in rectangular coordinates and in spherical coordinates. Of these, the spherical-coordinate representation is the more convenient one for describing nuclei which have spherical symmetry.

We rewrite the Hamiltonian, equation (21.28), in terms of the angular momentum operator L^2 (see chapter 12):

$$H = -\frac{\hbar^2}{2m}\left[\frac{1}{r^2}\frac{\partial}{\partial r}\left(r^2\frac{\partial}{\partial r}\right) - \frac{L^2}{\hbar^2 r^2}\right] + \frac{1}{2}m\omega^2 r^2 \tag{21.29}$$

And we seek a solution for the eigenfunctions as a product of the spherical harmonics (the angular momentum eigenfunctions) and a radial function R_{nl}:

$$\psi_{nlm}(r\theta\phi) = R_{nl}(r)Y_{lm}(\theta\phi) \tag{21.30}$$

It's just like we are solving the hydrogen atom except that the central potential is different. The radial function R (suspending its subscripts momentarily) must obey the equation

$$\left[\frac{1}{r^2}\frac{d}{dr}\left(r^2\frac{d}{dr}\right) + \frac{2m}{\hbar^2}\left(E - \frac{1}{2}m\omega^2 r^2 - \frac{\hbar^2 l(l+1)}{2mr^2}\right)\right]R(r) = 0 \tag{21.31}$$

The asymptotic form of the equation is obtained by ignoring those terms that $\rightarrow 0$ as $r \rightarrow \infty$. Calling the asymptotic solution R_∞, we have

$$\frac{d^2R_\infty}{dr^2} - \left(\frac{m\omega}{\hbar}\right)^2 r^2 R_\infty = 0$$

A permissible solution is

$$R_\infty \sim \exp(-m\omega r^2/2\hbar)$$

So now we try a solution for R that incorporates this asymptotic behavior. We also define a dimensionless new variable

$$\rho = (m\omega/\hbar)^{1/2} r \tag{21.32}$$

So we put

$$R(r) = \exp(-\rho^2/2)v(\rho) \tag{21.33}$$

Upon substitution into equation (21.31), we get the equation that must be satisfied by $v(\rho)$:

$$\frac{d^2v}{d\rho^2} + \left(\frac{2}{\rho} - 2\rho\right)\frac{dv}{d\rho} + \left(\lambda - 3 - \frac{l(l+1)}{\rho^2}\right)v = 0 \tag{21.34}$$

where we have introduced the dimensionless energy eigenvalue

$$\lambda = 2E/\hbar\omega \tag{21.35}$$

We now put

$$v(\rho) = \rho^s u(\rho) \tag{21.36}$$

with the purpose of taking out the singular part of equation (21.34). Substituting equation (21.36) into equation (21.34), we get

$$\begin{aligned} u[s(s-1) + 2s - l(l+1)]\rho^{s-2} + [2s+2]\,(du/d\rho)\rho^{s-1} \\ + [(d^2u/d\rho^2) - 2su + (\lambda - 3)u]\rho^s - 2(du/d\rho)\rho^{s+1} = 0 \end{aligned} \tag{21.37}$$

Divide by ρ^{s-2} and let $\rho \to 0$. This gives

$$s(s+1) - l(l+1) = 0$$

Consequently, s must be either l or $-l-1$, but the latter value is not permissible (why?). By substituting equation (21.36) with $s = l$ into equation (21.34), we get the equation satisfied by u

$$\rho \frac{d^2u}{d\rho^2} + (2l + 2 - 2\rho^2)\frac{du}{d\rho} + (\lambda - 3 - 2l)\rho u = 0 \tag{21.38}$$

Now we try the power series method of solution and assume the expansion

$$u = \sum_k a_k \rho^k \tag{21.39}$$

Substituting into equation (21.38) and realizing that the coefficient of each power of ρ must equal zero, we get the recursion relations between the coefficients a_k as

$$(2l + 2)a_1 = 0 \tag{21.40}$$

$$\cdots\cdot$$

$$a_{k+2} = \frac{2k + 3 + 2l - \lambda}{(k+2)(k+1) + (2l+2)(k+2)}\,a_k \tag{21.41}$$

Since the very first a_1 must be zero, all the odd coefficients vanish by virtue of equation (21.41). The even terms form a series for which

$$\frac{a_{k+2}}{a_k} \sim \frac{1}{k}$$

for large k. In this way we see that the power series of $u(\rho)$ diverges as $\exp(\rho^2)$ unless it terminates. The condition for termination is, of course,

$$2k + 3 + 2l - \lambda = 0 \tag{21.42}$$

as we can see from equation (21.41). This determines the eigenvalues as

$$\lambda = (2E/\hbar\omega) = 2k + 2l + 3$$

which can be written as

$$E = \hbar\omega(k + l + 3/2) \quad \text{with } k = 0,2,4,\ldots \tag{21.43}$$

Since E is quantized, at this point we introduce suitable subscripts for E, namely, nl, and write

$$E_{nl} = \hbar\omega(2n + l - \tfrac{1}{2}) \tag{21.44}$$

where we have defined n via the equation $k = 2(n - 1)$; $n = 1,2,3,\ldots$. Defined in this way, n represents the number of radial nodes of the wave function, including the one at 0 but excluding the one at ∞.

Now here is something interesting. If we solved the eigenvalue problem in rectangular coordinates, it is easy to see that an eigenvalue of $(\Lambda_i + \frac{1}{2})\hbar\omega$ is obtained for each degree of freedom, and consequently, the total eigenvalue comes out to be $(\Lambda_x + \Lambda_y + \Lambda_z + 3/2)\hbar\omega$, or, writing $\Lambda = \Lambda_x + \Lambda_y + \Lambda_z$, we have

$$E = (\Lambda + 3/2)\hbar\omega \tag{21.45}$$

where Λ is referred to as the major shell number. Comparing with equation (21.44), we can see that

$$\Lambda = 2(n - 1) + l \tag{21.46}$$

Figure 21.6 (the column on the left) shows the eigenvalue spectrum with its obviously degenerate shell structure.

The first shell $\Lambda = 0$ consists of one level, $1S(n = 1, l = 0)$, which can, according to the Pauli principle, take 2 protons and 2 neutrons. (Mind you, nuclear physics texts usually use lowercase letters to denote l for single-particle states.) So the first oscillator shell closure comes at $Z = 2$, $N = 2$, which is the nucleus ^{4}He. The next major shell, $\Lambda = 1$, also contains the single level $1P$, which can take 6 protons and 6 neutrons. So we get the next shell closure at $Z = 8$, $N = 8$, which is ^{16}O. The $\Lambda = 2$ major shell contains two degenerate levels, $2S$ and $1D$, and accommodates 12 protons and 12 neutrons. So the $\Lambda = 2$ shell closes at $Z = 20$, $N = 20$, ^{40}Ca. Interestingly, 2, 8, 20, are all examples of *magic numbers* for Z and N, for which nuclei are especially stable. Indeed, ^{4}He, ^{16}O, and ^{40}Ca are examples of *doubly closed shell nuclei*, similar in some ways to the atomic noble gases.

But then the jig is up. We can see from figure 21.6 that the next oscillator shell closes at $Z = 40$, $N = 40$, but the next experimental magic number occurs at $Z = 28$, $N = 28$. So the simple shell structure of the three-dimensional oscil-

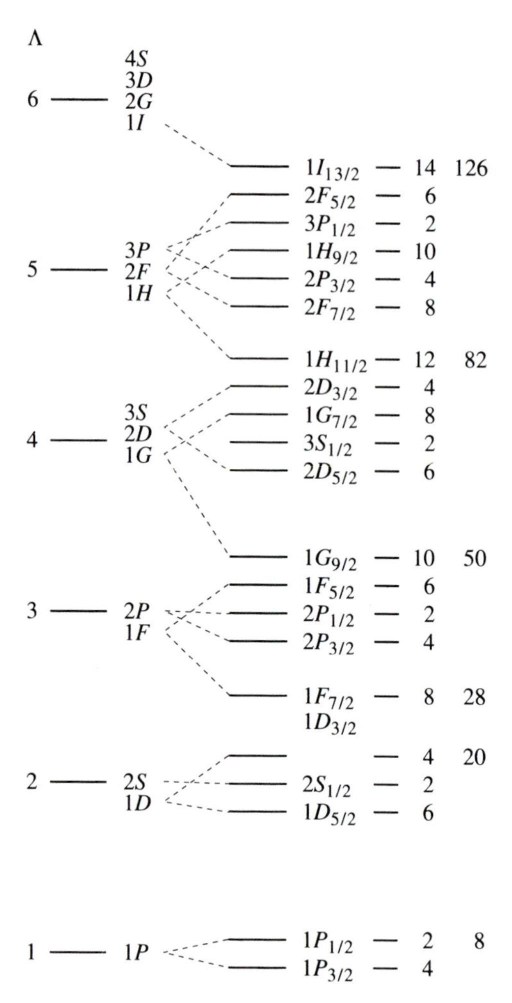

FIGURE 21.6
On the left are shown the three-dimensional harmonic oscillator energy levels and on the right the energy levels of harmonic oscillator plus the spin-orbit potential shell model that explains the nuclear magic numbers.

lator energy levels does not go all the way to explain the experimental shell structure found in nuclei. Something else has to be added. The new element is a spin-orbit potential that, however, has a very different origin from the atomic spin-orbit potential (see below).

Finally, the oscillator radial eigenfunctions, which can be determined from the recursion relations above, are customarily defined in terms of the now familiar Laguerre polynomials, but of half-integer order. The radial functions for $n = 1,2$ are

$$R_{1l}(r) = \left(\frac{m\omega}{\hbar}\right)^{3/4}\left[\frac{2}{\Gamma(l+3/2)}\right]^{1/2}\rho^l e^{-\rho^2/2}$$

$$R_{2l}(r) = \left(\frac{m\omega}{\hbar}\right)^{3/4}\left[\frac{2l+3}{\Gamma(l+3/2)}\right]^{1/2}\rho^l e^{-\rho^2/2}\left[1 - \frac{2}{2l+3}\rho^2\right] \tag{21.47}$$

where Γ is a gamma function.

The Nuclear Shell Model

It was the genius of M. G. Mayer and J. H. D. Jensen that discovered that if we add a spin-orbit potential to the three-dimensional oscillator, the nuclear magic numbers

$$Z, N = 2,\ 8,\ 20,\ 28,\ 50,\ 82,\ 126$$

can be explained. Interestingly, in contrast to atoms, the nuclear spin-orbit potential is an attractive potential

$$V_{ls}(r) = -U(r)\mathbf{L}\cdot\mathbf{S} \tag{21.48}$$

For our discussion, the form of $U(r)$ will not be necessary; in principle $U(r)$ can be determined from the empirical data on the two-nucleon interaction, which has a two-body spin-orbit component. The $\mathbf{L}\cdot\mathbf{S}$ potential causes the $j = l \pm \frac{1}{2}$ levels to split up, with the $j = l + \frac{1}{2}$ lowered in energy (fig. 21.6, right). The first-order energy shift is

$$\begin{aligned}\langle nljm|V_{ls}(r)|nljm\rangle &= -\langle U(r)\rangle_{nl}\tfrac{1}{2}[j(j+1) - l(l+1) - \tfrac{3}{4}]\hbar^2 \\ &= \tfrac{1}{2}(l+1)\hbar^2\langle U(r)\rangle_{nl} \quad \text{for } j = l - \tfrac{1}{2} \\ &= -\tfrac{1}{2}l\hbar^2\langle U(r)\rangle_{nl} \quad \text{for } j = l + \tfrac{1}{2}\end{aligned} \tag{21.49}$$

where $\langle U(r)\rangle_{nl}$ is the radial integral

$$\langle U(r)\rangle = \int_0^\infty r^2\, dr\, R_{nl}^2(r)U(r)$$

The splitting of the $j = l \pm \frac{1}{2}$ levels is

$$\Delta E_l = [(2l+1)/2]\hbar^2\langle U(r)\rangle_{nl} \tag{21.50}$$

The radial integral is only weakly sensitive to l. Therefore the spin-orbit splitting increases with l roughly as $(2l+1)$; the greater the orbital angular momentum, the more are the corresponding j-levels separated by the spin-orbit interaction.

Because of the spin-orbit interaction, the single-particle states of nucleons are now specified using the labels nlj; a complete state specification is $|nljm\rangle$. Therefore, each nlj level can take up to $2j + 1$ protons and neutrons, corresponding to the $2j + 1$ values of m.

So we go through the exercise of filling the shells again, but in the new scheme (fig. 21.6, right). The $1S$ level of the first major shell now is transformed to $1S_{1/2}$ [notice the notation: n(appropriate letter for l)$_j$], which is filled with $(2 \times \frac{1}{2} + 1) = 2$ protons and neutrons as before. The next level $1P$ now splits into two levels $1P_{3/2}$ and $1P_{1/2}$, $1P_{3/2}$ being the lower of the two. However, they are still close enough to count in the same shell. They each can take four and two protons and neutrons, respectively, and thus the shell still closes at Z, $N = 8$. The splittings in the next shell are similar; instead of $2S$ and $1D$, we now have $2S_{1/2}$, $1D_{5/2}$, and $1D_{3/2}$, but no matter, they all belong to the same shell; consequently, they still fill with the same number of neutrons and protons as before.

So the first significant change occurs for the levels of the next major shell, $\Lambda = 3$. Here the $1F$ level has a large enough l for the spin-orbit splitting between the $1F_{7/2}$ and $1F_{5/2}$ to be so large that $1F_{7/2}$ must be regarded as a lone-star shell by itself. Since the $1F_{7/2}$ level can take eight protons or neutrons, this now explains the magic number Z, $N = 28$. The remaining levels of the $\Lambda = 3$ major shell, $1F_{5/2}$, $2P_{3/2}$, and $2P_{1/2}$, can accommodate only 12 more particles, but their total shell capacity is boosted by another 10 via the lowering of the $1G_{9/2}$ level to their rank; the l of this last level is large enough to fit the bill (fig. 21.6). This explains the next magic number: $28 + 12 + 10 = 50$.

We can go on and on, but you get the picture. The next two magic numbers are explained the same way. You can easily reconstruct the next two shells with the help of figure 21.6, and this will be left as an exercise.

What is truly amazing in nuclear structure physics is that this simple basis of harmonic oscillator + spin-orbit potential explains not only the magic numbers but also the ground-state spins and magnetic moments of most near-closed-shell nuclei. For more complex nuclei with several particles outside of closed shells, we have to invoke the *residual interaction* between the nucleons for the shell model to have explanatory power, but that's a small price to pay. Incidentally, this language (residual interaction) is perfectly appropriate, because the harmonic oscillator + spin-orbit potential is indeed the nuclear analog of the Hartee-Fock atomic self-consistent potential.

21.3 BOSON GAS: PHOTON STATISTICS AND PLANCK'S LAW

What is the effect of exchange symmetry on the behavior of a system of identical bosons? Bosons, as you know, are particles whose wave function has to be symmetric under particle exchange—what does this symmetry correlation imply when we build a many-boson system? In this section, we will show that bosons tend to condense or bunch together, the opposite of fermion systems' antibunching "aufbau" behavior. Specifically, we will deal with the noninteracting photon gas and derive Planck's radiation law as a consequence of photon condensation.

Consider n identical noninteracting photons enclosed in a finite volume. Our job is to build a many-photon state starting with a state of zero photons, which we call the *vacuum* state. We will put photons—one at a time, starting from the vacuum state—into the confining volume. Let $|\psi_0\rangle$ designate the vacuum state and $|1\rangle$ the state of a single particle added to the vacuum. If H_1 is the interaction responsible for the transition $|\psi_0\rangle \rightarrow |1\rangle$, then the probability of the transition is given by the square of the matrix element of H_1 sandwiched between $|\psi_0\rangle$ and $|1\rangle$:

$$P_1 = |\langle 1|H_1|\psi_0\rangle|^2 = |a|^2 \tag{21.51}$$

calling the amplitude $\langle 1|H_1|\psi_0\rangle$, a.

Now suppose we introduce a second photon to make a two-particle state $|\psi_2\rangle$. We could write the two-particle state as

$$|\psi_2\rangle = |1\rangle|2\rangle$$

where $|2\rangle$ is again a single-particle state of particle 2 added to the vacuum. But that won't do because this implies that we add photon 2 on the top of the single-particle state of photon 1. But why can't we do this the other way around since photons are indistinguishable? This is how symmetry enters, of course; and we already know the answer: The two-particle state must be the symmetric combination of the two products of one-particle states. In other words

$$|\psi_2\rangle = (1/\sqrt{2})\,[|1\rangle|2\rangle + |2\rangle|1\rangle] \tag{21.52}$$

The amplitude for the transition $|\psi_0\rangle \rightarrow |\psi_2\rangle$ is

$$\langle\psi_2|H_1|\psi_0\rangle = (1/\sqrt{2})\,[\langle 2|H_1|\psi_0\rangle\langle 1|H_1|\psi_0\rangle + \langle 1|H_1|\psi_0\rangle\langle 2|H_1|\psi_0\rangle]$$

We will now invoke the fact that the photons themselves do not interact. Therefore,

$$\langle 1|H_1|\psi_0\rangle = \langle 2|H_1|\psi_0\rangle = a \tag{21.53}$$

Consequently, the probability of realizing a two-photon state upon starting with the vacuum is given as

$$P_2 = \tfrac{1}{2}[|a|^2 + |a|^2]^2 = 2[|a|^2]^2 \tag{21.54}$$

This is a radically new result, strictly quantum. This does not happen in classical physics where the particles are distinguishable. Classically, for distinguishable particles, the total probability is obtained by summing the individual probabilities (not the amplitudes) of the various possibilities and then dividing by the number of possibilities (in the quantum case, normalization accomplishes this last step); in other words, we average over the probabilities of the possibilities. We get

$$P_{2c} = \tfrac{1}{2}[\{|a|^2\}^2 + \{|a|^2\}^2] = \{|a|^2\}^2 \tag{21.55}$$

where the subscript c stands for classical.

You can see that in quantum physics, where particles are indistinguishable bosons, the probability of finding two particles inside the volume is twice as large as that for distinguishable particles of classical physics. Identical bosons tend to occupy the same space or condense in the same state. This is also completely opposite to the separatist exclusionary tendency of identical fermions.

Now let's consider putting three bosons into the volume. The three-boson symmetrized state is given as

$$|\psi_3\rangle = (1/3!)^{1/2}\Sigma_i p_i|1\rangle|2\rangle|3\rangle \tag{21.56}$$

where p_i denotes all the possible permutations of which there are 3! or 6. There are thus six possible ways to put three bosons in our box, and each of the amplitudes for the transition $|\psi_0\rangle \rightarrow |\psi_3\rangle$ for doing this is the product $(1/3!)^{1/2}a\cdot a\cdot a$. As before, we must add the amplitudes and then square them to obtain the quantum probability:

$$P_3 = (1/3!)|(3!)a\cdot a\cdot a|^2 = (3!)[|a|^2]^3 \tag{21.57}$$

Generalizing the result for n identical bosons, we have for the probability of obtaining an n photon state

$$P_n = (n!)[|a|^2]^n \tag{21.58}$$

which is $n!$ times the corresponding probability for distinguishable classical particles.

Suppose the system already is in an n photon state; what is the probability for adding one additional photon? It is the ratio P_{n+1}/P_n:

$$P_{n\rightarrow n+1} = P_{n+1}/P_n = (n+1)|a|^2 \tag{21.59}$$

The amplitude for the process then is given as (choosing it to be real)

$$\langle\psi_{n+1}|H_1|\psi_n\rangle = (n+1)^{1/2}a \tag{21.60}$$

Because of the rule backward amplitude = (forward amplitude)*, we can write

$$\langle \psi_n | H_1 | \psi_{n+1} \rangle = (n+1)^{1/2} a$$

which gives the amplitude of the process in which the number of photons decreases by 1 as

$$\langle \psi_{n-1} | H_1 | \psi_n \rangle = n^{1/2} a \tag{21.61}$$

Note that these matrix elements are similar to the ones obtained for the creation and destruction operators for the harmonic oscillator derived in chapter 7. It follows that photons (or any identical bosons) obey the same operator algebra as that of the one-dimensional harmonic oscillator, and their states can be represented the same way. Therefore, the vacuum state of the photons, $|\psi_0\rangle$, corresponds to the oscillator ground state, and the n photon state $|\psi_n\rangle$ is equivalent to the nth oscillator state.

Planck's Radiation Distribution Law

Now let's examine radiation in equilibrium in a cavity of volume Ω at a temperature T. The cavity, of course, constitutes the usual black body with its walls assumed perfectly absorbing and having only a tiny hole that does not affect the equilibrium. We will assume that the atoms of the wall of the cavity emit and absorb photons as a transition between two states $|c\rangle$ and $|d\rangle$ (fig. 21.7), where $E_d - E_c = \hbar\omega$, ω being the angular frequency of the photon. Let R_{ab} and R_{em} be the absorption and emission rates which must be equal at equilibrium:

$$R_{\text{ab}} = R_{\text{em}}$$

Moreover, each rate is equal to the product of the number of atoms in the respective states from which the transition takes place and the probability of transition per unit time:

$$R_{\text{ab}} = N_c P_{c\to d} = R_{\text{em}} = N_d P_{d\to c} \tag{21.62}$$

Now using Boltzmann's distribution to estimate N_c and N_d (this is justified because the energy states of atoms in thermal motion are continuous), we have

$$N_c/N_d = \exp[(E_d - E_c)/kT] = \exp(\hbar\omega/kT) = P_{d\to c}/P_{c\to d} \tag{21.63}$$

FIGURE 21.7
Transition between two atomic levels at energy E_d and E_c, where $E_d = E_c + \hbar\omega$.

Now realize that in the absorption of a photon, $c \to d$, the number of photons in the cavity decreases by 1; therefore, the transition probability is given by the square of the amplitude given by equation (21.61), except that we must replace n by n_{av}, the average number of photons in the cavity:

$$P_{c\to d} = n_{av}|a|^2 \tag{21.64}$$

On the other hand, in emission, for the transition $|d\rangle$ to $|c\rangle$, the number of photons in the cavity goes up, and we must use the square of the amplitude for creating a photon, equation (21.60), for the transition probability:

$$P_{d\to c} = (n_{av} + 1)|a|^2 \tag{21.65}$$

Consequently,

$$\frac{P_{d\to c}}{P_{c\to d}} = \frac{n_{av} + 1}{n_{av}} = \exp\left(\frac{\hbar\omega}{kT}\right) \tag{21.66}$$

where we have used equation (21.63). Solving for n_{av}, we obtain

$$n_{av} = \frac{1}{e^{\hbar\omega/kT} - 1} \tag{21.67}$$

From the analogy of the harmonic oscillator, it is clear that the energy of the n_{av} photons is $(n_{av} + \frac{1}{2})\hbar\omega$. However, since $n_{av} \gg 1$, we can ignore the zero-point energy and write for the energy distribution in the cavity

$$E(\omega) = n_{av}\hbar\omega = \frac{\hbar\omega}{e^{\hbar\omega/kT} - 1} \tag{21.68}$$

The energy spectrum of the photons obtained above is discrete; however, any observed distribution of frequencies will be continuous since a very large number of discrete frequencies can coexist in a cavity whose dimensions are much greater than the wavelength of the radiation. The number of states d^3n in a box in the phase space volume Ωd^3p from equation (21.1) is

$$d^3n = dn_x\,dn_y\,dn_z = (L/2\pi)^3\,dk_x\,dk_y\,dk_z = (\Omega/2\pi\hbar)^3\,d^3p$$

and the density of states per unit energy interval d^3n/dE is

$$\frac{d^3n}{dE} = \frac{\Omega}{(2\pi\hbar)^3}4\pi p^2\frac{dp}{dE} = \frac{d^3n}{d(\hbar\omega)} \tag{21.69}$$

where the factor 4π comes from the angular integration. For photons, $E = pc$ giving $dp/dE = 1/c$. There are also two states of photon polarization for each $E = \hbar\omega$. Consequently, we have

$$\frac{d^3n}{d\omega} = \frac{\Omega\omega^2}{\pi^2 c^3} \tag{21.70}$$

If we multiply equation (21.68) with the density of photon states calculated above, we get the continuous distribution of energy as a function of ω; dividing further by the volume Ω of the cavity, we get the distribution of the energy density $u(\omega)$:

$$u(\omega) = \frac{1}{\pi^2 c^3 \hbar^2} \frac{(\hbar\omega)^3}{e^{\hbar\omega/kT} - 1} \tag{21.71}$$

This is Planck's radiation law. To convert to the form of equation (1.4) in terms of the linear frequency $\nu = \omega/2\pi$, we note that

$$u(\omega)\, d\omega = u(\nu)\, d\nu$$

and thus

$$u(\nu) = u(\omega)\frac{d\omega}{d\nu} = \frac{8\pi h\nu^3}{c^3} \frac{1}{e^{h\nu/kT} - 1} \tag{21.72}$$

We began our exploration of quantum mechanics with Planck's law, and we finally have derived the law from purely quantum mechanical considerations—an important achievement, indeed.

PROBLEMS

1. The number density of free electrons is $8.5 \times 10^{22}\ \text{cm}^{-3}$ for copper. What is the Fermi energy in electron volts?
2. Consider a Fermi gas model of the nucleus consisting of independent proton and neutron Fermi gases. Derive expressions for the Fermi energy of both the neutron and proton gases consisting of N neutrons and Z protons, respectively, where $N + Z = A$, the mass number of the nucleus. If the radius of a nucleus is given by $R = 1.2\, A^{1/3}$ F, and $m_p \approx m_n = 1.6 \times 10^{-24}$ g, calculate the numerical values of the Fermi energies for ^{198}Au.
3. A one-dimensional periodic potential consists of a sequence of Dirac delta functions with a distance c between them (this is called the Dirac comb):

$$V(x) = V_0 \sum_{n=-\infty}^{n=\infty} \delta(x + nc)$$

Determine the energy eigenvalues (energy bands) of this potential.

4. Suppose the width c of the hills of the Kronig-Penny potential is allowed to $\to \infty$ without changing the width b of the wells or the depth V_0. For $E < V_0$, show that the energy bands become narrower and narrower, eventually contracting to the energy levels of the square well.
5. (a) The wave functions of a three-dimensional oscillator can be written as products of three one-dimensional oscillator eigenfunctions corresponding to each of the rectangular coordinates with quantum numbers Λ_x, Λ_y, and Λ_z, respectively. Find the energy, parity, and degeneracy of the lowest three distinct groups of energy levels.
(b) Now consider the solution for the three-dimensional oscillator in spherical coordinates that gives the same eigenvalues but different eigenfunctions. Using the knowledge of parity and degeneracy of the various states, deduce which values of l are associated with each group of levels enumerated in part (a).
6. A particle moves in a three-dimensional oscillator potential with Hamiltonian

$$H_{\text{osc}} = (\mathbf{p}^2/2m) + \tfrac{1}{2} m\omega^2 r^2$$

Additionally, the particle is charged ($e > 0$) and is in a uniform magnetic field $\mathbf{B}$ along the z-axis. Choosing the vector potential $\mathbf{A}(\mathbf{r}) = -\frac{1}{2}(\mathbf{r} \times \mathbf{B})$, show that the complete Hamiltonian of the charged particle can be written as

$$H = H_{\text{osc}} + H_1(\omega_L)$$

where $\omega_L = eB/2mc$ and H_1 is the sum of a term linear in ω_L and one quadratic in ω_L. Show that the stationary states of H can be determined exactly.
7. The energy spacing between the $1D_{5/2}$ and $1D_{3/2}$ levels for the nucleus ^{17}O is 5.0 MeV. What is the value of $\langle U(r) \rangle$ of the spin-orbit potential for the $1D$ state?
8. What do you expect the ground-state spin of ^{41}Ca nucleus to be according to the shell model? Calculate the magnetic moment of this state.

ADDITIONAL PROBLEMS

A1. The organic molecule octatetraen absorbs strongly at the wavelength $\lambda = 3020$ Å. A simple model looks at the molecule as a one-dimensional chain consisting of eight basic units with each unit contributing one free electron (free to move the entire length of the chain). Assume that the size of each unit is 1.39 Å; take the total effective length of the molecule as 7×1.39 Å.

Assume that the electronic states are given by those for an infinite one-dimensional-box potential. Use the Pauli exclusion principle, electron spin, and the aufbau principle to determine which energy levels E_n are occupied to form the ground state of the molecule. Determine the smallest energy involved in an absorption process and find the wavelength at which this absorption occurs.

A2. Write down the 4×4 matrix of the three-dimensional oscillator Hamiltonian in the truncated basis consisting of the lowest two energy levels. Now add the perturbation $C\mathbf{L}\cdot\mathbf{S} + D\mathbf{L}^2$, where C and D are constant and display the 4×4 matrix in terms of the parameters C and D.

A3. Consider the motion of a particle in the three-dimensional oscillator potential perturbed by the additional interaction $V' = Cr^4 \sin^4\theta \sin^4\phi$ where C is a constant. Use first-order perturbation theory to determine the change in the energy of the two lowest states of the oscillator.

A4. Consider the motion of a particle in a spherical harmonic oscillator of mass m and characteristic frequency ω. The motion is perturbed by the potential

$$V_1 = m\omega^2(x^2/10 + y^2/20 + z^2/30)$$

(a) What are the exact eigenvalues of the total Hamiltonian in the Cartesian representation, $E(n_1, n_2, n_3)$?

(b) Use perturbation theory to determine the ground state energy to the lowest nonzero order in the perturbation. Compare your answer with the exact answer.

REFERENCES

A. Das and A. C. Mellissinos. *Quantum Mechanics: A Modern Introduction.*
R. N. Liboff. *Introductory Quantum Mechanics.*
M. K. Pal. *Theory of Nuclear Structure.*
D. Park. *Introduction to the Quantum Theory.*
H. Smith. *Introduction to Quantum Mechanics.*

22

Time-Dependent Perturbation Theory and Application to Atomic Radiation and Scattering

We have avoided dealing with time-dependent Hamiltonians except for a brief foray with the ammonia maser and nuclear-magnetic resonance, where we were aided by the simplicity of 2×2 matrices. If H contains interaction terms that are explicitly time dependent in the general case of infinite dimensional matrices, how do we proceed?

Obviously, exact solution is out of the question, since in the presence of explicit time dependence of the Hamiltonian, we cannot reduce the Schrödinger equation to an eigenvalue problem. However, in many cases, the time-dependent part of the Hamiltonian can be regarded as a perturbation and we are able to treat it by means of a perturbation theory. This is called the time-dependent perturbation theory.

So in this chapter we will consider Hamiltonians of the form

$$H(t) = H_0 + H_1(t) \tag{22.1}$$

where H_0 is the unperturbed time-independent Hamiltonian solvable in terms of stationary states, and H_1, the time-dependent part, is only a small perturbation over and above H_0. The time-dependent interaction causes an admixture of the eigenstates of the stationary Hamiltonian H_0, and this admixture must be a function of time. So the pertinent question is this: If initially the system is in one of the stationary states $|i\rangle$ of H_0, what is the probability that at a later time t, the system will be found in another stationary state $|m\rangle$? In other words, what is the probability of transition from $i \rightarrow m$? In much of what follows, we will develop a simple but elegant answer to this question for the case of a harmonic time dependence of the perturbation. We will then illustrate our result with two examples: (1) interaction of atoms with time-dependent radiation fields and (2) elastic scattering of particles from a static potential.

We will, however, begin with formulating a perturbation theory for a general time dependence of H_1.

22.1 TIME-DEPENDENT PERTURBATION THEORY: GENERAL FORMULATION AND FIRST-ORDER THEORY

The first step in trying to find a perturbation solution of the Hamiltonian of equation (22.1) should be obvious. We expand the eigenstates $|\psi(t)\rangle$ of the full Hamiltonian in terms of the stationary eigenstates $|\phi_n\rangle$ of H_0, which form a complete set:

$$|\psi(t)\rangle = \sum_n c_n(t)|\phi_n\rangle \tag{22.2}$$

The question, "given $|\phi_n\rangle$, what is $|\psi(t)\rangle$?" now becomes the question, "given $c_n(0)$, what is $c_n(t)$?"

We know the answer to that last question if H_1 is zero. Then,

$$c_n(t) = c_n(0)\exp(-iE_n t/\hbar) \tag{22.3}$$

where E_n is the energy of the stationary state $|\phi_n\rangle$, $H_0|\phi_n\rangle = E_n|\phi_n\rangle$. Therefore, it seems wise to redefine the $c_n(t)$ to incorporate this knowledge. We write

$$|\psi(t)\rangle = \sum_n c_n(t)\exp(-iE_n t/\hbar)|\phi_n\rangle \tag{22.4}$$

If we substitute this expansion in the full Schrödinger equation

$$i\hbar\partial|\psi(t)\rangle\partial t = (H_0 + H_1)|\psi(t)\rangle \tag{22.5}$$

we obtain

$$\sum_n \left[i\hbar\frac{dc_n(t)}{dt} + E_n c_n(t)\right]e^{-iE_n t/\hbar}|\phi_n\rangle = \sum_n [E_n + H_1(t)]c_n(t)e^{-iE_n t/\hbar}|\phi_n\rangle$$

The E_n-terms cancel out. Now take the scalar product with $\langle\phi_m|$ from the left and use the orthonormality of the set $\langle\phi_m|\phi_n\rangle = \delta_{mn}$. This gives

$$i\hbar\frac{dc_m(t)}{dt} = \sum_n c_n(t)e^{i(E_m-E_n)t/\hbar}\langle\phi_m|H_1|\phi_n\rangle \tag{22.6}$$

So far everything is exact; we will now obtain the first-order theory.

First-Order Theory

We expand c_n as a perturbation series:

$$c_n = c_n^{(0)} + c_n^{(1)} + c_n^{(2)} + \cdots \tag{22.7}$$

Suppose initially the system is in a particular eigenstate $|\phi_i\rangle$ of H_0; that is,

$$c_n(0) = c_n^{(0)} = \delta_{ni} \tag{22.8}$$

Substituting equations (22.7) and (22.8) into equation (22.6), we obtain in first order (for $m \neq i$)

$$i\hbar\, dc_m^{(1)}(t)/dt = \exp[i(E_m - E_i)t/\hbar]\langle\phi_m|H_1|\phi_i\rangle$$

The solution is

$$c_m^{(1)}(t) = -\frac{i}{\hbar}\int_0^t dt'\, e^{i(E_m-E_i)t'/\hbar}\langle\phi_m|H_1(t')|\phi_i\rangle \tag{22.9}$$

Therefore, up to and including first order, we have

$$c_m(t) = \delta_{mi} - \frac{i}{\hbar}\int_0^t dt'\, e^{i(E_m-E_i)t'/\hbar}\langle\phi_m|H_1(t')|\phi_i\rangle \tag{22.10}$$

Note that this solution is reliable only if $|c_m^{(1)}(t)| \ll 1$.

According to the measurement postulate, the probability that a measurement of H_0 of the system at time t will give the value E_m is given as (provided no other measurement disturbed the system between $t = 0$ and $t = t$)

$$P_{i\to m} = |c_m(t)|^2 \tag{22.11}$$

The reason for the subscript on P is as follows: The result of the above measurement at time t leaves the system in a particular eigenstate $|\phi_m\rangle$; thus, this is the probability that the system has made a transition, a quantum jump, from $i \to m$; the probability that the system starts in state $|\phi_i\rangle$ and ends in state $|\phi_m\rangle$ at time t is given by the square of the expansion coefficient of $|\phi_m\rangle$ in the expansion of $|\phi_i\rangle$.

Example: Time-Dependent Perturbation of the Harmonic Oscillator

First, let's consider a purely pedagogic example with an easily integrable time-dependent interaction:

$$H_1 = -Cx\exp(-t^2/t_0^2)$$

Let H_1 act on a linear harmonic oscillator in its ground state from $t = -\infty$ to $t = +\infty$. We want the probability that at $t = +\infty$, the oscillator is found in the nth state $|u_n\rangle$.

Using equation (22.10), we have for $n \neq 0$,

$$c_n(\infty) = -\frac{i}{\hbar}\int_{-\infty}^{\infty} dt'\ (-C)\langle u_n|x|u_0\rangle e^{-t'^2/t_0^2}e^{in\omega t'}$$

where we substituted for the oscillator energies, $E_n = (n + \frac{1}{2})\hbar\omega$. Since x can connect the ground state with only the first excited state (why?), only c_1 is nonzero. Substituting the matrix element of x and performing the integration, we get

$$c_1(\infty) = \frac{iC}{\hbar}\left(\frac{\hbar}{2m\omega}\right)^{1/2}(\pi t_0^2)^{1/2}e^{-\omega^2 t_0^2/4}$$

In this way we find the probability of transition $|u_0\rangle \rightarrow |u_1\rangle$ as

$$P_{0\rightarrow 1} = |c_1(\infty)|^2 = (C^2\pi t_0^2/2m\hbar\omega)\exp(-\omega^2 t_0^2/2)$$

22.2 PERIODIC PERTURBATION: FERMI'S GOLDEN RULE

Consider a system subject to a perturbing interaction that varies periodically with time, for example, an atom being bathed in a monochromatic light beam. We have

$$\begin{aligned} H_1(t) &= 2H_1 \sin\omega t \\ &= (1/i)H_1[\exp(i\omega t) - \exp(-i\omega t)] \end{aligned} \tag{22.12}$$

where H_1 is a time-independent operator. Let's assume that the system comes in contact with the interaction at time $t = 0$. We can immediately calculate the first-order transition amplitude from $i \rightarrow m$, $m \neq i$ (writing $E_m - E_i = \hbar\omega_{mi}$):

$$\begin{aligned} c_m(t) &= -\frac{2i}{\hbar}\langle\phi_m|H_1|\phi_i\rangle\int_0^t dt' \sin\omega t' e^{i\omega_{mi}t'} \\ &= \frac{\langle\phi_m|H_1|\phi_i\rangle}{\hbar}\left[\frac{e^{i(\omega_{mi}-\omega)t} - 1}{\omega_{mi} - \omega} - \frac{e^{i(\omega_{mi}+\omega)t} - 1}{\omega_{mi} + \omega}\right] \end{aligned} \tag{22.13}$$

Since the interaction matrix element $\langle\phi_m|H_1|\phi_i\rangle$ is assumed to be small, the transition probability is appreciable only if one of the denominators approximately equals zero, in other words, only if a resonance condition is satisfied. We can have (1) absorption if

$$\omega_{mi} - \omega \approx 0, \quad \text{or} \quad E_m \approx E_i + \hbar\omega$$

or (2) emission if

$$\omega_{mi} + \omega \approx 0, \quad \text{or} \quad E_m \approx E_i - \hbar\omega$$

The interpretation of the situation is as follows. Suppose condition (1) is the case and $E_m > E_i$. As a result of the perturbation, $|c_m(t)|^2$ grows, but at whose expense? At the expense of $|c_i(t)|^2$, of course. And the expectation value for the energy of the system keeps on increasing as the shift occurs in the state. This is the meaning of absorption (fig. 22.1a). On the other hand, if condition (2) is satisfied and $E_m < E_i$, then the expectation of energy decreases and the system gives up energy to the perturbing field, which is emission; this is more appropriately called *stimulated emission* since the system is stimulated to emit a photon as opposed to spontaneous emission (fig. 22.1b).

Let's recognize that for a perturbation with one sharp frequency, only one of the two conditions can be satisfied at a time, not both. Therefore, we can calculate $|c_m(t)|^2$ from equation (22.13) without worrying about the interference effect between the two terms:

$$\begin{aligned} |c_m(t)|^2 &= \frac{|\langle\phi_m|H_1|\phi_i\rangle|^2}{\hbar^2}\left|\frac{e^{i(\omega_{mi}\pm\omega)t}-1}{\omega_{mi}\pm\omega}\right|^2 \\ &= \frac{|\langle\phi_m|H_1|\phi_i\rangle|^2}{\hbar^2}\frac{\sin^2[(\omega_{mi}\pm\omega)t/2]}{[(\omega_{mi}\pm\omega)/2]^2} \end{aligned} \tag{22.14}$$

Let's consider what we have gotten! The function

$$\frac{\sin^2[(\omega_{mi}\pm\omega)t/2]}{[(\omega_{mi}\pm\omega)/2]^2} \tag{22.15}$$

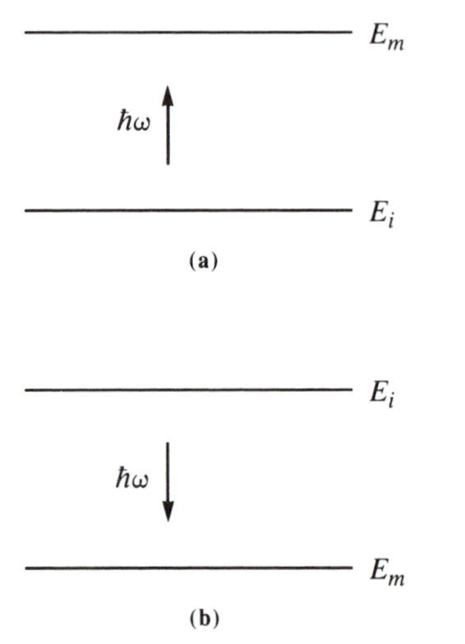

FIGURE 22.1
Time-dependent perturbation of a quantum system by a periodic potential V. (a) Absorption: the quantum system absorbs energy $\hbar\omega$ from H_1 and jumps up to an excited state. (b) Stimulated emission: the quantum system, initially in an excited state, surrenders energy $\hbar\omega$ to H_1 and deexcites.

as a function of $\beta = \omega_{mi} \pm \omega$ at a fixed t, is very interesting (fig. 22.2). It is sharply peaked around $\beta = \omega_{mi} \pm \omega = 0$, its value at this peak is t^2, and the width of the curve around the peak $\sim 1/t$, so that the area under the curve is $\sim t$. Thus if the "resonance" condition is exactly satisfied, the transition probability $|c_m(t)|^2$ grows quadratically with time. This aspect of $|c_m(t)|^2$ is quite counterintuitive. We expect the function to grow $\sim t$ in order to talk about a transition rate, as we did in chapter 21. Furthermore, note that the width of the central peak is large for small t; however, as t becomes large, the width practically goes to zero. The longer the perturbation is on, the less is the energy flexibility of the final state—in fact, the more the function (22.15) acts as a reminder of the conservation of energy in the transition. What does this mean?

We see why this last unappealing implausibility is occurring by realizing that we are assuming very sharp energy levels, whereas, in actuality, in the presence of a time-dependent perturbation, the energy eigenvalues are also modified to take on a width. This is equivalent to saying that there is an uncertainty ΔE in the value of the resonant β. We can think of the time that the perturbation is on as a time uncertainty, call it Δt. Therefore, all that variation of the width of the resonant curve with time is just a reflection of the time-energy uncertainty relation $\Delta E \cdot \Delta t \sim \hbar$ (see chapter 2).

And fortunately, if we take account of the actual experimental situation, we find that we have to introduce further elements in the derivation, the final

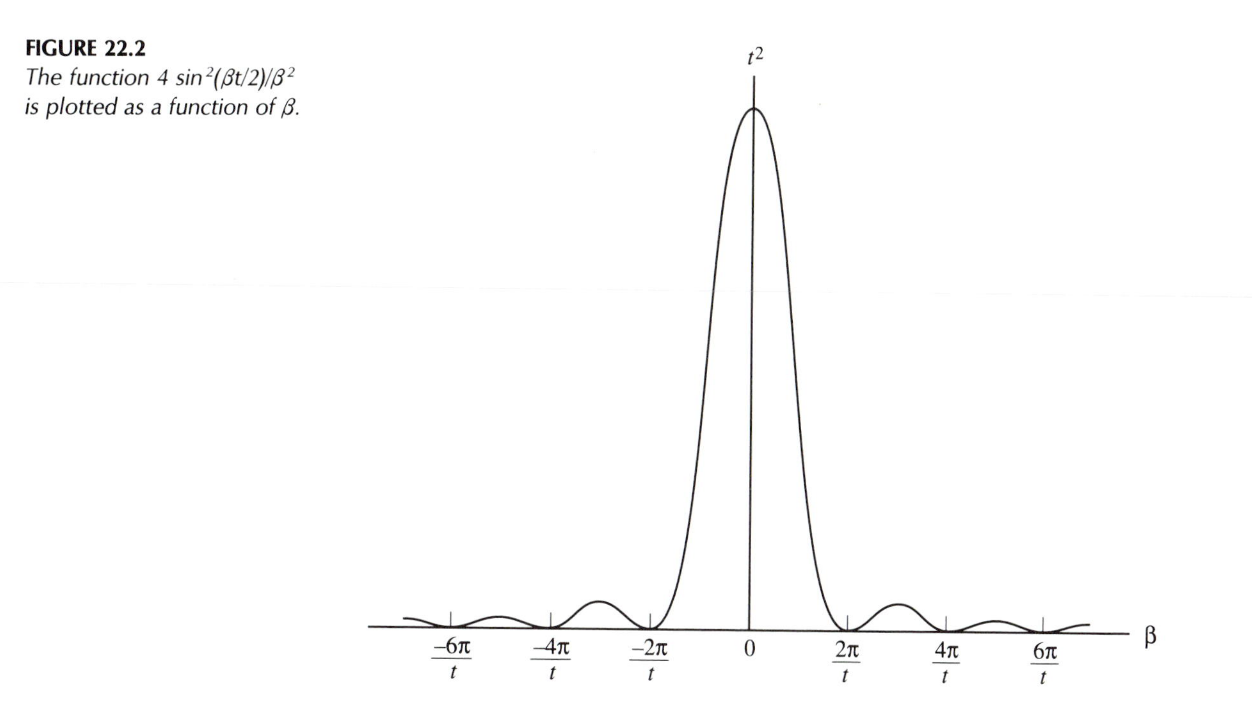

FIGURE 22.2
The function $4\sin^2(\beta t/2)/\beta^2$ is plotted as a function of β.

upshot of which is a formula for the transition probability that does make sense, that does vary in proportion to t.

First, we note that the highly peaked function

$$\frac{\sin^2[(\omega_{mi} \pm \omega)t/2]}{[(\omega_{mi} \pm \omega)/2]^2}$$

for large t goes over to the following delta function

$$2\pi t\delta(\omega_{mi} \pm \omega) = 2\pi\hbar t\delta(E_m - E_i \pm \hbar\omega) \tag{22.16}$$

This can be seen by noting that if $g(\beta)$ is a smooth function of β, we have for large t

$$\int_{-\infty}^{\infty} g(\beta)(4/\beta^2)\sin^2(\beta t/2)\,d\beta \approx g(0)\int_{-\infty}^{\infty} (4/\beta^2)\sin^2(\beta t/2)\,d\beta$$
$$= 2tg(0)\int_{-\infty}^{\infty} (1/u^2)\sin^2 u\,du = 2\pi tg(0)$$

Therefore, in the large t limit, the transition probability does vary as t, and we can define a transition probability per unit time

$$R_{i\to m} = 2\pi\hbar(1/\hbar^2)|\langle\phi_m|H_1|\phi_i\rangle|^2\delta(E_m - E_i \pm \hbar\omega)$$
$$= (2\pi/\hbar)|\langle\phi_m|H_1|\phi_i\rangle|^2\delta(E_m - E_i \pm \hbar\omega) \tag{22.17}$$

Admittedly, this derivation is not completely satisfactory. Applying a perturbation theoretic expression for "large t" is tricky! Moreover, we have ended up with a transition rate that is proportional to a delta function, a difficult function to give meaning to unless we integrate over it.

In practice, however, we have situations where the final (or the initial) state is one belonging to a group of states, all of which satisfy the resonant condition approximately (fig. 22.3), and we do have to integrate over these final (or initial) states. For example, we may be measuring the photoelectric effect where the final state consists of continuum electrons. In such a case, the transition rate that we measure is really the sum of the transition rates to an entire group of final states:

$$W_{i\to m} = \sum_{\Delta m} R_{i\to m} \tag{22.18}$$

Since the states $|m\rangle$ are continuously distributed, we should replace the sum above by an integral. To this end, we invoke the density of final states $\rho(E_m)$,

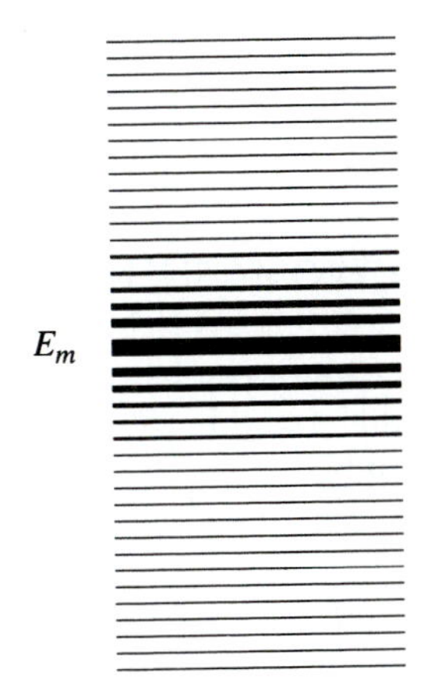

FIGURE 22.3
A group of final states around the state E_m.

the number of states per unit energy interval d^3n/dE introduced in chapter 21, and convert the sum in equation (22.18) into an integral to write

$$W_{i\to m} = \frac{2\pi}{\hbar}\int |\langle\phi_m|H_1|\phi_i\rangle|^2\rho(E_m)\,dE_m\delta(E_i - E_m \pm \hbar\omega)$$
$$= \frac{2\pi}{\hbar}\,\overline{|\langle\phi_m|H_1|\phi_i\rangle|^2\rho(E_m)|}_{E_m=E_i\pm\hbar\omega} \tag{22.19}$$

where the bar over the square of the H_1-matrix element indicates its average value over the final states. This is a most important result with many applications; hence the name *golden rule* was given it by Fermi.

Next we will apply the considerations that led to the golden rule to the calculation of the rate of electromagnetic transitions in atoms, which is a case of harmonic perturbation. This will further clarify the meaning of the density of final states and how to go about calculating it.

Although we have derived Fermi's golden rule using a harmonic perturbation, it is easy to see that the same formula applies for a constant (in time) potential, which is the special case of $\omega = 0$, noting that only one of the two terms in equation (22.12) contributes at a time to the final result. We will deal with the case of constant potential in a subsequent section as another application of the golden rule and also as an introduction to the subject of scattering theory.

However, a critique of the golden rule is also in order. The golden rule implies that the transition probability per unit time is a constant. But if this were true for all times, the probability would exceed unity at some point in time, which is not kosher. Fortunately, the reason for this apparent failure is not hard to find; it is due to our failure to account for the depletion of the initial wave function, which in turn is due to our failure to normalize the first-order wave function. The normalization of wave functions in perturbation theories always involves an order higher than the first. Therefore, the notion of the transition probability being equal to Wt holds only to lowest order in W, for small W (see later).

22.3 INTERACTION OF THE ELECTROMAGNETIC FIELD WITH ATOMS

The interaction of an electron with the radiation field, using the gauge condition $\nabla \cdot \mathbf{A} = 0$, was derived in chapter 14 and is given by the Hamiltonian (ignoring the small A^2-term in eq. [14.9]):

$$H = \frac{p^2}{2m} - e\phi + \frac{e}{mc}\mathbf{A}(\mathbf{r},t)\cdot\mathbf{p} \tag{22.20}$$

The last term is the time-dependent interaction. Suppose for $\mathbf{A}$ we take a plane wave; then the time dependence of $\mathbf{A}$ is

$$\begin{aligned}\mathbf{A} &= 2\mathbf{A}_0 \cos[\mathbf{k}\cdot\mathbf{r} - \omega t] \\ &= \mathbf{A}_0 \exp(i\mathbf{k}\cdot\mathbf{r})\exp(-i\omega t) + \mathbf{A}_0 \exp(-i\mathbf{k}\cdot\mathbf{r})\exp(i\omega t) \\ &= \mathbf{A}_0(\mathbf{r})\exp(-i\omega t) + \mathbf{A}_0^*(\mathbf{r})\exp(i\omega t)\end{aligned} \tag{22.21}$$

thus defining two new quantities $\mathbf{A}_0(\mathbf{r})$ and $\mathbf{A}_0^*(\mathbf{r})$. What do they mean? We know from section 22.2 that the second term $\sim\exp(i\omega t)$ takes part in induced emission of a photon, while the first term $\sim\exp(-i\omega t)$ is responsible for photon absorption. It is natural, then, to associate $\mathbf{A}_0^*(\mathbf{r})$ with the creation of a photon and $\mathbf{A}_0(\mathbf{r})$ with the annihilation of a photon. Here again is the analogy with the harmonic oscillator creation and annihilation operators that we noticed in chapter 21 (section 21.4). Indeed, if we quantize the electromagnetic field, these descriptions naturally follow.

We will, however, avoid a fully quantum mechanical description of the electromagnetic field and, instead, use arguments based on the correspondence principle to proceed further, which is to be able to calculate the matrix element of $\mathbf{A}_0(\mathbf{r})$ between appropriate initial and final states.

The electric and magnetic fields associated with the vector potential are given by

$$\boldsymbol{\mathcal{E}} = -(1/c)\partial\mathbf{A}/\partial t = (i\omega/c)\mathbf{A}_0 \exp[i(\mathbf{k}\cdot\mathbf{r} - \omega t)] + \text{complex conjugate}$$

$$\mathbf{B} = \nabla \times \mathbf{A} = i\mathbf{k} \times \mathbf{A}_0 \exp[i(\mathbf{k}\cdot\mathbf{r} - \omega t)] + \text{complex conjugate}$$

Using these relations, it is easy to express the energy density of the electromagnetic field in terms of $\mathbf{A}_0$:

$$(1/8\pi)(\boldsymbol{\mathcal{E}}^2 + \mathbf{B}^2) = (1/8\pi)[(2\omega^2/c^2)\mathbf{A}_0\cdot\mathbf{A}_0^* + 2(\mathbf{k}\times\mathbf{A}_0)\cdot(\mathbf{k}\times\mathbf{A}_0)^*]$$

where we have ignored the oscillating terms since they time-average out to zero. The gauge condition $\nabla \cdot \mathbf{A} = 0$ implies that

$$\mathbf{k} \cdot \mathbf{A}_0 = 0$$

which we now can use to simplify the second term above:

$$(\mathbf{k} \times \mathbf{A}_0) \cdot (\mathbf{k} \times \mathbf{A}_0^*) = k^2 \mathbf{A}_0 \cdot \mathbf{A}_0^* = (\omega^2/c^2) \mathbf{A}_0 \cdot \mathbf{A}_0^*$$

It follows that

$$(1/8\pi)(\mathcal{E}^2 + \mathbf{B}^2) = (\omega^2/2\pi c^2) \mathbf{A}_0 \cdot \mathbf{A}_0^*$$

Now enclose the system in a box of volume V, then the total energy of the electromagnetic field is given by the integral of the energy density calculated above over this volume:

$$\int d^3r \, \frac{1}{8\pi} (\mathcal{E}^2 + \mathbf{B}^2) = \frac{\omega^2 V}{2\pi c^2} |\mathbf{A}_0|^2$$

If we assume that this energy is carried by N photons, since the energy of each photon is $\hbar\omega$, we get

$$(\omega^2 V/2\pi c^2)|\mathbf{A}_0|^2 = N\hbar\omega$$

This gives the magnitude of $\mathbf{A}_0$. The direction of $\mathbf{A}_0$ is given by the polarization vector $\boldsymbol{\epsilon}$, which is a unit vector satisfying $\boldsymbol{\epsilon} \cdot \mathbf{k} = 0$. Finally, we have

$$\mathbf{A}_0(\mathbf{r}) = (2\pi c^2 N\hbar/\omega V)^{1/2} \boldsymbol{\epsilon} \exp[i\mathbf{k} \cdot \mathbf{r}] \tag{22.22}$$

To intuit the quantum mechanical modification, let's consult equations (21.60) and (21.61) of chapter 21. We can easily see that the following modifications are called for:

1. For the absorption of a photon by an electron from an initial state containing N photons to a final state containing $N - 1$ photons, according to equation (21.61), we must take $\mathbf{A}$ in the Hamiltonian equation (22.20) to be

$$\mathbf{A}(\mathbf{r}, t) = (2\pi c^2 N\hbar/\omega V)^{1/2} \boldsymbol{\epsilon} \exp[i(\mathbf{k} \cdot \mathbf{r} - \omega t)] \tag{22.23}$$

2. For the emission of a photon from an initial state containing N photons to a final state containing $N + 1$ photons, we must take $\mathbf{A}$ to be

(see eq. [21.60] and take note from eq. [22.21] that it is now $\mathbf{A}_0^*$ that comes into play)

$$\mathbf{A}(\mathbf{r},t) = [2\pi c^2(N+1)\hbar/\omega V)^{1/2}\boldsymbol{\epsilon}\exp[-i(\mathbf{k}\cdot\mathbf{r} - \omega t)] \tag{22.24}$$

What have we gained by invoking a quantum mechanical description of the radiation field, nonrigorous as it may be? Simply this. We now have a Hamiltonian for the interaction of a system with an electromagnetic field that not only applies to the case of absorption and stimulated emission of light, but also one that leads to the emission of a photon *even when there is no photon in the initial state*, $N = 0$; in other words, even when there is no incident electromagnetic wave. This *spontaneous* emission is a purely quantum effect and can be understood by remembering zero-point fluctuations that are characteristic of quantum systems. Even when there is no incident radiation, an atomic electron in an excited state continues to interact with the zero-point fluctuations of the radiation field and thus is able to emit a photon.

It should also be clear from equation (22.24) that if there is a large number of photons N of a certain frequency present, then the probability of stimulated emission increases enormously. The laser (*l*ight *a*mplification by *s*timulated *e*mission of *r*adiation) works on this principle (and also the maser). Many atoms are pumped up to a metastable excited state, causing *population inversion*. A few of these spontaneously emit photons of the "right" frequency; subsequently, these photons multiply via stimulated emission. The photons are entrapped between mirrors and thus provide the right environment for further lasing action (fig. 22.4).

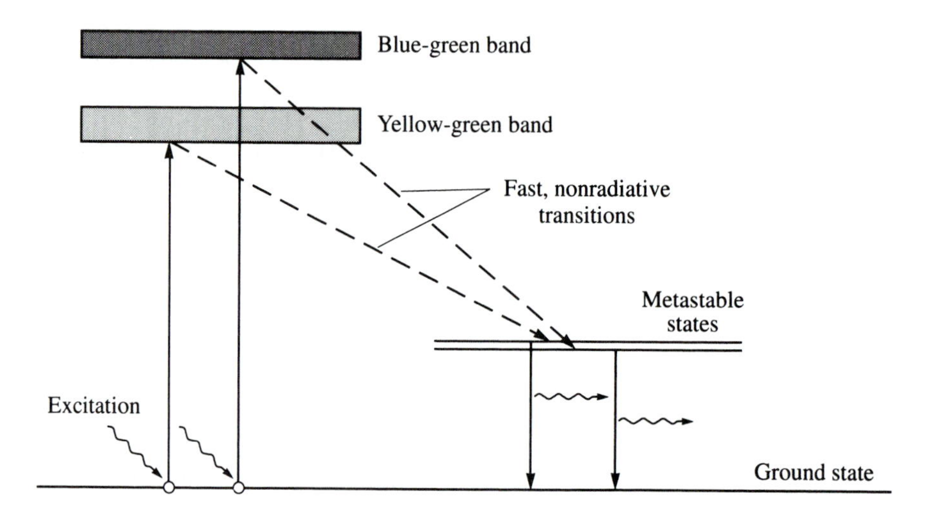

FIGURE 22.4
How a laser works.

22.4 CALCULATION OF TRANSITION RATE FOR SPONTANEOUS EMISSION

The interaction potential $e\mathbf{A}\cdot\mathbf{p}/mc$ in equation (22.20) for the case of spontaneous emission is therefore given as, using equation (22.24) with $N=0$ for $\mathbf{A}$,

$$H_1(t) = (e/mc)(2\pi c^2\hbar/\omega V)^{1/2}\boldsymbol{\epsilon}\cdot\mathbf{p}\exp[-i(\mathbf{k}\cdot\mathbf{r} - \omega t)] \tag{22.25}$$

From equation (22.17), the transition rate $R_{i\to m}$ for spontaneous emission of a photon from the initial state $|\phi_i\rangle$ to the final state $|\phi_m\rangle$ is given as

$$R_{i\to m} = \frac{2\pi}{\hbar}\frac{2\pi e^2\hbar}{m^2\omega V}|\langle\phi_m|e^{-i\mathbf{k}\cdot\mathbf{r}}\boldsymbol{\epsilon}\cdot\mathbf{p}|\phi_i\rangle|^2\delta(E_i - E_m - \hbar\omega) \tag{22.26}$$

We want, as before, to calculate the sum of these rates over a group of final states. In the present case, the final states consist of products of discrete atomic states and continuum photon states. The sum over the photonic states can be converted into an integral using the concept of density of states introduced earlier. We want the density of states $\rho(E)$ for which the emitted photon has momentum between $\mathbf{p}$ and $\mathbf{p} + d\mathbf{p}$; since the photons are contained in a box of volume V, we have

$$\begin{aligned} W_{i\to m} &= \sum \int dE\,\rho(E)R_{i\to m} \\ &= \sum \int dE\,(d^3\mathbf{n}/dE)R_{i\to m} \\ &= (V^{1/3}/2\pi\hbar)^3 \sum \int d^3\mathbf{p}\,R_{i\to m} \end{aligned} \tag{22.27}$$

where $\sum$ denotes the remaining sum over the final atomic states and where we have used the definition of $\rho(E)$,

$$\rho(E) = d^3\mathbf{n}/dE$$

and also the sense of equation (21.3). Now we can write

$$d^3\mathbf{p} = p^2\,dp\,d\Omega_p = (\hbar\omega/c)^2 d(\hbar\omega/c)\,d\Omega_p$$

The transition probability per unit time then becomes

$$W_{i\to m} = \sum \int \frac{4\pi^2 e^2}{m^2 \omega V} |\langle \phi_m | e^{-i\mathbf{k}\cdot\mathbf{r}} \boldsymbol{\epsilon}\cdot\mathbf{p} | \phi_i \rangle|^2 \, d\Omega_p \frac{V}{(2\pi\hbar)^3} \frac{\hbar^2\omega^2}{c^3}$$
$$\times\, d(\hbar\omega)\, \delta(E_i - E_m - \hbar\omega)$$
$$= \frac{\alpha}{2\pi} \sum \int d\Omega_p \frac{E_i - E_m}{\hbar} \left| \frac{1}{mc} \langle \phi_m | e^{-i\mathbf{k}\cdot\mathbf{r}} \boldsymbol{\epsilon}\cdot\mathbf{p} | \phi_i \rangle \right|^2 \quad (22.28)$$

In addition to the integral over the photon states carried out here, we must also sum over all the final atomic states that are relevant (see below). In addition, if the polarization is not measured, we must also sum the result over the two polarization states of the photon in order to obtain the desired transition rate.

The Dipole Approximation

The wave functions of the atomic states involved in the matrix element in equation (22.28) are well localized. If we take the center of the atom as the origin of the coordinate system, these wave functions spread no more than a few angstroms around the origin. In comparison, the wavelength of the emitted photons ranges from hundreds to tens of thousands of angstroms. Under these circumstances, $\exp(i\mathbf{k}\cdot\mathbf{r})$ can be expanded in a power series in the small quantity $(\mathbf{k}\cdot\mathbf{r})$:

$$\exp(i\mathbf{k}\cdot\mathbf{r}) = 1 + i\mathbf{k}\cdot\mathbf{r} + \tfrac{1}{2}(i\mathbf{k}\cdot\mathbf{r})^2 + \cdots$$
$$\approx 1 \quad (22.29)$$

If we keep only the first term, we can see that we have to calculate the matrix element

$$\boldsymbol{\epsilon}\cdot\langle \phi_m | \mathbf{p} | \phi_i \rangle$$

But this can be further simplified by noting that

$$[H_0, x] = -i\hbar p_x / m$$

Consequently, we have

$$\boldsymbol{\epsilon}\cdot\langle \phi_m | \mathbf{p} | \phi_i \rangle = (im/\hbar)\boldsymbol{\epsilon}\cdot\langle \phi_m | [H_0, \mathbf{r}] | \phi_i \rangle$$
$$= (im/\hbar)\boldsymbol{\epsilon}\cdot(E_m - E_i)\langle \phi_m | \mathbf{r} | \phi_i \rangle$$
$$= -im\omega_{im}\boldsymbol{\epsilon}\cdot\langle \phi_m | \mathbf{r} | \phi_i \rangle \quad (22.30)$$

where $\omega_{im} = (E_i - E_m)/\hbar$. So in this approximation, we have to calculate only the matrix element of the dipole operator $\sim\mathbf{r}$; therefore, it is called the dipole approximation.

Summing Over Polarization

Let's assume that both directions of the photon's polarization are measured. Then we have to sum over the two polarizations of the photons. Suppose the wave vector $\mathbf{k}$ of the photon is along $\mathbf{e}_z$ and the two polarizations $\boldsymbol{\epsilon}_1$ and $\boldsymbol{\epsilon}_2$ along $\mathbf{e}_x$ and $\mathbf{e}_y$, respectively (fig. 22.5). Calling $\langle \phi_m | \mathbf{r} | \phi_i \rangle = \mathbf{M}$, we find the sum to be of the form

$$\sum_{\text{pol}} |\boldsymbol{\epsilon}\cdot\mathbf{M}|^2 = |\boldsymbol{\epsilon}_1\cdot\mathbf{M}|^2 + |\boldsymbol{\epsilon}_2\cdot\mathbf{M}|^2 = |M_x|^2 + |M_y|^2$$

$$= |M|^2 - |M_z|^2$$

Since the direction of $\mathbf{k}$ is arbitrary, we have

$$\sum_{\text{pol}} |\boldsymbol{\epsilon}\cdot\mathbf{M}|^2 = |M|^2 - |\mathbf{k}\cdot\mathbf{M}|^2/k^2$$

$$= |\mathbf{M}|^2 - |\mathbf{M}|^2 \cos^2\theta = |\mathbf{M}|^2 \sin^2\theta$$

Now we can carry out the angular integration in equation (22.28); we have

$$\int \sin^2\theta \, d\Omega_p = 8\pi/3 \tag{22.31}$$

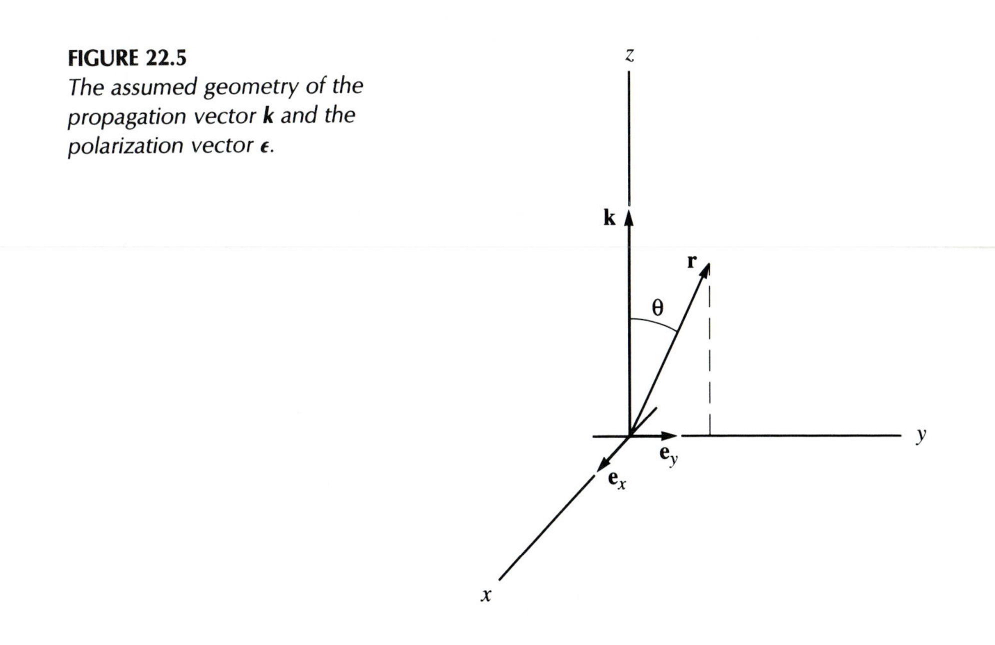

FIGURE 22.5
The assumed geometry of the propagation vector **k** *and the polarization vector* $\boldsymbol{\epsilon}$.

Combining equations (22.28), (22.29), and (22.30), we get for the transition rate the following expression:

$$W_{i\to m} = (4e^2\omega_{im}^3/3\hbar c^3)|\langle\phi_m|\mathbf{r}|\phi_i\rangle|^2 \tag{22.32}$$

This is the *Einstein A-coefficient.*

The result, equation (22.32), compares very well with that of classical radiation theory. The power radiated by an accelerated electron, according to the classical Larmor formula (see Jackson, *Classical Electrodynamics*), is given by

$$P = 2e^2a^2/3c^3$$

where a is the acceleration. For a circular orbit of radius r and angular velocity ω, $a = \omega^2 r$. Now we can argue that the lifetime of a state is the time τ it takes to radiate half the energy $\hbar\omega/2$ away. This gives

$$\frac{1}{\tau} = \frac{2P}{\hbar\omega} = \frac{1}{\hbar\omega}\frac{4}{3}\frac{e^2}{c^3}\omega^4 r^2 = \frac{4e^2\omega^3}{3\hbar c^3}r^2 \tag{22.33}$$

and $1/\tau$ is, of course, the same as the transition rate calculated in equation (22.32). This agreement is another spectacular example of classical correspondence.

The Dipole Selection Rules

The dipole selection rules arise from the integration over the angles of $\mathbf{r}$ of the matrix element

$$\langle\phi_m|\mathbf{r}|\phi_i\rangle$$

If the initial and final states of the atom are hydrogenic, characterized by the angular quantum numbers l, m and l', m', respectively, then the angular integral of the matrix element above consists of

$$\int d\Omega\, Y_{l'm'}^*(\theta\phi)\mathbf{e}_r Y_{lm}(\theta\phi) \tag{22.34}$$

where $\mathbf{e}_r$ is a unit vector along $\mathbf{r}$. Using the explicit representation of $Y_{1m}(\theta\phi)$ given in equation (11.43), only a little algebra is needed to express $\mathbf{e}_r$ in terms of Y_{1m}:

$$\mathbf{e}_r = \left(\frac{4\pi}{3}\right)^{1/2}\left[\frac{-\mathbf{e}_x + i\mathbf{e}_y}{\sqrt{2}}Y_{11} + \frac{\mathbf{e}_x + i\mathbf{e}_y}{\sqrt{2}}Y_{1-1} + \mathbf{e}_z Y_{10}\right] \tag{22.35}$$

Obviously the angular integral in equation (22.34) involves the integral of three spherical harmonics:

$$\int d\Omega \; Y^*_{l'm'}(\theta\phi) Y_{1M}(\theta\phi) Y_{lm}(\theta\phi)$$

Although the evaluation of this integral is beyond the mathematical level of this book, the result involves Clebsch-Gordan coefficients that are already familiar to the reader. Since the selection rules follow most transparently from the properties of these Clebsch-Gordan coefficients, we will use this mathematical result without proof (presumably you will not be too perturbed about a lack of proof; after all, an integral is an integral is an integral!):

$$\int d\Omega \; Y^*_{l'm'} Y_{LM} Y_{lm} = \left(\frac{(2l+1)(2L+1)}{4\pi(2l'+1)} \right)^{1/2} \langle lmLM | l'm' \rangle \langle l0L0 | l'0 \rangle \quad (22.36)$$

In the present case of the dipole matrix element, $L = 1$.

With these ideas in mind, the dipole selection rules are obtained as follows:

1. The Clebsch-Gordan coefficient $\langle lm1M | l'm' \rangle$ is zero unless

$$m' - m = M = 1, 0, -1 \quad (22.37)$$

This selection rule has been invoked previously in connection with the Zeeman effect.

2. The angular momentum selection rule for dipole radiation is

$$l' = l - 1, l, l + 1 \quad (22.38)$$

3. The Clebsch-Gordan coefficient is also zero if $l = l' = 0$; therefore, no $0 \to 0$ transition is possible.
4. The Clebsch-Gordan coefficient

$$\langle l0L0 | l'0 \rangle$$

vanishes unless

$$(-1)^{l+L-l'} = 1$$

(We can see this by changing the sign of all the m's and using eq. [17.30]). This is called the parity rule. For the electric dipole transition, $L = 1$, and the parity rule dictates that the initial and final atomic states be of opposite parity.

The Radial Integral: The $2P \to 1S$ Transition

Now we will give the complete calculation for a specific case: The $2P \to 1S$ transition in hydrogenic atoms. First let's calculate the radial integral. Substituting from equation (13.23) for $2P$ and $1S$ radial functions, we get

$$\int_0^\infty dr\, R_{10}^*(r) r^3 R_{21}(r) = \frac{1}{\sqrt{6}} (Z/a_0)^4 \int_0^\infty dr\, r^4 e^{-3Zr/2a_0}$$
$$= 4\sqrt{6}(2/3)^5(a_0/Z) \tag{22.39}$$

The angular integral

$$\left(\frac{4\pi}{3}\right)^{1/2} \int d\Omega\, Y_{00}^* \left[\frac{-\mathbf{e}_x + i\mathbf{e}_y}{\sqrt{2}} Y_{11} + \frac{\mathbf{e}_x + i\mathbf{e}_y}{\sqrt{2}} Y_{1-1} + \mathbf{e}_z Y_{10}\right] Y_{1m}$$

can be evaluated directly (since Y_{00} is just a number) or by using equation (22.36) (the Clebsch-Gordan coefficients can be found from table 17.1); the result is

$$(1/\sqrt{3})\,[\mathbf{e}_z \delta_{m,0} - \{(-\mathbf{e}_x + i\mathbf{e}_y)/\sqrt{2}\}\delta_{m,-1} - \{(\mathbf{e}_x + i\mathbf{e}_y)/\sqrt{2}\}\delta_{m,1}] \tag{22.40}$$

Combining equations (22.39) and (22.40), the square of the matrix element is

$$|\langle m|\mathbf{r}|i\rangle|^2 = 96(2/3)^{10}(a_0/Z)^2(1/3)\,[\delta_{m,0} + \delta_{m,1} + \delta_{m,-1}] \tag{22.41}$$

since the unit vectors $\mathbf{e}_x$, $\mathbf{e}_y$, and $\mathbf{e}_z$ are orthogonal. From equation (22.32), the transition rate is

$$W_{2P\to 1S} = (4/3)(e^2\omega^3/\hbar c^3)|\langle \phi_m|\mathbf{r}|\phi_i\rangle|^2$$
$$= (2^{17}/3^{11})(e^2\omega^3/\hbar c^3)(a_0/Z)^2[\delta_{m,0} + \delta_{m,1} + \delta_{m,-1}] \tag{22.42}$$

If the initial P-state is unpolarized (i.e., unaligned in any particular m-state), we must average the result, equation (22.42), over the m's. Since the summand is independent of m except for the Kronecker δ's, and since

$$(1/3)\sum_m (\delta_{m,0} + \delta_{m,1} + \delta_{m,-1}) = 1$$

we get

$$W_{2P\to 1S} = (2^{17}/3^{11})(e^2\omega^3/\hbar c^3)(a_0/Z)^2 \tag{22.43}$$

In order to get a numerical value, we need one more number. From the hydrogenic eigenvalue formula, equation (13.12), we have

$$\omega = (E_{2P} - E_{1S})/\hbar = (3/8)(mc^2/\hbar)Z^2\alpha^2$$

Substituting into equation (22.43), we get

$$W_{2P\to 1S} = (2/3)^8 (mc^2/\hbar)\alpha^5 Z^4$$
$$\approx 6 \times 10^8 Z^4 \text{ s}^{-1} \quad (22.44)$$

For the lifetime τ of the $2P$ state, this gives

$$\tau = 1/W = 1.6 \times 10^{-9} Z^{-4} \text{ s} \quad (22.45)$$

which agrees very well with the experimental data for atomic lifetimes.

There is profound meaning in the fact that quantum systems can make spontaneous transitions from excited energy levels to lower energy states: Excited states that decay cannot be, strictly speaking, stationary. At the most, if the decay lifetime is relatively long compared to the time it takes light to travel through the body (for an atom this is 10^{-16} s), as in the case of the dipole transition above, we can call the states *quasi-stationary*.

22.5 EXPONENTIAL DECAY LAW, LIFETIME, AND LINE WIDTH

Let's return to the little tarnish on the otherwise golden rule that we discussed before. The golden rule says that the transition rate is constant if we wait long enough, long compared to

$$\hbar/|(E_m - E_i + \hbar\omega)|$$

But on the other hand, if we wait too long, we must take into account the depletion of the initial state, which the first-order theory fails to do; there is no normalization correction in first order. We can try to correct the situation in the following way: The probability that has gone into all the final states must come from the initial state. Thus, the probability that the initial state remains intact must be

$$P_i(t) = 1 - \left(\sum_{f\neq i} W_{i\to f}\right) t \quad (22.46)$$

where the sum extends over all possible final states. Unfortunately, this expression becomes meaningless for large t, and the only way to interpret it is as the lowest-order approximation of a more general decay law. Let's note that if the decay law for quasi-stationary states is given by the well-known exponential form famous from the study of radioactivity,

$$P_i(t) = \exp\left[-t \sum_{f\neq i} W_{i\to f}\right] \quad (22.47)$$

equation (22.46) is indeed obtained as the first-order approximation. Not surprisingly, when we carry out a careful analysis on the subject, we find that under certain suitable assumptions, $P(t)$ is indeed given by equation (22.47). It is in the sense of this exponential decay law that we can speak of a lifetime of an initial state $\tau = 1/W$, which we introduced earlier.

However, the exponential decay must also be only approximate. This can be seen as follows. Suppose the state at $t = 0$, $|\psi(0)\rangle$, develops at time t into $|\psi(t)\rangle$. Suppose, and this is in accordance with the exponential law, that the form of the overlap is

$$\langle\psi(0)|\psi(t)\rangle = \exp(-\chi t/\hbar)$$

where $\chi = iE_0 + W/2$. Let's generalize the expression for all t. For $t < 0$, we have

$$\begin{aligned}\langle\psi(0)|\psi(t)\rangle &= \langle\psi(0)|\exp(-iHt/\hbar)|\psi(0)\rangle = \langle\psi(0)|\exp(iHt)|\psi(0)\rangle^* \\ &= \langle\psi(0)|\psi(-t)\rangle^* = \exp(\chi^* t/\hbar)\end{aligned}$$

Thus if we write

$$\langle\psi(0)|\psi(t)\rangle = \exp[-(iE_0 t + \tfrac{1}{2}W|t|)/\hbar] \tag{22.48}$$

the expression holds for all t. We now expand $|\psi(0)\rangle$ in terms of the eigenstates of the complete H:

$$|\psi(0)\rangle = \int g(E)|\psi(E)\rangle\, dE$$

where $g(E)$ is the expansion coefficient. This gives

$$\begin{aligned}\langle\psi(0)|\psi(t)\rangle &= \langle\psi(0)|\exp(-iHt/\hbar)|\psi(0)\rangle \\ &= \int |g(E)|^2 \exp(-iEt/\hbar)\, dE\end{aligned}$$

By taking an inverse Fourier transform, we can write

$$|g(E)|^2 = (2\pi\hbar)^{-1}\int_{-\infty}^{\infty} dt\, \langle\psi(0)|\psi(t)\rangle e^{iEt/\hbar} \tag{22.49}$$

Substituting equation (22.48) into the integral in equation (22.49) and carrying out the integration, we get

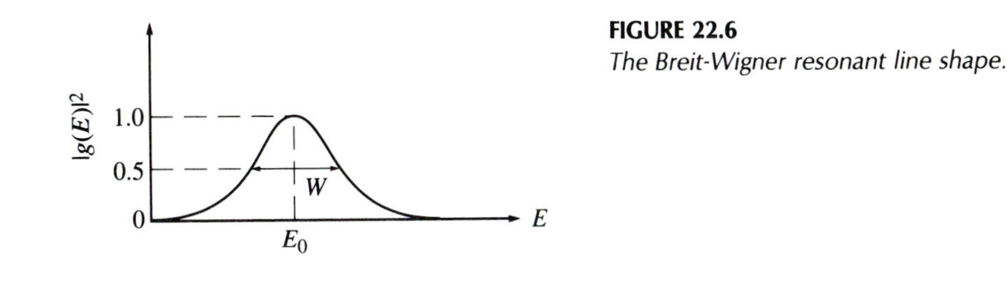

FIGURE 22.6
The Breit-Wigner resonant line shape.

$$|g(E)|^2 = \frac{W/2\pi}{(E - E_0)^2 + W^2/4} \tag{22.50}$$

This is strictly >0 for all E, which is not right behavior for it at all. The energy of a system always has a minimum value, hence $g(E)$ must go to zero for some value of $E \leq E_{\text{min}}$. So the exponential law of decay must be faulty, only approximate, and there must be deviations from it. But there is an interesting problem here. Exponential-decay law has no history. But if unstable particles show nonexponential decay, then in principle we could distinguish between them in terms of their age. This violates the principle of indistinguishability that we hold so dear in quantum mechanics. One way to get over this puzzle is to say, as Wigner does, that unstable particles are not elementary, and therefore they do not have to obey the principle of indistinguishability accurately!

Anyway, the function of equation (22.50) is the famous Breit-Wigner distribution of energy (fig. 22.6) and is widely used as the line shape (intensity of the line versus energy curve) for a wide range of quantum systems. As calculated above, W is called the natural line width; the actual line width may be further broadened due to other processes such as collision.

Phase-Space Factor for Decay Rates of Unstable Particles

Elementary particle physics is fraught with unstable particles that often decay into other particles in more than one way. For example, suppose an unstable elementary particle A can decay in two ways:

$$A \to B + C$$

$$A \to D + E \tag{22.51}$$

Can the first-order theory and the golden rule tell us which mode of decay is the preferred one?

Consider the decay of particle A into the two particles B and C. The first-order theory in the form of the golden rule (assuming it applies) gives us the decay rate as

$$W = (2\pi/\hbar)|\langle BC(E_0)|H_1|A\rangle|^2\rho(E_0) \tag{22.52}$$

where H_1, the interaction that causes the decay, has been assumed to be constant in time, and where $|BC(E_0)\rangle$ denotes the state of the final particles B and C with energy E_0, the initial energy (energy is conserved!), and $\rho(E_0)$ is the density of final states. It is appropriate to take the final state as the momentum eigenstate of the two particles (the product of their plane waves), in which case, in many theories, the dependence of the matrix element on the energy can be ignored. If this is the case, then the *phase-space factor*, the density $\rho(E_0)$, determines the decay rate enough to tell which of the two pathways in (22.51) is more probable.

Suppose A decays from rest. Then by the conservation of momentum, B and C have opposite momentum; take them as $\pm\mathbf{p}$. Note that the important thing is to realize that momentum conservation implies a constraint that eliminates one of the final momentum variables from being an independent variable. Since we have free particles, the energy eigenvalue of H_0 in the final state is given as (we have to use the relativistic formula, however)

$$E = E_B + E_C = (p^2c^2 + m_B^2c^4)^{1/2} + (p^2c^2 + m_C^2c^4)^{1/2} \tag{22.53}$$

The density of states $\rho(E)$, assuming that the decay particles are emitted in a box and have box-normalized states (call it the box of convenience), is

$$\rho(E) = d^3n/dE = (2\pi\hbar)^{-3}V\,d^3p/dE = (2\pi\hbar)^{-3}V\,d\Omega\,p^2\,dp/dE \tag{22.54}$$

Thus it boils down to the calculation of dp/dE. For this we use equation (22.53) and obtain

$$\frac{dp}{dE} = \left(\frac{dE_B}{dp} + \frac{dE_C}{dp}\right)^{-1} = \left(\frac{pc^2}{E_B} + \frac{pc^2}{E_C}\right)^{-1} = \frac{E_BE_C}{Epc^2} \tag{22.55}$$

But in the present case $E = E_0 = M_Ac^2$. We can also express the product E_BE_C as $\{(E_B + E_C)^2 - (E_B - E_C)^2\}/4$ with $E_B^2 - E_C^2 = (m_B^2 - m_C^2)c^4$, from which $E_B - E_C$ can be evaluated. The final result for $\rho(E = E_0)$ is

$$\rho(E_0) = (2\pi\hbar)^{-3}V\,d\Omega\,(p/4m_A^3)\,[m_A^4 - (m_B^2 - m_C^2)^2] \tag{22.56}$$

where p can be calculated from equation (22.53) to be

$$p = (c/2m_A)\,[(m_A + m_B + m_C)(m_A + m_B - m_C)(m_A - m_B + m_C) \times (m_A - m_B - m_C)]^{1/2} \tag{22.57}$$

Clearly then, the phase-space factor is greater for smaller masses of the final particles, the momentum p is greater. Thus we have proven a very important result: The decay rate is greater for the decay into smaller masses, or the decay that releases the most kinetic energy; and that's what determines the favorable pathway.

Incidentally, if you are worried about the volume of the box in the density of states, take note that there is also a $1/V$ factor coming from the box normalization of the wave function of the emitted particles in the matrix element; and thus, V does cancel, as it must (also, see below).

22.6 BORN APPROXIMATION AND INTRODUCTION TO SCATTERING

The elastic scattering of a particle projectile from a target can be looked upon as a transition from an initial momentum eigenstate into a final state of a different direction of momentum but the same energy. This situation is amenable to treatment by the time-dependent perturbation theory, in particular, the golden rule developed above.

If the initial and final momenta are denoted respectively by $\mathbf{p}_0 = \hbar\mathbf{k}_0$ and $\mathbf{p} = \hbar\mathbf{k}$, where $\mathbf{k}_0$ and $\mathbf{k}$ are the respective wave vectors, then the initial and final box-normalized wave functions are given as

$$\langle \mathbf{r}|\phi_i\rangle = L^{-3/2}\exp(i\mathbf{k}_0\cdot\mathbf{r}), \qquad \langle \mathbf{r}|\phi_m\rangle = L^{-3/2}\exp(i\mathbf{k}\cdot\mathbf{r})$$

where L is the dimension of the box. Then putting $H_1 = V(\mathbf{r})$ where V is the constant potential responsible for the scattering, we find for the perturbation matrix element the following equation:

$$\begin{aligned}\langle \phi_m|V|\phi_i\rangle &= L^{-3}\int d^3r\, e^{-i\mathbf{k}\cdot\mathbf{r}}Ve^{i\mathbf{k}_0\cdot\mathbf{r}} \\ &= L^{-3}\int d^3r\, V(\mathbf{r})e^{i\mathbf{q}\cdot\mathbf{r}}\end{aligned} \tag{22.58}$$

where $\mathbf{q} = \mathbf{k}_0 - \mathbf{k}$, and $\hbar\mathbf{q}$ is the *momentum transfer*. You will recognize the integral above as the three-dimensional Fourier transform of the potential $V(\mathbf{r})$, call it $V(\mathbf{q})$:

$$V(\mathbf{q}) = \int d^3r\, V(\mathbf{r})e^{i\mathbf{q}\cdot\mathbf{r}} \tag{22.59}$$

Consequently, the matrix element, equation (22.58), can be written as

$$\langle \phi_m|V|\phi_i\rangle = L^{-3}V(\mathbf{q}) \tag{22.60}$$

The density of states, as before, is given as

$$\rho(E) = (L/2\pi\hbar)^3\, d^3p/dE = (L/2\pi\hbar)^3\, d\Omega\, p^2\, dp/dE$$

Since $E(p) = p^2/2\mu$, where μ is the mass of the particle (we use the notation μ because the same treatment applies to the two-body scattering with μ regarded as the reduced mass of the equivalent one body in the center-of-mass system). We have $dE/dp = p/\mu$, giving $dp/dE = \mu/p$. We are thus left with

$$\rho(E) = \mu(L/2\pi\hbar)^3 p\, d\Omega \tag{22.61}$$

Substituting equations (22.60) and (22.61) into the golden rule, equation (22.19), we obtain

$$W_{i\to m} = (1/L^3)(1/4\pi^2\hbar^4)|V(\mathbf{q})|^2 \mu p\, d\Omega \tag{22.62}$$

But the expression in equation (22.62) still has an undesirable dependence on the volume L^3 of our box of convenience! Has the convenience of box normalization turned into an inconvenience? Not quite! The value of W obtained above is the number of particles scattered into the solid angle $d\Omega$ when there is one incident particle in the volume L^3. The corresponding incident flux is $L^{-3}v$, where $v = p/\mu$ is the velocity of the incident particle (which is the same as that of the scattered particle from energy conservation). We have to divide W by this flux to obtain the rate of transition per unit incident flux, which is what we measure, what is called the element of *differential cross section* $d\sigma$.

We have

$$\begin{aligned} d\sigma &= \frac{W}{p/\mu L^3} \\ &= \frac{\mu^2}{4\pi^2\hbar^4}\, d\Omega\, |V(\mathbf{q})|^2 \end{aligned}$$

In this way we obtain

$$\frac{d\sigma}{d\Omega} = \frac{\mu^2}{4\pi^2\hbar^4}\, |V(\mathbf{q})|^2 \tag{22.63}$$

This expression is called the *Born approximation* for the differential cross section; it was calculated by Born in the same paper in which he enunciated the celebrated probability interpretation of quantum mechanics.

The Born approximation for the differential cross section has been obtained from perturbation theory, and therefore it is valid only when the scattering potential is weak. We will develop a more satisfactory approach for scattering calculations in chapter 23 (where questions such as, How do we calculate the laboratory cross section from the center-of-mass cross section? will also be addressed); right now, let's consider an example of Born calculation.

The Case of the Yukawa or the Screened Coulomb Potential

We are confronted with calculating the three-dimensional Fourier transform of a spherically symmetric potential:

$$V(\mathbf{q}) = \int d^3r\, V(\mathbf{r}) e^{i\mathbf{q}\cdot\mathbf{r}} \tag{22.64}$$

with $\mathbf{q} = \mathbf{k}_0 - \mathbf{k}$ (fig. 22.7). Since $V(\mathbf{r}) = V(r)$, independent of angles, the angular integral is easy to perform. Choose the z-axis along $\mathbf{q}$. Then $\mathbf{q}\cdot\mathbf{r} = qr\cos\theta'$ (calling the polar angles of $\mathbf{r}$, $\theta'\phi'$). Note that q, the magnitude of $\mathbf{q}$, is given from

$$q^2 = \mathbf{k}^2 + \mathbf{k}_0^2 - 2kk_0\cos\theta = 2k^2(1 - \cos\theta) = 4k^2\sin^2\theta/2 \tag{22.65}$$

since $|\mathbf{k}| = |\mathbf{k}_0| = k$. Here θ is the scattering angle.

The integral in equation (22.64) is then

$$\int_0^\infty r^2\,dr\,V(r)\int_0^\pi \sin\theta'\,d\theta'\,e^{iqr\cos\theta'}\int_0^{2\pi} d\phi' = 4\pi\int_0^\infty dr\,r\,\frac{\sin qr}{q}\,V(r) \tag{22.66}$$

The Yukawa or screened Coulomb potential has the form

$$V(r) = C\exp(-r/a)/r \tag{22.67}$$

where a is to be interpreted as the range of the potential or the screening radius, as the case may be.

The Fourier transform, according to equation (22.66), is given as

$$\begin{aligned} V(q) &= 4\pi\int_0^\infty r\,\frac{\sin qr}{q}\,C\,\frac{e^{-r/a}}{r}\,dr = \frac{2\pi}{iq}\,C\int_0^\infty (e^{iqr} - e^{-iqr})e^{-r/a}\,dr \\ &= \frac{2\pi C}{iq}\left[\frac{1}{\frac{1}{a} - iq} - \frac{1}{\frac{1}{a} + iq}\right] = \frac{4\pi C}{\frac{1}{a^2} + q^2} \end{aligned} \tag{22.68}$$

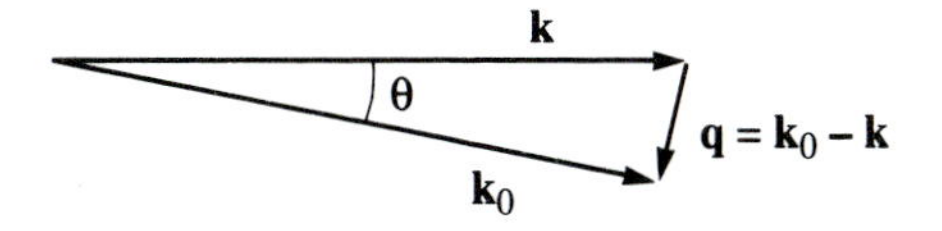

FIGURE 22.7
The geometry of the vectors $\mathbf{k}$, $\mathbf{k}_0$, *and* $\mathbf{q}$.

Substituting into equation (22.63), we find the Born differential cross section for the Yukawa potential:

$$\frac{d\sigma}{d\Omega} = \frac{\mu^2}{4\pi^2\hbar^4}|V(q)|^2 = \left[\frac{2\mu C}{\hbar^2\left(\frac{1}{a^2} + 4k^2\sin^2\frac{\theta}{2}\right)}\right]^2 \tag{22.69}$$

Note an important characteristic of the differential cross section calculated above: It sharply peaks in the forward direction as k^2 becomes large.

Using equation (22.69), we can now calculate the differential cross section even for the Coulomb potential by taking the limit $a \to \infty$ and putting $C = ZZ'e^2$. (Note that this limit could not be approached before because the $V(q)$-integral diverges.) We obtain

$$\left.\frac{d\sigma}{d\Omega}\right|_{\text{Coulomb/Born}} = \left[\frac{\mu ZZ'e^2}{2\hbar^2k^2\sin^2\theta/2}\right]^2 \tag{22.70}$$

This is identical (a result unique for the Coulomb potential) with the classical expression for scattering of a particle by the Coulomb field as first calculated by Rutherford (and a good thing, too, because the discovery of quantum mechanics via the Bohr pathway depended upon it).

As mentioned above, the Born treatment has limited validity. In chapter 23, we will treat scattering in a more complete fashion by treating the three-dimensional Schrödinger equation in the continuum in a proper manner.

PROBLEMS

1. A particle is in the ground state of the infinite one-dimensional potential box:

$$\begin{aligned} V(x) &= \infty, &&\text{for } x < -a \\ &= 0, &&\text{for } -a \le x \le a \\ &= \infty, &&\text{for } x > a \end{aligned}$$

A time-dependent perturbation

$$H_1 = H_{10}\cos(\pi x/a)\delta(t), \qquad H_{10} \text{ a constant}$$

acts on the particle. Calculate the transition probability to the first excited state.

2. Consider a one-dimensional harmonic oscillator in its ground state perturbed by the following time-dependent interaction:

$$H_1 = \delta\omega x^2 \cos ft \qquad \delta\omega \ll \omega$$

where ω is the oscillator frequency. Calculate the transition probability to the second excited state. Can there be a transition to any other excited state?

3. Show that in second order, the transition rate R obeys the same equation as the first-order theory (i.e., eq. [22.17]) except that instead of $\langle\phi_m|H_1|\phi_i\rangle$, we have

$$-\sum_{k\neq i} \frac{\langle\phi_m|H_1|\phi_k\rangle\langle\phi_k|H_1|\phi_i\rangle}{E_k - E_i}$$

The intermediate states $|\phi_k\rangle$ are called *virtual states.*

4. Calculate the $3D \rightarrow 2P$ transition rate for the hydrogen atom. Look up the needed Clebsch-Gordan coefficients from Condon and Shortley or some other book on angular momentum.

5. Calculate the $1P \rightarrow 1S$ transition rate for the three-dimensional oscillator. (*Hint:* Use the oscillator functions of eq. [21.47].)

6. Consider the decay of an unstable particle A into three particles:

$$A \rightarrow B + C + D$$

Use the relativistic energy-momentum relations to calculate the density of states, noting that there are only two independent momenta for the problem.

7. Calculate the Born approximation differential cross section for the following potentials:

(a) $V(r) = V_0 \exp(-r/a)$

(b) $V(r) = V_0 \exp(-r^2/a^2)$

(c) $V(r) = V_0 \qquad r \leq a$

$\qquad\quad = 0 \qquad r > a$

where V_0 is a constant.

8. Using Born approximation, obtain an expression for the differential cross section of the potential

$$V = -V_0(1 + P_r)$$

where P_r is the space-exchange operator and V_0 is a constant potential of range a.

9. Obtain an expression for the total cross section defined as

$$\sigma = \int d\Omega \; (d\sigma/d\Omega)$$

from the Born approximation expression for $d\sigma/d\Omega$ for the Yukawa potential obtained in the text. What happens to σ as $a \to \infty$? Why?

ADDITIONAL PROBLEMS

A1. Consider a gas of molecules with molecular velocities small compared to the speed of light. You can assume each molecule to be a rigid rotator with moment of inertia I.

(a) Write down the Hamiltonian and its eigenvalues for a molecule.

(b) Suppose a molecule makes a spontaneous transition from one of its rotational states to another. The center of mass of the molecule recoils as the photon carries away approximately the momentum $h\nu_R/c$, where $h\nu_R$ is the change in rotational energy of the molecule. Show that there is a recoil correction to the energy of the photon $h\nu_R$ given by

$$(-\hbar/Mc)\mathbf{k}\cdot\mathbf{n}\nu_R$$

where M is the mass of the molecule, $\hbar\mathbf{k}$ is the initial momentum of the center of mass, and $\mathbf{n}$ is the unit vector denoting the direction of the momentum of the emitted photon.

A2. Calculate the Born approximation differential cross section for the triangular potential shown in figure 22.8.

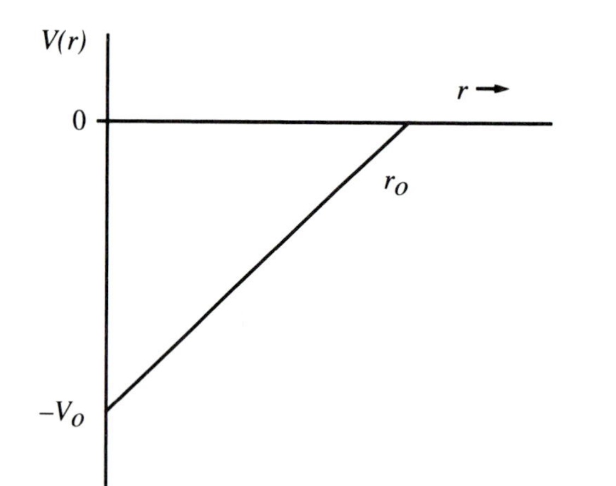

FIGURE 22.8

REFERENCES

S. Gasiorowicz. *Quantum Physics.*
D. Park. *Introduction to the Quantum Theory.*
R. Shankar. *Principles of Quantum Mechanics.*
A. Sudbury. *Quantum Mechanics and the Particles of Nature.*

23

Scattering Theory

Figure 23.1 shows a typical scattering experiment. On the left is a source of particles such as an accelerator. The particles are directed at a target that scatters some of them; the scattered particles are detected by counters placed at a distance from the target, at an angle to the incident beam.

However, as you know, all theoretical analyses of two-body problems such as this are done in quantum mechanics using the center-of-mass reference frame.

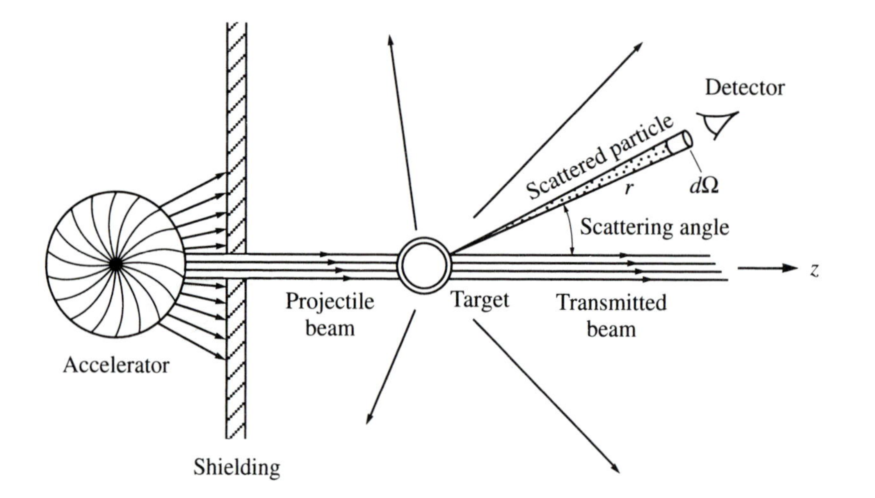

FIGURE 23.1
A scattering experiment (viewed classically). The scattered particles are detected at an angle θ to the incident beam within a solid angle $d\Omega$.

How does the above experiment look from the point of view of the center-of-mass system? In this frame, the two particle beams (projectile and target) come at one another but effectively we have a one-particle problem—a particle of reduced mass μ being scattered by a potential from an initial state to a final state. It is this one-particle scattering problem that we will analyze theoretically in most of this chapter; however, we will spend one section at the beginning establishing the connection between the center-of-mass quantities that we calculate and the laboratory quantities that we measure. (There are now some scattering experiments, using colliding beams, which are carried out in the center-of-mass frame, and the transformation from one set of coordinates to the other is rendered unnecessary.)

In order to develop theory in the simplest and most straightforward manner, we will make another assumption. The incident particle in the scattering experiment is most appropriately described by a wave packet, as is the final state consisting of the incident wave along with a scattered wave (fig. 23.2). However, the analysis of scattering in terms of wave packets is time-dependent; we have to take account of the spreading of the wave packets in time, and this becomes very complicated. So we resort to a time-independent approach.

Why do we do scattering experiments at all? One motivation is to learn about the interaction potential responsible for the scattering. However, our analysis is most straightforward with the added assumption that the potential is one of short range, as is the case in nuclear physics. We will not deal with the important Coulomb potential, because it needs special treatment.

We will begin with a discussion of what is measured and how the measurables are transformed to the center-of-mass system that our theory uses. Then we will develop a systematic solution of the Schrödinger equation for the scattering boundary conditions by the method of *partial waves*. And although most of the chapter is devoted to elastic scattering, some attention is given to inelastic scattering toward the end.

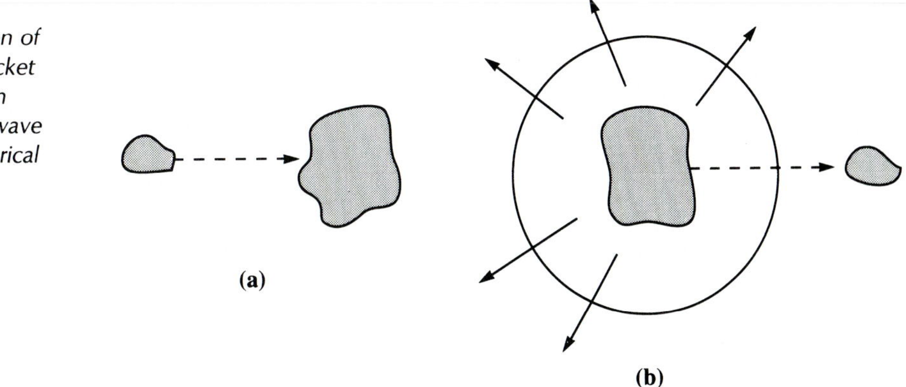

FIGURE 23.2
Wave packet description of scattering. (a) Wave packet of projectile incident on target; (b) transmitted wave packet along with spherical outgoing scattered wavefront.

23.1 SCATTERING AMPLITUDE, DIFFERENTIAL AND TOTAL CROSS SECTIONS, CENTER-OF-MASS, AND LABORATORY FRAMES

One constant challenge of quantum mechanics is to connect theory with experiment since we calculate quantities, wave functions, that are not directly measurable. So here we are in a scattering situation where the boundary conditions on the wave function are clear once we have made our basic assumptions. How do we relate these wave functions to the measurables, such as the differential cross section introduced in chapter 22?

Let's make the boundary conditions explicit. In the time-independent approach that we adopt, the incident particle is looked upon as a free particle and its wave function as a plane wave

$$\phi_k(\mathbf{r}) = \langle \mathbf{r} | \phi_k \rangle = \exp(ikz) \tag{23.1}$$

where $\hbar \mathbf{k}$ is the momentum of the incident particle, whose direction we have chosen as the z-axis, and where we have normalized the wave function so that there is just one particle per unit volume, which is the normalization we will choose for all our wave functions.

Now in the time-independent approach, we are looking at the scattering experiment of figure 23.1 in the way depicted in figure 23.3. If the scattering potential $V(\mathbf{r})$ is spherically symmetric (i.e., $V(\mathbf{r}) = V(|\mathbf{r}|)$), then asymptotically from the scattering source where we assume our detectors are, if we take the scattering source as the origin, we will have spherically outgoing scattered waves $\sim\exp(ikr)/r$ (see below for the justification of this nomenclature). The asymptotic form of the total wave function after the scattering by the potential is then represented by the sum of the incident and the scattered waves:

$$\begin{aligned}\psi_k(\mathbf{r}) \underset{r\to\infty}{\rightarrow} & \ \phi_k(\mathbf{r}) + \psi_k^{\text{sc}}(\mathbf{r}) \\ & = \exp(ikz) + f(\theta,\phi)\exp(ikr)/r\end{aligned} \tag{23.2}$$

where the function $f(\theta,\phi)$ keeps account of the angular distribution of the scattered objects and is called the *scattering amplitude*. From the definition, we can tell that it has the dimension of length.

We will show below that the scattered wave function

$$\psi_k^{\text{sc}}(\mathbf{r}) \underset{r\to\infty}{\rightarrow} f(\theta,\phi)\exp(ikr)/r \tag{23.3}$$

chosen in this fashion does satisfy the free-particle equation. In a sense this should be obvious to you except perhaps for the factor $1/r$, which is imperative so that the probability of particles being scattered into a solid angle $d\Omega$ does not depend on the radial distance.

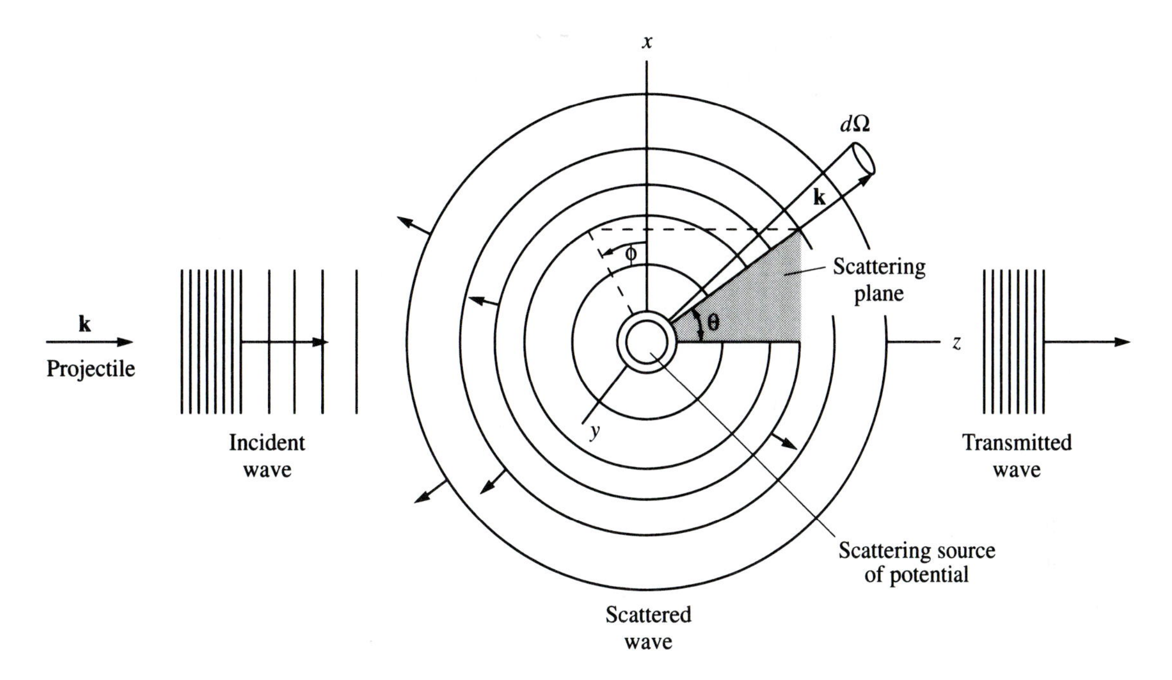

FIGURE 23.3
Idealized description of scattering: incident plane wave and spherical outgoing scattered waves. The scattering plane is formed by the incident and scattered wave vectors, **k** *and* **k**′; θ *is the scattering angle and* ϕ, *the azimuth. For elastic scattering,* $|\mathbf{k}'| = |\mathbf{k}| = k$.

Probability Currents for Elastic Scattering

Now let's calculate the probability currents associated with both the incident and the scattered waves above. The incident current is

$$\mathbf{j}_{\text{inc}} = \frac{\hbar}{2i\mu}\,[\phi_k^* \nabla \phi_k - (\nabla \phi_k^*)\phi_k]$$

$$= \frac{\hbar k}{\mu}\,\mathbf{e}_z \tag{23.4}$$

where, as before, μ denotes the reduced mass of the effective one-body problem in the center-of-mass system.

The current scattered by the target is given as

$$\mathbf{j}_{\text{sc}} = \frac{\hbar}{2i\mu}\,[\psi_k^{\text{sc}*} \nabla \psi_k^{\text{sc}} - (\nabla \psi_k^{\text{sc}*})\psi_k^{\text{sc}}] \tag{23.5}$$

For large distances, we must use the asymptotic form, equation (23.3), for ψ^{sc}. It turns out that only the radial component of the scattered current is relevant. To see this, let's calculate an angular current:

$$\mathbf{e}_\theta \cdot \mathbf{j}_{sc} = \frac{\hbar}{2i\mu}\left[\psi_k^{sc*}\frac{1}{r}\frac{\partial}{\partial\theta}\psi^{sc} - \text{complex conjugate}\right]$$

This is of the order of $\sim 1/r^3$ if we use the asymptotic form of ψ_k^{sc}, equation (23.3). At large distances, such a current, even after multiplied by the area factor $dS = r^2\, d\Omega$ of the sphere of radius r that collects the particles, goes to zero $\sim 1/r$.

In contrast, the radial current is easily seen to give a term that is independent of r when multiplied by $r^2\, d\Omega$:

$$(\mathbf{j}_{sc})_r = \mathbf{j}_{sc}\cdot\mathbf{e}_r = \frac{\hbar}{2i\mu}\left[\psi^{sc*}(r)\frac{\partial}{\partial r}\psi^{sc} - \text{complex conjugate}\right]$$
$$\underset{r\to\infty}{\longrightarrow} \frac{\hbar k}{\mu}\frac{|f(\theta,\phi)|^2}{r^2} \tag{23.6}$$

where we ignore the term $\sim 1/r^3$. Clearly,

$$\mathbf{j}_{sc}\cdot d\mathbf{S} = (\mathbf{j}_{sc})_r r^2\, d\Omega = (\hbar k/\mu)|f(\theta,\phi)|^2\, d\Omega \tag{23.7}$$

is independent of r.

Differential Cross Section

The most convenient and universal measure of how much scattering is taking place in a given situation is the scattering cross section, which we must define in a way that is independent of detecting equipment and all that.

First let's define the differential cross section for elastic scattering. If N is the number of incident particles per unit area per unit time, and ΔN is the number scattered by the target (assumed infinitesimally thin) into the detector (assumed infinitesimally small) at the angle (θ,ϕ) into the solid angle $\Delta\Omega$ in unit time, then the differential cross section is

$$\sigma(\theta,\phi) = \frac{d\sigma}{d\Omega} = \lim_{\Delta\Omega\to 0}\frac{1}{N}\frac{\Delta N}{\Delta\Omega} \tag{23.8}$$

where the notation $\sigma(\theta,\phi)$ is sometimes used for the differential cross section purely for convenience. In terms of the currents, $\Delta N = \mathbf{j}_{sc}\cdot d\mathbf{S}$, $N = \mathbf{j}_{inc}\cdot\mathbf{e}_z$, and after substituting from equations (23.4) and (23.7), we obtain

$$\sigma(\theta,\phi) = |f(\theta,\phi)|^2 \tag{23.9}$$

The differential cross section has the unit of area (per steradian). For low-energy nuclear scattering processes, the barn (1 b = 10^{-24} cm^2) is a convenient unit of area (because it should be as easy to hit a nuclear target with low energy neutrons as, say, "hitting the broad side of a barn").

If the potential causing the scattering is central, as assumed here, the scattering must be axially symmetric, and therefore $\sigma(\theta, \phi)$ is independent of the azimuthal angle ϕ and varies only with θ and with the incident energy of the projectile.

Total Cross Section

If we integrate out all angular dependence, we get the integrated elastic cross section, or more commonly, the *total cross section* denoted as σ_T:

$$\sigma_T = \int d\Omega \, \frac{d\sigma}{d\Omega} = \int_0^{2\pi} d\phi \int_0^{\pi} \sin\theta \, d\theta \, \frac{d\sigma}{d\Omega} \tag{23.10}$$

The total cross section still depends on energy, of course. It is a useful quantity, since it represents how the strength and nature of scattering changes as a function of energy.

Then, there is also inelastic scattering in which some flux is lost (absorbed). When such processes occur, we enlarge the concept of total cross section to include both the integrated elastic and inelastic differential cross sections:

$$\begin{aligned} \sigma_T &= \int d\Omega \left[\left(\frac{d\sigma}{d\Omega} \right)_{\text{el}} + \left(\frac{d\sigma}{d\Omega} \right)_{\text{inel}} \right] \\ &= \sigma_{Tel} + \sigma_{Tinel} \end{aligned} \tag{23.11}$$

Actually, the measurement of total cross section is easier than it seems from the above. In practice, all we have to measure is the depletion of the incident beam intensity upon making its journey through a target material which is of finite thickness (fig. 23.4). Suppose the target thickness is x and the density of

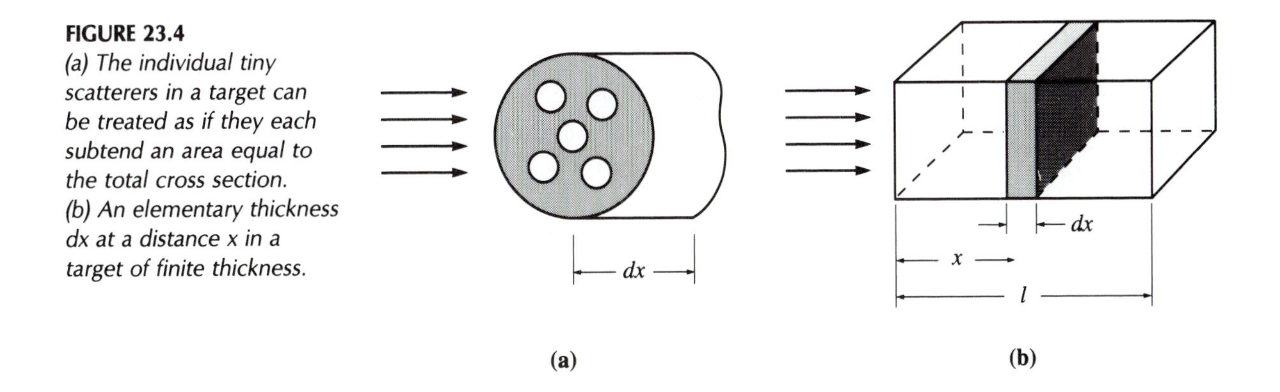

FIGURE 23.4
(a) The individual tiny scatterers in a target can be treated as if they each subtend an area equal to the total cross section.
(b) An elementary thickness dx at a distance x in a target of finite thickness.

the target material is ρ. Additionally, suppose that the incident beam intensity is I_0 and the intensity of the beam upon traveling a distance x of the target is $I(x)$. Then the depletion of the beam intensity corresponding to an infinitesimal thickness dx of the target is given as

$$dI = -\sigma_T \rho \, dx \, I \tag{23.12}$$

since ρ is the number of scattering centers per unit volume and σ_T is the effective area. As the beam traverses a finite thickness, the intensity decreases continuously. The final intensity $I(x)$ upon traversing a thickness x can consequently be obtained by integration of equation (23.12):

$$\int_{I(0)}^{I(x)} \frac{dI}{I} = -\int_0^x dx \, \sigma_T \rho$$

We obtain

$$\ln I(x) - \ln I(0) = -\sigma_T \rho x \tag{23.13}$$

Or

$$I(x) = I(0)\exp(-\sigma_T \rho x)$$

where we assume that ρ is uniform throughout the sample and that σ remains constant as well (that is, we are ignoring the change in the particle energies).

In using equation (23.13) for determining the total cross section, however, some practical considerations have to be given to things such as detector size, which makes it impossible not to count a few scattered particles while counting the transmitted ones, especially at small angles.

Transformation from the Center-of-Mass Frame to Laboratory Frame

As noted already, all the calculations in this chapter refer to the center-of-mass system; however, in this section we'll give the formulas necessary to convert center-of-mass quantities to laboratory quantities. Consider the scattering of two particles of mass m_1 and m_2 with m_2 being the stationary target and m_1 the projectile moving with velocity v_{1L} ($\ll c$) along the x-axis, where the subscript L denotes laboratory. The velocity of the center of mass (COM) of the system is along the x-axis and is equal to

$$v_{cm} = m_1 v_{1L}/(m_1 + m_2)$$

Let the collision take place in the xy-plane. Before collision, in the COM frame, the particles have velocities given by (the subscript c is used to distinguish the COM frame):

$$v_{1cx} = v_{1Lx} - v_{cm} = m_2 v_{1L}/(m_1 + m_2) \qquad v_{1cy} = 0$$

$$v_{2cx} = v_{2Lx} - v_{cm} = -m_1 v_{1L}/(m_1 + m_2) \qquad v_{2cy} = 0 \tag{23.14}$$

It is clear that in the COM system the two particles move with equal and opposite momenta before collision, and therefore, the same situation must persist, in this reference frame, after collision (fig. 23.5a). However, the situation in the laboratory is more complicated (fig. 23.5b).

If the center-of-mass scattering angle is θ_c and primes label velocities after collision, we have

$$v'_{1cx} = \frac{m_2}{m_1 + m_2} v_{1L} \cos\theta_c, \qquad v'_{1cy} = \frac{m_2}{m_1 + m_2} v_{1L} \sin\theta_c$$

$$v'_{2cx} = -\frac{m_1}{m_1 + m_2} v_{1L} \cos\theta_c, \qquad v'_{2cy} = -\frac{m_1}{m_1 + m_2} v_{1L} \sin\theta_c \tag{23.15}$$

Since $\mathbf{v}'_{1c} = \mathbf{v}'_{1L} - \mathbf{v}_{cm}$, equating x- and y-components of this equation and using equations (23.14) and (23.15), we get

$$v_{1c} \cos\theta_c = v'_{1L} \cos\theta_{1L} - v_{cm}$$

and

$$v_{1c} \sin\theta_c = v'_{1L} \sin\theta_{1L}$$

From these two equations, we get

$$\tan\theta_{1L} = \frac{\sin\theta_c}{\cos\theta_c + \gamma} \tag{23.16}$$

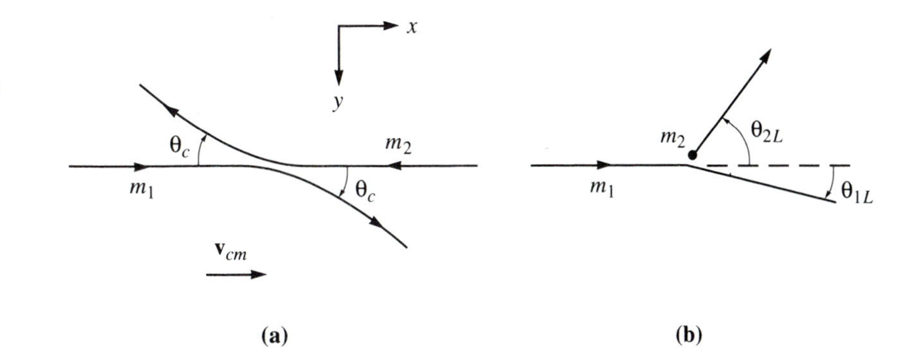

FIGURE 23.5
Scattering in (a) center-of-mass system and (b) in the laboratory system.

where

$$\gamma = v_{cm}/v_{1c} = m_1/m_2 \tag{23.17}$$

Similarly, it can easily be seen that

$$\tan \theta_{2L} = \cot \theta_c/2 \tag{23.18}$$

These equations can be generalized easily for inelastic collisions such as the reaction

$$m_1 + m_2 \rightarrow m_3 + m_4$$

If the amount of mass energy converted to kinetic energy of the final particles is Q (the Q-value of the reaction; $Q > 0$ for exothermic reactions, and $Q < 0$ for endothermic reactions), then γ obtains the value

$$\gamma = \left[\frac{m_1 m_3}{m_2 m_4} \frac{E}{E + Q}\right] \tag{23.19}$$

where E is the initial center-of-mass energy of m_1 and m_2, and equation (23.16) still holds.

The relation between the differential cross sections in the two frames is obtained by noting that the same number of particles that scatters into the solid angle $d\Omega_c$ at (θ_c, ϕ_c) goes into the solid angle $d\Omega_L$ at (θ_L, ϕ_L). Therefore,

$$\sigma_L(\theta_L, \phi_L) \sin \theta_L \, d\theta_L \, d\phi_L = \sigma_c(\theta_c, \phi_c) \sin \theta_c \, d\theta_c \, d\phi_c$$

In other words

$$\sigma_L(\theta_L, \phi_L) = \sigma_c(\theta_c, \phi_c) \, [d \cos \theta_c / d \cos \theta_L] \, d\phi_c / d\phi_L \tag{23.20}$$

Since we are assuming rotational symmetry, $\phi_c = \phi_L$. Using this and equations (23.16) and (23.18), we obtain the relation between the cross sections as (the details will be left as an exercise)

$$\sigma_L(\theta_{1L}, \phi_{1L}) = \frac{(1 + \gamma^2 + 2\gamma \cos \theta_c)^{3/2}}{|1 + \gamma \cos \theta_c|} \sigma_c(\theta_c, \phi_c) \tag{23.21}$$

$$\sigma_L(\theta_{2L}, \phi_{2L}) = 4 \cos \theta_{2L} \sigma_c(\theta_c, \phi_c) \tag{23.22}$$

And as far as the total cross sections are concerned, they must be the same for both coordinate systems and for either outgoing particle—the total number of scattering events has to be independent of the mode of description or reference frame used. You can also verify this using equations (23.21) and (23.22), which will be left as an exercise.

An important special case is $m_1 = m_2$ for which $\gamma = 1$, and the transformation equations between the reference frames simplify considerably. We get

$$\theta_{1L} = \frac{\theta_c}{2}, \qquad \theta_{2L} = \frac{\pi}{2} - \theta_{1L}$$

$$\sigma_L(\theta_{1L}, \phi_{1L}) = 4\cos\theta_{1L}\,\sigma_c(\theta_c, \phi_c), \qquad \sigma_L(\theta_{2L}, \phi_{2L}) = 4\cos\theta_{2L}\,\sigma_c(\theta_c, \phi_c) \tag{23.23}$$

23.2 CONTINUUM QUANTUM MECHANICS: PARTIAL WAVES

When we were calculating bound states back in chapter 12, we found it useful to incorporate rotational symmetry of the Hamiltonian into the solution and decompose the total wave function into eigenstates of orbital angular momentum. The same strategy of breaking up a wave into partial l waves will now be employed for the continuum problem starting with the case of a free particle.

Free-Particle Schrödinger Equation in Spherical Coordinates

The Schrödinger equation for a free particle

$$(\nabla^2 + k^2)\psi(r, \theta, \phi) = 0$$

upon substitution of the partial wave decomposition

$$\psi(r, \theta, \phi) = \sum_{lm} R_l(r) Y_{lm}(\theta\phi)$$

gives the following radial equation for R_l (see chapter 12, if you need a review; note that since we are dealing with continuum states, the quantum number n is dropped from the subscripts of R):

$$\left[\frac{d^2}{dr^2} + \frac{2}{r}\frac{d}{dr} - \frac{l(l+1)}{r^2}\right] R_l(r) + k^2 R_l(r) = 0 \tag{23.24}$$

Introducing a new variable $\rho = kr$, we rewrite equation (23.24) in the spherical Bessel form that you will recognize:

$$\left[\frac{d^2}{d\rho^2} + \frac{2}{\rho}\frac{d}{d\rho}\right] R_l(\rho) + \left[1 - \frac{l(l+1)}{\rho^2}\right] R_l(\rho) = 0 \tag{23.25}$$

The (unnormalized) solution of this equation, regular at the origin (which is the appropriate boundary condition here), is the spherical Bessel function

$$R_l(\rho) = R_l(kr) = j_l(kr) \tag{23.26}$$

You will recall from equation (12.48) that the asymptotic form of $j_l(kr)$ is given as

$$j_l(kr) \underset{r\to\infty}{\to} \frac{1}{kr}\sin\left(kr - \frac{l\pi}{2}\right) \tag{23.27}$$

Therefore, the asymptotic form of R_l is

$$R_l \sim -\frac{1}{2ikr}\left[e^{-i(kr-l\pi/2)} - e^{i(kr-l\pi/2)}\right] \tag{23.28}$$

In this way, we see that not only is the outgoing spherical wave $\exp(ikr)/r$ an asymptotic solution of the Schrödinger equation, something that we have wanted to prove, but also the "incoming" spherical wave $\exp(-ikr)/r$ is a solution. That this last wave is incoming is easily established by calculating its current, which is equal to and the negative of the current of the outgoing wave. It follows that the net flux is zero, which it must be since there is no source of flux anywhere.

What happens when there is a potential? Since we assume that $V(r)\to 0$ asymptotically faster than $1/r^2$, the asymptotic solution is still a free-particle solution (i.e., a linear combination of incoming and outgoing waves), but the linear combination must be appropriately chosen so that the asymptotic solution continuously matches the solution that is regular at the origin, a solution that must be determined with the potential "on." However, such a linear combination must still conserve the flux; the incoming and outgoing flux cannot differ.

In general then, we can parameterize the $V(r) \neq 0$ asymptotic solution of the Schrödinger equation in the form

$$R_l(r) \sim -\frac{1}{2ikr}\left[e^{-i(kr-l\pi/2)} - S_l(k)e^{i(kr-l\pi/2)}\right] \tag{23.29}$$

with the constraint that

$$|S_l(k)|^2 = 1 \tag{23.30}$$

which is the constraint of flux conservation. The S-function satisfying the constraint, equation (23.30), can always be written as

$$S_l(k) = \exp[2i\delta_l(k)] \tag{23.31}$$

where the functions $\delta_l(k)$ are real. They are called *phase shifts* because the asymptotic radial function, equation (23.29), can be rewritten in the form

$$R_l(r) \sim e^{i\delta_l} \frac{\sin[kr - l\pi/2 + \delta_l(k)]}{kr} \tag{23.32}$$

Apart from the phase factor in front, this is the same as the asymptotic form of the free particle wave function, equation (23.28), except that it is shifted in phase by δ_l.

You can understand the phase shift qualitatively as follows: If the potential is attractive, it accelerates the particle as it scatters it, and consequently, the wavelength of the particle is shortened in the scattering region and the phase shift is positive (fig. 23.6). Conversely, if the potential is repulsive, the particle is decelerated, the wavelength is lengthened, the wave tends to be pushed out of the scattering region, and the phase shift is negative (fig. 23.7).

Expansion of a Plane Wave into Partial Waves

Now we will expand the incident plane wave of our scattering scenario in terms of the infinite hoard of partial waves by using the complete set $j_l(kr)Y_{lm}$. Since a plane wave $\exp(ikz) = \exp(ikr\cos\theta)$ does not depend on the azimuth ϕ, it is clear that the wave does not possess any angular momentum component along the z-axis, $m = 0$. Since $Y_{l0}(\theta\phi) \sim P_l(\cos\theta)$, we write our expansion as

$$\exp(ikz) = \sum_{l=0}^{\infty} a_l j_l(kr) P_l(\cos\theta) \tag{23.33}$$

where the coefficients of the expansion a_l are yet to be determined. To this end, we multiply equation (23.33) by $P_{l'}(\cos\theta)$ and integrate over the solid angle. This gives

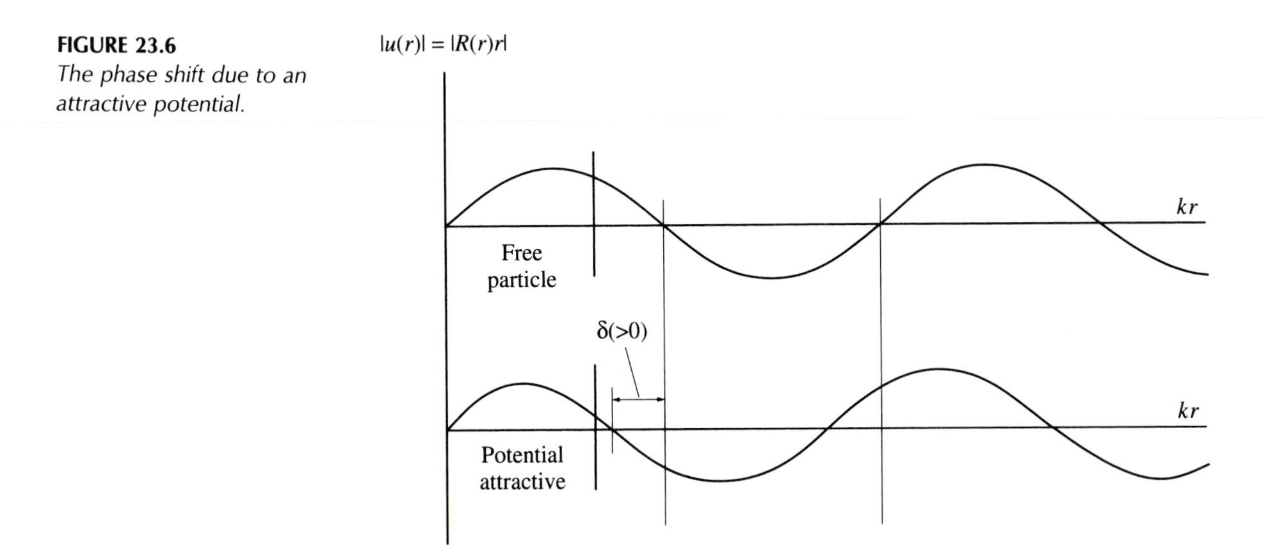

FIGURE 23.6
The phase shift due to an attractive potential.

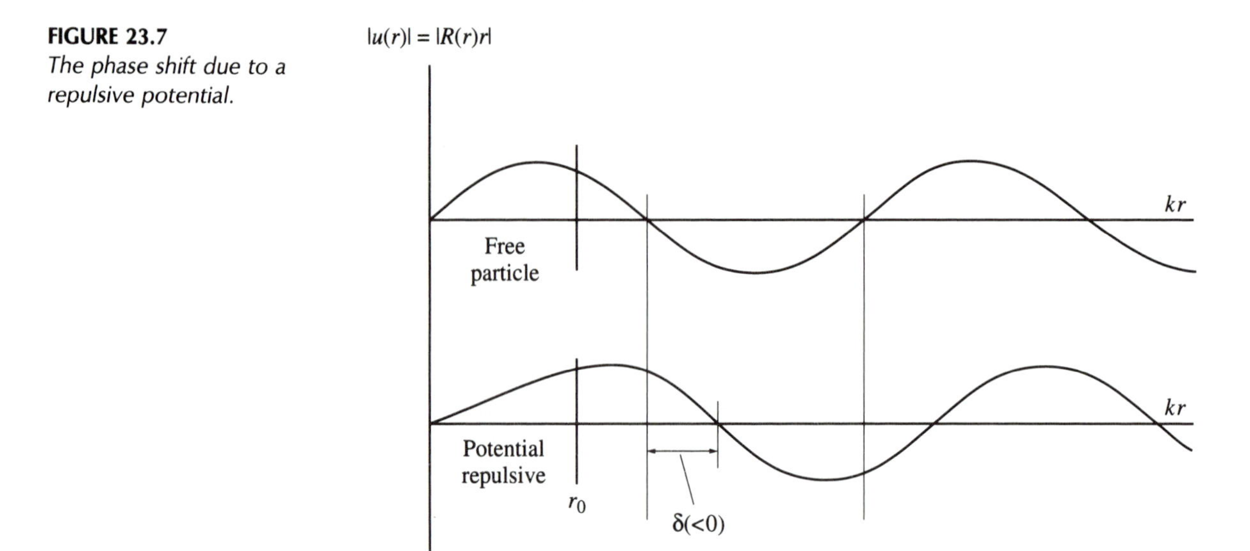

FIGURE 23.7
The phase shift due to a repulsive potential.

$$\int_0^\pi \sin\theta \, d\theta \, e^{ikr\cos\theta} P_{l'}(\cos\theta) = \sum_{l=0}^{\infty} a_l j_l(kr) \int_0^\pi \sin\theta \, d\theta \, P_l(\cos\theta) P_{l'}(\cos\theta)$$

But the last integral above is

$$(2/2l + 1)\delta_{ll'}$$

by virtue of the orthogonality of the Legendre polynomials. It also turns out that the integral on the left-hand side is a well-known integral representation of the spherical Bessel function (look it up!):

$$\int_0^\pi \sin\theta \, d\theta \, e^{ikr\cos\theta} P_l(\cos\theta) = 2i^l j_l(kr)$$

Substituting, we obtain

$$a_l = (2l + 1)i^l$$

Therefore, the partial wave expansion of a plane wave is given as

$$e^{ikz} = \sum_{l=0}^{\infty} i^l (2l + 1) j_l(kr) P_l(\cos\theta) \tag{23.34}$$

For asymptotic distances, we have

$$e^{ikz} \underset{r\to\infty}{\to} \sum_{l=0}^{\infty} i^l [(2l + 1)/kr] \sin(kr - l\pi/2) P_l(\cos\theta) \tag{23.35}$$

Partial Wave Expansion of the Scattering Amplitude

Strictly speaking, it makes sense to say only that a wave can be expanded in terms of partial waves; nonetheless, it is customary to talk about partial wave expansion of the scattering amplitude. How do we obtain such an expansion?

The trick is to invoke the change, the phase shift, that the wave function undergoes in the presence of the potential compared to the free particle. From equation (23.32) it is clear that the total wave function, after scattering, must have the asymptotic form

$$\psi_k(r) \underset{r\to\infty}{\to} \sum_{l=0}^{\infty} a_l i^l (2l+1) e^{i\delta_l(k)} \frac{\sin(kr - l\pi/2 + \delta_l)}{kr} P_l(\cos\theta) \tag{23.36}$$

where a_l is yet to be determined by comparing the expansion above with the scattering boundary condition we previously imposed upon ψ_k, namely,

$$\psi_k \underset{r\to\infty}{\to} \exp(ikz) + f(\theta,\phi)\exp(ikr)/r \tag{23.2}$$

Substituting for $\exp(ikz)$ from equation (23.35) and rearranging, we can write

$$f(\theta,\phi)\frac{e^{ikr}}{r} = \sum_{l=0}^{\infty} a_l \frac{2l+1}{kr} i^l e^{i\delta_l} \sin\left(kr - \frac{l\pi}{2} + \delta_l\right) P_l(\cos\theta) - \sum_{l=0}^{\infty} \frac{2l+1}{kr} i^l \sin\left(kr - \frac{l\pi}{2}\right) P_l(\cos\theta)$$

The key point to note is that our boundary condition for scattering excludes the incoming wave (why? because nobody ever saw a scattering event where waves converge onto a center instead of diverging from it!) from the left-hand side. Consequently, the incoming wave must cancel from the right-hand side, too. By writing

$$\sin\left(kr - \frac{l\pi}{2} + \delta_l\right) = (1/2i)\left\{\exp\left[i\left(kr - \frac{l\pi}{2} + \delta_l\right)\right] - \exp\left[-i\left(kr - \frac{l\pi}{2} + \delta_l\right)\right]\right\}$$

we can easily see that the above condition can be true only if $a_l = 1$. This also gives the partial wave expansion of the scattering amplitude:

$$\begin{aligned} f(\theta,\phi) &= \frac{1}{2ik}\sum_{l=0}^{\infty}(2l+1)(e^{2i\delta_l} - 1)P_l(\cos\theta) \\ &= \frac{1}{k}\sum_{l=0}^{\infty}(2l+1)e^{i\delta_l}\sin\delta_l P_l(\cos\theta) \\ &= \Sigma(2l+1)f_l P_l(\cos\theta) \end{aligned} \tag{23.37}$$

which defines f_l as the scattering amplitude for the lth partial wave.

Once we have the partial wave expansion for the scattering amplitude, the differential cross section follows from equation (23.9):

$$\sigma(\theta,\phi) = \frac{1}{k^2}\left|\sum_{l=0}^{\infty}(2l+1)e^{i\delta_l}\sin\delta_l P_l(\cos\theta)\right|^2 \tag{23.38}$$

What is the advantage of the partial wave expansion? The behavior of $j_l(kr)$ as $r \to 0$ is given as

$$j_l(kr) \sim (kr)^l \qquad (kr \to 0)$$

Therefore, each of the j_l's in the expansion of the plane wave, equation (23.34), is small until $kr \sim l$; this means that when a plane wave encounters a short-range scattering potential $V(r)$, some of its partial waves will not have any significant value inside the region of the potential and will thus be unaffected by it. Physically, it is the centrifugal potential that repels partial waves of large l from the region of the potential for low energy. At low energy (small k) then we need to include only a few partial waves in the expansion of the scattering amplitude, which is a great simplification.

The differential cross section, equation (23.38), determines the angular distribution of the scattered particles in the center-of-mass system. If only one partial wave l dominates (which is not unusual), the angular distribution is proportional to

$$|P_l(\cos\theta)|^2$$

At low energy, only the S-wave contributes, and the angular distribution is isotropic (fig. 23.8). As the energy of the incident particle increases, higher partial waves enter the picture, and the angular distribution reflects interference between the partial waves (fig. 23.9).

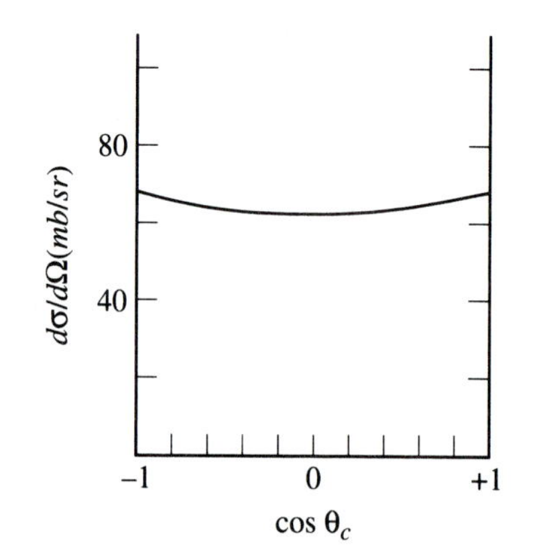

FIGURE 23.8
The COM differential cross section for neutron-proton scattering at a laboratory energy of 14.1 MeV (1 mb = 10^{-27} cm^2).

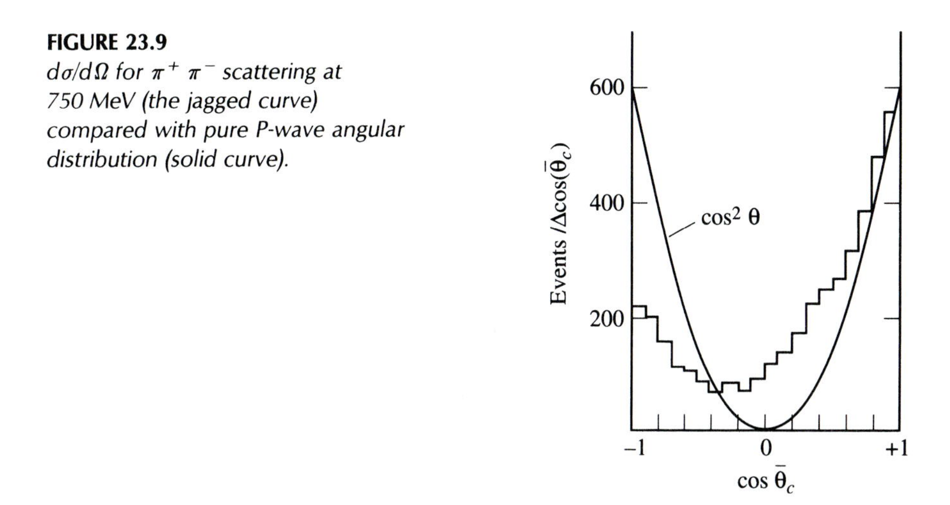

FIGURE 23.9
$d\sigma/d\Omega$ for $\pi^+ \pi^-$ scattering at 750 MeV (the jagged curve) compared with pure P-wave angular distribution (solid curve).

Total Cross Section and the Optical Theorem

The total cross section is obtained by integrating equation (23.38) over all solid angles:

$$\sigma_T = \int d\Omega\, \sigma(\theta,\phi) = \sum_{l=0}^{\infty} \sum_{l'=0}^{\infty} \frac{2l+1}{k} \frac{2l'+1}{k} e^{i(\delta_l - \delta_{l'})} \sin\delta_l \sin\delta_{l'} \times \int d\Omega\, P_l(\cos\theta) P_{l'}(\cos\theta)$$

Using the orthogonality of the Legendre polynomials once again, we obtain the following rather simple expression for σ_T:

$$\sigma_T = \frac{4\pi}{k^2} \sum_{l=0}^{\infty} (2l+1)\sin^2\delta_l \tag{23.39}$$

Now compare this with the value of the forward scattering amplitude, $f(\theta, \phi)$ for $\theta = 0$, which can easily be calculated from equation (23.37):

$$f(0) = \sum_{l=0}^{\infty} \frac{2l+1}{k} e^{i\delta_l} \sin\delta_l P_l(1)$$

But $P_l(1) = 1$, so we get

$$\operatorname{Im} f(0) = \sum_{l=0}^{\infty} \frac{2l+1}{k} \sin^2\delta_l = \frac{k}{4\pi} \sigma_T$$

In other words,

$$\sigma_T = \frac{4\pi}{k} \operatorname{Im} f(0) \tag{23.40}$$

This relationship between the total cross section and the forward scattering amplitude is called the *optical theorem*. Although we have derived it for elastic scattering, it remains true even in the presence of inelastic processes, as we will see later.

If it puzzles you a little how $f(0)$ can occur linearly in a relationship involving the total cross section, consider this. The total cross section represents removal of flux from the incident beam. Such removal is the result of destructive interference between the incident current and the elastically scattered current in the forward direction, and the latter is proportional to the imaginary part of $f(0)$.

23.3 SCATTERING BY A SQUARE-WELL POTENTIAL AT LOW ENERGIES

Many of the important aspects of scattering can be illustrated by taking the scattering potential as a square well and by restricting ourselves to low energy when only the S-wave contributes. The attractive square-well potential, as you recall, is given by

$$\begin{aligned} V(r) &= -V_0 \qquad r < a \\ &= 0 \qquad r > a \end{aligned}$$

Consequently, the S-wave radial equation for inside the well, $r < a$, is

$$\frac{d^2u}{dr^2} + k_{\text{in}}^2 u = 0, \qquad k_{\text{in}}^2 = \frac{2\mu}{\hbar^2}(E + V_0) \tag{23.41}$$

where, as in chapter 12, $u(r) = rR(r)$. The appropriate solution is the one that vanishes at $r = 0$; we have

$$u(r) = A \sin k_{\text{in}} r \tag{23.42}$$

For $r > a$, we have the free-particle radial equation

$$\frac{d^2u}{dr^2} + k^2 u = 0, \qquad k^2 = \frac{2\mu E}{\hbar^2} \tag{23.43}$$

The solution is phase shifted, however,

$$u(r) = B\sin(kr + \delta_0) \tag{23.44}$$

where the subscript 0 on δ denotes l, which is zero. (Note that the above is just another way of writing the most general solution of equation [23.43].)

Now we must ensure the continuity of the wave function and its derivative at the boundary $r = a$; the boundary conditions in their turn determine the phase shift δ_0. Also recall that the easiest way to incorporate the boundary conditions is to equate logarithmic derivatives of the outside and inside solutions at $r = a$:

$$k\cot(ka + \delta_0) = k_{in}\cot k_{in}a$$

or

$$(1/k)\tan(ka + \delta_0) = (1/k_{in})\tan k_{in}a \tag{23.45}$$

This is equivalent to

$$\frac{1}{k}\frac{\tan ka + \tan\delta_0}{1 - \tan ka\tan\delta_0} = \frac{1}{k_{in}}\tan k_{in}a$$

Rearranging,

$$\tan\delta_0(k_{in} + k\tan ka\tan k_{in}a) = k\tan k_{in}a - k_{in}\tan ka$$

or

$$\tan\delta_0 = \frac{(k/k_{in})\tan k_{in}a - \tan ka}{1 + (k/k_{in})\tan ka\tan k_{in}a} \tag{23.46}$$

To solve this for δ_0, define $\tan Ka$ such that

$$\tan Ka = (k/k_{in})\tan k_{ina}$$

We have

$$\tan\delta_0 = \frac{\tan Ka - \tan ka}{1 + \tan Ka\tan ka} = \tan(Ka - ka)$$

In this way we find δ_0,

$$\delta_0 = Ka - ka = \tan^{-1}\{(k/k_{in})\tan k_{in}a\} - ka \tag{23.47}$$

For low-energy scattering, $ka \ll 1$, $\tan ka \approx ka$; suppose additionally that the potential is shallow; then the denominator of equation (23.46) ≈ 1, and we get

$$\tan \delta_0 \approx \delta_0 \approx ka\left[\frac{\tan k_{\text{in}} a}{k_{\text{in}} a} - 1\right] \tag{23.48}$$

In this approximation, δ_0 is in the first quadrant. Now if we gradually increase the depth of the potential, at some point $k_{\text{in}} a$ will go through $\pi/2$. Recall from chapter 12 that this is the condition for the appearance of a bound state (at zero energy), that is, the potential is barely deep enough to bind the particle. So, what can we say now about the phase shift? From equation (23.46), since $\tan k_{\text{in}} a \to \infty$, we have

$$\tan \delta_0 = \cot ka \to \infty \tag{23.49}$$

This means that the phase shift δ_0 is going through $\pi/2$.

Suppose we increase the well depth again just a tad; now we are back to the same situation that led to equation (23.48); that is, $\tan \delta_0 \sim O(ka)$. The difference is that we must realize that δ_0 has to be in the third quadrant, where again the tangent is positive. We conclude that

$$\delta_0 \approx ka\left[\frac{\tan k_{\text{in}} a}{k_{\text{in}} a} - 1\right] \tag{23.48}$$

when there is no bound state, and

$$\delta_0 \approx \pi + ka\left[\frac{\tan k_{\text{in}} a}{k_{\text{in}} a} - 1\right] \tag{23.50}$$

when there is one bound state.

If we make the potential deeper still, a second bound state will appear when $k_{\text{in}} a$ goes through $3\pi/2$; then we will have to entertain a solution replacing equation (23.48) by

$$\delta_0 \approx 2\pi + ka\left[\frac{\tan k_{\text{in}} a}{k_{\text{in}} a} - 1\right]$$

and so forth. We are getting a glimpse of a general theorem of scattering theory known as *Levinson's theorem*:

$$\delta(E = 0) = (\text{number of bound states}) \times \pi \tag{23.51}$$

So what does all this mean? To investigate, let's look at the total cross section. When δ_0 is given by equation (23.48), the total cross section is given by

$$\sigma_T = \frac{4\pi}{k^2}\sin^2\delta_0 \approx \frac{4\pi}{k^2}\delta_0^2 \approx \frac{4\pi}{k^2}(ka)^2\left[\frac{\tan k_{\text{in}}a}{k_{\text{in}}a}\right]^2$$

$$= 4\pi a^2\left[\frac{\tan k_{\text{in}}a}{k_{\text{in}}a}\right]^2 \tag{23.52}$$

This is a constant. Of course, there is a $(ka)^2$ term here hiding behind all our approximations, so the total cross section is only approximately constant.

Now what happens when $\tan\delta_0 \to \infty$? Then

$$\sigma_T = \frac{4\pi}{k^2}\sin^2\delta_0 = \frac{4\pi}{k^2} \tag{23.53}$$

The total cross section approaches a maximum value (fig. 23.10). This is called a resonance (see below). If there is a bound state at zero energy and the phase shift goes through $\pi/2$, the cross section peaks, and we have a resonance.

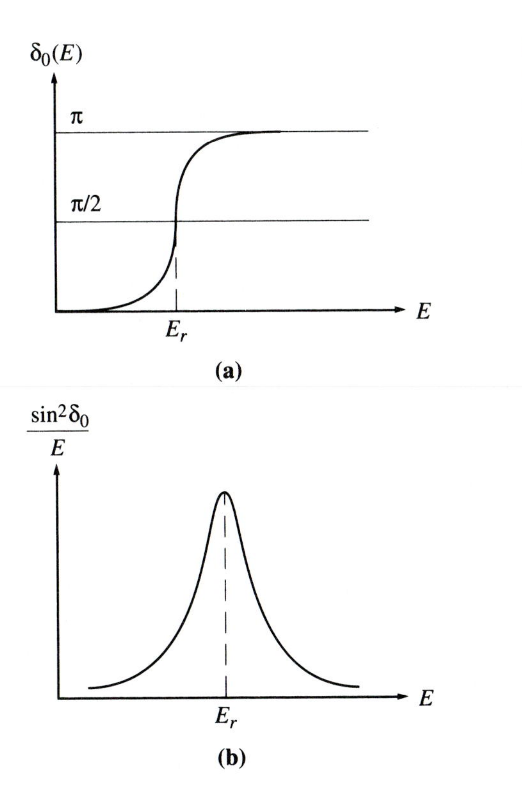

FIGURE 23.10
When the phase shift increases through $\pi/2$ (a), the corresponding partial wave cross section achieves a resonance (b).

Breit-Wigner Formula

Let's examine the behavior of a cross section near a resonance. At the resonant energy E_r, the phase shift increases through $\pi/2$, and we have

$$\cos\delta_0(E_r) = 0 \qquad \sin\delta_0(E_r) = 1$$

At $E \sim E_r$, we can expand $\sin\delta_0$ and $\cos\delta_0$ by means of Taylor series:

$$\begin{aligned}\sin\delta_0(E) &= \sin\delta_0(E_r) + \cos\delta_0(E)\left.\frac{d\delta_0(E)}{dE}\right|_{E=E_r}(E-E_r)\\ &= 1\end{aligned} \tag{23.54}$$

$$\begin{aligned}\cos\delta_0(E) &= \cos\delta_0(E_r) - \sin\delta_0(E)\left.\frac{d\delta_0(E)}{dE}\right|_{E=E_r}(E-E_r)\\ &= -\left.\frac{d\delta_0(E)}{dE}\right|_{E=E_r}(E-E_r) = -\frac{2}{\Gamma}(E-E_r)\end{aligned} \tag{23.55}$$

The last equation defines Γ, which we will interpret a little later.

The scattering amplitude is given by

$$f_0(\theta,E) = (1/k)\exp[i\delta_0(E)]\sin\delta_0(E) \tag{23.56}$$

The $1/k$-variation with energy is slow and uninteresting. The interesting, rapidly varying part comes from the phase shifts; let's call this part $f_0(\delta)$. We have

$$\begin{aligned}f_0(\delta) &= e^{i\delta_0(E)}\sin\delta_0(E) = \frac{\sin\delta_0(E)}{\cos\delta_0(E) - i\sin\delta_0(E)}\\ &\approx \frac{1}{-2(E-E_r)/\Gamma - i} = -\frac{\Gamma/2}{(E-E_r) + i\Gamma/2}\end{aligned} \tag{23.57}$$

where we have used the Taylor-expanded values of $\cos\delta_0$ and $\sin\delta_0$. The total cross section is now given by

$$\sigma_T = \frac{4\pi}{k^2}|f_0(\delta)|^2 = \frac{4\pi}{k^2}\frac{\Gamma^2/4}{(E-E_r)^2 + \Gamma^2/4} \tag{23.58}$$

which is the Breit-Wigner resonance formula (the resonance curve is shown in fig. 23.10b). Clearly, Γ represents the width of the resonant curve.

Low-Energy Neutron-Proton Scattering: Scattering Length and Effective Range

Suppose we consider the scattering from a square-well potential with enough depth that there is a bound state with a small binding energy. This is a model for neutron-proton scattering; the neutron-proton system does indeed have a bound state near zero energy, namely deuteron. The square-well treatment of deuteron was given in chapter 12, where we derived the following equation (changing the notation slightly):

$$k'_{\text{in}} \cot k'_{\text{in}} a = -k_B \tag{12.37}$$

with

$$k'^2_{\text{in}} = 2\mu(V_0 - |E|)/\hbar^2 \qquad k_B^2 = 2\mu|E|/\hbar^2$$

Since the binding energy $|E|$ is small, we can replace k'_{in} by $k_0 = [2\mu V_0/\hbar^2]^{1/2}$. Then we have

$$k_0 \tan(\pi/2 - k_0 a) = -k_B$$

Or, if $k_0 a$ is close to $\pi/2$, we have

$$\pi/2 - k_0 a = -k_B/k_0$$

giving

$$k_0 a = \frac{\pi}{2} + \frac{k_B}{k_0} \tag{23.59}$$

Coming back to low-energy scattering, the boundary conditions in the continuum case give us

$$(1/k)\tan(ka + \delta_0) = (1/k_{\text{in}})\tan k_{\text{in}} a \tag{23.45}$$

where now $k_{\text{in}}^2 = k^2 + k_0^2$. Since $ka \ll 1$, we have $k_{\text{in}} a \approx k_0 a$ as well. Consequently, equation (23.45) can be written as

$$k \tan k_0 a = k_0 \tan(ka + \delta_0)$$

Substituting for $k_0 a$ from equation (23.59) and expanding the tangent on the right-hand side, we get

$$k \tan\left(\frac{\pi}{2} + \frac{k_B}{k_0}\right) = k_0 \frac{\tan ka + \tan \delta_0}{1 - \tan ka \tan \delta_0}$$

Now

$$\tan(\pi/2 + k_B/k_0) = -\cot(k_B/k_0) = -[\tan(k_B/k_0)]^{-1} = -k_0/k_B$$

since $k_B/k_0 \ll 1$. Also $\tan ka \approx ka$. Substituting and rearranging, we get

$$-k(1 - ka \tan \delta_0) = k_B ka + k_B \tan \delta_0$$

Solving for $\tan \delta_0$, we obtain

$$\tan \delta_0 = \frac{k(k_B a + 1)}{k^2 a - k_B}$$

This last equation can be written as

$$k \cot \delta_0 = -\frac{k_B}{k_B a + 1} + \frac{a}{k_B a + 1} k^2 \tag{23.60}$$

And since $k_B a \ll 1$, we finally obtain

$$k \cot \delta_0 = -k_B + ak^2 \tag{23.61}$$

This is a special case of a very general result known as the *effective range expansion* for $k \cot \delta_0$ for low-energy potential scattering:

$$k \cot \delta_0 = -\frac{1}{\alpha} + \frac{1}{2} r_0 k^2 \tag{23.62}$$

The quantity α is called the *scattering length* and r_0, the *effective range.*

As $k \to 0$, we have from the effective range formula

$$\delta_0 \to -\alpha k \tag{23.63}$$

What happens to the scattering amplitude in this limit of zero energy? The near-zero energy ($l = 0$)-scattering amplitude is given as

$$f_0 = \frac{1}{k} e^{i\delta_0} \sin \delta_0 \underset{k \to 0}{\to} -\alpha \tag{23.64}$$

And the near-zero energy total cross section is given as

$$\sigma_T = (4\pi/k^2)\sin^2 \delta_0 \underset{k \to 0}{\to} 4\pi\alpha^2 \tag{23.65}$$

The meaning of effective range is self-suggesting. Let's try to give a geometric interpretation of the scattering length α. We begin by noting that at very low incident energy, the Schrödinger radial equation outside the range of the potential

$$\frac{d^2 u}{dr^2} + k^2 u = 0$$

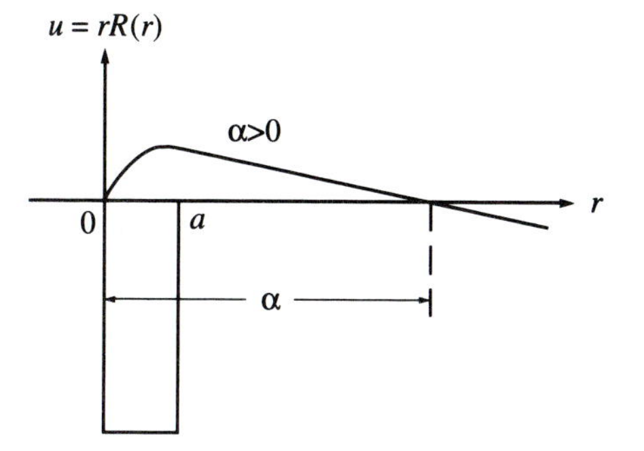

FIGURE 23.11
The scattering length α can be geometrically interpreted as the r-intercept of the asymptotic radial wave function u = rR(r).

becomes

$$\frac{d^2u}{dr^2} \approx 0$$

Clearly, the asymptotic solution must be a linear function of r. To find the function, we look at the asymptotic form of the wave function in the limit of $k \to 0$:

$$\begin{aligned} r\psi &= r\exp(i\mathbf{k}\cdot\mathbf{r}) + f_0\exp(ikr) \\ &\to r - \alpha \end{aligned}$$

as $k \to 0$. It is obvious that the scattering length is to be interpreted as the r-intercept of the asymptotic wave function $r\psi$ (fig. 23.11).

Scattering Lengths for Neutron-Proton Scattering

Coming back to the neutron-proton system, by comparing equations (23.60) and (23.62) we can see that

$$\alpha = \frac{k_B a + 1}{k_B} = a + \frac{1}{k_B}$$

For the deuteron binding energy of $|E| = 2.23$ MeV, noting that the reduced mass $\mu = m/2$, where m = nucleon mass = 940 MeV, we have

$$\begin{aligned} k_B^{-1} &= [\hbar^2/m|E|]^{1/2} = (\hbar/mc)[mc^2/|E|]^{1/2} \\ &\approx 4.3 \times 10^{-13} \text{ cm} \end{aligned}$$

where we have substituted for the proton Compton wavelength $\hbar/mc$. Taking the range of the potential $a = 1.2 \times 10^{-13}$ cm, we get

$$\alpha = 5.5 \times 10^{-13} \text{ cm} = 5.5 \text{ F}$$

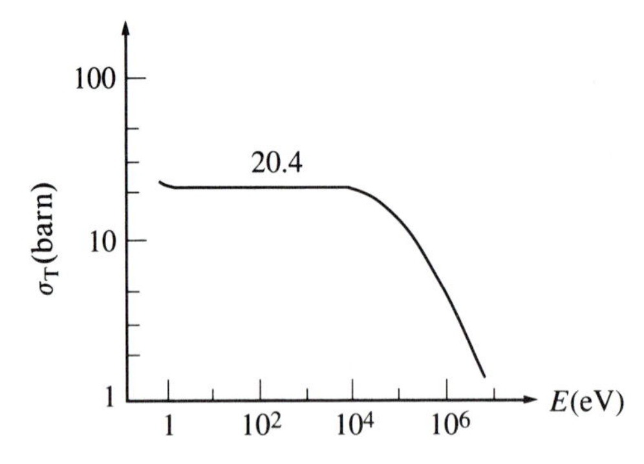

FIGURE 23.12
Experimental neutron-proton total scattering cross section at very low energies of 1 eV to 10 keV is constant; however, it is much greater in magnitude than the theoretical prediction. The answer is spin dependence of scattering.

The near-zero energy cross section is then given from equation (23.65) as

$$\sigma_T = 4\pi\alpha^2 \approx 4 \times 10^{-24}\ \text{cm}^2 = 4\ \text{barns}$$

Scattering cross section measurement with thermal neutrons gives a value of (fig. 23.12)

$$\sigma_T^{\text{exp}} \approx 20\ \text{barns}$$

This is a huge discrepancy!

The explanation is that the potential is different for the two different possible spin-states of the neutron-proton system, the singlet $S = 0$ and the triplet $S = 1$. When the spins are aligned, the neutrons scatter with a different cross section from the protons than when they are not. In a sense, this should be obvious since only the triplet potential is sufficiently deep to have a bound state, which is the deuteron ground state. The resultant cross section is, therefore, equal to the weighted sum of the triplet and singlet cross sections:

$$\sigma_T = 3\sigma_t/4 + \sigma_s/4 \tag{23.66}$$

where the subscripts t and s denote triplet and singlet, respectively. The 4 barns we estimated above are really the triplet cross section (since we used deuteron ground state for the estimate). So to obtain agreement with experiment, we must have

$$\sigma_s = 4\sigma_T - 3\sigma_t = 68\ \text{barns}$$

Such a large cross section suggests a resonance phenomenon. The wisdom from this is that the singlet potential barely misses producing a bound state at zero energy, hence the resonant enhancement of the near-zero energy cross section. The singlet wave function just misses turning over to have a node, and thus its

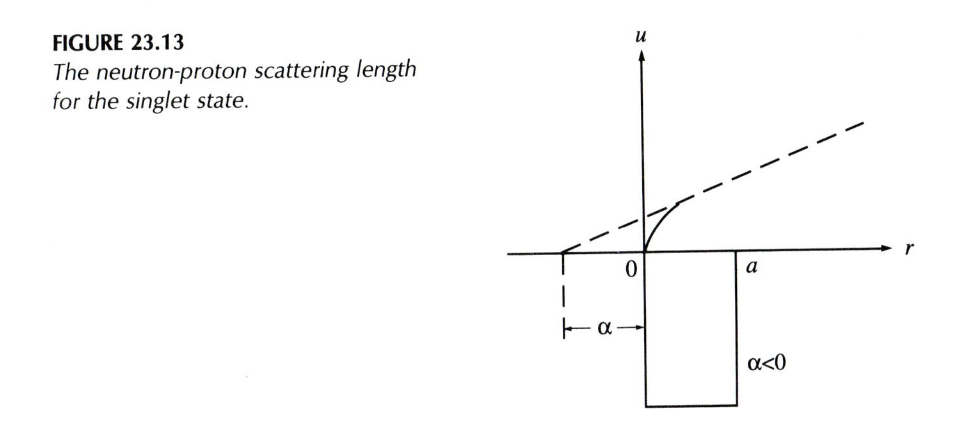

FIGURE 23.13
The neutron-proton scattering length for the singlet state.

r-intercept—the scattering length—is large and negative (fig. 23.13). The negative scattering length of the singlet n-p state has been verified by other means.

23.4 INELASTIC SCATTERING

Inelastic scattering occurs when the incident beam is robbed of flux, when there is a net loss of flux. Under this circumstance, the S-function introduced in equation (23.31) cannot be a pure phase, $\exp(2i\delta_l)$ with δ_l real. So how do we modify our formalism to account for the absorbed flux? The answer is to redefine the S-function:

$$S_l(k) = \eta_l(k)\exp[2i\delta_l(k)] \tag{23.67}$$

with $0 \leq \eta_l(k) \leq 1$. With this modification, the partial wave amplitude $f_l(k)$ is given by

$$\begin{aligned} f_l(k) &= \frac{S_l(k) - 1}{2ik} = \frac{\eta_l(k)\exp[2i\delta_l(k)] - 1}{2ik} \\ &= \frac{\eta_l \sin 2\delta_l}{2k} + i\,\frac{1 - \eta_l \cos 2\delta_l}{2k} \end{aligned} \tag{23.68}$$

The total elastic cross section is now easily obtained

$$\begin{aligned} \sigma_{Tel} &= 4\pi \sum_l (2l+1)|f_l(k)|^2 \\ &= \pi \sum_l (2l+1)\,\frac{1 + \eta_l^2 - 2\eta_l \cos 2\delta_l}{k^2} \end{aligned} \tag{23.69}$$

But there is also the job of calculating the cross section for the inelastic processes; fortunately, the total effect of all inelastic processes can be calculated without specifying precise mechanisms for inelasticity. We will calculate this total inelastic cross section by calculating the difference of the incoming flux and the outgoing flux and dividing the difference by the incident flux.

First note that the asymptotic total wave function can be written in terms of the $S_l(k)$'s as follows:

$$\psi_k(\mathbf{r}) \to \frac{i}{2k} \sum_l (2l+1) i^l \left(\frac{e^{-i(kr - l\pi/2)}}{r} - S_l(k) \frac{e^{i(kr - l\pi/2)}}{r} \right) P_l(\cos\theta) \quad (23.70)$$

where $S_l(k)$ is now given by equation (23.67). The radial flux $\int j_r r^2\, d\Omega$ of the incoming lth partial wave

$$(2l+1)(i/2k)\,[\exp(-ikr)/r]\,P_l(\cos\theta)$$

is given as

$$-(1/4k^2)(\hbar k/\mu) 4\pi (2l+1)$$

where we have used the results of the angular integration:

$$\int P_l^2(\cos\theta)\, d\Omega = \frac{4\pi}{2l+1} \int Y_{l0}^2\, d\Omega = \frac{4\pi}{2l+1}$$

Likewise, the radial flux of the outgoing part of the wave function, equation (23.70),

$$(2l+1)(iS_l/2k)\,[\exp(ikr)/r]\,P_l(\cos\theta)$$

is given as

$$[|S_l|^2/4k^2]\,(\hbar k/\mu) 4\pi (2l+1)$$

But $|S_l|^2 = \eta_l^2$. In this way we see that the net flux of the lth partial wave is given by

$$(2l+1)\,\frac{\pi}{k^2}\,\frac{\hbar k}{\mu}\,(\eta_l^2 - 1)$$

The negative of this quantity is the flux removed from the incident beam. Dividing that with the incident flux $\hbar k/\mu$, we get the contribution of the lth partial wave to the inelastic total cross section. Summing over l, we get σ_{Tinel}:

$$\sigma_{Tinel} = \frac{\pi}{k^2} \sum_{l=0}^{\infty} (2l + 1)(1 - \eta_l^2(k)) \tag{23.71}$$

The grand total of elastic and inelastic total cross sections is

$$\sigma_T = \sigma_{Tel} + \sigma_{Tinel} = (2\pi/k^2) \sum_l (2l + 1)(1 - \eta_l \cos 2\delta_l) \tag{23.72}$$

Now let's verify a comment made earlier that the optical theorem remains valid even when inelastic processes are included. Calculate $\operatorname{Im} f(0)$, the forward scattering amplitude, with the present form of f_l, equation (23.68),

$$\begin{aligned} \operatorname{Im} f(0) &= \sum_l (2l + 1) \operatorname{Im} f_l(k) \\ &= (1/2k) \sum_l (2l + 1)(1 - \eta_l \cos 2\delta_l) \\ &= (k/4\pi)\sigma_T \end{aligned}$$

It just works out this way.

Example: Scattering from a Black Disc

A black disc is a perfectly absorbing disc with all

$$\eta_l(k) = 0$$

Additionally, we will assume that only l-values up to some $l_{\max}$ need to be considered (this is equivalent to assuming that the disc has a sharp edge). We also consider the situation of large k, thus

$$l_{\max} = ka$$

With these caveats, the total inelastic cross section is given by the sum

$$\sigma_{Tinel} = \frac{\pi}{k^2} \sum_{l=0}^{ka} (2l + 1) = \frac{\pi}{k^2} (ka + 1)^2 \approx \pi a^2 \tag{23.73}$$

Perhaps it's a little surprising since we are considering total absorption, but equation (23.69) makes it clear that there is still elastic scattering going on; we have for the total elastic cross section

$$\sigma_{Tel} = \frac{\pi}{k^2} \sum_{l=0}^{ka} (2l + 1) \approx \pi a^2 \tag{23.74}$$

Consequently, the total cross section

$$\sigma_T = \sigma_{Tel} + \sigma_{Tinel} = 2\pi a^2 \tag{23.75}$$

The total cross section is twice the geometrical cross section of πa^2. This result, unexpected from a classical physics point of view, can be explained with the idea of *shadow scattering*. The point is that the shadow cast by the absorbing disc extends only up to a finite distance; far away we can't see a shadow, so the shadow must get filled in. But how so? Only if there is scattering of some of the wave at the edge of the disc. And this scattered flux must have the same magnitude as the flux that is taken out of the incident beam in order to do the job of filling up the shadow. You can see that the scattering and inelastic scattering cross sections would have to be the same, both must be πa^2, and their sum therefore exceeds the classically expected value by a factor of two.

23.5 OUTLOOK

We are going to end our discussion of scattering theory here. The main omission is the effect of spin degrees of freedom, best incorporated using a matrix formulation for the scattering amplitude, which is beyond the scope of this book.

With an exposition of the partial wave analysis of scattering, we have also formally treated most of the quantum phenomena of importance and most that is new. There are, of course, more surprises in store for the reader, for example, the relativistic equation of Dirac. And there are many more phenomena to calculate, some requiring more sophisticated mathematics than the level assumed in this book. But nobody expects any major revision of the basic principles of quantum mechanics introduced here.

Anyway, this is a good place to end the formal presentations of this book. There is, however, a final chapter, an unfinished one at that, which is an informal presentation of some of the ideas that may interest you. What is the meaning of the radical quantum principles? How do we interpret quantum mechanics? Is the ontological question (this is the one that turned on Einstein) of the philosophy of quantum mechanics answerable or worth answering?

PROBLEMS

1. Consider the scattering of particles of mass m from a target of objects of mass $M = 2m$. Suppose the beam has a laboratory kinetic energy of mc^2 and that we are observing scattering at an angle of 30°. (a) What is the relation of the scattering angles in the laboratory and the center-of-mass frames? (b) What is the relation between the differential cross sections?

2. Neutrons are scattered by protons at such energy that only S and P waves need be considered. Assume that the scattering potential is spherically symmetric.
   (a) Show that the differential cross section can be written in the form

$$\sigma(\theta, \phi) = a + b \cos\theta + c \cos^2\theta$$

   (b) What are the values of a, b, c in terms of the phase shifts?
   (c) What is the value of the total cross section in terms of a, b, and c? What is the value of the forward scattering amplitude?
3. Derive equations (23.21) and (23.22). Verify by direct integration of equations (23.21) and (23.22) that the total cross sections are the same for both laboratory and center-of-mass coordinate systems.
4. Consider S-wave neutron-proton scattering in the square-well potential model (depth V_0 and range a). (a) Draw graphs showing the variation of phase shift with energy for (i) $2\mu V_0 a^2/\hbar^2 < \pi^2/4$, (ii) $\pi^2/4 < 2\mu V_0/\hbar^2 < 9\pi^2/4$, and (iii) $2\mu V_0 a^2/\hbar^2 = \pi^2/4$. (b) Show that if there is a bound state at zero energy, the scattering length α becomes ∞ and the effective range $r_0 = a$.
5. Calculate the S-wave phase shift for a repulsive square-well potential. Discuss the limit of large and small depth of potential and the case of very low incident energy.
6. Calculate the S-wave phase shift for the *hard core* potential:

$$V(r) \to \infty \qquad r < c$$

$$V(r) = 0 \qquad r > c$$

   What is the value of the S-wave scattering length?
7. If the neutron-neutron potential is the same as the neutron-proton potential (for which the only bound state exists for l even state), give an argument against a dineutron bound state.
8. What is the S-wave phase shift for a repulsive hard-core (of radius c) plus attractive square-well potential (fig. 23.14)?
9. Consider the S-wave neutron scattering by the delta-function potential

$$2\mu V/\hbar^2 = -(s/c)\delta(r - c)$$

   Show that

$$\tan\delta_0(k) = \frac{s \sin^2 kc/kc}{1 - s \sin kc \cos kc/kc}$$

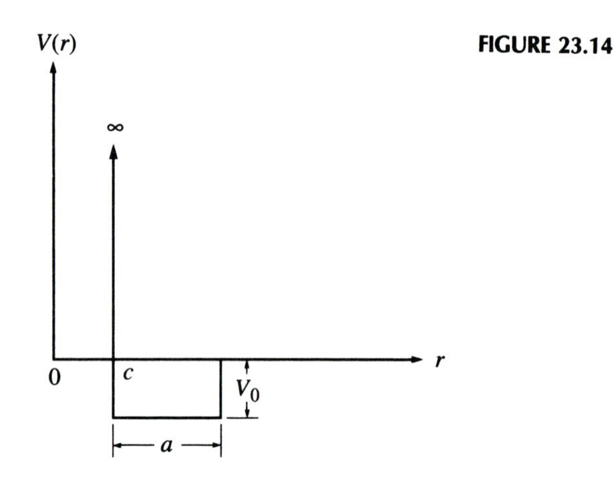

FIGURE 23.14

10. Consider the scattering of 6 MeV neutrons from a heavy black nucleus of diameter 5 F.
(a) What is the approximate number of partial waves that are affected?
(b) Calculate and make a simple plot of the elastic differential cross section $\sigma(\theta, \phi)$ as a function of θ.
(c) Calculate the total elastic, total inelastic, and total cross sections.

ADDITIONAL PROBLEMS

A1. Show that for a central potential $V(r)$ the scattering amplitude in the Born approximation can be written as

$$f_B(\theta) = -\frac{2\mu}{\hbar^2} \int_0^\infty r^2\, dr \frac{\sin qr}{qr} V(r)$$

where q is the momentum transfer as defined in figure 22.7.

A2. Using the formula

$$\frac{\sin qr}{qr} = \sum_l (2l + 1)[j_l(kr)]^2 P_l(\cos\theta)$$

in the Born expression above for the scattering amplitude, show that for small phase shifts δ_l (for which $\exp[2i\delta_l] - 1 \approx 2i\delta_l$), the Born approximation for the phase shifts is given as

$$\delta_l^B \approx \frac{2\mu}{\hbar^2} k \int_0^\infty V(r) j_l^2(kr) r^2\, dr$$

A3. Explain why for two identical spin-$\frac{1}{2}$ fermions, the scattering amplitude should be written in the form

$$f = [f_s(\theta) + f_s(\pi - \theta)] + [f_t(\theta) - f_t(\pi - \theta)]$$

where the subscripts s and t refer to singlet and triplet spin states, respectively.

REFERENCES

A. Das and A. C. Mellissinos. *Quantum Mechanics: A Modern Introduction.*
S. Gasiorowicz. *Quantum Physics.*

24

The Unfinished Chapter: The Meaning and Interpretation of Quantum Mechanics

Quantum mechanics works! But still, there are a couple of thorns in its side. It predicts wave functions that can be looked upon as coherent superpositions of contradictory facets, coherent superpositions of different eigenstates of an observable. Which slit did the electron really pass through in a double-slit experiment? Can it be that the electron's spin is both up and down at the same time? On one hand, a plane wave is an eigenstate of momentum; but reckoning in a different way, it is a coherent superposition of an infinite number of angular momentum eigenstates, the partial waves. Hence arises the problem of ontological meaning (the question of being, such as, Is there a quantum reality?) and interpretation. And when we observe, we always observe one facet, one eigenvalue of the observable we are measuring. The wave function, we assume, collapses upon measurement. But how so? Quantum epistemology (how we know things) seems to need a theory of measurement.

Then there is the question of nonlocal influences. In both the Aharonov-Bohm effect and the violation of Bell inequalities, it is clear that quantum-phase correlations imply a nonlocal influence. What is the implication of nonlocality for the interpretation of quantum mechanics?

We have been struggling for many decades to answer such questions. Some progress *has* occurred. Philosophical issues have sharpened. For example, it now appears to many authors that the Copenhagen interpretation, as explicated by Bohr, not only is ontologically impoverished (this was Einstein's big quibble with Bohr—see chapter 10) but is not even epistemologically satisfactory. Bohr tried to solve all the philosophical problems of quantum mechanics with one idea—complementarity. Implicit in Bohr's philosophy is a classical-quantum dichotomy that bogs down with paradoxes such as Schrödinger's cat; attempts to deal with such paradoxes through such linguistic gymnastics as censoring what you can and cannot ask in quantum mechanics seem repressive. In order to make Bohr's philosophy consistent, one must give a testable rationale for the classical-quantum dichotomy—for example, macrorealism (see chapter 14). It is also clear that no alternative to quantum mechanics can be found based on local hidden variables (see section 17.4). Furthermore, new experiments have shed increasing doubts on the validity of a statistical ensemble interpretation, an easy way out of the philosophical quandaries of quantum mechanics.

On the positive side, several different theories of quantum reality—ontologies, really (one of them is an extension of Heisenberg's version of the Copenhagen interpretation, which we have adapted in this book)—have been suggested that all seem able to resolve paradoxes of quantum measurement and nonlocality when properly applied. But each of these theories is "strange" in one way or another, and consequently, their universal acceptance will depend on further development, most likely experimental.

Therefore, although (or because!) the job remains unfinished, it seems worthwhile to delve into some of the issues that these ontologies, old and new, are raising (these are, after all, the most fascinating issues to most quantum initiates!). Some of the ontologies are mathematically sophisticated (John Bell has called them unromantic); others are philosophically rich, but mathematically underdeveloped (Bell calls them romantic). We will skip most of the mathematics, however.

Nonetheless, we will begin with a thorough mathematical discussion of von Neumann's measurement theory, which has raised many of the issues that the current ontologies are trying to resolve. Next we will discuss the ontologies.

The rest of the chapter will read more like a research paper than a chapter in a book. In that spirit, I will give numbered references to original work whenever appropriate and a list of references at the end of the chapter for your convenience. Also, there will be no problems to do at the end. Enjoy!

24.1 MEASUREMENT THEORY

In chapter 10, we introduced the concept of pure states and mixtures for ensembles of objects. For a pure state, the entire ensemble is described by the same state vector, which can be expressed as a coherent superposition

$$|\psi\rangle = \sum_i a_i |\psi_i\rangle \tag{24.1}$$

where the complete set $|\psi_i\rangle$ can be eigenstates of energy, for example. The square of the coefficients $P_i = |a_i|^2$ gives the probability that a particular eigenstate in "potentia" is realized upon measurement. A mixture, on the other hand, consists of different pure states

$$|\psi^{(1)}\rangle, |\psi^{(2)}\rangle \cdots |\psi^{(n)}\rangle$$

already realized in the ensemble, the probability or weights of their occurrence being $w_1, w_2, \ldots, w_n$, respectively. Of course,

$$\sum_n w_n = 1 \tag{24.2}$$

The probabilities w_i are probabilities in the classical sense.

What is the difference between the two cases? In the case of the pure state of the ensemble—that of the quantum mechanical description—if our knowledge is limited to the statistical distribution of energy among the objects of the ensemble, that is, only to the P_i's, the phases of the a_i remain indeterminate. It is this ignorance that, in some sense, is the bliss of quantum mechanics! In the mixture description, this ignorance does not occur.

The mixture description of an ensemble is the purport of the statistical interpretation of quantum mechanics, and we have already given some reason for looking at this model with disfavor.[1] But there is another reason for bringing up the concept of pure state and mixture. Measurements on a quantum ensemble routinely convert a pure state into a mixture—this is a consequence of the measurement postulate. Let's therefore invest a little more formal effort in the subject.

The Density Matrix

Can we find a concise, formal notation that describes statistical ensembles in complete generality? In particular, we are interested in developing a formalism in quantum mechanics that can deal with situations such as a mixture of two different pure state preparations.

For a mixture, the average value of a physical quantity F is easily defined:

$$\langle F\rangle = \sum_n w_n \langle\psi^{(n)}|F|\psi^{(n)}\rangle$$

or, if you prefer, introducing a particular basic set $|i\rangle$,

$$\langle F\rangle = \sum_n w_n \sum_{ij} \langle\psi^{(n)}|i\rangle\langle i|F|j\rangle\langle j|\psi^{(n)}\rangle$$

Define the density matrix ρ of the ensemble such that its matrix element in the basis $|i\rangle$ is given as

$$\rho_{ji} = \sum_n \langle j|\psi^{(n)}\rangle w_n \langle\psi^{(n)}|i\rangle = \sum_n w_n \langle\psi^{(n)}|j\rangle^* \langle\psi^{(n)}|i\rangle \tag{24.3}$$

Then

$$\langle F \rangle = \sum_{ij} F_{ij}\rho_{ji} = Tr(F\rho) = Tr(\rho F) \tag{24.4}$$

where Tr refers to trace—the sum of the diagonal elements of a matrix. Because the trace of a matrix is independent of any representations, we can use any basis whatsoever for evaluating $Tr(\rho F)$. Consequently, equation (24.4) is a very useful relationship.

We can also write the density matrix in the operator form (called the density or *statistical operator*) independent of any particular basis:

$$\rho = \sum_n |\psi^{(n)}\rangle w_n \langle \psi^{(n)}| \tag{24.5}$$

The density operator has two important properties: (1) it is hermitian and (2) it obeys the normalization condition

$$Tr\rho = 1 \tag{24.6}$$

The proofs will be left to the reader as an exercise.

The case of the pure state is included in this definition of the density matrix as the special case where all the w_n's are zero except one, for which it is unity. We have

$$\rho = |\psi^{(n)}\rangle\langle\psi^{(n)}| \tag{24.7}$$

for the pure case without the sum over n. Clearly,

$$\rho^2 = \rho \tag{24.8}$$

or

$$\rho(\rho - 1) = 0 \tag{24.9}$$

Therefore, for the pure case, we have not only $Tr\rho = 1$, but also

$$Tr(\rho^2) = Tr\rho = 1 \tag{24.10}$$

Furthermore, inserting in equation (24.9) a complete set of base states that diagonalizes the density matrix ρ, we can easily see that the eigenvalues of the hermitian matrix ρ are either zero or 1. The diagonal form of ρ is therefore of the form:

$$\begin{pmatrix} 1 & 0 & 0 & \cdots \\ 0 & 0 & 0 & \cdots \\ 0 & 0 & 0 & \cdots \\ \vdots & \vdots & \vdots & \vdots \end{pmatrix}$$

The density matrix for the pure state is sometimes called an *elementary matrix.* For a mixture, we never can have $\rho^2 = \rho$; the proof of this also will be left to the reader.

Perhaps you have enough formalism for now. Before we apply the density matrix formalism to measurement theory, let's consider another one of its useful applications—the description of a partially polarized beam of electrons.

Example of the Density Matrix Formalism: A Partially Polarized Beam

If we pass an electron beam through an S_z-Stern-Gerlach filter that orients all the outcoming spins in the $|S_z{=}{+}\frac{1}{2}\rangle$ state, we get a completely polarized beam in the pure state $|S_z{=}{+}\frac{1}{2}\rangle$. The density operator for such a state is given as

$$\rho = |S_z{=}{+}\tfrac{1}{2}\rangle\langle S_z{+}\tfrac{1}{2}|$$

for which the matrix representation is

$$\begin{pmatrix} 1 \\ 0 \end{pmatrix} (1 \quad 0) = \begin{pmatrix} 1 & 0 \\ 0 & 0 \end{pmatrix}$$

On the other hand, an unpolarized beam is described by a mixture of 50% $|S_z{=}{+}\frac{1}{2}\rangle$ and 50% $|S_z{=}{-}\frac{1}{2}\rangle$ states. The density operator for this case is given as

$$\rho = (0.5)|S_z{=}{+}\tfrac{1}{2}\rangle\langle S_z{=}{+}\tfrac{1}{2}| + (0.5)|S_z{=}{-}\tfrac{1}{2}\rangle\langle S_z{=}{-}\tfrac{1}{2}|$$

for which the matrix representation reads

$$\begin{pmatrix} \frac{1}{2} & 0 \\ 0 & \frac{1}{2} \end{pmatrix}$$

Indeed, it is easy to verify that the expectation value of S_z is zero in this state. From equation (24.4)

$$\langle S_z \rangle = Tr(\rho S_z) = 0$$

ρ (in this case) is proportional to the unit matrix, S_z is proportional to the Pauli matrix σ_z, and $Tr(\sigma_z) = 0$. Note that the expectation values $\langle S_x \rangle = \langle S_y \rangle = 0$ as well. [Question: Can you see the difference between an unpolarized beam (a mixture) and a 45° polarized beam (a pure state)?]

We have a partially polarized beam when we have an unequally weighted mixture of $|S_z=+\frac{1}{2}\rangle$ and $|S_z=-\frac{1}{2}\rangle$ states; let the weights be 60% and 40%, respectively. Then the matrix of ρ is given as

$$0.6\begin{pmatrix} 1 & 0 \\ 0 & 0 \end{pmatrix} + 0.4\begin{pmatrix} 0 & 0 \\ 0 & 1 \end{pmatrix} = \begin{pmatrix} 0.6 & 0 \\ 0 & 0.4 \end{pmatrix}$$

For this mixture, the expectation value of S_z is given by

$$\langle S_z \rangle = Tr(\rho S_z) = 0.1\hbar$$

The Interaction of a System and a Measuring Apparatus

The Schrödinger equation is as much a deterministic equation as the equations of classical physics. True, it gives coherent superpositions as the answer, but they time-evolve in a continuous, deterministic fashion. Using our new language, we should say that a pure case remains a pure case so long as the Schrödinger system is left alone.

This last statement can be proved in the following way. Suppose a density matrix at $t = 0$ is given by

$$\langle x|\rho|x'\rangle = \sum_n w_n \psi_n(x)\psi_n^*(x')$$

where we have used the position representation for the basis $|i\rangle$. If $\psi_n(x,t)$ time-evolves according to the Schrödinger equation

$$i\hbar\partial\psi_n/\partial t = H\psi_n(x,t)$$

with the initial condition $\psi_n(x,0) = \psi_n(x)$, then the time evolution of the density matrix is given by

$$\langle x|\rho(t)|x'\rangle = \sum_n w_n \psi_n(x,t)\psi_n^*(x',t)$$

From this, we can easily calculate $Tr\rho^2$. It is not necessary to write down the expression; all we need is to realize that the wave function enters $Tr\rho^2$ in the form of a scalar product, and that the scalar product of the wave function $\psi^*\psi$ is independent of time for hermitian Hamiltonians, and thus we must conclude that ρ^2 does not change with time. The upshot is this: Schrödinger time evolution can never change a pure state ($\rho^2 = \rho$) into a mixture ($\rho^2 \neq \rho$).

However, a problem arises, and hence the difference with classical physics, when the system interacts with a measuring apparatus—this brings in a change that has to be specially described as part of the fundamental postulates of the theory. Thus, argued von Neumann, quantum mechanics has two distinct laws of time evolution. The first is the Schrödinger equation, which gives a deterministic continuous prediction of the future states of the system if the initial state is known. The second is the reduction postulate, which operates whenever the system is subjected to measurement; now probability enters, for the reduction postulate is a probabilistic statement describing a discontinuous, acausal change in the system. Before measurement there is the coherent superposition, after measurement only the eigenstate of the measured observable. But this reduction cannot be described by the Schrödinger equation.

Let's be more explicit. Suppose we hold on to the idea, as seems proper, that quantum physics should apply to all physical systems, and this must include the measuring apparatus. Then let's examine the quantum mechanics of the measurement process to see if we can understand the relation of the reduction postulate and the Schrödinger equation. Let $|\psi_1\rangle$, $|\psi_2\rangle$ be two eigenstates of the system corresponding to two different eigenvalues and two different results of measurement. Suppose these measurements leave the apparatus in states $|\phi_i\rangle$ and $|\phi_2\rangle$, respectively. For example, $|\phi_1\rangle$ and $|\phi_2\rangle$ may correspond to two different readings of a pointer.

Assume that the apparatus is initially in the state $|\phi_0\rangle$. Now we make the measurement, the upshot of which is that if the system is in a state $|\psi_1\rangle$ before measurement, it will be in the same state after measurement while the state of the apparatus changes to $|\phi_1\rangle$. On the other hand, if the system is initially in the state $|\psi_2\rangle$, after measurement the apparatus will end up in the state $|\phi_2\rangle$. If we assume that the time evolution of the system is given by the operator $\exp(-iHt/\hbar)$, where H is the Hamiltonian that includes some kind of interaction between the system and the measurement apparatus, and t is the duration of the measurement, we must have

$$\exp(-iHt/\hbar)\,[|\psi_1\rangle|\phi_0\rangle] = \exp(i\theta_1)\,[|\psi_1\rangle|\phi_1\rangle]$$

$$\exp(-iHt/\hbar)\,[|\psi_2\rangle|\phi_0\rangle] = \exp(i\theta_2)\,[|\psi_2\rangle|\phi_2\rangle] \tag{24.11}$$

where θ_1 and θ_2 indicate phases introduced by the measurement. Quantum theory presents no problem here.

But now suppose that before measurement the quantum system is in a state of coherent superposition

$$|\psi_0\rangle = \alpha|\psi_1\rangle + \beta|\psi_2\rangle$$

so that the state of the system plus measuring apparatus is

$$[\alpha|\psi_1\rangle + \beta|\psi_2\rangle]|\phi_0\rangle \tag{24.12}$$

Then, after measurement, the system + apparatus will be in the state

$$\begin{aligned}\exp(-iHt/\hbar)\,[|\psi_0\rangle|\phi_0\rangle] &= \exp(-iHt/\hbar)\,[\alpha|\psi_1\rangle|\phi_0\rangle + \beta|\psi_2\rangle|\phi_0\rangle] \\ &= \exp(i\theta_1)\alpha|\psi_1\rangle|\phi_1\rangle + \exp(i\theta_2)\beta|\psi_2\rangle|\phi_2\rangle \quad (24.13)\end{aligned}$$

where we have used equation (24.11). This is a coherent superposition of two pointer readings, states that are supposed to be macroscopically distinguishable!

If we apply the reduction postulate to the state, equation (24.13), what do we get? If we measure the pointer position with, let's say, a camera, we will find the pointer in state $|\phi_1\rangle$ with a probability of $|\alpha|^2$; likewise, the probability is $|\beta|^2$ that the camera will show the pointer in state $|\phi_2\rangle$. In the former case, the state of the system is $|\psi_1\rangle$, in the latter it is $|\psi_2\rangle$. This is what the reduction postulate, applied directly to the system, also tells us.

Notice that the system is no longer in a state of coherent superposition; instead, its state is correlated with that of the apparatus. In a sense it has been reduced! But this has not solved the measurement problem, because the state of the combined system is still a coherent superposition, and we have to apply the reduction postulate to the combined (system + apparatus) in order to have an unambiguous answer as to the state of the system.

And this can go on ad infinitum. For example, if a second apparatus measures the first one, we get for the state of the total system

$$\sum_i c_i|\psi_i\rangle|\phi_i\rangle|\chi_i\rangle$$

which is again a coherent superposition ($|\chi_i\rangle$ denotes the state of the second measuring apparatus). This is the von Neumann chain, referred to in chapter 10. We cannot avoid discontinuous collapse, the reduction of the wave function, at some level or other. We can take some consolation from the fact that where we apply the reduction does not seem to matter (Bohr seemed to take some assurance from this fact), but that seems vain.

Von Neumann's own solution to the problem of quantum measurement was to suggest[2] (and later Wigner[3] supported the idea) that since the system never does collapse from within the manifold of quantum mechanical systems, the answer must be a jump out of the system. He said that our consciousness terminates the von Neumann chain, because we know that a conscious state is never dichotomic; consciousness must be outside the jurisdiction of Schrödinger evolution. This solution, too, bogs down with paradoxes, because faulty notions of consciousness and flawed ontology are applied (as we will see).

Any serious solution of the measurement problem seems to involve a supportive ontology. To see this more clearly, let's examine one of the "textbook" solutions of the measurement problem.[4]

24.2 THE PRINCIPLE OF MACROSCOPIC DISTINGUISHABILITY

The principle of macroscopic distinguishability asserts that the measurement of a state of a system can be carried out unambiguously because the states of the apparatus are macroscopically distinguishable. To examine the claim, let's be concrete and discuss the case of a Stern-Gerlach measurement of the magnetic quantum number of atomic states (fig. 24.1).

The measurement, as you recall, consists of the observation of the motion of the atoms through an inhomogeneous magnetic field in the z-direction along which we want to measure the component m of the angular momentum. The coordinate $\mathbf{R}$ of the center of mass of the atoms acts out the role of the pointer reading of the measuring apparatus. The internal coordinates of the atoms, $\mathbf{r}$, play the role of the coordinates of the system.

Before the atoms go through the inhomogeneous magnetic field (i.e., for small values of time t) their wave function is separable in the center-of-mass system and can be written as

$$\psi(t,\mathbf{R},\mathbf{r}) = v_0(\mathbf{R}) \sum_m c_m u_m(\mathbf{r}) \tag{24.14}$$

with $\sum_m |c_m|^2 = 1$. Note the analogy with equation (24.12); this is the wave function of system + apparatus before measurement.

When t is large, as the atoms pass through the magnetic field to the right side of the field region, Schrödinger time evolution implies, as before, that the total wave function of the system + apparatus becomes

$$\psi'(t,\mathbf{R},\mathbf{r}) = \sum_m c_m v_m(\mathbf{R},t) u_m(\mathbf{r}) \tag{24.15}$$

As before, we note that the special property of this wave function is that the internal state of the atom is uniquely correlated with the center-of-mass wave

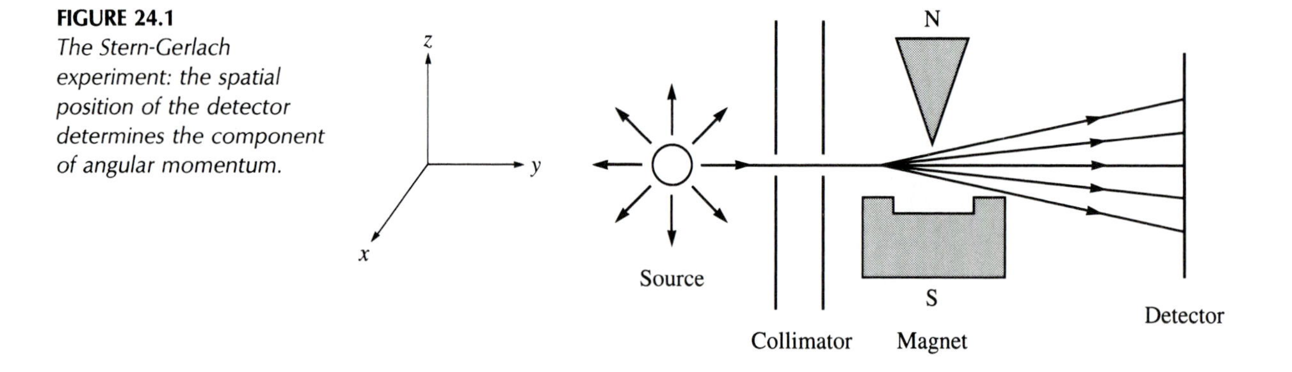

FIGURE 24.1 *The Stern-Gerlach experiment: the spatial position of the detector determines the component of angular momentum.*

function. A measurement of **R** uniquely fixes m, because each v_m is nonzero only within a small volume $V_m(t)$.

Now the density matrix corresponding to the pure state, equation (24.15), is given by

$$\langle \mathbf{r},\mathbf{R}|\rho(t)|\mathbf{r}',\mathbf{R}'\rangle = \sum_{mm'} c_m c_{m'}^* v_m(\mathbf{R},t)u_m(\mathbf{r})v_{m'}^*(\mathbf{R}',t)u_{m'}^*(\mathbf{r}') \tag{24.16}$$

The expectation value of an observable F in the state $\psi(\mathbf{r},\mathbf{R},t)$ is given as

$$\langle F\rangle = Tr(\rho F) = \int d^3r\, d^3r'\, d^3R\, d^3R' \langle \mathbf{r},\mathbf{R}|\rho|\mathbf{r}',\mathbf{R}'\rangle\langle \mathbf{r}',\mathbf{R}'|F|\mathbf{r},\mathbf{R}\rangle \tag{24.17}$$

It now can be argued that for large enough time, the regions $V_m(t)$ and $V_{m'}(t)$ where $v_m(\mathbf{R},t)$ and $v_{m'}(\mathbf{R}',t)$, respectively, are localized become macroscopically separated whenever $m \neq m'$. For all known physical observables, we know that

$$\langle \mathbf{r}',\mathbf{R}'|F|\mathbf{r},\mathbf{R}\rangle = 0$$

if $|\mathbf{R} - \mathbf{R}'|$ is macroscopic. Therefore, the $m \neq m'$ terms of the sum in equation (24.16) do not contribute to the expectation of F at all. Consequently, we have

$$Tr(\rho F) = Tr(\rho' F) \tag{24.18}$$

where ρ' is a reduced density matrix given by

$$\langle \mathbf{r},\mathbf{R}|\rho'(t)|\mathbf{r}',\mathbf{R}'\rangle = \sum_{m} |c_m|^2 v_m(\mathbf{R},t)u_m(\mathbf{r})v_m^*(\mathbf{R}',t)u_m^*(\mathbf{r}') \tag{24.19}$$

ρ' is the density matrix of a mixture (check it out, $Tr\rho^2 \neq 1$); but so long as equation (24.18) holds, it is impossible to distinguish experimentally if the density matrix in the final state is ρ or ρ'. In effect, then, we have arrived at a reduction quite independent of whether or not a conscious observer has looked at the pointer readings.

The difference between ρ and ρ' is the coherence effect—interference—introduced by the phases of c_m and $c_{m'}$. Thus we are able to say that it is *as if* phases are destroyed by the measurement process.

So have we solved the measurement problem? Strictly speaking, no. First of all, the equivalence of ρ and ρ' above is significant only in the framework of a statistical, ensemble interpretation of quantum mechanics, which is hard to uphold for other reasons.[1] The question of measurement of a single system is not solved by proving such equivalence.

For a single system, we must be able to argue that the phases are *really* destroyed by the measurement, not "as if" destroyed. The point is this. Suppose, instead of looking at the pointer in a Stern-Gerlach experiment, we make a record of the various values of **R** in a photographic emulsion. If the memory of the photograph is perfect, then there is a genuine solution of the measurement problem without invoking the reduction postulate. A perfect memory implies that there will never be any interference between the various m-states of the system; such a measurement has to destroy all phase coherence! But is any record in any way permanent? The answer is no, because there is no way to prove irreversibility staying within the context of quantum mechanics (see below).

The textbook resolution of the measurement problem holds water only if we back up the conclusion with a well-defined ontology or with daring ideas. For example, it has been proposed that quantum interference returns, but only after a time much longer than those observed in our laboratories.[5] All measurement of quantum effects, in this view, is temporary. If we don't want to go that far, we must resort to some ontology that goes beyond quantum mechanics.

24.3 REALISTIC ONTOLOGIES

For example, if we say that by nature quantum-measurement apparatuses are such that there is never any interference phenomenon at the macro level, then we have solved the measurement problem. This is the philosophy of macrorealism, which we have discussed before. The purpose of the SQUID program is to demonstrate experimentally that there is no quantum interference at the macro level.[6]

If macrorealism is correct, quantum mechanics does not hold in the realm of macroscopic measurement; there is a real quantum-classical dichotomy. If this solution turns out to be the right one, it may prove embarrassing to those theories that apply quantum mechanics to the whole cosmos—how a part of a quantum cosmos can behave truly classically would be a *real* mystery!

Similarly, one can make an ontological assumption about fundamental irreversibility in nature and then also the above resolution of the measurement problem begins to make sense. Let's discuss this in some detail.

Irreversibility

The problem of quantum measurement is the termination of the von Neumann chain. If irreversibility of the measuring apparatus is absolute, if it never returns to its initial state, even in principle, then the measurement process, as we saw above, terminates with the apparatus.

Consider that one way to distinguish quantum from classical is the quantum's ability of regeneration. A quantum interaction, if we don't measure it, leaves no record at all that it ever took place. But it is the job of the measuring apparatus to record the event, to make a memory. Anytime a measurement apparatus detects a quantum event, if the detection process produces perfect memory, we can never regenerate the event. Then the event can be said to have been measured.

Consider an example. An electron with spin in a direction $\mathbf{e}_n$ is a coherent superposition of S_z-spin up and down states. If the electron now passes the Stern-Gerlach field, it emerges at random either spin up (with probability $\cos^2 \theta/2$) or spin down (with probability $\sin^2 \theta/2$), as can be seen from pointer readings on a detector.

When we say quantum events can be regenerated, what we mean is this. Suppose we place a second Stern-Gerlach field in the path of the electron, but inverted, and before we detect it, the original electron (with spin along $\mathbf{e}_n$) is reconstructed back. Thus the Stern-Gerlach device alone is not enough to "measure" the electron; a detector is needed. But then the question arises, Is the detector enough?

From the argument that leads to the von Neumann chain, the answer is no. The detector becomes a coherent superposition of pointer readings. And the same is true for any subsequent measurement apparatus, giving us the von Neumann chain once again. Since these apparatuses have now taken on quantum behavior, it is in principle always possible to envisage a mechanism, similar in effect to the reversed Stern-Gerlach field, that will regenerate the original quantum event so that one cannot tell if it was ever measured. It may take a very, very long time to return all these macroapparatuses to their original situations; macrobodies have large regeneration times, but time is not of the essence here.

At first sight, it seems that memories *are* irreversible. But there really isn't any conclusive evidence for it because we have not waited long enough, and in principle it will never be long enough! Hence we have to depend on theory. And thus, more to the point is whether the equation that quantum objects obey, the Schrödinger equation, is time reversible or not.

Let's check this out because there is a source of slight confusion if we glibly make the transformation $t \to -t$ in the Schrödinger equation. The Schrödinger equation

$$i\hbar \partial \psi(t,\mathbf{r})/\partial t = H\psi(t,\mathbf{r}) \tag{24.20}$$

unlike Newton's, is first order in time. The problem is resolved when we take the complex conjugate of the equation and then change $t \to t' = -t$. We get

$$i\hbar \frac{\partial \psi^*(-t',\mathbf{r})}{\partial t'} = H\psi^*(-t',\mathbf{r}) \tag{24.21}$$

where we have utilized the fact that the Hamiltonian H given by

$$H = p^2/2m + V(\mathbf{r})$$

is time-reversal invariant (clearly, **r** is not affected by time reversal, and although **p** changes sign, its square does not). The time-reversal invariance of the Schrödinger equation follows once we recognize the proper operator T that transforms the wave function. It is such that

$$T\psi(t,\mathbf{r}) = \psi^*(-t,\mathbf{r}) \tag{24.22}$$

Verify that if ψ is a solution of equation (24.20), so is $T\psi$. An operator defined by equation (24.22) is *antilinear* and *antiunitary* because of the complex conjugation involved.

Because of the time-reversal invariance of our basic dynamics, most conventional wisdom has it that absolute irreversibility is impossible; the irreversibility that we see in nature has to do with the small probabilities that exist for retracing the path of evolution of a complex macrobody back to the initial configuration that has more relative order.

Thus it would seem that irreversibility is not the answer to the measurement problem; it cannot be invoked to save us from the von Neumann chain. Moreover, Wigner and others have argued against using irreversibility as a solution to the measurement problem on the basis of negative result measurements.[7] A simple example is to aim a telescope and a counter at only one of the slits in the double-slit experiment, not both, when we inquire which slit the electron is passing through. If the counter does not click, we must infer that the electron passed through the other slit; but in this case there is no interaction of the electron with the counter, and thus it is hard to see how irreversibility of the counter can be relevant.

It would seem that irreversibility is not the answer to the measurement problem, either; it cannot be invoked to save us from the von Neumann chain—unless we are ready to accept irreversibility, in the form of randomness, as more fundamental than quantum mechanics. Here is how that works, in brief. Suppose matter is fundamental in its randomness, and it is its random *stochastic* behavior that, through occasional fluctuations, generates approximate behavior that we may call quantum.[8] Quantum mechanics is an epiphenomenon, as is all other orderly behavior! Unfortunately, there is no experimental data to support such a theory.

This would be an ingenious solution to the measurement problem if it could be proven, but here again, there is an ontological assumption of an underlying medium that causes the randomness (such as the underlying random motion of molecules that causes the random "Brownian" motion of pollens that are easily seen under the microscope). Such an assumption, like the assumption of the hidden variables, runs afoul of Bell's theorem; it must accommodate nonlocality. But it is hard to accept nonlocal Brownian motion!

Nonlocal Hidden Variables

Although Bell's work has led to a revision of hidden-variable theories, nevertheless the nonlocal hidden-variable theory, as formulated by Bohm and his collaborators, remains a viable possible interpretation of quantum mechanics because of the mathematical and ontological clarity (although the mathematics is quite complex).[9]

In this ontology, physical realism (the idea that there is only one physical reality independent of the observer) is explicitly abandoned in favor of a two-level, realistic to be sure, ontology. The physical "explicate" world is described by the equations of classical physics except that the objects here move not according to classical laws of inertia and acceleration but are guided by quantum mechanical amplitudes or *pilot waves* of an "implicate" world via a quantum potential that enters the classical equation. It is a causal and deterministic theory—therefore, this would be a causal interpretation of quantum mechanics. The equations of the theory are based on some subtle modifications of the mathematics of quantum mechanics.

Experiments such as the double-slit and the delayed-choice experiments find causal explanations such as the particle going through one slit while the wave goes through both and modifies the particle's path. (Can we predict where an electron will land in the double-slit experiment? The answer is still no because, although the equations are causal, they are not manipulable in parts.) The theory avoids a quantum-classical dichotomy as well as the usual traps of a dualistic philosophy—namely, how do the two worlds interact?—since the pilot waves do not carry any energy, just information.

In a measurement situation where there is more than one possible outcome, the wave function (the pilot wave) breaks up into several separated packets. The physical particles enter one of these packets, and macroscopic distinguishability is achieved without the "embarrassment" of the reduction postulate. However, there is a different kind of embarrassment—that of empty wave packets that continue to exist as part of the total reality. This seems to be an unusually nonparsimonious way for nature to operate!

Many-Worlds Ontology

It is this existence of empty wave packets that the many-worlds theory successfully avoids since in this theory all possibilities become actuality—the world multiplies whenever there is an observation. The observers multiply with the multiplying worlds. Obviously, lack of parsimony multiplies, too.

On the face of it, the many-worlds theory seems to have the virtue that no mathematical extension of quantum mechanics is needed. However, this may be an illusion since the observer may be playing a special role (see below).

24.4 DOES CONSCIOUSNESS COLLAPSE THE WAVE FUNCTION?—IDEALIST EXTENSION OF HEISENBERG'S ONTOLOGY

So far the ontologies we have considered are all based on the philosophy of realism—the idea of some sort of independent objective reality. We now come back to von Neumann's own solution to the measurement problem—namely, conscious observation collapses the wave function or reduces the state vector of quantum objects and their measurement apparatuses.

Let's go back to the paradox of Schrödinger's cat, discussed in chapter 10. After the hour, the cat is half-dead and half-alive; as a measuring apparatus for the radioactive atom in the cage, it has taken on the Schrödinger time evolution and hence the dichotomy that originated with the atom. Von Neumann's resolution of the measurement problem is that when *we* look, the von Neumann chain terminates, because consciousness is a jump out of the system and is beyond quantum mechanics.

But questions have been raised. These are the issues of causal action of mind over matter (How can consciousness affect matter?) and a more subtle criticism that is called the paradox of Wigner's friend (see problem 4 in chapter 10). Let's examine these objections and see if the objections fade away when we adopt the correct ontological context for this resolution.

Heisenberg's Ontology and an Idealist Extension of the Copenhagen Interpretation

As an unspoken interpretation of the Copenhagen philosophy, many physicists today look upon quantum mechanics as purely instrumental, as merely a mathematical method to calculate the behavior of submicroscopic objects. But Niels Bohr, one of the founding fathers of the Copenhagen interpretation, also gave us the complementarity principle that hints at something much more. The complementarity principle declares that the wave and particle aspects of a quantum object are complementary; yet for a single quantum object, the wave nature never manifests; whenever we look, we always "see" a quantum object localized, as a particle. This opens the door to an idealist interpretation of complementarity—the wave aspect of a quantum object belongs in potentia, it exists in another domain of reality transcending the space-time reality of manifest appearance.

The transcendent domain of potentia is also reminiscent of Alfred Whitehead's process ontology. According to Whitehead, the events that characterize our world of appearance are not to be conceived as the motions of persistent material objects in given space and time. Instead, he said, "space and time [and material objects] must result from something in process which transcends objects."

The same ontology underlies the discussion of a quantum object's trajectory under the spell of the uncertainty principle. When we look, we find the center of mass of the moon exactly where we would expect it; yet we cannot define a trajectory for it. Where was it between observations? The straightforward answer is, in the domain of potentia.

The experimental demonstration of quantum nonlocality by Aspect et al.[10] adds further substance to the Heisenberg ontology of potentia in physics. According to Stapp, the message of quantum nonlocality is that "the fundamental processes of space-time lie outside of space-time but generate events which can be located in space-time."[11] Henceforth, we will use the term nonlocal in this "outside of space-time" in potentia sense (albeit "outside space-time" can be understood only paradoxically as both nowhere and everywhere in relation to space-time; paradox is always involved when we apply concepts to try to explain the idea of transcendent potentia).

One can still ask, can the ontology of potentia be incorporated within realism by postulating another order beyond the material order? In Bohm's hidden-variables theory above, there are two orders of reality, the implicate order (the nonlocal realm of the quantum pilot wave or hidden variables) and the explicate order (the world of appearance); however, there is causal continuity between the two worlds. This not only raises the old specter of determinism, but stops short of transcendence in idealism, which is causally discontinuous from the ordinary realm of appearance.

But there is another alternative for interpreting quantum mechanics—to embrace a monistic idealist view that the domain of potentia is a domain of consciousness, as is the empirical world of appearance. In this view, quantum objects are posited to exist in potentia (Heisenberg's basic ontology) until translated to the manifest world of appearance by the discontinuous act of conscious measurement (von Neumann's measurement theory). The only clarifying new element of the ontology is that all this happens in consciousness because nothing is outside consciousness. This monistic view of reality is crucial. Basically, this philosophy of monistic idealism is as old as realism; in the West it can be found to have originated in the writings of Plato.

The important aspect of quantum mechanics that makes it necessary to introduce consciousness in a physical theory is that here subjects and objects cannot be neatly separated. When we don't look, a quantum object expands as a wave (in potentia); but when *we* look, the wave collapses. Thus it is natural to expect that a theory of "us," subjects, would be connected with how we formulate our theory for the object.

Let's restate the von Neumann resolution of the paradox of Schrödinger's cat in the context of the above ontology based on monistic idealism: It is our consciousness whose observation of the cat resolves its alive-or-dead dichotomy. Coherent superpositions, the multifaceted quantum waves, exist in the transcendent order until consciousness brings them to the world of appearance through the act of observation. And in the process, consciousness chooses one facet out of two, or many, that are permitted by the mathematics of quantum mechanics, the Schrödinger equation; it is a limited choice, to be sure, subject to the overall probability constraint of quantum mathematics (i.e., consciousness is lawful).[12]

Since the act of conscious observation, in order to terminate the von Neumann chain, must be a jump out of the system, the subject of consciousness itself must act from outside of space-time, nonlocally. That consciousness must be nonlocal is further supported by the Aspect experiment.[10] According to the idealist proposition above, it is conscious observation that collapses the wave function of one of the two correlated photons in the Aspect experiment.[12] And as noted in chapter 17, the wave function of the correlated partner also collapses immediately. But a consciousness that can collapse the wave function of a photon at a distance instantly must itself be nonlocal.

Now the important question: Can the idealist ontology defined above answer the criticisms that are usually leveled against von Neumann's idea of consciousness collapsing the wave function?

Mind over Matter or What?

Explicitly, the concern of mind over matter in the paradox of Schrödinger's cat is this. If the cat is a coherent superposition (both live and dead) before we look, but has a unique state, dead or alive, after we look, then we must be *doing* something by just looking. Thus, Phillip Pearle[13] wonders, "It is hard to believe that a tiny peek at a cat would have a big effect on the physical state of the cat." If you look at this problem from the standpoint of the philosophy of dualism—mind separate from the body—you have to solve the problem by vainly trying to find evidence of psychokinesis, moving matter with the mind.[14] But in monistic idealism, objects are already in consciousness as primordial, possibility forms in potentia. *The collapse is not about doing something to objects via observing, but consists of choosing among the alternative possibilities that the wave function presents and recognizing the result of choice.*

The collapse of the wave function of a system, said Heisenberg, is a change in *our* knowledge of the system. That comment now begins to make sense, doesn't it?

Wigner's Friend

Next let's examine the paradox of Wigner's friend.[15] Suppose that instead of making the observation of the cat himself, Wigner asks a friend to do so. His friend opens the cage, finds the answer, and then reports it to Wigner. At this point, we can say, Wigner has just actualized the reality that includes his friend and the cat. But there is a paradox here. Was the cat alive or dead when Wigner's friend observed it, but before he reported the observation? To say that the state of the cat did not collapse when his friend observed the cat is to maintain that his friend remained in a state of suspended animation until Wigner asked him, that his friend's consciousness could not decide whether the cat was alive or dead without Wigner's prodding. This amounts to the philosophy of solipsism—the idea that only Wigner's consciousness is real; all other consciousness, including his friend's, is Wigner's imagination. But if we say Wigner's friend's consciousness collapses the cat's wave function, aren't we opening up a hornet's nest? What if Wigner and his friend look at the cat simultaneously—whose choice is going to count? The world would be in pandemonium if individual people were to decide the behavior of the objective world, because we know subjective impressions are often contradictory. Hence the argument of solipsism stands and is regarded as a fatal blow against von Neumann's resolution of Schrödinger's cat paradox.

However, there is an alternative to solipsism. Wigner's problem arises from his dualistic thinking, his consciousness separate from his friend's. The paradox disappears if there is only one subject—not separate subjects as we are used to thinking. This may seem strange, but as Schrödinger once said, "Consciousness is a singular for which there is no plural."[16] The antidote to the solipsism of Wigner's friend is a unitive subject-consciousness, and this is the way idealist philosophy portrays consciousness. If Wigner's friend's consciousness is in essence no different from Wigner's, if it is always one consciousness collapsing the wave function, there is no paradox.

When Is a Measurement?

There is, however, a subtle criticism that can be applied to unitive, nonlocal consciousness collapsing the wave function of quantum objects—such a consciousness is omnipresent. An omnipresent consciousness collapsing the wave function does not resolve the measurement paradox because we can ask, At what point is the measurement complete if consciousness is always looking? The answer is crucial. *The measurement is not complete without the inclusion of a self-referential mind-brain-awareness*! London and Bauer in a classic paper already suspected this.[17] Indeed, this agrees with our empirical observation that there is no experience of a material object without the presence of a concomitant mental object such as the thought, "I see this object," or at least, awareness.

There is a causal circularity in the proposition that is underscored above, and it is this: Awareness is needed to complete the measurement, but without the completion of measurement, there is no awareness. This causal circularity is crucial for self-reference, the introspective self that we experience. Space does not permit going into the details here, but a central success for the idealist resolution of the quantum-measurement problem has been to resolve the causal circularity above and to understand how self-reference in the form of a personal I emerges in us as an epiphenomenon of experience.[18]

The Watched Pot Does Boil

There is a question of Schrödinger's cat itself being a conscious being that is now worth considering: Why can't the cat collapse its own wave function? In fact, we can make the issue even more acute by putting a human being inside the cage, radioactive atom and all. Suppose we open the cage after the hour, and if he is still alive, ask him if he experienced a half-alive half-dead state. Of course not! he will say. Are we getting into trouble here with the idealist interpretation? But what if we ask him if he experienced being alive all the time? Of course! he will say, maybe a little puzzled. But he would be misrepresenting his case, because the fact is that we are not conscious of our body all the time; the continuity of the stream of our consciousness is only illusory, similar to how we see a motion picture—our brain-mind cannot discern the individual still pictures when they are paraded before our eyes at a speed of 24 frames per second. Thus, what seems to be continuity to a human observer watching himself is really a mirage, fraught with many discontinuous collapses.

It is sometimes argued that the idea of consciousness collapsing the quantum wave function and bringing it to the world of appearance must be wrong, because then we could prevent change of any object by staring at it. At first glance it seems that such an object would never have the opportunity for time evolution; its wave function would remain collapsed all the time. This paradox is sometimes called "the watched pot never boils" phenomenon. But the above argument about the limits of our perceptual apparatus also nullifies this criticism. The world comes into being discontinuously, event to event. Any continuity, such as watching a pot continuously, is fabricated by the mind-brain, but there is no reality to it. Thus, the watched pot does boil.

Careful experimentation, on the other hand, has demonstrated that measurement interruptions do affect the lifetime of an unstable object, because the transition probability after such interruption can be shown to vary for a short while as the square of the time and not time itself.[19] However, detecting such a "quantum Zeno effect" is tricky; in particular, the question of continuous observation of decay by a Geiger counter is fraught with the same kind of confusion discussed above. So watch it when you read the literature.

Reconciliation with the Many-Worlds Interpretation and the Cosmological Question

We will now argue that the idealist interpretation can easily incorporate the many-worlds idea with a slight reinterpretation of the latter. Actually, if we carefully examine the many-worlds theory (see chapter 1), we find that conscious observation plays a subtle role in this theory. For example, how does Everett define *when* a branching of the universe occurs?[20] E. J. Squares[21] has proposed a more explicit role for consciousness in Everett's original many-worlds theory—it is conscious participation that decides the branching; this is commendable, but we can do even better.

According to the idealist interpretation, coherent superpositions exist in a transcendent domain as formless potentia. Suppose the parallel universes of the many-worlds theory are not material but potential in content; suppose they are potential universes in the transcendent domain of potentia. Then instead of saying that each observation splits off a branch or branches of the material universe that decouple from this one, we can say that each observation makes a causal pathway in the fabric of possibilities in the domain of potentia; once the choice is made, all except one of the possibilities are excluded (decoupled) from the world of appearance.

This way of reinterpreting the many-worlds formalism gets rid of the ugly (and costly) proliferation of material universes—universes in potentia are cheap!

One of the attractive features of the many-worlds theory is that the existence of many worlds makes it a little more palatable in applying quantum mechanics to the entire cosmos. Since quantum mechanics is a probabilistic theory, one justifiably feels uncomfortable thinking about a wave function for the entire universe: Can one ascribe meaning to such a wave function if there is only one of a kind? Many worlds, even ones in potentia, come in handy in this situation.

Now we can see the idealist resolution of the cosmological paradox, How came the cosmos before conscious beings evolved in the world? Very simple. The cosmos never appeared in concrete and never stays laid out in concrete; there isn't any place where past universes, one after the other, can be seen like paintings on a canvas that are uncovered with time. (Although if we think about it, this is how many realists picture the universe!) That would be as bizarre and as uneconomical as the many-worlds theory. But look at the situation with the idea that the universe exists as formless potentia, as myriads of possible branches in the transcendent domain, and becomes manifest in the world of appearance as and when observed by conscious beings. It is these observations that chart out the universe's causal history, choosing among the myriads of parallel alternatives that never found their way to the physical world of appearance.

This way of interpreting our cosmological history may help solve one of the most puzzling aspects of the evolution of life and consciousness—namely, the very low probability for the evolution of life from prebiotic matter and the similar low probability of mutation. But once we recognize that biological mutation (which includes the mutation of prebiotic molecules) is a quantum event, we realize that the universe bifurcates in every such event (in the domain of potentia), becoming many branches, until in one of the branches there is a sentient being that can look with self-referential awareness and complete a quantum measurement. It is at this point that a causal pathway is chosen, a pathway that includes the sentient being. John Wheeler calls this kind of scenario the closure of the meaning circuit.[22] This idealist interpretation of cosmology thus incorporates Wheeler's idea of a participatory universe—a real play of the observer and the observed. It is also easy to see that the present view supports a strong anthropic principle[23]—conscious beings are an essential component of the universe.

Summary of the Idealist Interpretation of Quantum Measurement

Let's summarize the idealist interpretation:

1. What is being measured? A quantum object that exists as a coherent superposition in a transcendent domain of potentia.
2. What collapses the quantum wave function? A nonlocal, unitive consciousness.
3. When is a measurement completed? The measurement is completed only when a conscious being (presumably a mind-brain) looks with self-referential awareness at the macroapparatus involved with the event of measurement.
4. What is the role of the measurement apparatus? The macro measuring apparatus is needed to amplify and record a quantum event. The point is this. A measuring apparatus is classical only in the sense of the correspondence principle; although ultimately quantum in nature, through its complexity it loses the practically instant regenerativity that a simple quantum object has. Because of this long regeneration time, a measuring apparatus can make a record (although only temporary) after its wave function is collapsed and the measurement is completed.

The macrobodies that we conventionally call "classical" are distinguished from the micro ("quantum") in that they acquire continuity, habitual behavior, and temporary memory because of the large regeneration time. All these characteristics make the macroworld a suitable reference point for experience.

But if there is no ultimate irreversibility in the motion of matter, how does the idealist interpretation handle the question of time's arrow? In the idealist interpretation, time is a two-way street in the domain of potentia, showing signs of only approximate irreversibility for motion of more and more complex objects, as explained above. But when consciousness collapses the wave function of the mind-brain, it manifests the subjective one-way time that we observe. There is something irreversible in the process of collapse itself, in quantum measurement, as Szilard suspected long ago.[24] The details of what this irreversibility entails will require further investigation.

It can be asked, How can mind have any significant effect on atoms (besides, no such effect has been observed)? But the idealist interpretation as proposed here implies no such effect as far as atoms are concerned, no effect of mind over atoms beyond what the Copenhagen interpretation already assumes. The atomic probability calculus is quite fixed by material interactions via the Schrödinger equation, as experiments amply bear out. The situation with the brain may be altogether different, however. Here there is fertile ground for new quantum theory[25] (modified quantum mechanics) because the existing theories based on classical physics are severely inadequate.

Furthermore, if the consciousness-resolution of the quantum theory of measurement turns out to be correct, then the door opens for the application of quantum principles to the problems of the brain-mind.

24.5 A FINAL OUTLOOK

I hope you get the gist of the chapter. With quantum mechanics, we have not only a brilliant machinery for calculations in physics with enormous technological applications but also the possibility of truly understanding what reality is all about. If we can discern experimentally between the realistic and idealist ontologies (most of the ontologies have verifiable consequences),[26] then not only will a millennia-old philosophical debate end, but also perhaps the road will be paved toward an understanding of the human being itself.

REFERENCES

1. See the review by P. Gibbins, *Particles and Paradoxes* (Cambridge, U.K.: Cambridge University Press, 1987).
2. J. von Neumann, *Mathematical Foundations of Quantum Mechanics* (Princeton, N.J.: Princeton University Press, 1955), Ch. VI.
3. E. P. Wigner, in *The Scientist Speculates*, I. J. Good, ed. (Kindswood, Surrey, U.K.: The Windmill Press, 1962).
4. K. Gottfried. *Quantum Mechanics.*
5. P. L. Csonka, *Phys. Rev.* **D4**, 1607 (1971).
6. A. J. Leggett, in *The Lesson of Quantum Theory*, J. De Boer, E. Dal, and O. Ulfbeck, eds. (Amsterdam: North-Holland, 1986), p. 35.

7. J. M. Jauch, E. P. Wigner, and M. M. Yanase, *Nuovo Cimento* **48B**, 144 (1967); M. Renninger, *Z. Phys.* **157**, 417 (1960).
8. I. Progogine, *From Being to Becoming* (San Francisco: Freeman, 1980).
9. D. Bohm, B. J. Hiley, and P. N. Kaloyerou, *Physics Reports* **144**, 323 (1987).
10. A. Aspect, J. Dalibard, and G. Roger, *Phys. Rev. Lett.* **49**, 1804 (1982).
11. H. P. Stapp, *Nuovo Cimento* **40B**, 191 (1977).
12. A. Goswami, *Physics Essays* **2**, 385 (1989).
13. P. Pearle, in *The Wave Particle Dualism*, S. Diner et al., eds. (Dordrecht: Riedel, 1984).
14. R. D. Mattuck and E. H. Walker, in *The Iceland Papers: Experimental and Theoretical Explorations into the Relation of Consciousness and Physics*, A. Puharich, ed. (Amherst, Wis.: Essentia Research Associates, 1979), p. 111.
15. E. Wigner, *Symmetries and Reflections* (Bloomington: Indiana University Press, 1967).
16. E. Schrödinger, *What Is Life? and Mind and Matter* (London: Cambridge University Press, 1969).
17. F. London and E. Bauer, in *Quantum Theory and Measurement*, J. A. Wheeler and W. Zurek, eds. (Princeton, N.J.: Princeton University Press, 1983), p. 217.
18. A. Goswami, *Journal of Mind and Behavior* **11**, 75 (1990); A. Goswami, *The Self-Aware Universe* (New York: Tarcher/Putnam, 1993).
19. W. M. Itano, D. J. Heinzen, J. J. Bollinger, and D. J. Wineland, *Phys. Rev.* **A41**, 2295 (1990).
20. H. Everett III, *Rev. Mod. Phys.* **29**, 454 (1957), and in *The Many-Worlds Interpretation of Quantum Mechanics*, B. Dewitt and N. Graham, eds. (Princeton, N.J.: Princeton University Press, 1973).
21. E. J. Squares, *European J. Phys.* **8**, 171 (1987).
22. J. A. Wheeler, in *New Techniques and Ideas in Quantum Measurement Theory*, D. M. Greenberger, ed. (New York: N.Y. Academy of Science, 1986), p. 304.
23. For a discussion of the strong anthropic principle, see J. D. Burrow and F. J. Tipler, *The Anthropic Cosmological Principle* (New York: Oxford University Press, 1986).
24. L. Szilard, *Z. Phys.* **53**, 840 (1929).
25. Many authors have suggested the idea that mind-brain should contain a macro quantum system in addition to the classical neuronal system. See, for example, H. P. Stapp, *Foundations of Physics* **12**, 963 (1982).
26. See references 6, 9, 12, and 18. Some new experimental ideas have been suggested even for the many-worlds interpretation by D. Deutsch, *Int. J. of Theor. Phys.* **24**, 1 (1985).

Appendix: The Delta Function

The Dirac delta function can be defined by

$$\delta(x - x_0) = \frac{1}{2\pi} \int_{-\infty}^{\infty} dk \, e^{ik(x-x_0)} \tag{A.1}$$

But this is a most peculiar function; for clearly it must be zero when $x \neq x_0$, and it must $\to \infty$ when $x = x_0$, since the range of integration in equation (A.1) is infinitesimal. Indeed, it is best to think of the delta function not as a function but as a *distribution* or, if you prefer, as a *generalized* function, which (and the properties of which) does not have any meaning by itself, but only when it appears in the form

$$\int dx \, \delta(x - x_0) f(x) \equiv f(x_0) \tag{A.2}$$

where the function $f(x)$ is assumed to be sufficiently smooth in the range defined by the arguments of the delta function. That is, we should accept equation (A.2) as the definition of the delta function, and regard equation (A.1) as one of the many *representations* of the function. With this firmly understood, in this appendix we will present some of the useful properties and representations of the delta function.

Properties of the Delta Function

We will first prove some of the important properties of the delta function.

1. The delta function is normalized

$$\int dx\, \delta(x - x_0) = 1 \tag{A.3}$$

2.

$$\int dx\, \delta(x) f(x) = f(0) \tag{A.4}$$

These properties follow from the definition, equation (A.2).

3. The delta function is real

$$\delta^*(x) = \delta(x) \tag{A.5}$$

4. The delta function is an even function

$$\delta(-x) = \delta(x)$$

Properties 3 and 4 are easily proven from equation (A.4); proceed by applying the equation for the case of a real function $f(x) = f^*(x)$, and for an even function $f(-x) = f(x)$, respectively.

5. Let's now prove the following important property:

$$\delta(ax) = (1/a)\delta(x) \qquad a > 0 \tag{A.6}$$

Using the normalization condition (A.3), we have

$$\int dx\, \delta(ax) = \int (1/a) d(ax) \delta(ax) = \int (1/a)\, du\, \delta(u) = 1/a$$

where we have substituted $u = ax$. Now compare with

$$\int dx\, \delta(x) = 1$$

Equation (A.6) follows.

6. The following property also follows immediately:

$$\delta(ax) = (1/|a|)\delta(x) \tag{A.7}$$

7.
$$f(x)\delta(x-a) = f(a)\delta(x-a) \tag{A.8}$$

This follows directly from the definition, equation (A.2).

8. It also follows that

$$\delta(x^2 - a^2) = (1/2|a|)\,[\delta(x-a) + \delta(x+a)] \tag{A.9}$$

First note that the argument of the delta function is 0 for both $x = a$ and $x = -a$. Therefore, we must be able to write

$$\begin{aligned}\delta(x^2 - a^2) &= \delta[(x-a)(x+a)] \\ &= (1/|x+a|)\delta(x-a) + (1/|x-a|)\delta(x+a) \\ &= (1/2|a|)\,[\delta(x-a) + \delta(x+a)]\end{aligned}$$

where, in obtaining the last two steps, we have used equations (A.7) and (A.8), respectively.

Now let's enumerate without proof one more property of the delta function (you may find it interesting to find the proof yourself):

9.
$$\int \delta(x-b)\delta(a-x)\,dx = \delta(a-b) \tag{A.10}$$

Representations of the Delta Function

As already noted, equation (A.1) is a representation of the delta function; but this is not the only representation. Many other representations have been found. We will consider a few of the simple ones.

First, one of the simplest representations is

$$\delta(x) = \lim_{g\to\infty} \frac{\sin(gx)}{\pi x} \tag{A.11}$$

It can be proven by writing equation (A.1) in the form

$$\delta(x) = \frac{1}{2\pi} \lim_{g\to\infty} \int_{-g}^{g} dk\, e^{ikx}$$

and then carrying out the integration. We get

$$\begin{aligned}\delta(x) &= \frac{1}{2\pi} \lim_{g\to\infty} \frac{e^{igx} - e^{-igx}}{ix} \\ &= \lim_{g\to\infty} \frac{\sin(gx)}{\pi x}\end{aligned}$$

Two other useful representations of the delta function invoke the idea that in principle, any peaked function, if it is suitably normalized so that the area under it is one, will tend to a delta function in the limit that the width of the function $\rightarrow 0$. These representations are

$$\delta(x) = \lim_{a \to 0} \frac{1}{\pi} \frac{a}{x^2 + a^2} \tag{A.12}$$

$$\delta(x) = \lim_{a \to \infty} \frac{a}{\sqrt{\pi}} e^{-a^2 x^2} \tag{A.13}$$

The proofs will be left to the reader.

Does the derivative of a delta function make sense? Since the function invariably appears under an integral sign, always in the company of a smooth function, the derivative of a delta function obtains meaning. For example, we can easily prove the following relationship:

$$\int dx\, \delta'(x) f(x) = -f'(0) \tag{A.14}$$

where the prime denotes differentiation with respect to the argument.

Finally, for the integral of a delta function we have

$$\begin{aligned} \int_{-\infty}^{x} dy\, \delta(y - a) &= 0, \qquad \text{for } x < a \\ &= 1, \qquad \text{for } x > a \\ &\equiv \theta(x - a) \end{aligned} \tag{A.15}$$

where $\theta(x - a)$ is called a *step function*. Differentiating, we get one more relationship for the delta function—it is the derivative of a step function:

$$\delta(x - a) = d\theta(x - a)/dx \tag{A.16}$$

Bibliography

E. E. Anderson. *Modern Physics and Quantum Mechanics*, Philadelphia: Saunders, 1971.

P. W. Atkins. *Molecular Quantum Mechanics*, New York: Oxford University Press, 1983.

G. Baym. *Lectures on Quantum Mechanics*, New York: W. A. Benjamin, 1969.

D. Bohm. *Quantum Theory*, Englewood Cliffs, N.J.: Prentice Hall, 1951.

———. *Wholeness and Implicate Order*, London: Routledge and Kegan-Paul, 1980.

A. Z. Capri. *Nonrelativistic Quantum Mechanics*, Menlo Park, Calif.: Benjamin/Cummings, 1985.

M. Chester. *Primer of Quantum Mechanics*, New York: Wiley, 1987.

C. Cohen-Tannoudji, B. Diu, and F. Laloe. *Quantum Mechanics*, New York: Wiley, 1977.

E. U. Condon and G. H. Shortley. *The Theory of Atomic Spectra*, Cambridge: Cambridge University Press, 1959.

A. Das and A. C. Melissinos. *Quantum Mechanics: A Modern Introduction*, New York: Gordon & Breach, 1986.

A. S. Davydov. *Quantum Mechanics*, Reading, Mass.: Addison-Wesley, 1965.

B. d'Espagnat. *In Search of Reality*. New York: Springer-Verlag, 1983.

R. H. Dicke and J. P. Witke. *Introduction to Quantum Mechanics*, Reading, Mass.: Addison-Wesley, 1960.

P. A. M. Dirac. *The Principles of Quantum Mechanics*, Oxford: Clarendon Press, 1958.

Fayyazuddin and Riazuddin. *Quantum Mechanics*, Singapore: World Science Publishers, 1990.

E. Fermi. *Nuclear Physics*, Chicago: University of Chicago Press, 1950.

R. P. Feynman, R. B. Leighton, and M. Sands. *The Feynman Lectures in Physics*, vol 3. Reading, Mass.: Addison-Wesley, 1965.

P. Fong. *Elementary Quantum Mechanics*, Reading, Mass.: Addison-Wesley, 1962.

S. Gasiorowicz. *Quantum Physics*, New York: Wiley, 1974.

K. Gottfried. *Quantum Mechanics*, vol. 1, New York: W. A. Benjamin, 1966.

W. Heisenberg. *The Physical Principles of the Quantum Theory*, New York: Dover, 1930.

———. *Physics and Philosophy*, New York: Harper Torchbooks, 1958.

N. Herbert. *Quantum Reality*, New York: Doubleday, 1985.

J. D. Jackson. *Classical Electrodynamics*, New York: Wiley, 1975.

M. Jammer. *The Philosophy of Quantum Mechanics*, New York: Wiley, 1974.

———. *The Conceptual Development of Quantum Mechanics*, Woodbury, N.Y.: American Institute of Physics, 1989.

T. F. Jordan. *Quantum Mechanics in the Simplest Matrix Form*, New York: Wiley, 1986.

A. M. Leggett. In *The Lessons of Quantum Theory*, J. De Boer, E. Dal, and O. Ulfbeck (eds.), Amsterdam: North Holland, 1986.

R. N. Liboff. *Introductory Quantum Mechanics*, Reading, Mass.: Addison-Wesley, 1987.

E. Merzbacher. *Quantum Mechanics*, New York: John Wiley, 1970.

J. R. Oppenheimer. *Science and Common Understanding*, New York: Simon & Schuster, 1954.

M. K. Pal. *Theory of Nuclear Structure*, New Delhi: Affiliated East-West Press, 1982.

D. Park. *Introduction to the Quantum Theory*, New York: McGraw-Hill, 1964.

L. Pauling and E. Wilson. *Introduction to Quantum Mechanics*, New York: McGraw-Hill, 1935.

J. L. Powell and B. Crasemann. *Quantum Mechanics*, Reading, Mass.: Addison-Wesley, 1961.

A. I. M. Ray. *Quantum Mechanics*, New York: Wiley, 1981.

J. J. Sakurai. *Modern Quantum Mechanics*, Menlo Park, Calif.: Benjamin/Cummings, 1985.

J. A. Schumacher. In *Fundamental Questions in Quantum Mechanics*, L. M. Roth and A. Inomata (eds.), New York: Gordon and Breach, 1984.

L. I. Schiff. *Quantum Mechanics*, New York: McGraw-Hill, 1955.

R. Shankar. *Principles of Quantum Mechanics*, New York: Plenum, 1980.

H. Smith. *Introduction to Quantum Mechanics*, World Scientific, Singapore: 1991.

A. Sudbery. *Quantum Mechanics and the Particles of Nature*, New York: Cambridge University Press, 1986.

J. von Neumann. *The Mathematical Foundations of Quantum Mechanics*, Princeton: Princeton University Press, 1955.

J. A. Wheeler and W. H. Zurek. *Quantum Theory and Measurement*, Princeton: Princeton University Press, 1983.

R. L. White. *Basic Quantum Mechanics*, New York: McGraw-Hill, 1966.

E. H. Wichmann. *Quantum Physics*, New York: McGraw-Hill, 1971.

E. P. Wigner. *Symmetries and Reflections*, Bloomington: Indiana University Press, 1967.

F. A. Wolf. *Taking the Quantum Leap*, San Francisco: Harper & Row, 1981.

Index

Absorption of radiation, 455, 468
Accidental degeneracy, 189
Addition, of angular momenta, 353ff
  of spin and orbital angular momentum, 355ff
  of two spins, 357ff
Adiabatic approximation. *See* Born-Oppenheimer approximation
Aharanov-Bohm effect, 292ff
Airy function, 94-95, 173, 175, 262-63
  zeros, 95
Alpha decay, 79ff
Ammonia maser, 335ff
Ammonia molecule, 330ff, 387ff
  in electric field, 335ff, 387ff
  in time-dependent electric field, 338ff
Angular distribution, 502
Angular momentum, 6, 11, 192ff, 218ff, 322ff, 353ff
  commutation relations, 220
  condition in Bohr atom, 6, 11
  conservation, 243, 245ff
  coupling, 353ff
  eigenfunctions, 229ff
  eigenvalue problem, 223ff, 234ff
  matrices, 322ff
  operator, 192, 225
Angular momentum–angle uncertainty relation, 193
Anomalous Zeeman effect, 289, 401ff
Antisymmetrization of wave function, 197, 405, 411, 420, 433
Aspect experiment, 373
Associated Laguerre polynomials, 275
Associated Legendre polynomials, 228
Atomic structure, 421ff
  periodic table, 421ff
  shell structure, 421ff
Aufbau principle, 422, 453
Averaging over initial states, 476

Balmer series, 7
Band structure, 445
Barn, 493
Barrier penetration, 76ff, 178
Basis, 119ff, 302ff
Bell inequality, 370ff, 520
Bell's theorem, 370
Berry's phase correlation, 294
Bessel functions, 199 (problem), 253-54
  spherical, 253ff
Black-body radiation, 2ff
  energy density, 4
Black disc scattering, 515ff
Block diagonal matrix, 325
Bohm, 209, 532
Bohr, 1, 5ff, 25 (problem), 45, 46, 101, 103ff, 106, 162, 203, 212ff, 520-21
Bohr atomic model, 5ff
Bohr complementarity principle. *See* Complementarity principle
Bohr correspondence principle. *See* Correspondence principle
Bohr-Einstein debate. *See* Einstein-Bohr debate
Bohr radius, 7
Boltzmann factor, 344, 455

Born, 1, 13, 46, 482
Born approximation, 481ff
  for Coulomb scattering, 482ff
  limitations, 482
Born-Oppenheimer approximation, 425ff
Born probability interpretation, 1, 13
Boson, 198
Boson condensation, 198, 453
Bound states in potential well, 84
Box normalization, 62
Box potential. *See* Particle in a box
Bra, 119
Bracket, 119
Breit-Wigner formula, 479, 508

Center-of-mass reference frame, 194-95, 244
  (relation of) center-of-mass and laboratory cross sections, 494ff
Centrifugal barrier, 252
Classical limit, 43ff, 168
Clebsch-Gordan coefficients, 361, 402, 475
  table, 365
Coherent superposition, 22
Collapse of the wave function, 45ff, 68, 69, 207, 214, 368ff, 520, 535
Collision theory. *See* Scattering theory
Column matrix, 309
Commutation relations, 55-56, 220
  for angular momentum, 220
  for p and x, 55-56
Complementarity principle, 46, 109ff, 208
Completeness rule, 119
Complete set of commuting observables, 127ff
Complete set of eigenfunctions, 67
Conservation
  of angular momentum, 243, 245ff
  of energy, 70-71 (problem)
  of momentum, 166-67
  of probability, 22ff
Constant of the motion, 166
Continuity equation, 23, 60
Continuum quantum mechanics, 497ff
Coordinate representation. *See* Position representation
Copenhagen interpretation of quantum mechanics, 46-47, 107, 202-3, 208ff, 296, 520, 534
Correspondence principle, 5, 169ff, 474
Coulomb potential, 268ff
Creation operator, 155, 455
Cross section
  for black disc, 515
  differential, 492-93
  inelastic, 493
  optical theorem, 503-4, 515
  in terms of phase shifts, 503
  total, 493ff
  total elastic, 493
Cylindrical coordinates, 191ff

Davisson-Germer diffraction experiment, 9
De Broglie, 1, 8ff
De Broglie wavelength relation, 10ff
Degeneracy, 62
  accidental, 189
  for central potentials, 245ff
  for Coulomb potential, 273
  exchange, 195
  removal in atoms, 398ff, 403
  symmetry, 189
Degenerate eigenfunctions, 62
Degenerate perturbation theory, 386ff
Delayed choice experiment, 113ff
Delta function, 39-40, 543ff
  potential, 91-92
Density matrix, 522ff
Density operator, 523
Density of states, 456, 466, 482
Destruction operator, 155, 455
Deuteron, 257ff
Diatomic molecule, 244-45
Differential scattering cross section, 482ff
Diffraction, single slit, 103-4
Dipole selection rules, 474ff
Dirac, 1, 13, 108, 117
Dirac delta function. *See* Delta function
Dirac notation, 119ff
Double slit experiment, 107ff
Dualism, 204, 215, 536

EPR paradox. *See* Einstein-Podolsky-Rosen (EPR) paradox
Effective potential, 252
Effective range formula, 510
Ehrenfest theorem, 165ff, 168
Eigenfunctions, 60
  orthogonality, 63-64, 65-66
Eigenstates, 124
Eigenvalue, 60
Eigenvalue equation, 59ff
  in matrix form, 328ff
Einstein, 1, 6, 8, 103ff, 195, 520
Einstein-Bohr debate, 103ff
Einstein equivalence principle, 106
Einstein-Podolsky-Rosen (EPR) paradox, 195, 199, 204ff, 368ff
Einstein's A-coefficient, 474
Einstein's box experiment, 105-6
Electric dipole approximation, 472ff
  selection rules, 474ff
  transition rate for $2P \rightarrow 1S$, 476ff
Electric dipole moment, 335, 385
Electric polarizability, 385ff
Electromagnetic energy density, 468
Electron configurations of elements (table), 426ff
Electron in constant magnetic field, 287ff
Electron spin. *See* Spin
Emission of radiation, 455, 464, 468, 471ff
Energy bands in solids, 445
Energy matrix, 321, 331ff
Energy operator. *See* Hamiltonian operator
Energy shift
  first order, 380
  helium excited states, 411ff
  helium ground state, 404ff
  second order, 381
Ensemble interpretation of quantum mechanics, 107, 211, 529. *See also* Statistical interpretation of quantum mechanics
Equation of motion, 162ff, 165-66
Equivalence principle, 106
Everett interpretation. *See* Many-worlds interpretation
Exchange degeneracy, 195
Exchange effect in helium, 411ff
Exchange integral, 412, 434
Exchange operator, 197
Exchange symmetry, 195ff
Expansion postulate, 66-67

Expectation value, 16
  for powers of r in hydrogen, 280
Exponential decay law, 477ff

Fermi energy, 439
Fermi gas, 438ff
Fermi's golden rule, 463ff
Fermions, 198
Feynman, 15, 25, 109, 115 (problem), 300
Fine structure of atomic spectra, 397, 398
Fine structure constant, 272
Flux, 23
Flux conservation in radial equation, 498
Flux quantization. *See* Magnetic flux quantization
Fourier integral, 28, 39
Fourier transform, 28, 39-40
Free particle
  radial equation, 497ff
  Schrödinger equation, 12ff, 58ff

g factor. *See* Gyromagnetic ratio
Gauge invariance of Schrödinger equation, 291ff
Gauge transformation, 286
Gaussian wave packets, 14-15, 28ff, 39ff
Generators of symmetry transformations, 247
Golden rule for transition rate. *See* Fermi's golden rule
Ground state, 6, 20
Group velocity, 33-34
Gyromagnetic ratio, 299
  for electron, 299
  for proton, 342

Hamiltonian operator, 57ff
  for electron in external electromagnetic field, 287ff
Harmonic oscillator, 136ff, 189ff, 446ff
  classical correspondence, 169ff
  eigenfunctions, 143ff
  eigenvalue, 143ff
  matrix elements, 154ff
  operator method, 149
  three-dimensional, 446ff
  two-dimensional, 189ff
Hartree equation for atoms, 417ff
Hartree-Fock method, 420
Heisenberg, 1, 7, 36-37, 45, 46, 101, 203, 208, 325, 413, 521, 536
Heisenberg-Bohr microscope, 101ff, 212ff
Heisenberg equation of motion, 164ff
Heisenberg explanation of ferromagnetism, 413
Heisenberg ontology, 208, 534ff
Heisenberg picture, 163ff
Heisenberg uncertainty relation. *See* Uncertainty relations
Heitler-London method for molecules. *See* Valence bond method
Helium atom, 404ff
  exchange interaction, 411
  excited states, 411
  first-order level shifts, 405ff
  screening of nuclear charge, 408
  variational calculation, 409ff
Hermite polynomial, 143ff
Hermitian conjugate, 124, 309, 315
Hermitian matrices, 315
Hermitian operators, 57, 65, 124ff
Hidden variables, 209, 369ff, 532-33
Hilbert space, 119
Hund's Rules, 424-25
Hydrogen atom
  Bohr model, 5ff
  fine structure, 398ff
  radial equation, 269ff
  relativistic effects, 398ff
  spectrum, 273
  spin-orbit coupling, 398
  wave functions, 274ff
Hydrogen molecule, 433ff
Hyperfine structure, 397

Idealism, 203-4, 534
Idealistic interpretation of quantum mechanics, 533ff
Identical particles, 195ff
Induced absorption and emission, 464
Inelastic scattering, 493, 496, 513ff
Infinite box. *See* Particle in a box
Inner product. *See* Scalar product
Interpretation of expansion coefficients, 68
Interpretation of quantum mechanics, 45ff, 101ff, 202ff, 520ff
Interpretation of wave function, 13ff
Invariance
  under discrete displacements, 441-42
  rotation, 245ff
  time reversal, 531-32
  translation (problem), 265
Inverse Fourier transform, 39
Irreversibility, 530ff

j-j coupling, 423

K^0-$\bar{K}$ system (problem), 349
Ket, 119
Klein-Gordon equation (problem), 26
Kramer's rule, 280
Kronecker delta, 63-64
Kronig-Penny model, 442ff

L-S coupling, 424
Laboratory frame, 494ff
Lagrange multiplier in variational principle, 418
Laguerre polynomials, 275ff
Lamb shift, 401
Laplacian operator, 248
Larmor frequency, 289
Laser, 470
Legendre polynomials, 227ff
Levinson theorem, 506
Lifetime, 474, 477ff
Linear operators, 54-55, 124
Linear potential, 92ff, 262ff
Line width, 36, 465, 479
Logical positivism, 204

Macrorealism, 215, 296, 530
Macroscopic distinguishability, 528ff
Magic numbers, 446, 449, 451ff
Magnetic dipole moment, 280ff, 288, 298-99, 342
  of spin, 299, 342
Magnetic flux quantization, 295
Magnetic quantum number, 192
Magneton
  Bohr, 299
  nuclear, 343

Many-worlds interpretation, 211-12, 533, 538-39
Matrices, 156-57, 307ff
  diagonalization, 315, 317ff, 332-33, 388ff
  hermitian, 315
  multiplication, 308
  representation of operators, 307-8
  unitary, 320
Matrix element, 124
Matrix mechanics, 1, 8, 117, 321-22, 328ff
Maxwell's equations, 285ff
Measurement postulate, 67-68
Measurement theory, 525ff
Meissner effect, 295
Molecular orbital method, 430ff
Molecules
  ammonia, 330ff
  Born-Oppenheimer approximation, 425ff
  orbitals, 431
  rotation, 233-34, 429
  structure, 430ff
  types of motion, 429ff
  valence bond method, 433ff
  vibration, 429
Momentum conservation, 166-67
Momentum operator, 52ff, 186
  eigenfunctions, 62ff
  hermiticity, 56-57
Momentum representation, 121ff
Momentum transfer, 481
Monistic idealism, 203-4, 535

N-particle system, 199, 417
  Hamiltonian, 417ff
n-representation, 153
Neutron-proton scattering, 508ff
  spin dependence, 512
Nodes, connection with energy, 143
Nonlocality, 207-8, 294, 373, 520, 532-33, 535
  nonlocal hidden variables, 208-9, 373, 532-33
  nonlocal phase relationship, 294, 373
Normalization
  of eigenfunctions, 14, 22
  of wavepackets, 30, 53-54
Normalization of momentum eigenfunctions, 62-63
Nuclear magnetic resonance, 342ff
Number operator, 152
Number representation, 153

Observables, 54, 57
Ontology, 106-7, 203ff, 208, 521, 530ff, 533ff
Operator, 54ff, 123ff, 307ff
  displacement, 441
  hermitian, 57, 65, 124
  linear, 54-56, 124
  method for eigenvalue problems, 149ff, 234ff
  particle exchange, 197
  raising and lowering, 150ff, 234ff
  rotation, 247, 265 (problem), 350 (problem)
  space exchange, 360
  spin exchange, 360
  time evolution, 131ff
  unitary, 320, 361
Oppenheimer, 129
Optical theorem, 504, 515
Orbital angular momentum, 237
Orbitals, 421
Orthogonality, 63, 65-66
Orthohelium, 396
Orthonormality condition, 63
Overlap integral, 431

Paired electrons, 295-96
Parahelium, 396
Parity, 88ff, 140, 232
  selection rule, 475
Parseval's theorem, 54
Partial wave expansion of the scattering amplitude, 500ff
Partially polarized beam, 524-25
Particle in a box, 17ff, 64
  in three dimensions, 252ff
  in two dimensions, 187ff
Particle on a ring, 190ff
Particle on a sphere, 218ff
Pauli exclusion principle, 198, 360
Pauli spin matrices, 316ff
Periodic boundary conditions, 56-57
Periodic potential, 440ff
Periodic table, 421ff
Permanent dipole moment, 390
Perturbation theory
  convergence, 378
  degenerate, 386ff
  time dependent, 460ff
  time independent, 377ff
  treatment of ammonia molecule, 387ff
Pfleegor-Mandel experiment (problem), 116
Phase shift for radial solution, 499
  at resonance, 508
  for square well, 504ff
Phase space, 177, 479
  factor for multiparticle decay, 479ff
Phase velocity, 33
Phase of wave function, 21ff, 24, 295, 349, 373
Phonons, 143
Photon absorption and emission, 454-55, 468
Photon gas, 452ff
Photons, 6, 452ff
Pion Compton wavelength, 259
Planck's constant, 3, 10
Planck radiation formula, 4
  derivation, 2ff
  quantum mechanical derivation, 452ff
Plane wave expansion in spherical harmonics, 500
Polarizability. *See* Electric polarizability
Polarization of photon, 469
  summed over, 473-74
Polarized electron beams, 304, 524
Population inversion, 470
Position
  operator, 51, 56, 125ff
  representation, 122ff
Positronium (problem), 282
Postulates of quantum mechanics, 64ff
Potentia, 45, 68, 208, 534ff
Potential
  centrifugal, 252
  Coulomb, 268ff
  delta function, 91-92
  hard core (problem), 517
  linear, 92ff, 262ff
  periodic, 440-41
  self-consistent, 419
  square barrier, 76ff
  square well, 82ff, 255ff
  step (problem), 95
Potential barrier, 76ff
Potential scattering in Born approximation, 481ff

Potential scattering and phase shift, 499, 504ff
Precession of spin, 345ff
Principle of macroscopic distinguishability, 528ff
Probability conservation, 22
Probability current, 22ff, 24-25, 77, 280ff, 491ff
Probability density, 23
Probability distribution, 13-14
Probability interpretation of expansion coefficients, 68
Probability interpretation of wave function, 13ff
Probability of transition. *See* Transition probability

Quantization of angular momentum, 6, 192, 220ff
Quantization of energy, 3, 7, 18
Quantum conditions, 6
Quantum cosmology, 539
Quantum electrodynamic form of vector potential, 469-70
Quantum jump, 6, 7, 45, 79, 208
Quark interaction, 262ff
Quasi-stationary states, 477

Radial equation, 249ff
  bound state solution, 250ff
  continuum solution, 497ff
  solution for hydrogen, 269ff
  solution for linear potential, 262ff
  solution for spherical box, 252ff
  solution for square well, 255ff, 504ff
Radiation of atoms, 476ff
Radiative transitions, matrix element, $2P \rightarrow 1S$ rate, 476ff
Raising operators, 151
Ramsauer-Townsend effect, 84
Range of nuclear force, 258
Rare earths, 422
Rayleigh-Jeans black body radiation law, 2
Rayleigh-Ritz variational principle, 390ff
Realism, 203ff
Reality of expectation values, 57
Reduced mass, 194, 244
Reduction postulate, 68
Reduction of wave packet. *See* Collapse of the wave function
Reflection by potential step (problem), 95
Relativistic corrections to hydrogen spectrum, 400
Relaxation mechanism, 344
Representation, 118ff
  energy, 153
  momentum, 121ff
  number (n), 153
  position, 122ff
Residual interaction, 416, 419, 452
Resonance, 84, 465, 507-8, 512
Resonant scattering, 84, 508, 512
  Breit-Wigner formula, 508
Ritz variational principle. *See* Rayleigh-Ritz variational principle
Rotational motion of molecules, 233-34, 429-30
Rotational symmetry, 245ff
Rotator, 233-34
Row matrix, 309
Rutherford cross section for Coulomb scattering, 484
Rutherford model of the atom, 5
Rydberg, 272

S function, 498, 513-14
S wave scattering, 504ff
Scalar product, 119, 123
Scattering, spin dependence, 512
Scattering amplitude, 490ff
Scattering length, 510, 511, 513
Scattering theory, 481ff, 488ff
Schmidt orthogonalization, 66, 73 (problem)
Schrödinger, 1, 10, 12ff, 45
Schrödinger equation, 12
  cylindrical coordinates, 191
  derivation, 131ff
  as eigenvalue-eigenfunction equation, 51ff
  finite representation, 328-29
  free particle, 12ff, 61ff, 497ff
  in momentum space, 92ff
  for N particles, 199
  particle in potential, 60
  separation of angular coordinates, 245ff
  separation of center of mass motion, 194ff, 244-45
  separation of time dependence, 59
  in spherical coordinates, 245ff
  in three dimensions, 198
  time dependent, 59
  time independent, 60
Schrödinger picture, 163ff
Schrödinger's cat, 199, 209ff, 535
Schwartz inequality, 130
Screened Coulomb potential, 483
Screening of nuclear charge, 408ff
Secular equation, 315ff
Selection rules, 474-75
  for orbital angular momentum, 475
  for parity change, 475
  role in Zeeman effect, 289, 403
  for z-component of angular momentum, 475
  zero-zero transitions, 475
Self-consistent potential, 419
Semiconductors, 446
Shadow scattering, 516
Shell model of nucleus, 446ff
Shell structure
  atoms, 421
  nuclei, 449-50
Similarity transformation, 321
Simultaneous eigenfunctions, 62, 90, 127-28
  conditions on operators, 127-28
Singlet state, 360
  behavior under particle exchange, 360
  in EPR paradox, 368
Slater determinant, 420
Sommerfeld-Wilson quantization rule, 177
Space-exchange operator, 360
Spectroscopic notation, 423ff
Spectroscopic term value, 423ff
  table, 426
Spectrum of hydrogen, 273-74
  modifications, 398ff
Spherical Bessel functions, 253
  zeros, 254
Spherical Hankel functions, 261
Spherical harmonic addition theorem, 406
Spherical harmonics, 230ff
  integration over three spherical harmonics, 475
Spherical Neumann functions, 261
Spherical waves, 490, 497-98

Spin, 197-98, 237, 299, 302ff, 345ff
  analogy with light polarization, 311ff
Spin 1/2 operators, 307, 314ff
Spin-dependence of helium spectrum, 412
Spin-dependence of scattering lengths, 513
Spin exchange operator, 360
Spin matrices, 314ff
Spin-orbit coupling, 355
Spin-orbit interaction
  in atoms, 398, 399ff
  in nuclei, 451ff
Spinors, 309ff, 311ff, 348-49
Spin precession in magnetic field, 345ff
Spin singlet wave functions. *See* Singlet state
Spin-statistics connection, 197-98
Spin triplet wave function. *See* Triplet state
Spontaneous emission, 470, 471ff
Spreading of wave packet, 39ff, 47
Square integrable functions, 22, 61
Square well
  bound states in one dimension, 84ff
  bound states in three dimensions, 258ff
  resonant scattering, 84
  scattering in one dimension, 82ff
  in three dimensions, 504ff
SQUID, 216, 296ff
Stark effect, 383
Statistical interpretation of quantum mechanics, 529ff. *See also* Ensemble interpretation of quantum mechanics
Stefan-Boltzmann law, 5
Step function, 546
Stern-Gerlach experiment, 297ff, 302, 528ff, 531
Stern-Gerlach filters, 303
Stimulated emission, 464, 470
Superconductors, 295ff
Superposition principle, 14
Superposition of waves. *See* Wave packets
Symmetrization of wave function, 197, 454
Symmetry degeneracy, 189
Symmetry of Hamiltonian, 90, 243

Term value. *See* Spectroscopic term value
Thomas precession effect, 398
Time dependence of expectation values, 61
Time dependence of wave functions, 60
Time-energy uncertainty relation, 35-36, 105-6, 465
Time evolution of decaying state, 477ff
Time evolution operator, 131ff
Time evolution of systems, 68-69, 131ff
Time reversal, 531-32
Trace of a matrix, 324 (problem), 523
Transformation matrix, 317ff
Transition probability, 471-72
  relation to lifetime, 474, 477-78
Transmission coefficient for square well, 78ff
Transmission coefficient in WKB approximation, 178
Transpose of a matrix, 315
Triplet state, 359
  behavior under particle exchange, 359
Tunneling, 78-79
Two-particle system, 193ff, 244-45
Two-state systems, 328ff

Uncertainty relations, 15ff, 30ff, 34ff, 109 129ff, 143, 148, 202, 208, 272, 465
  between angular momentum and angle, 193
  dispersion, 43
  general proof, 129ff
  between position and momentum, 15ff, 30ff
  between time and energy, 35-36, 105-6, 465
  use for estimates, 34
Unitary transformation, 125, 320, 362
Unsold closure principle, 386

Vacuum, 453
Valence bond method, 433ff
Variational method, 390ff
  for helium, 409ff
  for many electron atoms, 417ff
  for molecules, 430ff
Vector model, 229
Vector potential, 286
  for constant magnetic field, 287
  for emission and absorption of photons, 469-70
Vibrational states of molecules, 429
Virial theorem (problems), 158, 283
Virtual states (problem), 485
von Neumann chain, 214-15, 527
von Neumann measurement theory, 526ff
von Neumann-Wigner resolution of quantum measurement paradox, 215, 535ff

Watched pot never boils paradox, 537-38
Wave equation. *See* Schrödinger equation
Wave mechanics, general structure, 117ff
Wave packets, 28ff
  gaussian, 14-15, 29ff
  in scattering, 489
  spreading, 39ff
  and uncertainty relations, 30ff
Wave-particle duality, 42, 45ff
Weak objectivity (problem), 216
Wentzel-Kramers-Brillouin (WKB) approximation, 162, 171ff
Wheeler, 114
Wheeler's participatory universe, 539
Wigner's friend, 534, 536

Yukawa potential, 483ff

Zeeman effect
  anomalous, 289, 401ff
  normal, 288ff
  for strong fields, 404
Zeno's paradox, 36-37
Zero-point energy, 23, 143
Zero-zero transition selection rule, 475